Transporters and Drug-Metabolizing Enzymes in Drug Toxicity

Transporters and Drug-Metabolizing Enzymes in Drug Toxicity

Edited by Albert P. Li

In Vitro ADMET Laboratories, Inc.
Columbia, MD
USA

WILEY

This edition first published 2021

Registered Office
John Wiley & Sons, Inc., 111 River Street, Hoboken, NJ 07030, USA

Editorial Office
111 River Street, Hoboken, NJ 07030, USA

For details of our global editorial offices, customer services, and more information about Wiley products visit us at www.wiley.com.

Wiley also publishes its books in a variety of electronic formats and by print-on-demand. Some content that appears in standard print versions of this book may not be available in other formats.

Library of Congress Cataloging-in-Publication Data applied for
ISBN: 9781119170846

Cover design by Wiley
Cover image: Courtesy of Albert P. Li

MIX
Paper from
responsible sources
FSC
www.fsc.org
FSC® C013604

Contents

Preface

A major goal of this book is to provide information to aid the advancement of experimental approaches to ensure drug safety in drug development. It should be of interest to students and researchers in drug metabolism, transport, and toxicology; practitioners in drug development; and governmental regulatory scientists.

The most challenging aspect of drug development is the selection of drug candidates with appropriate safety and efficacy to ensure regulatory approval and market acceptance. The paradigm of demonstration of safety and efficacy in preclinical animal models followed by human clinical trials needs to be refined. The inadequacy of animal models to predict human safety and efficacy is clearly illustrated by the estimated >90% clinical trial failure rate for candidates selected based on results of preclinical trials. Let us ponder this for a minute, despite the extensive time and resources spent in preclinical evaluation, 9 out of 10 candidates selected for clinical trial fail due to a host of factors, with the major ones being unexpected toxicity and/or lack of efficacy. Furthermore, numerous marketed drugs have been withdrawn or have their use limited due to severe, often idiosyncratic, adverse drug toxicity.

This book is intended to present information to overcome this challenge. Failure of preclinical studies to predict human safety and efficacy can be attributed to species differences in drug properties. The inability of clinical trials to eliminate drugs with idiosyncratic drug toxicity is likely due to the inadequate number of subjects employed in regulatory clinical trials to identify drugs causing severe idiosyncratic drug toxicity with an incidence of <1/5000. It is, therefore, of utmost importance to fully understand the key determinants of drug toxicity and apply the knowledge to identify drug candidates with human toxicity as well as to identify the at-risk human populations.

In this book, drug metabolism enzymes, uptake transporters, and efflux transporters are identified as key contributors of drug toxicity. Species differences in these key parameters result in the failure in the selection of drug candidates with the appropriate properties based on results of preclinical trials, resulting in clinical trial failure. Individuals may succumb to life-threatening toxicity of marketed drugs due to the chance co-occurrence of genetic and/or environmental risk factors which may exacerbate drug toxicity. Identification of drug candidates with toxicity exacerbated by known genetic and environmental risk factors must be an approach to be considered in drug development and management of drug administration, hopefully leading to the ultimate elimination of idiosyncratic drug toxicity.

I am grateful to the expert chapter authors who, as a result of their painstaking research, have provided significant progress toward the understanding of the contribution of drug metabolism and transport in drug toxicity.

Albert P. Li, Ph.D.
Columbia, MD

List of Contributors

Michael Arand
Institute of Pharmacology and Toxicology
University of Zürich
Zürich
Switzerland

Kristin Ashby
Division of Bioinformatics and Biostatistics
National Center for Toxicological Research
U.S. Food and Drug Administration
Jefferson
AR
USA

Kim L.R. Brouwer
Division of Pharmacotherapy and Experimental Therapeutics
UNC Eshelman School of Pharmacy
University of North Carolina at Chapel Hill
Chapel Hill
NC
USA

Gian Camenisch
ADME Department of PK Sciences
Translational Medicine
Novartis Institutes for BioMedical Research
Basel
Switzerland

Minjun Chen
Division of Bioinformatics and Biostatistics
National Center for Toxicological Research
U.S. Food and Drug Administration
Jefferson
AR
USA

Ann K. Daly
Translational and Clinical Research Institute
Newcastle University
Newcastle upon Tyne
UK

Mark P. Grillo
MyoKardia, Inc.
Brisbane
CA
USA

Paavo Honkakoski
Division of Pharmacotherapy and Experimental Therapeutics
UNC Eshelman School of Pharmacy
University of North Carolina at Chapel Hill
Chapel Hill
NC
USA
School of Pharmacy
Faculty of Health Sciences
University of Eastern Finland
Kuopio
Finland

Tao Hu
Department of Pharmaceutical Sciences
University of Maryland School of Pharmacy
Baltimore
MD
USA

Felix Huth
ADME Department of PK Sciences
Translational Medicine
Novartis Institutes for BioMedical Research
Basel
Switzerland

Vlasia Kastrinou-Lampou
ADME Department of PK Sciences
Translational Medicine
Novartis Institutes for BioMedical Research
Basel
Switzerland
Institute of Pharmacology and Toxicology
University of Zürich
Zürich
Switzerland

Stefanie Kennon-McGill
Department of Environmental and Occupational Health
Fay W. Boozman College of Public Health
University of Arkansas for Medical Sciences
Little Rock
AR
USA

Gerd A. Kullak-Ublick
Department of Clinical Pharmacology and Toxicology
University Hospital Zürich
University of Zurich
Zürich
Switzerland
Mechanistic Safety
CMO & Patient Safety
Global Drug Development
Novartis
Basel
Switzerland

Albert P. Li
In vitro ADMET Laboratories, Inc.
Columbia
MD
USA

Stacey Line
Department of Pharmacology and Therapeutics
University of Manitoba
Winnipeg
MB
Canada

Chuang Lu
Sanofi US
Waltham
MA
USA

Kazuya Maeda
Laboratory of Molecular Pharmacokinetics
Graduate School of Pharmaceutical Sciences
The University of Tokyo
Bunkyo-ku Tokyo
Japan

Mitchell R. McGill
Department of Environmental and Occupational Health
Fay W. Boozman College of Public Health
University of Arkansas for Medical Sciences
Little Rock
AR
USA

Donald Miller
Department of Pharmacology and Theurapetics
Max Rady College of Medicine
University of Manitoba
Winnipeg
MB
Canada

William A. Mturphy
Division of Pharmacotherapy and Experimental Therapeutics
UNC Eshelman School of Pharmacy
University of North Carolina at Chapel Hill
Chapel Hill
NC
USA

Birk Poller
ADME Department of PK Sciences
Translational Medicine
Novartis Institutes for BioMedical Research
Basel
Switzerland

Nur A. Safa
Department of Pharmacology and Therapeutics
University of Manitoba
Winnipeg
MB
Canada

Chitra Saran
Division of Pharmacotherapy and Experimental Therapeutics
UNC Eshelman School of Pharmacy, and Department of Pharmacology
UNC School of Medicine
University of North Carolina at Chapel Hill
Chapel Hill
NC
USA

Hongbing Wang
Department of Pharmaceutical Sciences
University of Maryland School of Pharmacy
Baltimore
MD
USA

Qingping Wang
Sanofi US
Waltham
MA
USA

Yue Wu
Division of Bioinformatics and Biostatistics
National Center for Toxicological Research
U.S. Food and Drug Administration
Jefferson
AR
USA

Yan Zhang
DMPK
Nuvation Bio, Inc.
New York
NY
USA

Part I

Overview

1

Overview

Drug Metabolism, Transporter-Mediated Uptake and Efflux, and Drug Toxicity

Albert P. Li

In vitro ADMET Laboratories, Inc., 9221 Rumsey Road, Suite 8, Columbia, MD, 21045, USA

1.1 Drug Toxicity as a Challenge in Drug Development

Drug discovery and development is critical to human health. In recent decades, there have been numerous exciting breakthroughs in the development of novel approaches to cure previously difficult to manage diseases, alleviate pain and discomfort, as well as scientifically acceptable approaches to prevent or delay the onset of common but deadly disorders. The explosion in knowledge of the human genome has been instrumental in the development of novel therapeutic targets in the past decade, allowing the development of small molecules, biologics, and gene therapy to retard or completely abolish the progression of diseases via the modulation of key pathways. Human life span has been extended in most developed countries due to enhancements in health care approaches.

Unexpected human-specific drug toxicity has been, and continues to be a major challenge in drug development (1–5). Pharmaceuticals by nature are biologically active chemicals designed to interact with biological pathways. Key uncertainties are the unintended biological effects which may lead to damage. Unintended drug toxicity is one of the major reasons for clinical trial failures as well as withdrawal or greatly limited use of drugs that have received regulatory approval. Unintended and unexpected drug toxicities are responsible for the high costs and time required for drug development. The most recent estimation of the cost for development of a new drug is over $1 billion USD, with the average time span from discovery to market of over 10 years (6, 7).

In drug development, approaches to develop drugs with acceptable safety margin to patients involve extensive preclinical evaluation in multiple species of laboratory animals. Only after the preclinical data support the safety of a drug

Transporters and Drug-Metabolizing Enzymes in Drug Toxicity, First Edition. Edited by Albert P. Li.

candidate can it be submitted for approval by regulatory agencies for phase 1, 2, and 3 clinical trials. A drug is approved for marketing after the clinical trials support its safety and efficacy. The occurrence of unexpected toxicity for drug candidates with acceptable safety profile from preclinical safety studies is one of the major reasons for clinical trial failure, illustrating that nonhuman animals may not provide adequate information for the assessment of human drug toxicity. However, numerous drugs approved for marketing based on clinical trial demonstration of safety have been withdrawn from the market or have been restricted in use due to unacceptable adverse events leading to deaths. The occurrence of unacceptable drug toxicity after a drug is marketed to the human population illustrates that clinical trials do not have the statistical power to assure safety when millions of patients are exposed to a new drug. The paradigm for drug safety evaluation of preclinical safety in laboratory animals followed by regulatory human clinical trials therefore is not always adequate.

I strongly advocate the transition of the practice of toxicology from an empirical to a mechanistic discipline. A thorough mechanistic understanding of the onset of the toxic events is necessary for the identification of risk factors and the estimation of the probability of the patient population with the risk factors based on genetic polymorphism, coadministered prescribed and non-prescribed medications, disease status, and life style factors such as diet, substance abuse, alcohol consumption, and the use of unregulated diet supplements and herbal medicines.

1.2 Fate of an Orally Administered Drug

The following are the events likely to occur with an orally administered drug.

Enteric events: An orally administered drug undergoes the following events:

1) Metabolism by intestinal microflora.
2) Entry into enterocytes via diffusion or transporter-mediated uptake.
3) Metabolism by drug-metabolizing enzymes in the enterocytes.
4) Efflux to the intestinal lumen via efflux transporters.
5) Entry of the drug and its metabolites into the portal circulation.

Hepatic events: Upon entry of a drug and its enteric metabolites into the portal circulation, the following events occur:

1) Entry into hepatocytes via diffusion or transporter-mediated uptake.
2) Metabolism by drug-metabolizing enzymes in the hepatocytes.
3) Excretion of parent drug and/or metabolites to the intestine as bile via diffusion or efflux transporters followed by excretion into the feces or reentry into the portal circulation (enterohepatic recirculation) via diffusion or uptake transporters.
4) Entry into the systemic circulation.

Extrahepatic events: Upon entry of a drug and its hepatic metabolites into the systemic circulation, the following events occur:

1) Entry into extrahepatic target cells via diffusion or transporter-mediated uptake.
2) Metabolism by drug-metabolizing enzymes in the target cells.
3) Exit of parent drug and/or metabolites from the target cells to the systemic circulation.
4) Excretion via urinary, perspiratory, respiratory pathways.

The manifestation of toxicity of the drug in question, either due to the parent molecule or its metabolites, is dependent on the concentration of the toxic moiety at the ultimate target. Metabolism (metabolic clearance, activation, and detoxification), uptake transport, and efflux transport can play critical roles on whether the toxic moiety will reach the critical concentration leading to the onset of toxicity.

1.3 The Multiple Determinant Hypothesis for Idiosyncratic Drug Toxicity

A perplexing observation in drug toxicity is the phenomenon of idiosyncratic drug toxicity which occurs at an incidence of less than 1 per 5000 patient-years, thereby cannot be detected with conventional regulatory clinical trials but would be clearly identified after marketing of a drug with a large patient population being exposed (1–5). To explain this phenomenon and to provide a possible mitigating strategy, I have proposed the multiple determinant hypothesis (3) which states that idiosyncratic drug toxicity occurs due to a confluence of multiple discrete events in the individual succumbing to drug toxicity (Figure 1.1).

I would like to illustrate the role of drug transport and metabolism in the manifestation of idiosyncratic drug toxicity based on the Multiple Determinant Hypothesis.

This hypothetical drug, T, is a substrate of an uptake transporter. Upon uptake into the hepatocytes, T is metabolized to a reactive metabolite, T*. T* is the ultimate toxicant which reacts with key cellular proteins, leading to cell death. T* also can form protein conjugates, leading to the formation of a neoantigen which would elicit a cytotoxic inflammatory event, leading to liver failure. However, T also undergoes metabolic clearance via glucuronidation and sulfation, with the conjugates excreted via transporter-mediated efflux. T* can be detoxified via glutathione *S*-transferase (GST)-mediated glutathione (GSH) conjugation.

The following are key determinants and the respective probability (p) in the patient population for the manifestation of liver toxicity, resulting in liver failure:

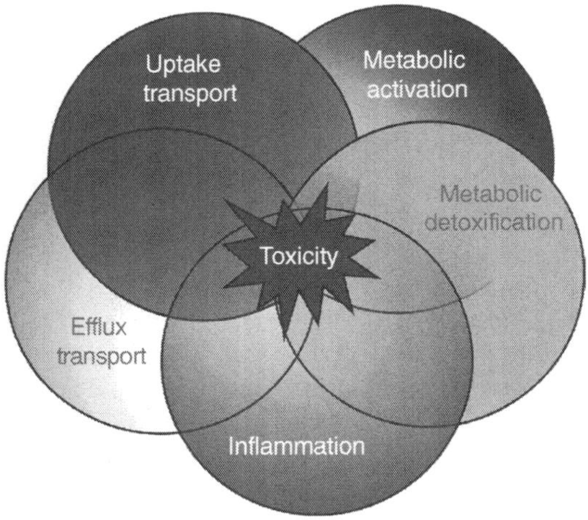

Figure 1.1 Schematic illustration of the Multiple Determinant Hypothesis of Idiosyncratic Drug Toxicity. Each circle represents a key determinant due to genetic and/or environmental factors. The hypothetical situation shown here is the confluence of the five determinants leading to severe liver toxicity in a patient at the time of drug administration: 1. Increased transporter-mediated uptake, leading to increased liver drug concentration. 2. Increased metabolic activation due to the co-exposure to enzyme inducers, leading to higher rate of formation of toxic/reactive metabolites. 3. Compromises metabolic detoxification due to co-exposure to GSH depleting agents and/or inhibitors of phase 2 conjugating pathways. 4. Sensitized inflammatory response due to diseases or exposure to sensitizing agents. 5. Compromised efflux of bile salt due to genetic polymorphism of BSEP or co-exposure to BSEP inhibitors.

1) *Determinant 1 (p1)*: Elevated uptake transporter activity, leading to higher than normal intracellular concentration.
2) *Determinant 2 (p2)*: Elevated drug metabolizing enzyme activity due to prior exposure to inducers or genetic polymorphism, resulting in higher than normal T* formation.
3) *Determinant 3 (p3)*: Suppressed UDP-glucuronosyltransferase (UGT) and sulfotransferase (SULT) detoxification activity.
4) *Determinant 4 (p4)*: Suppressed GST activity and/or reduced cellular GSH level.
5) *Determinant 5 (p5)*: Hypersensitized inflammatory reactions due to previous exposure to the same drug and the associated neoantigen.
6) *Determinant 6 (p6)*: Reduced efflux transporter activity of glucuronide and sulfate conjugates, leading to feedback inhibition of the detoxifying pathways.

In this hypothetical example, Determinant 2 (metabolic activation) is the most crucial, while the other determinants would exacerbate the hepatotoxic events. The patient who will succumb to liver failure is one with the confluence of risk factors – simultaneous occurrence of the various determinants of drug toxicity at the time of drug administration (Figure 1.1).

Based on the theory that co-occurrence of multiple determinants are required for the manifestation of severe drug-induced liver injuries (DILIs), the incidence of DILI in the patient population will be a product of the incidence of patients with each of the determinants at the time of drug administration as shown in the following equation:

$$I(\text{DILI}) = I(\text{Determinant } 1) \times I(\text{Determinant } 2) \times I(\text{Determinant } 3)\ldots \times I(\text{Determinant } n)$$

Considering a hypothetical case that the co-occurrence of four determinants is required for DILI, and that each determinant has an incidence of 10% of the patient population, I(DILI) will be $10\% \times 10\% \times 10\% \times 10\%$ or 1 in 10 000 patients. This event therefore cannot be readily detected in the regulatory clinical trials due to the limited number of patients, but will be manifested when the drug is administered to the general patient population upon marketing.

1.4 Concluding Remarks

Transporter-mediated drug uptake and efflux are clear determinants of intracellular concentrations of drugs and metabolites which are substrates of the transporters. Individual variations in the expression of drug metabolism enzymes and transporters due to genetic polymorphism, exposure to drugs, foods and food supplements, and various diseases have been well-established. A thorough understanding of the role of drug metabolism and transport in the disposition of a potentially toxic drug will aid the assessment of its toxic potential and the identification of the critical determinants of events leading to the formation and accumulation of the toxic moiety. Routine evaluation of the toxicological consequence of drug metabolism and transport may therefore aid the development of drugs with lower potential of causing idiosyncratic toxicity and may eventually lead to the identification of at-risk populations.

1.4.1 A Comprehensive Approach to Safety Evaluation in Drug Development

While the conventional preclinical and clinical safety evaluation are essential to the assessment of drug safety, additional considerations are needed to avoid the occurrence of idiosyncratic drug toxicity. I have previously proposed that drug safety assessment requires a comprehensive, multidisciplinary approach (8, 9) with input from pharmacologists, toxicologists, epidemiologists, population geneticists, and especially drug metabolism/pharmacokinetics (DMPK) experts to provide insight toward the possible risk factors discussed above.

Key parameters to be considered for drug safety evaluation based on this comprehensive approach include the following: (i) Pharmacology: Possible toxicity due to drug–target interactions, including interactions with unintended molecular targets, or with molecular targets in unintended organs. (ii) Chemistry: Chemical scaffolding and side chains with safety concerns. (iii) Toxicology: Toxicity in animals *in vivo*, and in relevant animal and human cells in culture. (iv) DMPK: Safety concerns due to toxification or detoxification, organ distribution, clearance, and pharmacokinetic drug–drug interactions. (v) Risk factors: Physiological, environmental, and genetic factors that may enhance a patient's susceptibility. It is proposed that this integrated, multidisciplinary approach to safety evaluation may enhance the accuracy of the prediction of drug safety and thereby the efficiency of drug development (8, 9).

1.4.2 The Dose Makes the Poison – Paracelsus Updated

A major doctrine in toxicology is the Paracelsus hypothesis:

All things are poison, and nothing is without poison; only the dose makes a thing not a poison.

When Philippus Aureolus Theophrastus Bombastus von Hohenheim (10, 11) made this statement which has become the cornerstone of the field of toxicology, he meant dose to individuals. An often cited example is that drinking too much water, a universally accepted non-poison, can lead to fatalities due to a phenomenon known as hyponatremia, lowering of plasma sodium concentrations due to over dilution of plasma, leading to hypo-osmolality-related fatal events (12, 13).

Based on the Multiple Determinant Hypothesis, I would like to propose a modernization to the Paracelsus doctrine for drug toxicity:

The dose makes the poison, whilst individuals make the dose.

References

1 Mosedale M, Watkins PB. "Understanding idiosyncratic toxicity: lessons learned from drug-induced liver injury". *J Med Chem* 2020; 63(12): 6436–6461.

2 Stepan AF, Walker DP, Bauman J, Price DA, Baillie TA, Kalgutkar AS, et al. "Structural alert/reactive metabolite concept as applied in medicinal chemistry to mitigate the risk of idiosyncratic drug toxicity: a perspective based on the critical examination of trends in the top 200 drugs marketed in the United States". *Chem Res Toxicol* 2011; 24(9): 1345–1410.

3 Li AP. “A review of the common properties of drugs with idiosyncratic hepatotoxicity and the ‘multiple determinant hypothesis’ for the manifestation of idiosyncratic drug toxicity”. *Chem Biol Interact* 2002; 142(1–2): 7–23.

4 Collins JM. “Idiosyncratic drug toxicity”. *Chem Biol Interact* 2002; 142(1–2): 3–6.

5 Park BK, Kitteringham NR, Powell H, Pirmohamed M. “Advances in molecular toxicology-towards understanding idiosyncratic drug toxicity”. *Toxicology* 2000; 153(1–3): 39–60.

6 DiMasi JA, Grabowski HG, Hansen RW. “The cost of drug development”. *N Engl J Med* 2015; 372(20): 1972.

7 Wouters OJ, McKee M, Luyten J. “Estimated research and development investment needed to bring a new medicine to market, 2009–2018”. *JAMA* 2020; 323(9): 844–853.

8 Li AP. “A comprehensive approach for drug safety assessment”. *Chem Biol Interact* 2004; 150(1): 27–33.

9 Li AP. “An integrated, multidisciplinary approach for drug safety assessment”. *Drug Discov Today* 2004; 9(16): 687–693.

10 Mann RD. “Famous names in toxicology. Paracelsus--born 500 years ago”. *Adverse Drug React Toxicol Rev* 1993; 12(2): 81–82.

11 Wilks MF. “Bringing chemistry to medicine-the contribution of Paracelsus to modern toxicology”. *Chimia (Aarau)* 2020; 74(6): 507–508.

12 Oh RC, Malave B, Chaltry JD. “Collapse in the heat-from overhydration to the emergency room-three cases of exercise-associated hyponatremia associated with exertional heat illness”. *Mil Med* 2018; 183(3–4): e225–e228.

13 O’Brien KK, Montain SJ, Corr WP, Sawka MN, Knapik JJ, Craig SC. “Hyponatremia associated with overhydration in U.S. Army trainees”. *Mil Med* 2001; 166(5): 405–410.

2

Transporter, Drug Metabolism, and Drug-Induced Liver Injury in Marketed Drugs

Minjun Chen, Kristin Ashby, and Yue Wu

Division of Bioinformatics and Biostatistics, National Center for Toxicological Research, U.S. Food and Drug Administration, Jefferson, Arkansas, USA

2.1 Introduction

Drug-induced liver injury (DILI) can lead to serious clinical outcomes, including acute liver failure. Current research on DILI suggests that it is mostly idiosyncratic. It is a significant concern for drug developers and agencies, and the current testing strategies provide an unacceptably high number of false negatives [1]. In past decades over 50 approved drugs have either been withdrawn from the market or have become the focus of regulatory actions, including the addition of boxed warnings and other product labeling modifications due to DILI [2, 3]. Hepatotoxicity also is cited as one of the leading reasons for drug development failure, accounting for over 20% of failed clinical trials due to drug safety issues [4]. In 2009, the US FDA published guidance on the clinical evaluation of DILI before premarketing approval, which aimed to address the collection and evaluation of laboratory measurements that signal the potential for DILI during clinical trials [5].

The liver is the major organ for drug metabolism and receives over 80% of its blood flow from the gastrointestinal tract. It contains several enzymes with biotransformation capacity for a large variety of foreign compounds, including therapeutic medications (Figure 2.1). P450 enzymes in phase I metabolism catalyze lipophilic drug molecules into water-soluble metabolites in order to eliminate them from the body. However, the biotransformation of these drug molecules can also form toxic reactive metabolites (RMs), which covalently bind and modify biological macromolecules such as enzymes, proteins and DNA, affecting their functions and potentially leading to toxicity. The physiological role of phase II reactions is to form conjugated products that are more water soluble than the original xenobiotic or active phase I metabolites; nevertheless, it also generates reactive conjugations that could result in toxicity.

Transporters and Drug-Metabolizing Enzymes in Drug Toxicity, First Edition. Edited by Albert P. Li.

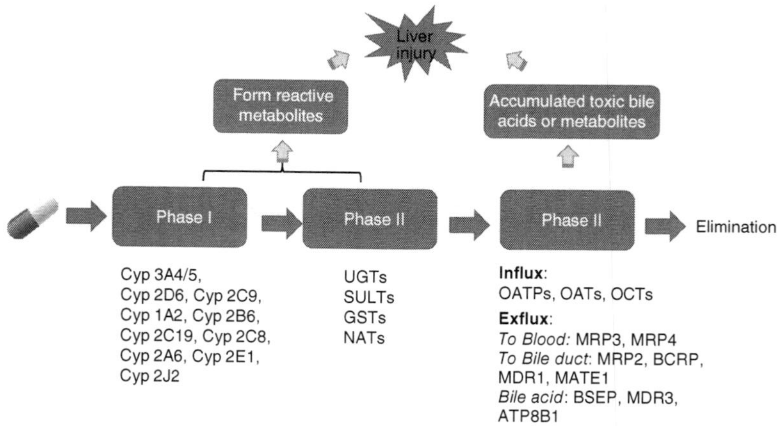

Figure 2.1 The physiological process of drug metabolism and transport and the role of enzymes in formation of reactive metabolites and accumulation of toxic bile acids, two proven mechanisms leading to drug-induced liver injury.

The inhibition of hepatic transporters such as the bile salt export pump (BSEP) presumably could cause toxic bile acids to accumulate in the liver and also is considered an important mechanism leading to DILI. In this chapter we will focus on drug metabolism and hepatic transporters, and their relationships to DILI. We will briefly introduce the enzymes involved in phase I and phase II hepatic metabolism and their roles in forming chemically RMs. Next, we will review the role of RMs in toxicity and how to detect and measure them using in silico and experimental approaches, as well as strategies for mitigating the risk of RMs in the drug discovery phase. We will also discuss the role of hepatic transporters and their relationship to hepatotoxicity. Finally, we will summarize the genetic variants of drug metabolism enzymes and hepatic transporters and their impact on pharmacokinetic behavior and drug safety.

2.2 Hepatic Metabolism

The liver's primary mechanism for metabolizing drugs is comprised of two reaction phases: biotransformation/detoxification (phase I) and conjugation (phase II). In phase I, reactive or polar groups are introduced into xenobiotics. In phase II, reactions these modified compounds are conjugated to polar compounds, which are easier to eliminate from the body.

2.2.1 Phase I Metabolism

Within phase I metabolism a series of biotransformation reactions, including oxidation, hydroxylation, and other reactions, are triggered. These reactions add functional hydroxyl, carboxyl, amino, or thiol groups to lipophilic drugs and convert them to a more hydrophilic status. This process is primarily mediated by the cytochrome P450 (CYP) enzymes, which account for about 75% of the total metabolism in the liver. CYP is a large family of enzymes containing heme as a cofactor. They function as monooxygenases localized primarily in the membrane of the endoplasmic reticulum (ER) in hepatocytes. In humans approximately 60 genes coding for the various CYP enzymes have been identified. The most frequent human P450 isoforms involved in the metabolism of drug molecules are 3A4/5, 2D6, 2C9, 1A2, 2B6, 2C19, 2C8, 2A6, 2E1, and 2J2 (Figure 2.2) [6]. CYP3A4/5 and 2D6 are the most abundant of all CYP450 enzymes in the human liver and metabolize half of medications, although their expression levels vary widely among individuals.

The expression level of these CYP enzymes controls the rate at which many drugs are metabolized. Usually a drug is metabolized by multiple CYP enzymes. Each of these has a limited capacity to metabolize drugs, and therefore can become overloaded when the drug level in blood is too high. The activities of the CYP enzymes significantly vary among individuals, and their gene expression and enzyme activities are tightly regulated by members of the nuclear receptor (NR) family of ligand-modulated transcription factors, such as the pregnane X receptor (PXR), farnesoid X receptor, vitamin D receptor, and hepatocyte nuclear factor 4 alpha.

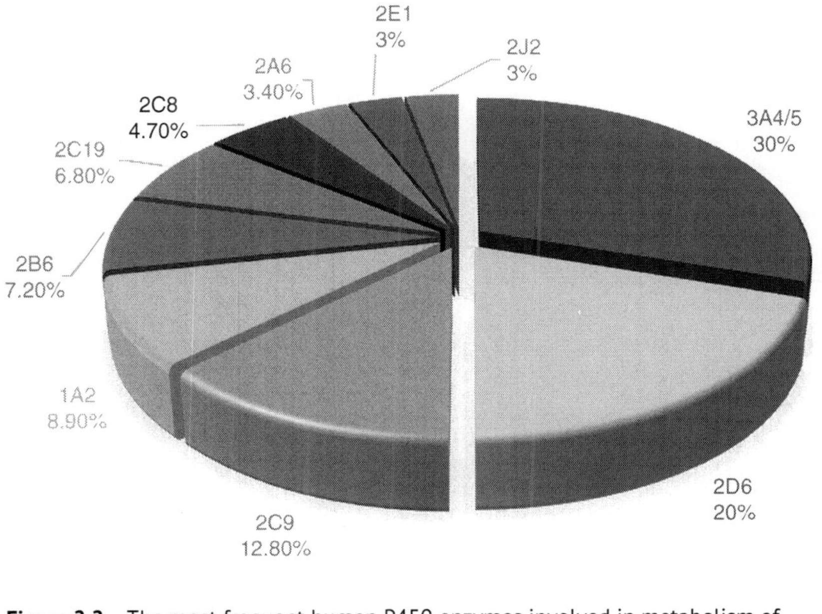

Figure 2.2 The most frequent human P450 enzymes involved in metabolism of clinically-used drugs. Numbers represent the fractions of drugs metabolized by the P450 enzyme. *Source:* Data from Zanger and Schwab [6].

2.2.2 Phase II Metabolism

Phase II metabolism is composed of a series of reactions, including glucuronidation, sulfoconjugation, glutathione (GSH) S-conjugation, acetylation, and methylation. They introduce glucuronic acid, sulfate, amino acids, or GSH molecules to phase I products and form conjugates, which increase water solubility and decrease pharmacologic activity; and finally, enhance detoxification of the compounds.

Glucuronidation accounts for the conjugation of about 40–70% of marketed drugs in humans, and UDP-glucuronosyltransferases (UGTs) are the key enzymes metabolizing various exogenous and endogenous compounds.

Sulfoconjugation (or sulfonation), generally described as a detoxification pathway for many xenobiotics, is mediated by a supergene family of enzymes called sulfotransferases (SULTs). The phase I active molecules can form a water-soluble compound by adding the sulfonate moiety, which can easily be removed from the body.

GSH S-conjugation plays a critical defensive role against oxidative stress. The family of glutathione S-transferases (GSTs) helps detoxify reactive intermediates formed by other metabolizing enzymes derived from drugs, chemicals, and environmental carcinogens.

Acetylation reactions are mostly driven by two *N*-acetyltransferase isoenzymes (NATs), NAT1, and NAT2, in humans. Most drugs in acetylation reactions either are hydrazines or aromatic amines.

Methylation is a minor pathway of xenobiotic biotransformation, which does not dramatically change the solubility of substrates. It is primarily involved in the metabolism of small endogenous compounds and certain drugs.

2.3 Reactive Metabolite Formation and Assessment

The physiological role of drug metabolism is detoxification, i.e. the biotransformation of lipophilic compounds into stable water-soluble metabolites, which are more readily eliminated from the body. However, drug metabolism can also lead to toxicity by the biotransformation of foreign compounds into metabolites which could be intrinsic to chemical reactivity for macromolecules. Such metabolites typically are unstable and undetectable in plasma, and usually have half-lives of less than one minute. They can be electron-deficient electrophiles or free radicals containing an unpaired electron [7], but most are electrophilic in nature [8]. Also, they are reactive with electron-rich nucleophiles such as biological macromolecules in which the functional side chains of arginine, lysine, histidine, cysteine, aspartic acid, glutamic acid, and tyrosine are potent nucleophiles in the unprotonated state [9].

RMs can irreversibly bind to and modify biological macromolecules such as enzymes, mitochondrial proteins, RNA, and DNA, and some of these mechanisms can eventually cause hepatocellular damage. For example, RMs can lead to an immune response by forming hepatic protein adducts. The overloading of RMs can deplete GSH, and the lack of GSH in the liver results in oxidative stress. They can also directly inhibit the BSEP, which may disrupt liver function and lead to hepatic injury. Free radicals also can bind lipids, initiating a chain reaction leading to lipid peroxidation, oxidative stress, or modification of biological macromolecules.

2.3.1 Metabolism and Reactive Metabolites

Many enzymes can generate and release RMs. Particularly, cytochromes P450, together with the reactivity of their oxygen intermediates, are the main drivers in catalyzing the formation of relatively inactive substrates into diverse chemically reactive species. These P450 enzymes are present in the ER and their hepatic expression is concentrated in the liver's centrilobular region; this is consistent with the typical histological findings of centrilobular necrosis in many acute DILI cases. Many drugs can form RMs, which are catalyzed by P450 enzymes and cause hepatotoxicity [10]. For example, acetaminophen is primarily metabolized via glucuronidation and sulfonation pathways, and the regular metabolites are rapidly excreted in urine. However, a small proportion of acetaminophen is metabolized by CYP3A4, CYP2E1, and CYP1A2, and undergoes bioactivation to form the RM, *N*-acetyl-*p*-benzoquinoneimine (NAPQI). At therapeutic doses, NAPQI can be rapidly detoxified by conjugation to GSH and safely eliminated from the liver. Acetaminophen overdosing, on the other hand, can result in GSH depletion and NAPQI accumulation, eventually leading to oxidative stress reactions, dysfunction of mitochondria, and DNA damage.

Phase II enzymes undoubtedly play an important role in the detoxification of various xenobiotics, but some, including UGTs, GSTs, NATs, and SULTs, also could catalyze the formation of RMs. Hepatotoxicity is considered a class effect of nonsteroidal anti-inflammatory drugs (NSAIDs) and the carboxylic acid functional groups they contain can undergo bioactivation to form acyl glucuronides by hepatic UGT isoforms. Benoxaprofen was an NSAID withdrawn from market in 1982 following several reports of fatal cholestatic jaundice. It contained a carboxylic acid moiety that can form a reactive acyl glucuronide and results in covalent protein adducts as reported from *in vitro* incubations with rat liver microsomes and from *in vivo* studies [11]. In FDA's guidance for industry on safety testing drug metabolites [12], additional safety assessments are required for products that potentially could form reactive acylglucuronides.

2.3.2 Dose and Reactive Mtabolites

The idiosyncratic nature of DILI suggests that it is independent of dose; however, many DILI cases have occurred with doses of drugs >100 mg per day, and cases for drugs given at doses of 10 mg per day or lower have been reported only rarely [13]. Furthermore, the average daily dose of drugs reported to cause hepatotoxicity was higher compared to drugs not associated with this [13]. Moreover, extensive hepatic metabolism has been associated with a higher risk of hepatic adverse events from oral medications [14]. High daily dose and extensive drug metabolism in the liver are associated with high risk of DILI, supporting the hypothesis that the formation and accumulation of RMs beyond a critical threshold is the precondition triggering the development of liver injury.

The quantitative relationship between daily dose, formation of RMs and risk of DILI in humans was reported by Chen et al. [15]. By considering $N = 354$ FDA-approved oral medications, the authors defined an algorithm for assigning a DILI score by factoring the relative contribution of daily dose, logP and RM, which permitted a quantitative assessment of clinical DILI risk:

$$\text{DILIscore} = 0.608 * \log_e(\text{daily dose / mg}) + 0.227 * \log P + 2.833 * \text{RM}$$
$$\text{DILIscore} > 6,\ \text{high risk}$$
$$3-6,\ \text{moderate risk}$$
$$<3,\ \text{low risk}$$

The three parameters in the DILIscore formula (i.e. daily dose/Cmax, logP, and RM formation) contributed significantly to DILI risk with the order of RM > daily dose > logP. The relationship between calculated DILI scores and DILI risk was validated by three independent datasets retrieved from the literature, and the score model demonstrates its correlation with the severity of clinical outcome by applying to $N = 159$ clinical cases collected from the NIH's LiverTox database (https://www.ncbi.nlm.nih.gov/books/NBK547852/).

2.3.3 Structural Alerts for Avoiding Reactive Metabolites

A compound's potential to form electrophiles and chemically RMs is determined by its chemical structure. Certain functional groups, referred to as structural alerts, are molecular fragments with high chemical reactivity, or which can be transformed into fragments with high chemical reactivity through bioactivation [16]. These structural alerts can be present in the parent compound or in its metabolites.

Structural alerts regularly have been employed for screening potential RM formation from candidate compounds in drug discovery, in order to limit undesirable toxic effects. General avoidance of certain functional groups related to RM formation is considered a normal practice at the lead optimization/candidate selection stage. Sometimes, if the structural alert must be carried into lead optimization, mitigation strategies learned from industry experience may be suggested to attenuate or avoid toxicity, including: (i) use substitutes resistant to metabolism to replace potential structural alerts; (ii) decrease electronic density; and (iii) introduce a structural element to redirect metabolism potential [16].

An increasing concern, however, is that the value of incorporating structural alerts for screening drug molecule-associated safety hazards may have been overstated. In a study examining the role of RMs, about 50% of the top 200 drugs marketed in the U.S. and ranked by prescription and sales were found to have one or more alerts in their chemical architecture [17]. Meanwhile, the absence of structural alerts cannot serve as a guarantee of drug safety. For example, ximelagatran did not present any alerts in its chemical structure but was recalled as a thrombin inhibitor with idiosyncratic hepatotoxicity [17].

Figure 2.3 lists some typical structural alerts found in the literature. Since the relationships among these alerts in generating RMs and toxicity are largely

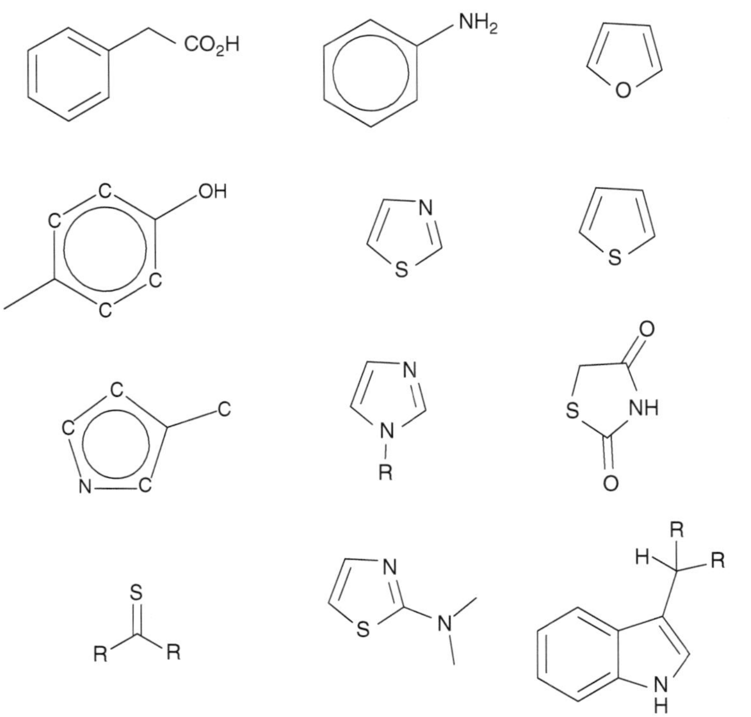

Figure 2.3 Some typical structure alerts for formation of reactive metabolites found in the literature.

unclear, most institutes and companies maintain their own libraries of structural alerts. Several review papers have summarized a comprehensive list of alerts. Enoch et al., for example, reported many structural alerts, each with an outline of their associated mechanistic chemistry of adduct formation. These alerts have been incorporated into the Organization for Economic Co-operation and Development (OECD) (Q)SAR Toolbox, a freely available software tool for analyzing structural alerts in the public domain [18]. Claesson and Minidis summarized 63 alerts with detailed SMART names, SMART pictures and examples of drugs, which could be the basis for an in-house structural alert application system. [19]

2.3.4 Experimental Approaches for Assessing Reactive Metabolites

The structural alerts for formation of RMs sometimes do not imply actual formation of RMs. In this circumstance, experimental evidence is needed to assess whether a drug candidate is bioactivated to RMs. We discuss some selected experimental approaches here, including covalent binding assay, electrophile trapping experiment and time-dependent inactivation of CYP450.

2.3.4.1 Covalent Binding Assay

The earliest approach for assessing protein covalent binding is using radiolabeled compounds with human liver microsomal or hepatocyte incubations. It is the gold-standard approach for measuring RM formation and can quantify the extent of RMs covalently binding to proteins. The original experimental protocol for determining covalent binding properties of radioactive drugs was developed by Evans et al. [20]. In their protocol *in vitro* human liver microsomes are used as the metabolizing enzyme source and incubated with radioactive drug and cofactors for oxidative metabolism. Quantification of covalent binding is measured through counting the radioactivity of the protein pellet.

Although the covalent binding assay is reliable and widely used by pharmaceutical companies, the use of radiolabeled drugs makes running high throughput screening difficult, and it is an expensive option for assessing whether a compound is likely to undergo bioactivation. A more common and economical screening approach is using electrophile trapping experiments that do not require radiolabeled drugs.

2.3.4.2 Electrophile Trapping Experiments

Electrophile trapping is the primary approach for screening RMs in the early phase of drug discovery without involving a radioactive drug. Normally it requires the addition of trapping agents (i.e. GSH) to microsomal incubation followed by liquid chromatography mass spectrometry (LC-MS) analysis. This approach also helps indicate which metabolite structure is responsible for this reactivity and characterizes the mechanism of covalent binding. It generates the stable trapping adducts and/or conjugates of electrophilic RMs using nucleophilic trapping reagents or *in vitro* incubation, along with hepatic microsome and cellular, animal, and human studies. The biological nucleophile GSH is the most frequently-used nucleophile reagent, commonly used for trapping soft electrophiles (e.g. epoxides, quinones, and quinone methides). For certain hard electrophiles with high charge density and high polarization, hard nucleophiles such as cyanide anion (CN^-) are preferred because the GSH adducts are unstable and GSHs have limited trapping efficiency.

Following *in vitro* incubations, various analytical approaches such as LC-MS, fluorescence or radiochemical detection could be used to analyze the trapped adducts and/or conjugates of RMs. Mass spectrometric methods are the most common approach. Taking advantage of specific fragmentation behavior of the peptide moiety of nucleophilic trapping reagents, stable trapping adducts, and/or conjugates of electrophilic RMs can be detected and characterized. For example, the neutral loss of 129 Da is commonly used by positive ion electrospray-tandem mass spectrometry to provide a generic endpoint for GSH trapping and measurement. Without using the radiolabeled drugs, this protocol can provide semi-quantitative estimates of adduct formation for RMs.

2.3.4.3 Time Dependent Inactivation of CYP450 Enzymes

Occasionally the formed RMs (electrophiles) are so highly reactive that they cannot flow out from the active site of the P450 enzymes that catalyzed their formation. Electrophilic intermediates derived from drug molecules are so reactive that they can covalently bind directly to an active site amino acid residue in the CYP enzyme itself. Covalent modification in the active site of a P450 enzyme may lead to the loss of enzyme activity over time via the modification of the P450 apoprotein. Therefore, time dependent inactivation (TDI) of CYP450 enzymes and other drug-metabolizing enzymes also indicates the generation of RMs during the drug metabolism process. Additionally, drug–drug interactions could be caused by this irreversible P450 inhibition. Covalent modification of P450 enzymes can also result in a neoantigen formation and trigger an autoimmune response in DILI.

The TDI assay is a two-step assay in which the drug of interest is preincubated with a source of P450 enzymes, such as human liver microsomes. This method can detect two scenarios for TDI: (i) time dependent loss of CYP activity following incubation at a single compound concentration; and (ii) an IC50 shift after preincubation at multiple concentrations. In the first step the tested compound is preincubated at multiple concentrations in the *in vitro* incubation of human liver microsome in the presence and absence of β-nicotinamide adenine dinucleotide

2′-phosphate reduced tetrasodium salt (NADPH) for a preset time period (usually ~30 minutes). The second step is the measurement of CYP450 activity using a probe CYP substrate. The IC50 values upon preincubation were calculated in the presence and absence of NADPH cofactor, and a decrease in IC50 will suggest the presence of TDI of CYP450 enzymes.

Electrophile trapping experiments are typically used for the stable conjugates of RMs. If a certain amount of formed RMs cannot escape the active site of the P450 enzymes, the trapping assays usually are not able to detect the RMs associated with the inactivation of P450 enzymes. The TDI assay therefore can supplement the electrophile trapping assay in detecting RM formation. Nakayama et al. [21] demonstrated that the combination of TDI assays and GSH trapping assays significantly correlated with the extent of covalent binding assay ($r = 0.77, P < 0.0001$), but both alone are not correlated, suggesting the combination of these two assays provides an alternative to the covalent binding assay for identifying RM formation by the drug molecules.

2.4 Hepatic Transporters

Hepatic transporter proteins facilitate drug metabolism and elimination by regulating the movement of drugs across hepatocyte cell membranes. They also are essential in maintaining bile acid homeostasis. Drug–drug interactions and genetic variants that impair the function of hepatic drug transporters can result in bile acid accumulation, which can lead to toxicity and DILI.

Hepatic transporters are generally described as influx or efflux transporters (Figure 2.4). Influx transporters located on the basolateral membrane of hepatocytes remove drugs and other compounds from blood in the liver sinusoidal space and transport into hepatocytes. Influx transporters include members of the solute carrier (SLC) superfamily, such as the sodium-independent organic anion transporters (OATs) and organic anion transporting polypeptides (OATPs), the organic cation transporters (OCTs), and the sodium-dependent sodium taurocholate cotransporting polypeptide (NTCP) transporter. The OATP transporters OATP1B1 (*SLCO1B1*), OATP1B3 (*SLCO1B3*), and OATP2B1 (*SLCO2B1*) are most abundant in the liver [22] and are responsible for the uptake of many endogenous compounds, as well as statins, methotrexate, fexofenadine, bosentan, and rifampicin, among other drugs. OATP1B1 and OAT1B3 are also involved in bilirubin and bile acid uptake; however, the NTCP is primarily responsible for bile acid uptake. The OATP transporters are associated with various drug–drug interactions; for example, OATP1B1 and OATP1B3 are inhibited by atazanavir and ritonavir, cyclosporine, and gemfibrozil [23], and OATP1B1 is inhibited by the tyrosine kinase inhibitors pazopanib and nilotinib [24]. Drugs such as cyclosporin and rifampin

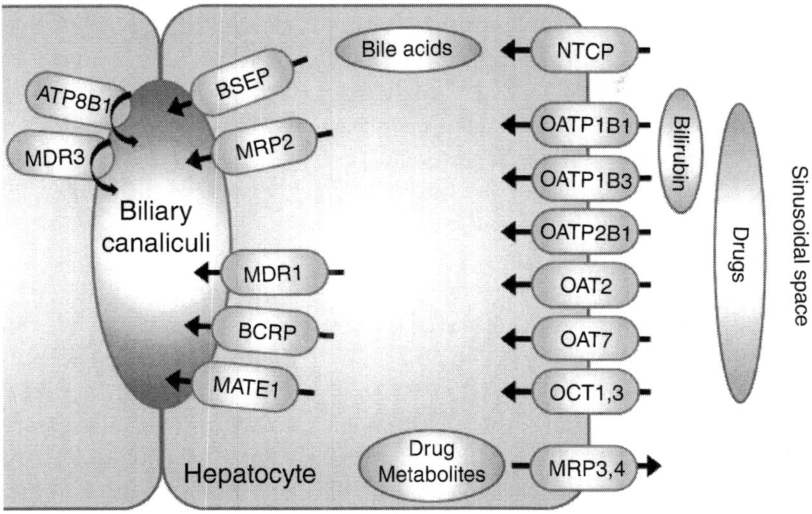

Figure 2.4 Hepatic proteins involved in drug, bilirubin, and bile acid transport. Drugs are removed from the blood by influx transporters on the basolateral membrane. These influx transporters include organic anion-transporting polypeptides (OATP), organic anion transporters (OAT), and organic cation transporters (OCT), all of which belong to the solute carrier superfamily. OATP1B1 and OAT1B3 also are involved in bilirubin and bile acid uptake; however, the sodium taurocholate co-transporting polypeptide (NTCP) is primarily responsible for transporting bile acids into the hepatocyte from the sinusoidal space. The bile salt export pump (BSEP) transports bile salts from the hepatocyte into the biliary canaliculi, while the multidrug resistance-associated protein 2 (MRP2) transports sulphated bile salts as well as drug metabolites into the biliary canaliculi. Multidrug resistance protein 3 (MDR3) and ATPase Phospholipid Transporting 8B1 (APT8B1) translocate phospholipids that, together with bile acids, form mixed micelles which protect the biliary tree from the detergent-like effects of bile salts. Located on the basolateral membrane, multidrug resistance associated proteins MRP3 (ABCC3) and MRP4 (ABCC4) return drug metabolites and bile acids to the liver sinusoids. If cholestatic conditions exist, MRP3 and MRP4 are upregulated to protect hepatocytes from toxic bile acid accumulation. Drug metabolites are exported into the biliary canaliculi by multidrug resistance protein 1 (MDR1), breast cancer resistance protein (BCRP), multidrug and toxin extrusion protein 1 (MATE1), and multidrug resistance-associated protein 2 (MRP2). ATP8B1, ATPase Phospholipid Transporting 8B1; BCRP, breast cancer resistance protein (BCRP); BSEP, bile salt export pump; MATE, multidrug and toxin extrusion protein; MDR, multidrug resistance protein; MRP, multidrug resistance-associated protein; NTCP, sodium taurocholate co-transporting polypeptide; OAT, organic anion transporter; OATP, organic anion-transporting polypeptide; OCT, organic cation transporter.

that inhibit both OATP1B1 and OATP1B3 disrupt bilirubin uptake and can cause hyperbilirubinemia [25, 26]. Rotor type hyperbilirubinemia occurs when genetic mutations render both OATP1B1 and OATP1B3 deficient, thereby inhibiting bilirubin uptake [27].

Efflux transporters are located either on the basolateral membrane or the canalicular membrane and belong to the ATP-binding cassette (ABC) transporter superfamily. Multidrug resistance-associated proteins MRP3 (*ABCC3*) and MRP4 (*ABCC4*) are located on the basolateral membrane and return drug metabolites and bile acids from liver cells to the sinusoid blood. Under cholestatic conditions, MRP3 and MRP4 are upregulated in order to protect hepatocytes from bile acid accumulation [28, 29].

Transporters on the canalicular membrane are responsible for the export of drug metabolites and bile components. Drug metabolites are exported into the biliary canaliculi by multidrug resistance protein 1 (MDR1), breast cancer resistance protein (BCRP), multidrug and toxin extrusion protein 1 (MATE1) and multidrug resistance-associated protein 2 (MRP2). In addition to drug metabolites, MRP2 also transports sulphated bile salts and conjugated bilirubin into the biliary canaliculi. The BSEP transports bile salts from the hepatocyte into the biliary canaliculi. Multidrug resistance protein 3 (MDR3) and ATPase Phospholipid Transporting 8B1 (APT8B1) translocate phospholipids that, together with bile acids, form mixed micelles which protect the biliary tree from the detergent-like effects of bile salts.

Efficient export of bile salts and acids is essential, as their accumulation can be toxic for hepatocytes due to their detergent-like effects. Drug interactions or genetic defects that impair hepatic transporters involved in bile salt or phospholipid secretion can result in intrahepatic cholestasis (Table 2.1). For example, mutations in *ABCB3*, which encodes the phospholipid transporter MDR3, can cause progressive familial intrahepatic cholestasis 3, a condition that usually presents as cholestasis in early childhood and may progress to end-stage liver disease [31]. MDR3 is inhibited by azole antifungals, including ketoconazole, itraconazole, and posaconazole [33]. Loss-of-function mutations in *ABCB11*, the gene encoding the BSEP, cause progressive familial intrahepatic cholestasis 2 [31] and drugs that inhibit the BSEP, such as bosentan, cyclosporin A, and troglitazone, can induce cholestatic DILI [34, 35].

Increasing evidence has shown that *in vitro* inhibition of the BSEP is one of the essential mechanisms leading to DILI and is also a recognized risk factor for DILI [34, 36]. However, it has been suggested that BSEP inhibition alone might not be a good predictor of DILI risk [37] as drugs inhibiting the BSEP often inhibit other transporters as well, and it may be the cumulative effect that causes toxic bile acid accumulation [38]. In support of this theory, hepatoxicity induced by isoniazid and rifampicin may be due to the inhibition of both NTCP and BSEP [39]. Additionally, the discontinued HER2 tyrosine kinase inhibitor CP-724,714 inhibits efflux transporters BSEP and MDR1, which contributes to the toxic accumulation of drug and bile acids [40].

Table 2.1 Genetic diseases related to hepatic transporters.

Gene/protein	Function	Related disease	Clinical phenotypes	References
Basolateral influx transporters				
SLCO1B1 organic-anion transporting polypeptide 1B1 (OAT1B1) *SLCO1B3* Organic-anion transporting polypeptide 1B3 (OATP1B3)	Sodium-independent uptake of drugs and endogenous compounds, including bilirubin, bile acids, and statins	Rotor syndrome, caused by deficiency in both OATP1B1 and OATP1B3	Conjugated hyperbilirubinemia, coproporphyrinuria	[27]
SLC10A1 sodium taurocholate co-transporting polypeptide (NTCP)	Sodium-dependent bile acid uptake transporter. Also transports statins	NTCP deficiency	Extreme elevated levels of bile salts	[30]
Canalicular efflux transporters				
ATP8B1 ATPase phospholipid transporting 8B1	ATP-dependent translocation of aminophospholipids into bile lumen	Progressive familial intrahepatic cholestasis 1	Cholestasis in children	[31]
ABCB4 multidrug resistance protein 3 (MDR3)	ATP-dependent translocation of phosphatidylcholine into bile lumen	Progressive familial intrahepatic cholestasis 3	Cholestasis in children	[31]
ABCB11 bile salt export pump (BSEP)	ATP-dependent bile acid transport into bile lumen	Progressive familial intrahepatic cholestasis 2	Cholestasis in children	[31]
ABCC2 MRP2: multidrug resistance associated protein 2	ATP-dependent transport bilirubin and drug metabolites transport into bile lumen	Dubin–Johnson syndrome	Conjugated hyperbilirubinemia, jaundice	[32]

2.5 Genetic Variants and Their Impact for Pharmacokinetic Behavior and Safety

Genetic variation can result in decreased, unchanged, or increased enzymatic function [41, 42], and therefore could significantly impact drug safety by altering pharmacokinetic characteristics in certain individuals [43]. Based on decades of pharmacogenetic research, the association between genetic variation and drug metabolism has been established for many drugs [44]. The FDA has published a list of pharmacogenetic associations with sufficient scientific evidence [45]. However, often contradictory evidence for a given drug is reported, usually due to the limited statistical power in many studies. This is of particular importance in establishing associations between genotypes and DILI, as DILI has a very low incidence (~20 per 100 000 inhabitants annually) and clinical diagnosis is challenging [46]. The association between genotype and hepatotoxicity has been suggested for CYP2C9-Bosentan, NAT2-isoniazid and others (Table 2.2).

For the major CYPs the single nucleotide polymorphism (SNP) has been estimated to be 14–127 per kilobase pair [62]. Of these SNPs, 9–41% are nonsynonymous variants. Only a limited number of SNPs (6–30%) have been cataloged or functionally annotated by the Pharmacogene Variation (PharmVar) Consortium, a public central repository for pharmacogene (PGx) variation (www.pharmvar.org). In addition to SNPs, variations such as insertions and deletions are present. The polymorphism phenotype can be categorized into four types; i.e. poor metabolizer, intermediate metabolizer, extensive metabolizer, and ultra rapid metabolizer. Poor metabolizer usually refers to individuals carrying homozygous or compound heterozygous null alleles (no function). In intermediate metabolizers one allele is partially defective while the other either is partially defective or null. For extensive metabolizers at least one allele is fully functional. Individuals with gain-of-function polymorphisms, e.g. carrying extra gene copies and mutations that enhance enzyme activity, belong to the ultra rapid metabolizer category.

2.5.1 *CYP3A4*

More than 34 SNPs of *CYP3A4* have been included in the PharmVar, but generally they do not translate to significant interindividual variability [41, 42]. The observed variability could not be explained by the SNP variations in the coding and 5′-flanking regions of *CYP3A4* [63]. Instead, the variations that affect *CYP3A4* expression are more relevant to drug metabolism. The *cis* genetic variant *CYP3A4*22* has a 15389C>T substitution in intron 6, which results in decreased hepatic *CYP3A4* expression. It has been suggested that *CYP3A4*22* might be of clinical relevance to statin therapy, as the plasma concentration of the statin drugs (e.g. simvastatin and lovastatin) largely is dependent on CYP3A4 activity [64]. In

Table 2.2 Selected reports for the association between gene variations and drug-induced liver injury.

Gene	Variant	Function	Affected subgroups	Drug	DILI risk association	References
CYP2B6	*1H and *1J (-2320T>C)	Increased expression	Ultrarapid metabolizers	Ticlopidine	Increased risk	[47]
CYP2B6	*6	Decreased function	Poor metabolizers	Efavirenz	Increased risk	[48]
CYP2C9	*2	Decreased function	Poor metabolizers	Bosentan	Increased risk	[49, 50]
CYP2E1	c1/c1	Increased expression	Ultrarapid metabolizers	Isoniazid	Increased risk	[51]
NAT2	*4	Without active alleles	Slow acetylators	Isoniazid	Increased risk	[51–53]
UGT1A6	A528G	Silent mutation		Tolcapone	Increased risk	[54]
UGT2B7	*2	Decreased function	Poor metabolizers	Diclofenac	Increased risk	[55, 56]
GSTM1 and GSTT1	Double null genotype	Decreased function	Poor metabolizers	Troglitazone	Increased risk	[57]
ABCB1	3435C>T	Decreased expression		Nevirapine	Decreased risk	[58, 59]
ABCB11	D676Y, G885R and V444A	Decreased function		Ethinylestradiol/ gestodene	Increased risk	[60]
ABCC2	C-24T	Decreased expression		Diclofenac	Increased risk	[55]
ABCC2	C-24T	Decreased expression		Deferasirox	Increased risk	[61]

a small cohort study the patients carrying a *CYP3A4*22* allele required a significantly lower dose to reach target levels of low-density lipoprotein-cholesterol [65]. Moreover, *CYP3A4* expression is also affected by the *trans* genetic variations. Polymorphisms in *CYP3A4* regulators, including *FoxA2/A3*, *HNF4a*, *PXR*, and *MDR1*, have been suggested to explain ~24% of the interindividual variability in hepatic *CYP3A4* expression [66]. Although CYP3A4 is the most abundant CYP in the liver and catalyzes the oxidation of a broad spectrum of drugs, little association has been established so far between *CYP3A4* polymorphisms and DILI [6].

2.5.2 *CYP3A5*

CYP3A5 expression is highly polymorphic. The most common loss-of-function variant is *CYP3A5*3*, which harbors a 6986A>G mutation in the third intron [67]. This causes aberrant alternative splicing and a premature stop in protein translation. The resulting truncated protein is defective in enzymatic function. *CYP3A5* genotype-based dosing has been suggested for the immunosuppressant tacrolimus, used in patients who received kidney [68, 69]. The therapeutic index of tacrolimus is narrow, and adverse reactions such as nephrotoxicity, hypertension, and diabetes can occur at concentrations slightly above or within the therapeutic range [70]. Therefore, therapeutic drug monitoring and dose adjustment of tacrolimus are clinically important. *CYP3A5*3* reportedly is associated with lower tacrolimus clearance, and poor metabolizer patients with *CYP3A5*3* require lower doses to achieve target concentrations [71]. In a cohort study, 39% of the variability in the blood concentration-to-dose ratio of tacrolimus could be explained by *CYP3A5*3* alone [72]. Whether *CYP3A5* genotype-based dosing is beneficial to kidney transplant recipients requires further investigation; however, it is considered an option in clinical practice.

2.5.3 *CYP2D6*

The *CYP2D6* polymorphism has been studied extensively, and 139 major variants have been cataloged in the PharmVar. The most common null allele in the Caucasian population is *CYP2D6*4*, accounting for the majority (70–90%) of the poor metabolizer phenotype [41, 73]. *CYP2D6*4* carriers all have a 1847G>A mutation, causing splicing defect and no protein expression in the liver. The *CYP2D6*10* allele is most commonly found in Asians [74]. Double SNP mutations 100C>T (P34S) and 4181G>C (S486T) render the protein unstable. In the African population the *CYP2D6*17* allele has the highest frequency [75]. Triple mutations, 1022C>T (T107I), 2851C>T (R296C), and 4181G>C (S486T), lead to a structural change in the active site, which alters the substrate specificity. The pharmacogenetic impact of *CYP2D6* on opioid drugs has been demonstrated [76]. Tramadol, for example, is bioactivated by CYP2D6 into active metabolite

O-desmethyltramadol [77]. Poor metabolizers are insensitive to opioids and have much lower analgesic response [78]. In contrast, ultrarapid metabolizers with duplicated *CYP2D6* genes show significantly increased efficiency in converting opioid prodrugs such as tramadol and codeine into active metabolites, which could cause opioid overdose [79]. The impact of *CYP2D6* polymorphism is not limited to opioid drugs but covers a spectrum of clinical drugs. It might be a contributing factor to the hepatoxicity of perhexiline and trazodone [80, 81]. The case of a CYP2D6 poor metabolizer with perhexiline-associated DILI has been reported. Significantly higher plasma Cmax of *m*-chloro, 4-phenylpiperazine, a hepatotoxic metabolite of trazodone, was observed in poor metabolizers in comparison to extensive metabolizers. However, the pharmacogenetic association has not been confirmed.

2.5.4 *CYP2C9*

Genetic variation in *CYP2C9* has been recognized as a host factor for adverse drug reactions because many of the CYP2C9 substrates have a narrow therapeutic index [41]. The most studied variants are *CYP2C9*2* (3608C>T, R144C) and *CYP2C9*3* (42614A>C, I359L), both of which have decreased enzyme activity. They are more prevalent in Caucasians than in Asians and Africans [82]. The most investigated case for CYP2C9 pharmacogenetics involves the vitamin K antagonist warfarin. Although warfarin contains both *R*- and *S*-enantiomers, the *S*-enantiomer is three to five times more effective in anticoagulation than the *R*-enantiomer [83]. *S*-warfarin metabolism is highly dependent on CYP2C9. Both *CYP2C9*2* and *CYP2C9*3* carriers have decreased warfarin biotransformation and thus could be considered for dose adjustment, especially when genetic variation in *VKORC1* (vitamin K epoxide reductase, complex I) is also present [84, 85]. *CYP2C9* polymorphism also affects the metabolism of some NSAIDs, such as diclofenac and ibuprofen. A potential pharmacogenetic association of CYP2C9 with diclofenac hepatotoxicity was proposed [86], but was negated by pharmacokinetics studies [87, 88]. *CYP2C9*2* has been reported to be associated with bosentan-related DILI, as the metabolism of bosentan is substantially reduced in *CYP2C9*2* carriers [49, 50]. *CYP2C9*3* was implicated to hepatotoxicity of leflunomide in a single case [89].

2.5.5 *CYP2C19*

The null allele of *CYP2C19* is common, especially in Asians, with a frequency of 13–25% [41]. *CYP2C19*2* and *CYP2C19*3* are the two most important nonfunctional variants, which occur mainly in Caucasians and Asians, respectively [90]. *CYP2C19*2* has a splicing defect while *CYP2C19*3* has a premature stop codon. The bioactivation of the antiplatelet drug clopidogrel is dependent largely on

CYP2C19 activity [23]. Poor metabolizer patients reportedly are associated with a decreased anticoagulant response and an increased risk of adverse cardiovascular events [91–93]. An isolated number of cases suggested a possible association of CYP2C19 variants with hepatotoxicity of Atrium (phenobarbital, febarbamate, and difebarbamate) [94] and troglitazone [95].

2.5.6 *CYP2B6*

The most common variant is *CYP2B6*6*, which harbors two SNP mutations (15631G>T and 18053A>G), causing two amino acid substitutions (Q172H and K262R). *CYP2B6*6* is prevalent in different populations with variable frequency [96]. *CYP2B6* has been suggested to have clinical implications in HIV therapy [97]. The antiretroviral medication efavirenz is metabolized by CYP2B6. Patients with *CYP2B6*6* have decreased CYP2B6 activity and are associated with high plasma levels of efavirenz, which could increase the risk of central neuron system adverse events [92] and hepatic adverse events [48]. *CYP2B6*1H and *1J* are associated with ticlopidine hepatotoxicity in Japanese populations, possibly due to increased *CYP2B6* expression [47].

2.5.7 *UGT1A1*

Ten of the UGT1A isoforms (1–10) are encoded by the *UGT1A* gene locus [98]. Currently 113 UGT1A1 variants have been cataloged by the UGT nomenclature committee. The clinically-relevant variant *UGT1A1*28* has an extra TA pair inserted into the $(TA)_6$ repeats in the TATA box [99]. This disrupts the transcriptional regulation of *UGT1A1*, and thus reduces protein expression. *UGT1A1*28* carriers exhibit impaired glucuronidation of lipophilic molecules, including bilirubin, resulting in decreased bilirubin clearance (Gilbert's syndrome) [98, 100].). It has been associated with the toxicity of the anticancer drug irinotecan [101]. Irinotecan is a prodrug that undergoes bioactivation into active metabolite SN-38, followed by inactivation by glucuronidation. The therapeutic window of irinotecan is rather small. Patients with *UGT1A1*28* alleles could be at increased risk of adverse drug reactions due to its less effective clearance. Some SNPs in the *UGT1A* locus (e.g. C908G in the 3′-UTR region) have been reported to be associated with the hepatic toxicity of tolcapone, all of which were in significant linkage disequilibrium with the silent mutation *UGT1A6-A528G* [54].

2.5.8 NAT2

NAT2 is expressed predominantly in the liver. More than 100 NAT2 alleles are being cataloged by the Arylamine *N*-acetyltransferase Gene Nomenclature Committee. Pharmacogenetic studies of NAT2 largely have focused on the anti-tuberculosis drug isoniazid. NAT2 is critical to the clearance of isoniazid and its metabolites, including acetylhydrazine and hydrazine, which have been associated with hepatotoxicity [102, 103]. Patients carrying NAT2 variants with decreased enzymatic function (slow acetylator) have been reported to have higher isoniazid exposure and increased risk of DILI during anti-tuberculosis treatment [52, 104, 105]. This suggests that genotype-based dosing for isoniazid might be of clinical relevance. In addition to the NAT2 pathway, an alternative pathway comprised of CYP2E1 and GST plays a role in the clearance of isoniazid. *CYP2E1* polymorphisms also have a suggested association with isoniazid hepatotoxicity [51, 106], however, with inconsistent evidence [104, 107]. The pharmacogenetic association remains uncertain.

2.5.9 Hepatic Transporters

Certain genetic polymorphisms in drug transporter genes have been associated with increased DILI risk. In a study comparing 33 DILI patients and 95 European controls, Lang et al. sequenced *ABCB11* and *ABCB4*, which encode BSEP and MDR3, and found four nonsynonymous mutations significantly associated with DILI, one of which resulted in a nonfunctional protein (*ABCB11*, exon 21: 2563G>A→G855R) [60]. Ciccacci et al. identified a variant in *MDR1* (c.3435C>T) associated with nevirapine-induced hepatoxicity in a study of 78 nevirapine-induced hepatoxicity cases and 78 patients without hepatoxicity in Mozambique [108]. Another study found evidence that certain *MDR1* polymorphisms can influence the basal CYP3A4 expression or function, i.e. individuals homozygous for *MDR1* 2677T (Ser893) had a higher hepatic expression or function of CYP3A4 than those homozygous for 2677G (Ala893) [109]. Daly et al. found that a variant in *ABCC2*, which encodes MRP2, was significantly associated with diclofenac-induced hepatoxicity in a European population [55]. Interestingly, this same variant, *ABCC2* C-24T, was associated with a higher clearance rate and a shorter half-life of deferasirox in the Chinese population [61]. A study of 94 drug- and herb-induced hepatoxicity cases identified several polymorphisms in *ABCC2* that were associated with susceptibility to liver injury in the Korean population [110].

2.6 Summary

Increasing evidence suggests that drug metabolism and hepatic transporters play essential roles in the development of hepatotoxicity. The formation of chemically-RMs and the abnormal accumulation of toxic bile acids or drug metabolites are two proven mechanisms causing DILI. The formation of RMs is a byproduct of physiological biotransformation of phases I and II metabolism enzymes, while bile acid accumulation results from abrupted functions of the hepatic transporters. Some of these drug-metabolizing enzymes and hepatic transporters have significantly varied expression levels among individuals, which can lead to outcomes other than the development of DILI. In addition to the gene variations summarized in this chapter, many environmental and host factors; such as age, gender, co-medications, comorbidities, and life-style all affect the susceptibility of individuals through the disturbance of drug-metabolizing enzymes and hepatic transporters. The contribution of these factors has been well-reviewed in the literature [111, 112] and the nature of the co-factors in these events has led to the "multiple determinants hypothesis" for idiosyncratic drug toxicity [113]. Further investigations are still needed to establish the association between interactions of these factors and their contributions to the development of DILI.

Acknowledgment

The authors thank Joanne Berger with the FDA Library for manuscript editing assistance.

Disclaimer

The views expressed in this manuscript do not necessarily represent those of the U.S. Food and Drug Administration.

References

1 Olson H, Betton G, Robinson D, Thomas K, Monro A, Kolaja G, et al. Concordance of the toxicity of pharmaceuticals in humans and in animals. *Regulatory Toxicology and Pharmacology* 2000; 32(1): 56–67.

2 Chen M, Suzuki A, Thakkar S, Yu K, Hu C, Tong W. DILIrank: the largest reference drug list ranked by the risk for developing drug-induced liver injury in humans. *Drug Discovery Today* 2016; 21(4): 648–53.

3 Chen M, Vijay V, Shi Q, Liu Z, Fang H, Tong W. FDA-approved drug labeling for the study of drug-induced liver injury. *Drug Discovery Today* 2011; 16 (15–16): 697–703.

4 Watkins P. Drug safety sciences and the bottleneck in drug development. *Clinical Pharmacology & Therapeutics* 2011; 89(6): 788–90.

5 US FDA. Drug-Induced Liver Injury: Premarketing Clinical Evaluation; 2009. Available from: https://www.fda.gov/regulatory-information/search-fda-guidance-documents/drug-induced-liver-injury-premarketing-clinical-evaluation.

6 Zanger UM, Schwab M. Cytochrome P 450 enzymes in drug metabolism: regulation of gene expression, enzyme activities, and impact of genetic variation. *Pharmacology & Therapeutics* 2013; 138(1): 103–41.

7 Uetrecht JP, editor *Myeloperoxidase as a Generator of Drug Free Radicals. Biochemical Society Symposia*; 1995: Portland Press Limited.

8 Srivastava A, Maggs J, Antoine D, Williams D, Smith D, Park B. Role of reactive metabolites in drug-induced hepatotoxicity. In: Uetrecht JP, editor *Adverse Drug Reactions*; 2010: Springer. p. 165–94.

9 Bischoff R, Schlüter H. Amino acids: chemistry, functionality and selected non-enzymatic post-translational modifications. *Journal of Proteomics* 2012; 75(8): 2275–96.

10 Walgren JL, Mitchell MD, Thompson DC. Role of metabolism in drug-induced idiosyncratic hepatotoxicity. *Critical Reviews in Toxicology* 2005; 35(4): 325–61.

11 Dahms M, Spahn-Langguth H. Covalent binding of acidic drugs via reactive intermediates: detection of benoxaprofen and flunoxaprofen protein adducts in biological material. *Die Pharmazie* 1996; 51 (11): 874–81.

12 FDA TU. Safety Testing of Drug Metabolites Guidance for Industry; 2020. Available from: https://www.fda.gov/media/72279/download.

13 Uetrecht J. Prediction of a new drug's potential to cause idiosyncratic reactions. *Current Opinion in Drug Discovery & Development* 2001; 4(1): 55–9.

14 Lammert C, Bjornsson E, Niklasson A, Chalasani N. Oral medications with significant hepatic metabolism at higher risk for hepatic adverse events. *Hepatology* 2010; 51(2): 615–20.

15 Chen M, Borlak J, Tong W. A model to predict severity of drug-induced liver injury in humans. *Hepatology* 2016; 64(3): 931–40.

16 Limban C, Nuţă DC, Chiriţă C, Negreş S, Arsene AL, Goumenou M, et al. The use of structural alerts to avoid the toxicity of pharmaceuticals. *Toxicology Reports* 2018; 5: 943–53.

17 Stepan AF, Walker DP, Bauman J, Price DA, Baillie TA, Kalgutkar AS, et al. Structural alert/reactive metabolite concept as applied in medicinal chemistry to mitigate the risk of idiosyncratic drug toxicity: a perspective based on the critical examination of trends in the top 200 drugs marketed in the United States. *Chemical Research in Toxicology* 2011; 24(9): 1345–410.

18 Enoch S, Ellison C, Schultz T, Cronin M. A review of the electrophilic reaction chemistry involved in covalent protein binding relevant to toxicity. *Critical Reviews in Toxicology* 2011; 41(9): 783–802.

19 Claesson A, Minidis A. Systematic approach to organizing structural alerts for reactive metabolite formation from potential drugs. *Chemical Research in Toxicology* 2018; 31(6): 389–411.

20 Evans DC, Watt AP, Nicoll-Griffith DA, Baillie TA. Drug– protein adducts: an industry perspective on minimizing the potential for drug bioactivation in drug discovery and development. *Chemical Research in Toxicology* 2004; 17(1): 3–16.

21 Nakayama S, Takakusa H, Watanabe A, Miyaji Y, Suzuki W, Sugiyama D, et al. Combination of GSH trapping and time-dependent inhibition assays as a predictive method of drugs generating highly reactive metabolites. *Drug Metabolism and Disposition* 2011; 39(7): 1247–54.

22 Zaïr ZM, Eloranta JJ, Stieger B, Kullak-Ublick GA. Pharmacogenetics of OATP (SLC21/SLCO), OAT and OCT (SLC22) and PEPT (SLC15) transporters in the intestine, liver and kidney. *Pharmacogenomics*, 2008, 9(5):597–624.

23 Dong AN, Tan BH, Pan Y, Ong CE. Cytochrome P 450 genotype-guided drug therapies: an update on current states. *Clinical and Experimental Pharmacology and Physiology* 2018; 45 (10): 991–1001.

24 Khurana V, Minocha M, Pal D, Mitra AK. Inhibition of OATP-1B1 and OATP-1B3 by tyrosine kinase inhibitors. *Drug Metabolism and Drug Interactions* 2014; 29(4): 249–59.

25 Campbell SD, de Morais SM, Xu JJ. Inhibition of human organic anion transporting polypeptide OATP 1B1 as a mechanism of drug-induced hyperbilirubinemia. *Chemico-Biological Interactions* 2004; 150(2): 179–87.

26 Chiou WJ, de Morais SM, Kikuchi R, Voorman RL, Li X, Bow DA. *in vitro* OATP1B1 and OATP1B3 inhibition is associated with observations of benign clinical unconjugated hyperbilirubinemia. *Xenobiotica* 2014; 44(3): 276–82.

27 van de Steeg E, Stránecký V, Hartmannová H, Nosková L, Hřebíček M, Wagenaar E, et al. Complete OATP1B1 and OATP1B3 deficiency causes human Rotor syndrome by interrupting conjugated bilirubin reuptake into the liver. *The Journal of Clinical Investigation* 2012; 122(2): 519–28.

28 Denk GU, Soroka CJ, Takeyama Y, Chen W-S, Schuetz JD, Boyer JL. Multidrug resistance-associated protein 4 is up-regulated in liver but down-regulated in kidney in obstructive cholestasis in the rat. *Journal of Hepatology* 2004; 40(4): 585–91.

29 Borst P, de Wolf C, van de Wetering K. Multidrug resistance-associated proteins 3, 4, and 5. *Pflügers Archiv-European Journal of Physiology* 2007; 453(5): 661–73.

30 Vaz FM, Paulusma CC, Huidekoper H, de Ru M, Lim C, Koster J, et al. Sodium taurocholate cotransporting polypeptide (SLC10A1) deficiency: conjugated hypercholanemia without a clear clinical phenotype. *Hepatology* 2015; 61(1): 260–7.

31 Davit-Spraul A, Gonzales E, Baussan C, Jacquemin E. Progressive familial intrahepatic cholestasis. *Orphanet Journal of Rare Diseases* 2009; 4(1): 1.

32 Paulusma CC, Kool M, Bosma PJ, Scheffer GL, ter Borg F, Scheper RJ, et al. A mutation in the human canalicular multispecific organic anion transporter gene causes the Dubin–Johnson syndrome. *Hepatology* 1997; 25(6): 1539–42.

33 Mahdi ZM, Synal-Hermanns U, Yoker A, Locher KP, Stieger B. Role of multidrug resistance protein 3 in antifungal-induced cholestasis. *Molecular Pharmacology* 2016; 90(1): 23–34.

34 Dawson S, Stahl S, Paul N, Barber J, Kenna JG. *in vitro* inhibition of the bile salt export pump correlates with risk of cholestatic drug-induced liver injury in humans. *Drug Metabolism and Disposition* 2012; 40(1): 130–8.

35 Fattinger K, Funk C, Pantze M, Weber C, Reichen J, Stieger B, et al. The endothelin antagonist bosentan inhibits the canalicular bile salt export pump: a potential mechanism for hepatic adverse reactions. *Clinical Pharmacology & Therapeutics* 2001; 69(4): 223–31.

36 Kis E, Ioja E, Rajnai Z, Jani M, Méhn D, Herédi-Szabó K, et al. BSEP inhibition–*in vitro* screens to assess cholestatic potential of drugs. *Toxicology in vitro* 2012; 26(8): 1294–9.

37 Chan R, Benet LZ. Measures of BSEP inhibition *in vitro* are not useful predictors of DILI. *Toxicological Sciences* 2017; 162(2): 499–508.

38 Watkins PB. The DILI-sim initiative: insights into hepatotoxicity mechanisms and biomarker interpretation. *Clinical and Translational Science* 2019; 12(2): 122–9.

39 Guo YX, Xu XF, Zhang QZ, Li C, Deng Y, Jiang P, et al. The inhibition of hepatic bile acids transporters Ntcp and Bsep is involved in the pathogenesis of isoniazid/rifampicin-induced hepatotoxicity. *Toxicology Mechanisms and Methods* 2015; 25(5): 382–7.

40 Feng B, Xu JJ, Bi Y-A, Mireles R, Davidson R, Duignan DB, et al. Role of hepatic transporters in the disposition and hepatotoxicity of a HER2 tyrosine kinase inhibitor CP-724, 714. *Toxicological Sciences* 2009; 108(2): 492–500.

41 Klein K, Zanger UM. Pharmacogenomics of cytochrome P450 3A4: recent progress toward the "missing heritability" problem. *Frontiers in Genetics* 2013; 4: 12.

42 Amacher DE. The primary role of hepatic metabolism in idiosyncratic drug-induced liver injury. *Expert Opinion on Drug Metabolism & Toxicology* 2012; 8(3): 335–47.

43 Madian AG, Wheeler HE, Jones RB, Dolan ME. Relating human genetic variation to variation in drug responses. *Trends in Genetics: TIG* 2012; 28 (10): 487–95.

44 Pachkoria K, Lucena MI, Molokhia M, Cueto R, Carballo AS, Carvajal A, et al. Genetic and molecular factors in drug-induced liver injury: a review. *Current Drug Safety* 2007; 2(2): 97–112.

45 FDA TU. Table of Pharmacogenetic Associations; 2020. Available from: https://www.fda.gov/medical-devices/precision-medicine/table-pharmacogenetic-associations?utm_campaign=2020-02-20%20Pharmacogenetic%20Associations%3A%20Scientific%20Evidence%20Underlying%20Gene-Drug%20Interactions&utm_medium=email&utm_source=Eloqua.

46 Sgro C, Clinard F, Ouazir K, Chanay H, Allard C, Guilleminet C, et al. Incidence of drug-induced hepatic injuries: a French population-based study. *Hepatology (Baltimore, MD)* 2002; 36(2): 451–5.

47 Ariyoshi N, Iga Y, Hirata K, Sato Y, Miura G, Ishii I, et al. Enhanced susceptibility of HLA-mediated ticlopidine-induced idiosyncratic hepatotoxicity by CYP2B6 polymorphism in Japanese. *Drug Metabolism and Pharmacokinetics* 2010; 25(3): 298–306.

48 Yimer G, Amogne W, Habtewold A, Makonnen E, Ueda N, Suda A, et al. High plasma efavirenz level and CYP2B6*6 are associated with efavirenz-based HAART-induced liver injury in the treatment of naïve HIV patients from Ethiopia: a prospective cohort study. *The Pharmacogenomics Journal* 2012; 12(6): 499–506.

49 Markova SM, De Marco T, Bendjilali N, Kobashigawa EA, Mefford J, Sodhi J, et al. Association of CYP2C9*2 with bosentan-induced liver injury. *Clinical Pharmacology & Therapeutics* 2013; 94(6): 678–86.

50 Seyfarth H-J, Favreau N, Tennert C, Ruffert C, Halank M, Wirtz H, et al. Genetic susceptibility to hepatoxicity due to bosentan treatment in pulmonary hypertension. *Annals of Hepatology* 2014; 13(6): 803–9.

51 Lee, SW, Chung, L, Huang, HH, Chuang, TY, Liou, YH, and Wu, L. NAT2 and CYP2E1 polymorphisms and susceptibility to first-line anti-tuberculosis drug-induced hepatitis. *The International Journal of Tuberculosis and Lung Disease* 2010; 14: 622–626.

52 Azuma J, Ohno M, Kubota R, Yokota S, Nagai T, Tsuyuguchi K, et al. NAT2 genotype guided regimen reduces isoniazid-induced liver injury and early treatment failure in the 6-month four-drug standard treatment of tuberculosis: a randomized controlled trial for pharmacogenetics-based therapy. *European Journal of Clinical Pharmacology* 2013; 69(5): 1091–101.

53 Cho H-J, Koh W-J, Ryu Y-J, Ki C-S, Nam M-H, Kim J-W, et al. Genetic polymorphisms of NAT2 and CYP2E1 associated with antituberculosis drug-induced hepatotoxicity in Korean patients with pulmonary tuberculosis. *Tuberculosis (Edinburgh, Scotland)* 2007; 87(6): 551–6.

54 Acuña G, Foernzler D, Leong D, Rabbia M, Smit R, Dorflinger E, et al. Pharmacogenetic analysis of adverse drug effect reveals genetic variant for susceptibility to liver toxicity. *The Pharmacogenomics Journal* 2002; 2(5): 327–34.

55 Daly AK, Aithal GP, Leathart JB, Swainsbury RA, Dang TS, Day CP. Genetic susceptibility to diclofenac-induced hepatotoxicity: contribution of UGT2B7, CYP2C8, and ABCC2 genotypes. *Gastroenterology* 2007; 132(1): 272–81.

56 Lazarska KE, Dekker SJ, Vermeulen NPE, Commandeur JNM. Effect of UGT2B7*2 and CYP2C8*4 polymorphisms on diclofenac metabolism. *Toxicology Letters* 2018; 284: 70–8.

57 Watanabe I, Tomita A, Shimizu M, Sugawara M, Yasumo H, Koishi R, et al. A study to survey susceptible genetic factors responsible for troglitazone-associated hepatotoxicity in Japanese patients with type 2 diabetes mellitus. *Clinical Pharmacology & Therapeutics* 2003; 73(5): 435–55.

58 Ritchie MD, Haas DW, Motsinger AA, Donahue JP, Erdem H, Raffanti S, et al. Drug transporter and metabolizing enzyme gene variants and nonnucleoside reverse-transcriptase inhibitor hepatotoxicity. *Clinical Infectious Diseases: An Official Publication of the Infectious Diseases Society of America* 2006; 43(6): 779–82.

59 Haas DW, Bartlett JA, Andersen JW, Sanne I, Wilkinson GR, Hinkle J, et al. Pharmacogenetics of nevirapine-associated hepatotoxicity: an Adult AIDS Clinical Trials Group collaboration. *Clinical Infectious Diseases: An Official Publication of the Infectious Diseases Society of America* 2006; 43(6): 783–6.

60 Lang C, Meier Y, Stieger B, Beuers U, Lang T, Kerb R, et al. Mutations and polymorphisms in the bile salt export pump and the multidrug resistance protein 3 associated with drug-induced liver injury. *Pharmacogenetics & Genomics* 2007; 17(1): 47–60.

61 Cao K, Ren G, Lu C, Wang Y, Tan Y, Zhou J, et al. ABCC2 c.-24 C> T single-nucleotide polymorphism was associated with the pharmacokinetic variability of deferasirox in Chinese subjects. *European Journal of Clinical Pharmacology* 2020; 76(1): 51–9.

62 Zanger UM, Klein K, Thomas M, Rieger JK, Tremmel R, Kandel BA, et al. Genetics, epigenetics, and regulation of drug-metabolizing cytochrome p450 enzymes. *Clinical Pharmacology and Therapeutics* 2014; 95(3): 258–61.

63 Hirota T, Ieiri I, Takane H, Maegawa S, Hosokawa M, Kobayashi K, et al. Allelic expression imbalance of the human CYP3A4 gene and individual phenotypic status. *Human Molecular Genetics* 2004; 13(23): 2959–69.

64 Neuvonen PJ, Kantola T, Kivistö KT. Simvastatin but not pravastatin is very susceptible to interaction with the CYP3A4 inhibitor itraconazole. *Clinical Pharmacology and Therapeutics* 1998; 63(3): 332–41.

65 Wang D, Guo Y, Wrighton SA, Cooke GE, Sadee W. Intronic polymorphism in CYP3A4 affects hepatic expression and response to statin drugs. *The Pharmacogenomics Journal* 2011; 11(4): 274–86.

66 Lamba V, Panetta JC, Strom S, Schuetz EG. Genetic predictors of interindividual variability in hepatic CYP3A4 expression. *Journal of Pharmacology and Experimental Therapeutics* 2010; 332(3): 1088–99.

67 Kuehl P, Zhang J, Lin Y, Lamba J, Assem M, Schuetz J, et al. Sequence diversity in CYP3A promoters and characterization of the genetic basis of polymorphic CYP3A5 expression. *Nature Genetics* 2001; 27(4): 383–91.

68 Elens L, Bouamar R, Hesselink DA, Haufroid V, van der Heiden IP, van Gelder T, et al. A new functional CYP3A4 intron 6 polymorphism significantly affects tacrolimus pharmacokinetics in kidney transplant recipients. *Clinical Chemistry* 2011; 57 (11): 1574–83.

69 Hesselink DA, van Schaik RHN, van der Heiden IP, van der Werf M, Gregoor PJHS, Lindemans J, et al. Genetic polymorphisms of the CYP3A4, CYP3A5, and

MDR-1 genes and pharmacokinetics of the calcineurin inhibitors cyclosporine and tacrolimus. *Clinical Pharmacology and Therapeutics* 2003; 74(3): 245–54.

70 Andrews LM, Li Y, De Winter BCM, Shi Y-Y, Baan CC, van Gelder T, et al. Pharmacokinetic considerations related to therapeutic drug monitoring of tacrolimus in kidney transplant patients. *Expert Opinion on Drug Metabolism & Toxicology*. 2017; 13 (12): 1225–36.

71 Staatz CE, Goodman LK, Tett SE. Effect of CYP3A and ABCB1 single nucleotide polymorphisms on the pharmacokinetics and pharmacodynamics of calcineurin inhibitors: part I. *Clinical Pharmacokinetics* 2010; 49(3): 141–75.

72 Birdwell KA, Grady B, Choi L, Xu H, Bian A, Denny JC, et al. The use of a DNA biobank linked to electronic medical records to characterize pharmacogenomic predictors of tacrolimus dose requirement in kidney transplant recipients. *Pharmacogenetics and Genomics* 2012; 22(1): 32–42.

73 Sachse C, Brockmöller J, Bauer S, Roots I. Cytochrome P450 2D6 variants in a Caucasian population: allele frequencies and phenotypic consequences. *American Journal of Human Genetics* 1997; 60(2): 284–95.

74 Johansson I, Oscarson M, Yue QY, Bertilsson L, Sjöqvist F, Ingelman-Sundberg M. Genetic analysis of the Chinese cytochrome P4502D locus: characterization of variant CYP2D6 genes present in subjects with diminished capacity for debrisoquine hydroxylation. *Molecular Pharmacology* 1994; 46(3): 452–9.

75 Masimirembwa C, Persson I, Bertilsson L, Hasler J, Ingelman-Sundberg M. A novel mutant variant of the CYP2D6 gene (CYP2D6*17) common in a black African population: association with diminished debrisoquine hydroxylase activity. *British Journal of Clinical Pharmacology* 1996; 42(6): 713–9.

76 Owusu Obeng A, Hamadeh I, Smith M. Review of opioid pharmacogenetics and considerations for pain management. *Pharmacotherapy* 2017; 37(9): 1105–21.

77 Rodieux F, Vutskits L, Posfay-Barbe KM, Habre W, Piguet V, Desmeules JA, et al. When the safe alternative is not that safe: tramadol prescribing in children. *Frontiers in Pharmacology* 2018; 9: 227–13.

78 Stamer UM, Musshoff F, Kobilay M, Madea B, Hoeft A, Stuber F. Concentrations of tramadol and O-desmethyltramadol enantiomers in different CYP2D6 genotypes. *Clinical Pharmacology and Therapeutics* 2007; 82(1): 41–7.

79 Gasche Y, Daali Y, Fathi M, Chiappe A, Cottini S, Dayer P, et al. Codeine intoxication associated with ultrarapid CYP2D6 metabolism. *The New England Journal of Medicine* 2004; 351 (27): 2827–31.

80 Morgan MY, Reshef R, Shah RR, Oates NS, Smith RL, Sherlock S. Impaired oxidation of debrisoquine in patients with perhexiline liver injury. *Gut* 1984; 25 (10): 1057–64.

81 Barbhaiya RH, Buch AB, Greene DS. Single and multiple dose pharmacokinetics of nefazodone in subjects classified as extensive and poor metabolizers of dextromethorphan. *British Journal of Clinical Pharmacology* 1996; 42(5): 573–81.

82 Johnson JA, Caudle KE, Gong L, Whirl-Carrillo M, Stein CM, Scott SA, et al. Clinical pharmacogenetics implementation consortium (CPIC) guideline for pharmacogenetics-guided warfarin dosing: 2017 update. *Clinical Pharmacology & Therapeutics* 2017; 102(3): 397–404.

83 Kaminsky LS, Zhang ZY. Human P450 metabolism of warfarin. *Pharmacology & Therapeutics*. 1997; 73(1): 67–74.

84 Sanderson S, Emery J, Higgins J. CYP2C9 gene variants, drug dose, and bleeding risk in warfarin-treated patients: a HuGEnet systematic review and meta-analysis. *Genetics in Medicine: Official Journal of the American College of Medical Genetics* 2005; 7(2): 97–104.

85 Consortium IWP, Klein TE, Altman RB, Eriksson N, Gage BF, Kimmel SE, et al. Estimation of the warfarin dose with clinical and pharmacogenetic data. *The New England Journal of Medicine* 2009; 360(8): 753–64.

86 Morin S, Loriot MA, Poirier JM, Tenneze L, Beaune PH, Funck-Brentano C, et al. Is diclofenac a valuable CYP2C9 probe in humans? *European Journal of Clinical Pharmacology* 2001; 56 (11): 793–7.

87 Yasar U, Eliasson E, Forslund-Bergengren C, Tybring G, Gadd M, Sjoqvist F, et al. The role of CYP2C9 genotype in the metabolism of diclofenac *in vivo* and *in vitro*. *European Journal of Clinical Pharmacology* 2001; 57 (10): 729–35.

88 Aithal GP, Day CP, Leathart JB, Daly AK. Relationship of polymorphism in CYP2C9 to genetic susceptibility to diclofenac-induced hepatitis. *Pharmacogenetics* 2000; 10(6): 511–8.

89 Sevilla-Mantilla C, Ortega L, Agúndez JAG, Fernández-Gutiérrez B, Ladero JM, Díaz-Rubio M. Leflunomide-induced acute hepatitis. *Digestive & Liver Disease: Official Journal of the Italian Society of Gastroenterology and the Italian Association for the Study of the Liver* 2004; 36(1): 82–4.

90 de Morais SM, Wilkinson GR, Blaisdell J, Meyer UA, Nakamura K, Goldstein JA. Identification of a new genetic defect responsible for the polymorphism of (S)-mephenytoin metabolism in Japanese. *Molecular Pharmacology* 1994; 46(4): 594–8.

91 Hulot J-S, Bura A, Villard E, Azizi M, Remones V, Goyenvalle C, et al. Cytochrome P450 2C19 loss-of-function polymorphism is a major determinant of clopidogrel responsiveness in healthy subjects. *Blood* 2006; 108(7): 2244–7.

92 Collet J-P, Hulot J-S, Pena A, Villard E, Esteve J-B, Silvain J, et al. Cytochrome P450 2C19 polymorphism in young patients treated with clopidogrel after myocardial infarction: a cohort study. *Lancet (London, England)* 2009; 373 (9660): 309–17.

93 Mega JL, Close SL, Wiviott SD, Shen L, Hockett RD, Brandt JT, et al. Cytochrome p-450 polymorphisms and response to clopidogrel. *The New England Journal of Medicine* 2009; 360(4): 354–62.

94 Horsmans Y, Lannes D, Pessayre D, Larrey D. Possible association between poor metabolism of mephenytoin and hepatotoxicity caused by Atrium, a fixed combination preparation containing phenobarbital, febarbamate and difebarbamate. *Journal of Hepatology* 1994; 21(6): 1075–9.

95 Farrell GC, Liddle C. Drugs and the liver updated, 2002. *Seminars in Liver Disease* 2002; 22(2): 109–13.

96 Lamba V, Lamba J, Yasuda K, Strom S, Davila J, Hancock ML, et al. Hepatic CYP2B6 expression: gender and ethnic differences and relationship to CYP2B6 genotype and CAR (constitutive androstane receptor) expression. *Journal of Pharmacology and Experimental Therapeutics* 2003; 307(3): 906–22.

97 Ward BA, Gorski JC, Jones DR, Hall SD, Flockhart DA, Desta Z. The cytochrome P450 2B6 (CYP2B6) is the main catalyst of efavirenz primary and secondary metabolism: implication for HIV/AIDS therapy and utility of efavirenz as a substrate marker of CYP2B6 catalytic activity. *Journal of Pharmacology and Experimental Therapeutics* 2003; 306(1): 287–300.

98 Strassburg CP. Pharmacogenetics of Gilbert's syndrome. *Pharmacogenomics* 2008; 9(6): 703–15.

99 Bosma PJ, Chowdhury JR, Bakker C, Gantla S, de Boer A, Oostra BA, et al. The genetic basis of the reduced expression of bilirubin UDP-glucuronosyltransferase 1 in Gilbert's syndrome. *The New England Journal of Medicine* 1995; 333 (18): 1171–5.

100 Zhang D, Zhang D, Cui D, Gambardella J, Ma L, Barros A, et al. Characterization of the UDP glucuronosyltransferase activity of human liver microsomes genotyped for the UGT1A1*28 polymorphism. *Drug Metabolism and Disposition* 2007; 35 (12): 2270–80.

101 Tukey RH, Strassburg CP, Mackenzie PI. Pharmacogenomics of human UDP-glucuronosyltransferases and irinotecan toxicity. *Molecular Pharmacology* 2002; 62(3): 446–50.

102 Metushi IG, Cai P, Zhu X, Nakagawa T, Uetrecht JP. A fresh look at the mechanism of isoniazid-induced hepatotoxicity. *Clinical Pharmacology & Therapeutics* 2011; 89(6): 911–4.

103 Tostmann A, Boeree MJ, Aarnoutse RE, de Lange WCM, van der Ven AJAM, Dekhuijzen R. Antituberculosis drug-induced hepatotoxicity: concise up-to-date review. *Journal of Gastroenterology & Hepatology* 2008; 23(2): 192–202.

104 Fukino K, Sasaki Y, Hirai S, Nakamura T, Hashimoto M, Yamagishi F, et al. Effects of N-acetyltransferase 2 (NAT2), CYP2E1 and Glutathione-S-transferase (GST) genotypes on the serum concentrations of isoniazid and metabolites in tuberculosis patients. *The Journal of Toxicological Sciences* 2008; 33(2): 187–95.

105 Saukkonen JJ, Cohn DL, Jasmer RM, Schenker S, Jereb JA, Nolan CM, et al. An official ATS statement: hepatotoxicity of antituberculosis therapy. *American Journal of Respiratory and Critical Care Medicine* 2006; 174(8): 935–52.

106 Vuilleumier N, Rossier MF, Chiappe A, Degoumois F, Dayer P, Mermillod B, et al. CYP2E1 genotype and isoniazid-induced hepatotoxicity in patients treated for latent tuberculosis. *European Journal of Clinical Pharmacology* 2006; 62(6): 423–9.

107 Ashrafian H, Horowitz JD, Frenneaux MP. Perhexiline. *Cardiovascular Drug Reviews* 2007; 25(1): 76–97.

108 Ciccacci C, Borgiani P, Ceffa S, Sirianni E, Marazzi MC, Altan AMD, et al. Nevirapine-induced hepatotoxicity and pharmacogenetics: a retrospective study in a population from Mozambique. *Pharmacogenomics* 2010; 11(1): 23–31.

109 Lamba J, Strom S, Venkataramanan R, Thummel KE, Lin YS, Liu W, et al. MDR1 genotype is associated with hepatic cytochrome P450 3A4 basal and induction phenotype. *Clinical Pharmacology & Therapeutics* 2006; 79(4): 325–38.

110 Choi JH, Ahn BM, Yi J, Lee JH, Lee JH, Nam SW, et al. MRP2 haplotypes confer differential susceptibility to toxic liver injury. *Pharmacogenetics and Genomics* 2007; 17(6): 403–15.

111 Chalasani N, Björnsson E. Risk factors for idiosyncratic drug-induced liver injury. *Gastroenterology* 2010; 138(7): 2246–59.

112 Chen M, Suzuki A, Borlak J, Andrade RJ, Lucena MI. Drug-induced liver injury: interactions between drug properties and host factors. *Journal of Hepatology* 2015; 63(2): 503–14.

113 Li AP. A review of the common properties of drugs with idiosyncratic hepatotoxicity and the “multiple determinant hypothesis” for the manifestation of idiosyncratic drug toxicity. *Chemico-Biological Interactions* 2002; 142 (1–2): 7–23.

3

Drug-Metabolism Enzymes and Transporter Activities as Risk Factors of Selected Marketed Drugs Associated with Drug-Induced Fatalities

Albert P. Li

In vitro ADMET LaboratoriesInc., 9221 Rumsey Road, Suite 8, Columbia, MD, USA

3.1 Introduction

Severe drug toxicity, especially drug-induced liver toxicity (DILI), is a major challenge in drug development. The occurrence of severe drug toxicity for marketed drugs is a clear indication that the current paradigm for drug safety evaluation during development, namely, preclinical evaluation in multiple species of laboratory animals followed by phases I, II, and III clinical trials, does not always provide adequate information allowing the assessment of human drug safety. The role of metabolism and transport in the toxicity of 12 marketed drugs which have been withdrawn from the market or received black box warning concerning their toxicity are reviewed here to provide evidence of their contribution to drug toxicity. The intent of this chapter is to emphasize the importance of the incorporation of mechanistic studies, at least starting with understanding the role of metabolism and transport in drug toxicity, to hopefully enhance the accuracy of human drug safety evaluation before the exposure of a new drug to the patient population.

3.2 Acetaminophen

Acetaminophen (APAP) (paracetamol) is an over-the-counter (OTC) analgesic drug (NSAID) used extensively by the US population for the treatment of pain and fever. It is often used in combination with other OTC products for the relief of pain and cold symptoms [1–7]. It is also one of the oldest drugs (>50 years) that remains on the US market in spite of its association with up to 50% of drug-induced liver failures in the US, leading to deaths or a need for liver transplantation [8–10]. APAP hepatotoxicity is often due to intentional or unintentional overdosage, with liver toxicity reported to be observed in patients administered

Transporters and Drug-Metabolizing Enzymes in Drug Toxicity, First Edition. Edited by Albert P. Li.

chronic doses of >5 g per day or acute doses of >7 g [11]. However, there are also cases of APAP-induced liver failure in patients who used the recommended therapeutic doses [10], with alcohol intake as a major exacerbating factor [12–20]. Hepatotoxicity can be attenuated upon intravenous treatment with *N*-acetyl cysteine (NAC) within eight hours of APAP ingestion. The effectiveness of NAC treatment is a function of plasma APAP level as well as the time interval between NAC administration and APAP ingestion [21, 22].

3.2.1 Drug Metabolism and Toxicity

APAP represents a prototypical protoxicant in which metabolic activation is a key mechanism of its hepatotoxicity. Its metabolic activation pathways have been extensively evaluated in both human *in vitro* systems as well as *in vivo* in laboratory animals. The generally accepted metabolic activation scheme is that APAP undergoes oxidative metabolism by P450 to *N*-acetyl-*p*-benzoquinone imine (NAPQI) which is a highly reactive metabolite that is either detoxified via glutathione (GSH) conjugation or undergoes covalent binding with cellular macromolecules, leading to hepatocellular necrosis. Inflammatory response elicited by cellular injuries and the subsequent Increased oxygen/nitrogen stress mediated by Kupffer cell activation [23] and recruitment of neutrophils [24] may also be involved in APAP hepatotoxicity.

In vitro studies using recombinant human P450 isoforms show that, in comparison to the other human P450 isoforms (CYP1A2, CYP2E1, and CYP2D6), CYP3A4 is the most effective in the metabolism of APAP to NAPQI, with Km equal to plasma drug concentration at therapeutic doses [25]. The association of alcohol consumption with APAP hepatotoxicity has led to the postulation that alcohol enhancement of CYP2E1 activity via increased gene expression [26] and/or enzyme stabilization [27], may be responsible for the increased toxicity [28, 29]. An alternative explanation is that APAP is metabolically activated by CYP3A4 while alcohol consumption depletion of cellular GSH [30–32], leading to compromised detoxification of NAPQI and thereby enhanced toxicity.

3.2.2 Transporters and Toxicity

APAP undergoes direct glucuronidation and sulfation, and the reactive metabolite NAPQI subjected to GSH conjugation. The various conjugates are excreted via transporter-mediated efflux into the bile duct. Recently, using membrane vesicles, APAP–GSH was found to be substrates of MRP1, MRP2, and MRP [33]. Hepatobiliary efflux transporters MRP1 and MRP4 mRNA levels were elevated in livers from patients after toxic APAP ingestion [34]. This upregulation of transporter expression by APAP is also observed in rodent models of APAP hepatotoxicity [34–36]. Treatment of animals with clodronate liposomes to deplete Kupffer cells abolished Mrp4 upregulation by APAP as well as increasing its hepatotoxicity, suggesting that transporter upregulation is protective, and that the upregulation may be a result of Kupffer cell activation [37].

3.2.3 Risk Factors

As APAP is metabolically activated by the inducible P450 isoforms, especially CYP1A2 and CYP3A4, to NAPQI which in turn is detoxified by GSH conjugation, genetic or environmental factors leading to increased CYP3A4 activity and depletion of hepatocellular GSH concentration are likely risk factors of its hepatotoxicity [38]. Omeprazole induction of CYP1A2 in CYP2C19 poor metabolizers have been postulated to be a risk factor of APAP toxicity [39]. Induction of CYP3A4 as a risk factor has been substantiated in a clinical study where a patient coadministered with phenytoin, a CYP3A4 inducer, experienced APAP hepatotoxicity at therapeutic doses [40, 41]. There is also evidence that concurrent inflammatory events may enhance APAP toxicity, suggesting that bacterial or viral infection may be risk factors [23, 42, 43]. As efflux transporters may have protective roles toward APAP toxicity, environmental and genetic factors leading to compromised efflux transporter activities may also be risk factors.

Life style risk factors for APAP toxicity include obesity [44] and chronic alcohol consumption [18, 45, 46]. It is interesting to note that alcohol consumption has been reported to lead to GSH depletion in human subjects [47], thereby supporting GSH as a key risk factor for APAP toxicity.

3.3 Cerivastatin

Cerivastatin belongs to a class of 3-hydroxy-3-methylglutaryl coenzyme A (HMG-CoA) reductase inhibitors commonly known as statins. Statins are one of the most prescribed drugs indicated for the lowering of plasma cholesterol. Cerivastatin was introduced by Bayer into the U.S. market in June 1997 and was withdrawn in August 2001 due to 31 deaths due to rhabdomyolysis [48–53]. Cerivastatin was introduced to compete with an existing statin, atorvastatin. Statins are commonly coadministered with fibrates for the treatment of patients with mixed hyperlipidemia [54]. Pharmacokinetic drug interactions between cerivastatin and fibrates, leading to elevated plasma levels of cerivastatin, is believed to be the major cause of rhabdomyolysis.

3.3.1 Drug Metabolism and Toxicity

Compromised hepatic clearance is believed to be the major mechanism of cerivastatin toxicity. Cerivastatin-induced rhabdomyolysis occurred mostly in patients coadministered with fibrate, suggesting that the toxicity is a result of interaction between these two administered drugs. Cerivastatin is metabolized by CYP2C8 to hydroxylated metabolites [55], and CYP3A4 to desmethylcerivastatin which is further glucuronidated [56]. Gemfibrozil was found to inhibit the formation of CYP2C8-mediated metabolite but not the CYP3A4 metabolites in human liver microsomes [57, 58]. Gemfibrozil coadministration was found to greatly enhance plasma cerivastatin concentrations in human patients [59] while potent CYP3A4 inhibitors such as erythromycin and itraconazole had no significant effects [60]. These observations suggest that CYP2C8 plays a more important role than CYP3A4 in the hepatic clearance of cerivastatin.

Rhabdomyolysis is also observed in patients administered cerivastatin alone. Association of CYP2C8 genetic polymorphism and cerivastatin toxicity, however, is not as definitive as that observed with fibrate inhibition of CYP2C8 activity [61]. Patients with rhabdomyolysis administered only cerivastatin and not coadministered with fibrate have been found to have a higher frequency of CYP2C8 polymorphic genes [62], leading to CYP2C8 activity both higher and lower than that for wildtype. Genetic polymorphism leading to lower CYP2C8 activity would lead to slower hepatic clearance, resulting in higher cerivastatin activity, thereby enhancing the likelihood of rhabdomyolysis. Higher CYP2C8 activity should lead to lower plasma cerivastatin concentrations and therefore would not lead to toxicity. It is possible that for these patients, the role of uptake transporter plays a greater role in hepatic clearance than CYP2C8 (see below).

3.3.2 Transporter and Toxicity

Cerivastatin is one of the drugs where toxicity can be a function of both drug metabolism and transport. Besides being a CYP2C8 inhibitor, Gemfibrozil and its glucuronide are inhibitors of cerivastatin uptake by the organic anion transporting polypeptide 2 (OATP2/OATP1B1:SLC21A6) [58, 63]. Inhibition of transporter-mediated uptake of cerivastatin into hepatocytes leads to lower hepatic metabolic clearance. Physiologically based pharmacokinetic (PBPK) models have been developed to evaluate the complex interaction between cerivastatin and gemfibrozil, taking into account the inhibitory effects of gemfibrozil on cerivastatin uptake and metabolism [64, 65].

3.3.3 Risk Factors

The toxicity of cerivastatin is a function of plasma drug concentration which is determined by hepatic uptake and metabolism. Environmental and genetic conditions leading to reduction of hepatic CYP2C8 and OATP transporter activities are likely to be risk factors for cerivastatin-induced rhabdomyolysis. Co-exposure to CYP2C8 inhibitor as a major key factor has already been demonstrated with the high incidence of rhabdomyolysis in patients coadministered gemfibrozil. Genome wide association evaluation of 185 rhabdomyolysis patients in a case control study has identified genetic variants of OATP1B1(SLCO1B1), a nonsynonymous coding variant rs4149056 variant, as a statistically significant risk factor for patients administered cerivastatin without gemfibrozil coadministration [61]. Transfection of the variant gene into an *in vitro* cell culture system shows reduced cerivastatin uptake when compared to wild type, strongly suggesting that patients with the variant OATP1B1 had a similar reduction in cerivastatin uptake. In the same study, no association of CYP2C8 variants with rhabdomyolysis was observed. The results appear to suggest that uptake transporter may be a more important risk factor for rhabdomyolysis for patients administered with cerivastatin alone.

3.4 Felbamate

Felbamate is an anticonvulsant with a dual mechanism of action both as an agonist of gamma-aminobutyric acid (GABA) receptors and also an antagonist of *N*-methyl-D-aspartate receptor (NMDA) receptors [66]. It was approved by FDA in August 1993 for the treatment of seizures associated with Lennox–Gastaut syndrome in children, but was withdrawn within a year in August 1994 due to its association with aplastic anemia [67]. Felbamate was provided a redemption in September 1994 for limited use in patients with refractory epilepsy [68], and with additional warning due to its association with rare incidence of acute liver failure, including cases with fatalities [67, 69–71].

3.4.1 Drug Metabolism and Toxicity

Association of felbamate administration and aplastic anemia was the key reason for the initial market withdrawal of felbamate [72]. Upon its redemption, felbamate was also known to be associated with idiosyncratic acute liver failures [73].

The mechanism for felbamate toxicity is not yet fully defined. While aplastic anemia was the initial reason for the market withdrawal of felbamate, most research activities were focused on its hepatotoxicity. Felbamate has been reported to be subjected to P450-mediated oxidation, with CYP3A4 and CYP2E1 identified to be the major isoforms involved in its metabolism [74], an observation that was made *in vitro* but was not confirmed in clinical studies with CYP3A4 inhibitors [75]. The reported reactive metabolites generated upon hepatic metabolism include alpha, beta-unsaturated aldehyde, atropaldehyde (ATPAL) [76, 77], and 2-phenylpropenal (a metabolite formed by hydrolysis of felbamate to 2-phenyl-1,3-propandiol monocarbamate [MCF], followed by oxidation to 3-carbamoyl-2-phenylpropionaldehyde [CBMA], and spontaneous loss of carbon dioxide and ammonia) [78–80]. An ongoing hypothesis is the oxidative stress and conjugation of macromolecules by reactive metabolites of felbamate, leading to the onset of a cascade of events which ultimately result in acute liver failure [81]. It is interesting to note that felbamate is an inhibitor of CYP2C19 [75] and a heteroactivator (not inducer) of CYP3A4 [74].

3.4.2 Transporters and Toxicity

Treatment of rats with verapamil, a Pgp inhibitor, was found to increase felbamate brain accumulation, suggesting that felbamate is a substrate for the efflux transporter [82]. However, no difference in felbamate distribution in the central nervous system was observed between wild type and multidrug resistance-associated protein 2 (MRP2)-deficient rats [83], suggesting that felbamate is not a substrate for this transporter. There are no reports on transporters for felbamate uptake.

3.4.3 Risk Factors

Based on current knowledge, genetic and environmental factors that would enhance the formation of reactive metabolites (CYP3A4 induction), reduction of detoxification (GSH depletion), and increased intracellular accumulation (Pgp inhibition) are potential risk factors for felbamate hepatotoxicity.

3.5 Flucloxacillin

Flucloxacillin, an isoxazolyl penicillin, is a narrow spectrum antibiotic of the penicillin class that has a broad range of uses in the treatment of Gram-positive bacterial infections of skin and soft tissue [84, 85], lung [86], urinary tract [87], meningitis [88, 89] and is used as a prophylaxis during surgery [90–92]. Flucloxacillin treatment has been associated with severe hepatotoxicity, resulting in liver failure [93, 94]. A genome wide association study (GWAS) showed an association of flucloxacillin-induced liver toxicity with the HLA-B*5701 genotype [95–97].

3.5.1 Drug Metabolism and Toxicity

Flucloxacillin has been shown to be metabolized primarily by CYP3A4, CYP3A7, and CYP2C9 in hepatocytes to 5-hydroxymethyl flucloxacillin [98, 99], a metabolite cytotoxic to biliary epithelial cells, resulting in hepatotoxicity [99–101]. Flucloxacillin has been shown to be an inducer of CYP3A4 both *in vitro* in hepatocytes, and in clinical studies [102], suggesting that prolonged treatment of patients may lead to increased CYP3A4 induction, resulting in a higher rate of formation of the cytotoxic metabolite. Plasma from patients exposed to flucloxacillin has been found to contain human serum albumin with modified lysine residues as a result of the covalent binding with 5-hydroxymethyl flucloxacillin, leading to the hypothesis that immune reaction to the neoantigens may be one of the key mechanism of hepatotoxicity [103].

3.5.2 Transporters and Toxicity

The membrane transporter MRP2 has been found to mediate the binding of flucloxacillin proteins localized in bile canaliculi regions [104] which has been hypothesized to be one of the determinants of its hepatotoxicity.

3.5.3 Risk Factors

The association of flucloxacillin-induced liver toxicity with the HLA-B*5701 genotype [95–97] represents the most successful application of GWAS in the identification of an at-risk population based on genotype. Unfortunately, the association cannot be extended to other DILI drugs. Based on the metabolism of flucloxacillin to the cytotoxic metabolite 5-hydroxymethyl flucloxacillin, patient populations with enhanced CYP3A4, CYP3A7, and CYP2C9 activities due to environmental and genetic factors, may be at risk of its hepatotoxicity. The findings with MRP2-mediation of the localization of flucloxacillin in biliary cells suggest that increased MRP2 activity may also be a risk factor.

3.6 Nefazodone

Nefazodone is a phenylpiperazine antidepressant which received approval from the Food and Drug Administration for treatment of major depressive disorder in 1994. Nefazodone enhances serotonin (5-hydroxytryptamine [5-HT]) synaptic transmission by acting as a potent antagonist of 5-HT2 receptors and as a 5-HT uptake inhibitor (SARI) as well as a weak serotonin-norepinephrine-dopamine uptake inhibitor (SNDRI) [105–109]. Nefazodone is efficacious toward the treatment of depression with depression-related anxiety symptoms [106]. Nefazodone administration has been found to be associated with the occurrence of idiosyncratic liver failure resulting in deaths and a need for liver transplantation [110–115]. Nefazodone was withdrawn from the US market in 2004 [116, 117]. A survey of

Canadian patients shows that a majority of the patients with nefazodone-induced liver injuries occurred within eight months of administration [118].

3.6.1 Drug Metabolism and Toxicity

in vitro studies suggest that the hepatotoxicity of nefazodone is associated with metabolic activation of the parent drug to highly reactive and toxic metabolites. Human liver microsome studies show that nefazodone undergoes hydroxylation and sulfydryl conjugation occurred on the 3-chlorophenylpiperazine-ring. This leads to the initial formation of *p*-hydroxynefazodone which is transformed to the reactive quinone–imine intermediate. The reactive intermediate is believed to be responsible for felbamate hepatotoxicity as well as for time-dependence of CYP3A4, the P450 isoform responsible for nefazodone metabolism [119]. Nefazodone and its metabolites are found to cause mitochondrial damage, leading to reactive oxygen species (ROS) formation as demonstrated by reduced GSH depletion [120]. An *in vitro* human hepatocyte assay for the identification of drugs associated with liver failures using ROS/cellular adenosine triphosphate (ATP) ratio as an endpoint routinely used nefazodone as a positive control representing DILI drugs [121].

3.6.2 Transporters and Toxicity

in vitro studies with human hepatocytes suggest that inhibition of the transporter ABCB11 (bile salt export pump [BSEP]), leading to intracellular accumulation of the cytotoxic bile salts, may be a mechanism of nefazodone hepatotoxicity [122, 123]. Nefazodone was reported to inhibit the efflux transporters MDR1 and MRP2, demonstrating that the drug and its metabolites may interact with efflux transporters *in vivo* [124]. While BSEP inhibition has been associated with DILI as a result of bile salt accumulation, the involvement of MDR1 and MRP2 in hepatotoxicity has not yet been clearly established. One possibility is the increased accumulation of the parent drug or its toxic metabolites which are efflux transporter substrates.

3.6.3 Risk Factors

The current knowledge on nefazodone metabolism suggest that increased metabolic activation (e.g. enhanced CYP3A4 activity due to environmental and genetic factors) as well as compromised detoxification pathways (e.g. decreased glutathione S-transferase (GST) activity and depletion of GSH by environmental agents) are likely risk factors of nefazodone hepatotoxicity. As nefazodone hepatotoxicity may be related to bile salt accumulation via BSEP inhibition, environmental, and genetic factors leading to compromised BSEP functions may also be a risk factor for its hepatotoxicity.

3.7 Obeticholic Acid

Obeticholic acid (6a-ethyl-chenodeoxycholic acid (OCA) is a farnesoid X receptor (FXR) agonist approved by FDA in May 2016 under the accelerated approval program for the treatment of primary biliary cholangitis (PBC) [125]. PBC is a chronic autoimmune cholestatic liver disease that predominantly affects women in early to middle age. It is typically associated with autoantibodies to mitochondrial antigens and results in immune-mediated destruction of small and medium-sized intrahepatic bile ducts leading to cholestasis, hepatic fibrosis and may progress to cirrhosis or hepatic failure and, in some cases, hepatocellular carcinoma [126–129].

FXR is an orphan nuclear hormone receptor found in the nucleus of cells in the liver, intestine, kidney, and adrenal glands [130–132] with bile acids – chenodeoxycholic acid, lithocholic acid, and deoxycholic acid as ligands [131]. OCA activation of FXR leads to reduction of hepatic bile acids via two mechanisms: (i) Reduction of the synthesis of the bile acid precursor, cholesterol, via downregulation of CYP7A1, the key enzyme for cholesterol synthesis [132, 133], and (ii) Upregulation of the efflux transporters of bile acids, BSEP [134], OST-a, and OST-b [135, 136].

OCA is used either as a single therapy or in combination with ursodeoxycholic acid (UDCA) in adult patients, for the 40% of PBC patients with an inadequate response to UDCA, and for the 10% of PBC patients who is unable to tolerate UDCA. As an agonist of FXR, OCA treatment reduces bile acid synthesis. The clinical efficacy of OCA was clearly demonstrated in PBC patients using plasma biomarkers for liver, alkaline phosphatase, gamma-glutamyl transpeptidase, and alanine aminotransferase as biomarkers of PBC progression [126–129, 137–142].

OCA administration has been associated with liver damage [143]. As a result of the occurrence of 19 deaths associated with OCA treatment, the FDA released not only a Drug Safety Communication on 21 September 2017, warning of serious liver injury caused by OCA due to incorrect dosing but also a Boxed Warning. FDA is also requiring a Medication Guide for patients to inform them about this issue (see weblink: FDA Adds Black Box Warning to Intercept's Liver Disease Medicine | 2018-02-06 | FDANews).

3.7.1 Drug Metabolism and Toxicity

As a relatively new drug, there are limited numbers of publications on OCA metabolism. Its metabolism is likely to be similar to that known for bile acids. Bile acids are subjected to glucuronidation by the UDP-glucuronosyltransferases (UGTs) 1A1, 2B4, and 2B7 and sulfation by sulfotransferases SULT2A1 and SULT2A8. The conjugated bile acids are secreted into bile via the canalicular bile salt export pump (BSEP), into the gastrointestinal tract. Some bile acids are passively absorbed in the upper intestine, but most are reabsorbed in the ileum as free bile acids upon deconjugation by the intestinal flora [144–146].

Five clinical drug–drug interaction studies have been reported on the effects of oral daily doses of 10 or 25 mg OCA on single-dose plasma pharmacokinetics of specific probe substrates for enzymes CYP1A2 (caffeine, R-warfarin), CYP3A (midazolam, R-warfarin), CYP2C9 (S-warfarin), CYP2D6 (dextromethorphan), CYP2C19 (omeprazole), and drug transporters, BCRP/OATP1B1/OATP1B3 (rosuvastatin), and P-gp (digoxin). OCA was found to have no interactions with S-warfarin, digoxin, and dextromethorphan, and weak interactions with caffeine, omeprazole, rosuvastatin, and midazolam [147]. The clinical findings therefore suggest that OCA does not significantly inhibit CY2C9, CYP2D6, CYP2C19, and CYP3A4.

The roles of OCA metabolism with its association of liver failures have not yet been defined. Primary cultured human hepatocytes were found to reproduce OCA activation of FXR leading to CYP7A suppression, but without associated cytotoxicity upon treatment for 72 hours with 0.1–100 μM of OCA, thereby suggesting that the clinical hepatotoxicity of OCA may not be a result of its direct cytotoxic effects [148, 149].

3.7.2 Transporters and Toxicity

FXR regulates transporter-mediated efflux via upregulation of BSEP [134], OST-a, and OST-b [135, 136]. OCA, as an agonist of FXR, has been shown to increase the mRNA levels of the bile salt export pump (BSEP), and the basolateral efflux heterodimer transporters, organic solute transporter alpha (OSTalpha) and OSTbeta in primary cultured human hepatocytes [148]. However, as of this writing, there the roles of transporters in OCA toxicity, if any, is not yet defined.

3.7.3 Risk Factors

As the mechanism of OCA toxicity is not yet elucidated, risk factors are not yet defined. Based on clinical findings, patients with preexisting liver cirrhosis may have a higher risk toward OCA hepatotoxicity. This "risk factor", however, is yet to be confirmed (https://livertox.nih.gov/ObeticholicAcid.htm).

3.8 Sitaxentan

Sitaxentan is a benzodioxole drug marketed for the treatment of pulmonary arterial hypertension (PAH) [150, 151]. It also has been found to be an effective treatment of digital ulcers, easing pain, and preventing the formation of new ulcers [152]. It is a competitive antagonist of endothelin-1 which has been found to be elevated in PAH patients [153]. Binding of endothelin-1 to its receptors, endothelin-A (ET-A) and endothelin-B (ET-B), leads to pulmonary vasoconstriction [153]. Inhibition of endothelin-1 binding to its receptors by Sitaxentan thereby inhibits the vasoconstriction effects of endothelin, leading to decreases in pulmonary vascular resistance [150, 151]. Sitaxentan was found to be associated with idiosyncratic acute liver failures [154–157], resulting in deaths and a need for liver transplantation. Sitaxentan was withdrawn from the market worldwide in 2010.

3.8.1 Drug Metabolism and Toxicity

As for most marketed drugs that are found to be associated with idiosyncratic liver toxicity, Sitaxentan has been extensively evaluated in preclinical safety tests in multiple animal species, including single- and multiple-dose studies in mice, rats, and dogs. Preclinical findings were non-remarkable. While signs of liver toxicity including liver hypertrophy were observed, the preclinical safety study results show that Sitaxentan has acceptable safety margins [158].

As of this writing, there is only one study on the metabolic fate of Sitaxentan. Using *in vitro* hepatic experimental systems including human hepatocytes, human liver microsomes, and complementary DNA (cDNA) expressed microsomes, the 1,3-benzodioxole ring of Sitaxentan is found to undergo enzymatic demethyleneation to an ortho-catechol metabolite that can further oxidize to a reactive ortho-quinone metabolite which undergoes GSH conjugation. CYP3A4 is found to be the major P450 isoform involved in the metabolic activation of Sitaxentan and is also inactivated during the metabolic process, further confirming the formation of highly reactive metabolites. Metabolic activation of Sitaxentan by CYP3A4 to reactive quinone metabolites which formed GSH conjugates is concluded to be the most likely mechanism for its hepatotoxicity [159].

3.8.2 Transporters and Toxicity

There are no reports on the interaction of Sitaxentan with uptake or efflux transporters.

3.8.3 Risk Factors

The major risk factors associated with Sitaxentan are likely to be environmental and genetic factors that would lead to enhanced formation of the toxic reactive metabolites (e.g. elevation of CYP3A4 activity due to exposure to inducers) or compromised detoxification pathways (e.g. decreased cellular GSH contents due to co-exposure to GSH depleting agents).

3.9 Sorivudine

Sorivudine (SRV; 1-beta-D-arabinofuranosyl-*E*-5-[2-bromovinyl] uracil) is a synthetic analog of thymidine intended as an antiviral drug for the treatment of varicellar zoster virus [160–162]. The antiviral activity of SRV is a result of its phosphorylation by thymidine kinase followed by incorporation into viral DNA, inhibiting DNA synthesis and thereby viral replication. SRV is also effective as an antiviral agent toward herpes simplex type 1 virus, and Epstein–Barr virus, and has been proposed as a potential treatment of viral outer retinal necrosis [163, 164]. SRV has high oral bioavailability and superior antiviral activity compared to the available antivirals at the time of its introduction [160, 165]. SRV was approved for marketing in Japan in 1993 and was withdrawn within 40 days of marketing due to its association with 23 cases of severe toxicity, leading to 15 deaths. The deaths and severe gastrointestinal toxicity were observed in cancer patients coadministered SRV and 5-fluorouracil (5-FU) prodrugs [166, 167]. Because of the severe toxicity associated with SRV, the drug did not receive regulatory approvals in any other countries, including the United States.

3.9.1 Drug Metabolism and Toxicity

Toxicity observed in patients coadministered SRV and 5-FU prodrugs is a result of unexpected elevation of plasma 5-FU levels. The toxicity observed included bone marrow damage, intestinal membrane mucosa atrophy, decreases in white blood cells and platelets, diarrhea with bloody flux, and severe anorexia. Both laboratory animal and clinical results confirm that SRV is metabolized by intestinal flora to (*E*)-5-(2-bromovinyl)uracil (BVU), a suicide inhibitor of dihydropyrimidine dehydrogenase (DPD), the key enzyme for 5-FU metabolism and elimination, leading to retardation of 5-FU metabolic clearance, resulting in its accumulation to toxic levels [166–168] (Figure 3.1).

3.9.2 Transporters and Toxicity

There is no known involvement of transporters in the uptake of SRV.

3.9.3 Risk Factors

Coadministration with 5-FU is the most important risk factor for the observed lethality.

3.10 Tacrine

Tacrine hydrochloride (1,2,3,4-tetrahydro-9-acridinamine monohydrochloride), an acetylcholinesterase inhibitor [169], was one of the first drugs marketed to combat Alzheimer's disease based on the cholinergic theory of the disease

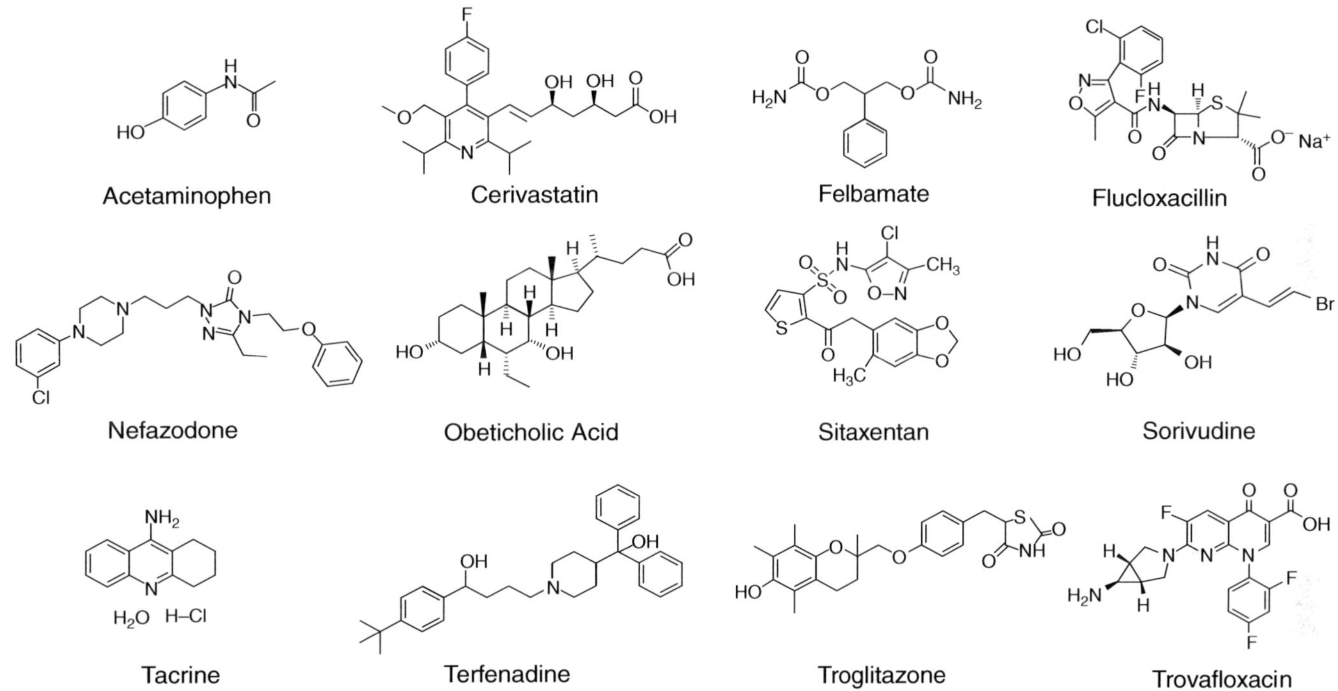

Figure 3.1 Chemical structure of the 12 toxic drugs reviewed.

mechanism. Tacrine was approved for use by US FDA in 1993 to treat Alzheimer's disease patients with mild to moderate symptoms of dementia. Tacrine was withdrawn from the U.S. market in May 2012 due to liver toxicity and the availability of alternative acetylcholinesterase inhibitor drugs with lower frequency of liver enzyme elevation.

Hepatotoxicity is a major complication of tacrine. It has been estimated that nearly half of the patients that received tacrine had reversible elevation of the clinical marker of liver toxicity, plasma alanine aminotransferase activity. This hallmark of liver toxicity appeared to be reversible in some patients as it was resolved by discontinuation of administration, and in some cases, without discontinuation. However, a small number of patients eventually developed severe and fatal liver toxicity in spite of frequent monitoring of serum liver enzyme activities [170–172].

3.10.1 Drug Metabolism and Toxicity

Toxic metabolite formation: In vitro results suggest that metabolism of tacrine, mainly by CYP1A2 [173], to highly reactive metabolites is a plausible mechanism of its hepatotoxicity. Incubation of tacrine with human liver microsomes has been reported to lead to the formation of mono- and dihydroxylated metabolites. Oxidative metabolism to hydroxylated metabolites appeared to involve highly reactive quinone methides that form covalent adducts with macromolecules, leading to hepatotoxicity. This hypothesis was further substantiated by the attenuation of protein binding by tacrine metabolites via supplementation of reduced L-GSH to human liver microsomes [174–176].

Species differences in metabolism: In vivo [177] and *in vitro* [178] evaluation of metabolite profiles showed similar metabolite profiles but apparent quantitative differences in tacrine metabolism. A key observation is that human liver microsomes formed the highest levels of reactive metabolites, followed by dog liver microsomes (intermediate) and rat liver microsomes (lowest) [178].

3.10.2 Transporters and Toxicity

As of this writing, there is no evidence that tacrine hepatotoxicity is associated with its interaction with transporters. However, results with rats show that tacrine distribution into rat brains across the blood brain barrier is a function of transporter-mediated uptake involving organic cation transporters [179].

3.10.3 Risk Factors

High CYP1A2 activities in combination with low GSH levels could be risk factors for tacrine hepatotoxicity. The caffeine breath test, a clinical bioassay for CYP1A2 activity in patients, however, was not successful in the identification of patients with high susceptibility to tacrine hepatotoxicity [180]. On the other hand, there appeared to be correlation between GST polymorphism and susceptibility to tacrine hepatotoxicity [181, 182]. For instance, multivariate Cox hazards model showed that the GST M1-T1 null genotype was an independent risk factor of tacrine hepatotoxicity [181].

3.11 Terfenadine

Terfenadine was marketed in 1985 in the United States as the first nonsedating antihistamine for the treatment of seasonal allergic rhinnitis. It was available as an OTC drug and was found to be associated with lethal cases of torsades de pointes, a form of ventricular arrythmia, in patients who were also taking macrolide antibiotics such as erythromycin [183–185], and the antifungal ketoconazole [184, 186–188], and in patients with compromised liver functions [189]. A black box warning for terfenadine was issued by FDA in 1992 and the drug was eventually removed from the market in 1997. Terfeanadine was replaced by fexofenadine, the pharmacologically active metabolite of terfenadine without cardiotoxicity [190].

3.11.1 Drug Metabolism and Toxicity

The legacy of terfenadine drug interactions: Drug interactions with terfenadine have historical significance. It was the first clear indication that a P450 inhibiting perpetrator drug could increase the plasma concentration of a coadministered victim drug that is a substrate of the inhibited P450 isoform, resulting in toxicity. As a result, US FDA published the first drug–drug interaction guidance document proposing a mechanistic-based approach to identify perpetrator and victim drugs of drug–drug interactions. A major advance catalyzed by FDA's proposed strategy is the recognition of species differences in drug metabolism. The FDA 1997 guidance document proposed that before approval for marketing, a drug needs to be investigated using *in vitro* human drug metabolizing enzyme-containing systems, with focus on human P450 isoforms, on whether its metabolism can be affected by the existing drugs that are inhibitors or inducers of drug metabolism, as well as the drug's potential to inhibit or induce drug metabolism of P450 isoforms that are responsible for the metabolism of marketed drugs [191]. The recommended *in vitro* human hepatic metabolic systems include human liver microsomes, cDNA-expressed P450 isoforms, and human hepatocytes [192]. This strategy is now globally recognized by international regulatory agencies for drug approval, has been continuously been improved, with the most recent US FDA guidance involving also uptake and efflux drug transporters [193]. The validity of this new approach is supported by known clinical cases of drug–drug interactions based on transporter activities [194].

Terfenadine metabolism and toxicity: Terfenadine causes prolongation of the QT interval due to suppression of specific delayed rectifier ventricular K+ currents as a result of the blockade of the hERG-IKr channel [195–198]. The lack of cardiotoxicity in patients administered terfenadine alone is a result of its rapid metabolic clearance at the therapeutic dose, as clearly demonstrated in a clinical study where detectable unmetabolized terfenadine was only observed in patients coadministered ketoconazole, and not in patients administered only terfenadine [187]. CYP3A4 is the major P450 isoform responsible for the metabolic clearance of terfenadine, resulting in hydroxy-terfenadine which is further oxidized to fexofenadine, the carboxylic acid metabolite of terfenadine.

3.11.2 Transporter and Toxicity

Results of an *in vitro* study with Caco2 cells show that terfenadine is a substrate of Pgp. The authors suggest that inhibition of intestinal Pgp efflux transport (e.g. by ketoconazole) may contribute to a higher plasma concentration [199]. This suggestion is not yet substantiated by clinical findings.

3.11.3 Risk Factors

The major risk factor associated with terfenadine cardiotoxicity is its coadministration with drugs that are potent inhibitors of CYP3A4 such as ketoconazole, fluconazole, itraconazole, erythromycin, clarithromycin, cimetidine, and troleandomycin [200–204]. Patients with low CYP3A4 activities due to genetic polymorphism or liver diseases may also be at high risk due to lower hepatic clearance of terfenadine [189]. There is evidence that grapefruit juice, a known inhibitor of intestinal CYP3A4 and the enteric efflux transporter Ppg, would increase plasma terfenadine concentration to detectable levels with QT prolongation effects [205–208]. The clinical significance of the grapefruit juice effects is yet to be substantiated.

3.12 Troglitazone (Rezulin®)

Troglitazone (2,4-thiazolidinedione) was the first peroxisome proliferator-activated receptor (PPAR) antagonist insulin sensitizer developed for the treatment of type 2 diabetes. It received FDA approval for marketing in 1997, and was withdrawn from the market in 2000 after reports of at least 12 cases of severe liver injuries [209], leading to deaths or a need for liver transplantation [209–211]. Troglitazone was one of the first drugs associated with idiosyncratic hepatotoxicity [212] generally classified as drugs with severe hepatotoxicity at a low incidence than cannot be readily detected in regulatory clinical trials. The reported incidences are 14 [213] and 19 [214] per 100 000 patients. Upon introduction of troglitazone to the market in March 1997 to its market withdrawal in March 2000, 83 cases of liver failure were reported with an estimated exposed population of 1.92 million [215]. Signs of troglitazone hepatotoxicity were observed during prospective clinical trials, with 1.9% of the patients exhibiting signs of liver injuries, leading to a recommendation of frequent monitoring of liver enzyme chemistries for the administered patients before the official withdrawal of the drug from the market [209]. Liver enzyme elevation would resolve with or without stoppage of drug administration. Routine monitoring of liver enzymes, however, did not prevent the onset of troglitazone-induced liver failures [215].

3.12.1 Drug Metabolism and Toxicity

As one of the first drugs associated with liver failures, extensive research has been performed with troglitazone to elucidate the key events associated with its toxicity. Investigations with human liver microsomes, recombinant P450s, and human hepatocytes show that troglitazone is metabolized by P450, especially CYP3A4, to highly reactive *o*-quinone methide and quinone epoxide metabolites that form conjugates with GSH and *N*-acetylcysteine [216, 217]. Troglitazone is also found to be metabolized to glucuronide and sulfate conjugates [218]. *In vitro* studies with human hepatocytes suggest that the parent drug is responsible for the cytotoxicity of troglitazone, with conjugative pathways as detoxifying [219]. Reviews of troglitazone hepatotoxicity have suggested that reactive metabolite formation is not likely to be responsible for its hepatocellular cytotoxicity [220, 221]. Reactive metabolites and the associated oxidative stress [121, 222, 223] and cytotoxic inflammatory responses to metabolite-protein conjugates are believed to be key events leading to hepatic failure [24, 224, 225].

3.12.2 Transporter and Toxicity

Troglitazone represents one of the first hepatotoxic drugs where the accumulation of intracellular bile salts to cytotoxic levels due to BSEP inhibition is a likely mechanism of its toxicity. The initial discovery of troglitazone-mediated BSEP inhibition was observed in rats *in vivo* and in rat liver plasma membrane preparations. Troglitazone was found to induce cholestasis in male and female rats. Troglitazone was found to inhibit BSEP-mediated taurocholate transport with an apparent *K*(*i*) value, 1.3 μM, while its conjugated metabolite, troglitazone sulfate, was a more potent inhibitor, with an apparent *K*(*i*) value of 0.23 μM [226, 227].

3.12.3 Risk Factors

While the mechanism of hepatotoxicity for troglitazone has not been fully established, experimental investigations, and correlation studies show that reactive metabolite formation and BSEP inhibition are the most likely mechanism for troglitazone-induced liver injuries. The potential risk factors are genetic and environmental conditions that would lead to enhanced formation of metabolites (e.g. P450 induction), compromised detoxification pathways (e.g. reduced cellular GSH levels or GST activities), and sensitized cytotoxic inflammatory responses. Null mutations of GSTM1 and GSTT1 have been associated with increases in liver enzymes in troglitazone patients, suggesting the involvement reactive metabolite formation in troglitazone hepatotoxicity [228]. However, a correlation analysis between P450 isoform activities and EC50 with primary human hepatocytes from 27 donors showed that the best correlation was made not with any of the individual isoforms, but with the activity product of CYP3A4 × UGT)/SULT [229], suggesting that troglitazone is directly cytotoxic, with CYP3A4 and UGT as detoxifying and sulfotransferase (SULT) as activating activities. This correlation is consistent with the hypothesis of BSEP inhibition by troglitazone sulfate conjugates, leading to accumulation of intracellular bile salts in hepatocytes, as a key mechanism of hepatotoxicity [230]. Further, experimental evidence exists suggesting that inflammatory events may exacerbate troglitazone hepatotoxicity [231]. Troglitazone-induced liver failures represent an example where occurrence of idiosyncratic liver toxicity is a result of simultaneous occurrence of multiple sensitizing events as stated by Li's Multiple Determinant Hypothesis [232], with the sensitizing events being enhanced metabolism of reactive metabolites (e.g. induced CYP3A4 activity), compromised GSH conjugation of reactive metabolites (e.g. reduced cellular GSH levels, reduced GST activities), enhanced sulfate formation (e.g. enhanced SULT activity), and sensitized cytotoxic immune response.

3.13 Trovafloxacin

Trovafloxacin is a fluoroquinolone (FQ) with broad spectrum antibiotic properties. Trovafloxacin inhibits two key enzymes for bacterial cell replication: (i) DNA gyrase, an essential enzyme that is involved in the replication, transcription, and repair of bacterial DNA; and (ii) Topoisomerase IV, an enzyme known to play a key role in the partitioning of the chromosomal DNA during bacterial cell division [233, 234]. FQ drugs have the potential to be effective toward bacteria that are resistant to commonly used antibiotics (e.g. penicillins, cephalosporins, aminoglycosides, macrolides, and tetracyclines) that have different mechanism of action.

Trovafloxacin was approved by FDA in 1997. In 1999, FDA placed limits on its application due to its association with 140 reports of hepatic events [235–237] and, including 14 cases of hepatic failure, leading to 6 deaths [235].

3.13.1 Metabolism and Toxicity

While the exact mechanism for the hepatotoxic properties of trovafloxacin is not yet fully elucidated, several lines of evidence suggest that enzymatic oxidation of the cyclopropylamine moiety to reactive metabolites is involved. Via synthesizing a drug model of trovafloxacin which contains the cyclopropylamine substructure, Sun et al. showed that chemical oxidants could oxidize the drug model to a reactive alpha, beta-unsaturated aldehyde [238]. The same laboratory also showed that CYP1A2 and myeloperoxidase could oxidize the drug model to the same reactive aldehyde which could conjugate reduced GSH and form protein adducts [239]. Shaw et al. showed that hepatotoxicity could be induced with a nonhepatotoxic dose of trovafloxacin upon coadministration of nonhepatotoxic doses of lipopolysaccharide (LPS), with significant elevation of TNFa proceeding the onset of liver injuries. Inhibition of TNFa transcription by pentoxifylline or inhibition of TNFa activity by etanercept were found to reduce trovafloxacin/LPS-induced hepatotoxicity [240–242]. The results suggest that inflammatory insults may exacerbate trovafloxacin hepatotoxicity, with TNFa as a possible mediator for the initiation of the toxic events.

3.13.2 Transporters and Toxicity

While there are no reports on the role of transporters on hepatic uptake and efflux, it has been reported that trovafloxacin could bioaccumulate in cultured human polymorphonuclear leukocytes (PMNs), human peritoneal macrophages, and tissue-cultured epithelial cells (McCoy cells), albeit in a nonsaturable manner and was enhanced at 4 °C, thereby eliminating the involvement of uptake transporters [243].

3.13.3 Risk Factors

The current information suggests that environmental and genetic factors that could enhance the formation of reactive metabolites (e.g. inducers of CYP1A1 and myeloperoxidase), reduction of cellular GSH, and inflammatory insults are likely risk factors for trovafloxacin hepatotoxicity.

3.14 Conclusions

A review on the uptake, efflux, and metabolism of the 12 toxic drugs show that drug–drug interactions, metabolic activation, and BSEP inhibition represent the key mechanism of toxicity (Table 3.1). With our current extensive mechanistic understanding of metabolic enzyme and transporter induction and inhibition, and the resulting effective *in vitro*/*in vivo* experimental approaches, toxicity due to drug–drug interactions with prescription drugs can be effectively managed. The next frontier for toxicity resulting from drug-interactions is health supplements which have not been systematically evaluated for drug interaction potential.

Toxicity affected by metabolic activation and detoxification remains challenging. The analysis of potential risk factors suggests that toxicity may be enhanced as a result of genetic and environmental factors that would lead to co-occurrence of multiple events: (i) Increased body burden of the toxic drug due to enhanced absorption resulting from enhanced enteric uptake transport and/or decreased efflux transport. (ii) Increased liver burden resulting from enhanced hepatic uptake transport and/or decreased efflux transport. (iii) Elevated toxic parent drug/metabolites due to enhanced metabolic activation and decreased metabolic

Table 3.1 A summary of the key mechanisms of drug-induced toxicity for the 12 toxic drugs.

	Mechanism of drug-induced toxicity		
Drug	**BSEP inhibition**	**Reactive metabolite formation**	**Metabolic drug interactions**
Acetaminophen	—	X	—
Cerivastatin	—	—	X
Felbamate	—	X	—
Flucloxacillin	—	X	—
Nefazodone	—	X	—
Obeticholic acid	X	—	—
Sitaxentan	—	X	—
Sorivudine	—	—	X
Tacrine	—	X	—
Terfenadine	—	—	X
Troglitazone	X	X	—
Trovafloxacin	—	X	—

detoxification. Severe drug toxicity resulting from the co-occurrence of these multiple risk enhancing factors is the basis for the Multiple Determinant Hypothesis of idiosyncratic drug toxicity.

References

1. Honig S, Murray KA. Postsurgical pain: zomepirac sodium, propoxyphene/-acetaminophen combination, and placebo. *J Clin Pharmacol* 1981;21 (10):443–8.
2. Pircio AW, Buyniski JP, Roebel LE. Pharmacological effects of a combination of butorphanol and acetaminophen. *Arch Int Pharmacodyn Ther* 1978;235(1):116–23.
3. Hopkinson JH, 3rd, Blatt G, Cooper M, Levin HM, Berry FN, Cohn H. Effective pain relief: comparative results with acetaminophen in a new dose formulation, propoxyphene napsylate-acetaminophen combination, and placebo. *Curr Ther Res Clin Exp* 1976;19(6):622–30.
4. Diamond S. Treatment of migraine with isometheptene, acetaminophen, and dichloralphenazone combination: a double-blind, crossover trial. *Headache* 1976;15(4):282–7.
5. Walker JM. Value of an acetaminophen-chlorzoxazone combination (parafon forte) in the treatment of acute musculoskeletal disorders. *Curr Ther Res Clin Exp* 1973;15(5):248–52.
6. Steele RW, Young FS, Bass JW, Shirkey HC. Oral antipyretic therapy. Evaluation of aspirin-acetaminophen combination. *Am J Dis Child* 1972;123(3):204–6.
7. Cameron JS, Specht PG, Wendt GR. Effects of placebo and an acetaminophen-salicylamide combination on moods, emotions, and motivations. *J Psychol* 1967;67(2):257–62.
8. Lee WM. The case for limiting acetaminophen-related deaths: smaller doses and unbundling the opioid-acetaminophen compounds. *Clin Pharmacol Ther* 2010;88(3):289–92.
9. Robinson AE, Sattar H, McDowall RD, Holder AT, Powell R. Forensic toxicology of some deaths associated with the combined use of propoxyphene and acetaminophen (paracetamol). *J Forensic Sci* 1977;22(4):708–17.
10. Amar PJ, Schiff ER. Acetaminophen safety and hepatotoxicity--where do we go from here? *Expert Opin Drug Saf* 2007;6(4):341–55.
11. Bailey BO. Acetaminophen hepatotoxicity and overdose. *Am Fam Physician* 1980;22(1):83–7.
12. Fercovic A, Serra I, Serra L. Acetaminophen hepatotoxicity in an alcohol addicted student. Case report. *Rev Med Chil* 1999;127(2):202–5.
13. Himmelstein DU, Woolhandler SJ, Adler RD. Elevated SGOT/SGPT ratio in alcoholic patients with acetaminophen hepatotoxicity. *Am J Gastroenterol* 1984;79(9):718–20.

14 Johnson MW, Friedman PA, Mitch WE. Alcoholism, nonprescription drug and hepatotoxicity. The risk from unknown acetaminophen ingestion. *Am J Gastroenterol* 1981;76(6):530–3.

15 Johnston SC, Pelletier LL, Jr. Enhanced hepatotoxicity of acetaminophen in the alcoholic patient. Two case reports and a review of the literature. *Medicine (Baltimore)*. 1997;76(3):185–91.

16 McClain CJ, Kromhout JP, Peterson FJ, Holtzman JL. Potentiation of acetaminophen hepatotoxicity by alcohol. *JAMA* 1980;244(3):251–3.

17 Poulsen HE, Lerche A, Pedersen NT. Potentiated hepatotoxicity from concurrent administration of acetaminophen and allyl alcohol to rats. *Biochem Pharmacol* 1985;34(6):727–31.

18 Schmidt LE, Dalhoff K, Poulsen HE. Acute versus chronic alcohol consumption in acetaminophen-induced hepatotoxicity. *Hepatology* 2002;35(4):876–82.

19 Wootton FT, Lee WM. Acetaminophen hepatotoxicity in the alcoholic. *South Med J* 1990;83(9):1047–9.

20 Zimmerman HJ, Maddrey WC. Acetaminophen (paracetamol) hepatotoxicity with regular intake of alcohol: analysis of instances of therapeutic misadventure. *Hepatology* 1995;22(3):767–73.

21 Chomchai S, Chomchai C, Anusornsuwan T. Acetaminophen psi parameter: a useful tool to quantify hepatotoxicity risk in acute acetaminophen overdose. *Clin Toxicol (Phila)* 2011;49(7):664–7.

22 Chomchai S, Lawattanatrakul N, Chomchai C. Acetaminophen Psi Nomogram: a sensitive and specific clinical tool to predict hepatotoxicity secondary to acute acetaminophen overdose. *J Med Assoc Thai* 2014;97(2):165–72.

23 Roth RA, Ganey PE. Intrinsic versus idiosyncratic drug-induced hepatotoxicity--two villains or one? *J Pharmacol Exp Ther* 2010;332(3):692–7.

24 Proctor WR, Chakraborty M, Fullerton AM, Korrapati MC, Ryan PM, Semple K, et al. Thymic stromal lymphopoietin and interleukin-4 mediate the pathogenesis of halothane-induced liver injury in mice. *Hepatology* 2014;60(5):1741–52.

25 Laine JE, Auriola S, Pasanen M, Juvonen RO. Acetaminophen bioactivation by human cytochrome P450 enzymes and animal microsomes. *Xenobiotica* 2009;39(1):11–21.

26 Takahashi T, Lasker JM, Rosman AS, Lieber CS. Induction of cytochrome P-4502E1 in the human liver by ethanol is caused by a corresponding increase in encoding messenger RNA. *Hepatology* 1993;17(2):236–45.

27 Badger TM, Ronis MJ, Ingelman-Sundberg M, Hakkak R. Pulsatile blood alcohol and CYP2E1 induction during chronic alcohol infusions in rats. *Alcohol* 1993;10(6):453–7.

28 Ueshima Y, Tsutsumi M, Takase S, Matsuda Y, Kawahara H. Acetaminophen metabolism in patients with different cytochrome P-4502E1 genotypes. *Alcohol Clin Exp Res* 1996;20 (1 Suppl):25A–8A.

29 Thummel KE, Slattery JT, Ro H, Chien JY, Nelson SD, Lown KE, et al. Ethanol and production of the hepatotoxic metabolite of acetaminophen in healthy adults. *Clin Pharmacol Ther* 2000;67(6):591–9.

30 Moss M, Guidot DM, Wong-Lambertina M, Ten Hoor T, Perez RL, Brown LA. The effects of chronic alcohol abuse on pulmonary glutathione homeostasis. *Am J Respir Crit Care Med* 2000;161 (2 Pt 1):414–9.

31 Kostrubsky VE, Szakacs JG, Jeffery EH, Wood SG, Bement WJ, Wrighton SA, et al. Role of CYP3A in ethanol-mediated increases in acetaminophen hepatotoxicity. *Toxicol Appl Pharmacol* 1997;143(2):315–23.

32 Liang Y, Yeligar SM, Brown LA. Chronic-alcohol-abuse-induced oxidative stress in the development of acute respiratory distress syndrome. *Sci World J* 2012;2012:740308.

33 Koenderink JB, van den Heuvel J, Bilos A, Vredenburg G, Vermeulen NPE, Russel FGM. Human multidrug resistance protein 4 (MRP4) is a cellular efflux transporter for paracetamol glutathione and cysteine conjugates. *Arch Toxicol* 2020, 94, 3027–3032.

34 Barnes SN, Aleksunes LM, Augustine L, Scheffer GL, Goedken MJ, Jakowski AB, et al. Induction of hepatobiliary efflux transporters in acetaminophen-induced acute liver failure cases. *Drug Metab Dispos* 2007;35 (10):1963–9.

35 Ghanem CI, Gomez PC, Arana MC, Perassolo M, Ruiz ML, Villanueva SS, et al. Effect of acetaminophen on expression and activity of rat liver multidrug resistance-associated protein 2 and P-glycoprotein. *Biochem Pharmacol* 2004;68(4):791–8.

36 Lima RA, Candido EB, de Melo FP, Piedade JB, Vidigal PV, Silva LM, et al. Gene expression profile of ABC transporters and cytotoxic effect of ibuprofen and acetaminophen in an epithelial ovarian cancer cell line *in vitro*. *Rev Bras Ginecol Obstet* 2015;37(6):283–90.

37 Campion SN, Johnson R, Aleksunes LM, Goedken MJ, van Rooijen N, Scheffer GL, et al. Hepatic Mrp4 induction following acetaminophen exposure is dependent on Kupffer cell function. *Am J Physiol Gastrointest Liver Physiol* 2008;295(2):G294–304.

38 Kwan D, Bartle WR, Walker SE. Abnormal serum transaminases following therapeutic doses of acetaminophen in the absence of known risk factors. *Dig Dis Sci* 1995;40(9):1951–5.

39 Sarich T, Kalhorn T, Magee S, al-Sayegh F, Adams S, Slattery J, et al. The effect of omeprazole pretreatment on acetaminophen metabolism in rapid and slow metabolizers of S-mephenytoin. *Clin Pharmacol Ther* 1997;62(1):21–8.

40 Cook MD, Williams SR, Clark RF. Phenytoin-potentiated hepatotoxicity following acetaminophen overdose? A closer look. *Dig Dis Sci* 2007;52(1):208–9.

41 Suchin SM, Wolf DC, Lee Y, Ramaswamy G, Sheiner PA, Facciuto M, et al. Potentiation of acetaminophen hepatotoxicity by phenytoin, leading to liver transplantation. *Dig Dis Sci* 2005;50 (10):1836–8.

42 Maddox JF, Amuzie CJ, Li M, Newport SW, Sparkenbaugh E, Cuff CF, et al. Bacterial- and viral-induced inflammation increases sensitivity to acetaminophen hepatotoxicity. *J Toxicol Environ Health A* 2010;73(1):58–73.
43 Tukov FF, Maddox JF, Amacher DE, Bobrowski WF, Roth RA, Ganey PE. Modeling inflammation-drug interactions *in vitro*: a rat Kupffer cell-hepatocyte coculture system. *Toxicol in vitro* 2006;20(8):1488–99.
44 Chomchai S, Chomchai C. Being overweight or obese as a risk factor for acute liver injury secondary to acute acetaminophen overdose. *Pharmacoepidemiol Drug Saf* 2018;27(1):19–24.
45 Makin A, Williams R. Paracetamol hepatotoxicity and alcohol consumption in deliberate and accidental overdose. *QJM* 2000;93(6):341–9.
46 Banda PW, Quart BD. The effect of mild alcohol consumption on the metabolism of acetaminophen in man. *Res Commun Chem Pathol Pharmacol* 1982;38(1):57–70.
47 Farinati F, Cardin R, de Maria N, Lecis PE, Della Libera G, Burra P, et al. Zinc, iron, and peroxidation in liver tissue. Cumulative effects of alcohol consumption and virus-mediated damage--a preliminary report. *Biol Trace Elem Res* 1995;47 (1–3):193–9.
48 Ozdemir O, Boran M, Gokce V, Uzun Y, Kocak B, Korkmaz S. A case with severe rhabdomyolysis and renal failure associated with cerivastatin-gemfibrozil combination therapy--a case report. *Angiology* 2000;51(8):695–7.
49 Ravnan SL, Locke C, Yee WP, Haase K. Cerivastatin-induced rhabdomyolysis: 11 case reports. *Pharmacotherapy* 2002;22(4):533–7.
50 Rodriguez ML, Mora C, Navarro JF. Cerivastatin-induced rhabdomyolysis. *Ann Intern Med* 2000;132(7):598.
51 Simpson S. Case reports of rhabdomyolysis associated with cerivastatin therapy. *Arch Intern Med* 2001;161 (21):2630–1.
52 SoRelle R. Baycol withdrawn from market. *Circulation* 2001;104(8):E9015–6.
53 Wooltorton E. Bayer pulls cerivastatin (Baycol) from market. *CMAJ* 2001;165(5):632.
54 Shek A, Ferrill MJ. Statin-fibrate combination therapy. *Ann Pharmacother* 2001;35 (7–8):908–17.
55 Kaspera R, Naraharisetti SB, Tamraz B, Sahele T, Cheesman MJ, Kwok PY, et al. Cerivastatin *in vitro* metabolism by CYP2C8 variants found in patients experiencing rhabdomyolysis. *Pharmacogenet Genomics* 2010;20 (10):619–29.
56 Muck W. Clinical pharmacokinetics of cerivastatin. *Clin Pharmacokinet* 2000;39(2):99–116.
57 Wang JS, Neuvonen M, Wen X, Backman JT, Neuvonen PJ. Gemfibrozil inhibits CYP2C8-mediated cerivastatin metabolism in human liver microsomes. *Drug Metab Dispos* 2002;30 (12):1352–6.
58 Shitara Y, Hirano M, Sato H, Sugiyama Y. Gemfibrozil and its glucuronide inhibit the organic anion transporting polypeptide 2 (OATP2/

OATP1B1:SLC21A6)-mediated hepatic uptake and CYP2C8-mediated metabolism of cerivastatin: analysis of the mechanism of the clinically relevant drug-drug interaction between cerivastatin and gemfibrozil. *J Pharmacol Exp Ther* 2004;311(1):228–36.

59 Backman JT, Kyrklund C, Neuvonen M, Neuvonen PJ. Gemfibrozil greatly increases plasma concentrations of cerivastatin. *Clin Pharmacol Ther* 2002;72(6):685–91.

60 Muck W. Rational assessment of the interaction profile of cerivastatin supports its low propensity for drug interactions. *Drugs* 1998;56 Suppl 1:15–23; discussion 33.

61 Marciante KD, Durda JP, Heckbert SR, Lumley T, Rice K, McKnight B, et al. Cerivastatin, genetic variants, and the risk of rhabdomyolysis. *Pharmacogenet Genomics* 2011;21(5):280–8.

62 Lucas RA, Weathersby BB, Rocco VK, Pepper JM, Butler KL. Rhabdomyolysis associated with cerivastatin: six cases within 3 months at one hospital. *Pharmacotherapy* 2002;22(6):771–4.

63 Shitara Y, Itoh T, Sato H, Li AP, Sugiyama Y. Inhibition of transporter-mediated hepatic uptake as a mechanism for drug-drug interaction between cerivastatin and cyclosporin A. *J Pharmacol Exp Ther* 2003;304(2):610–6.

64 Yao Y, Toshimoto K, Kim SJ, Yoshikado T, Sugiyama Y. Quantitative analysis of complex drug-drug interactions between cerivastatin and metabolism/transport inhibitors using physiologically based pharmacokinetic modeling. *Drug Metab Dispos* 2018;46(7):924–33.

65 Varma MV, Lin J, Bi YA, Kimoto E, Rodrigues AD. Quantitative rationalization of gemfibrozil drug interactions: consideration of transporters-enzyme interplay and the role of circulating metabolite gemfibrozil 1-O-beta-glucuronide. *Drug Metab Dispos* 2015;43(7):1108–18.

66 Burdette DE, Sackellares JC. Felbamate pharmacology and use in epilepsy. *Clin Neuropharmacol* 1994;17(5):389–402.

67 Zupanc ML, Roell Werner R, Schwabe MS, O'Connor SE, Marcuccilli CJ, Hecox KE, et al. Efficacy of felbamate in the treatment of intractable pediatric epilepsy. *Pediatr Neurol* 2010;42(6):396–403.

68 Heyman E, Levin N, Lahat E, Epstein O, Gandelman-Marton R. Efficacy and safety of felbamate in children with refractory epilepsy. *Eur J Paediatr Neurol* 2014;18(6):658–62.

69 Thakkar K, Billa G, Rane J, Chudasama H, Goswami S, Shah R. The rise and fall of felbamate as a treatment for partial epilepsy--aplastic anemia and hepatic failure to blame? *Expert Rev Neurother* 2015;15 (12):1373–5.

70 Dieckhaus CM, Thompson CD, Roller SG, Macdonald TL. Mechanisms of idiosyncratic drug reactions: the case of felbamate. *Chem Biol Interact* 2002;142 (1–2):99–117.

71 Pellock JM, Brodie MJ. Felbamate: 1997 update. *Epilepsia* 1997;38 (12):1261–4.

72 Pennell PB, Ogaily MS, Macdonald RL. Aplastic anemia in a patient receiving felbamate for complex partial seizures. *Neurology* 1995;45 (3 Pt 1):456–60.

73 Shah YD, Singh K, Friedman D, Devinsky O, Kothare SV. Evaluating the safety and efficacy of felbamate in the context of a black box warning: a single center experience. *Epilepsy Behav* 2016;56:50–3.

74 Egnell AC, Houston B, Boyer S. *in vivo* CYP3A4 heteroactivation is a possible mechanism for the drug interaction between felbamate and carbamazepine. *J Pharmacol Exp Ther* 2003;305(3):1251–62.

75 Glue P, Banfield CR, Perhach JL, Mather GG, Racha JK, Levy RH. Pharmacokinetic interactions with felbamate. *in vitro–in vivo* correlation. *Clin Pharmacokinet* 1997;33(3):214–24.

76 Kapetanovic IM, Torchin CD, Strong JM, Yonekawa WD, Lu C, Li AP, et al. Reactivity of atropaldehyde, a felbamate metabolite in human liver tissue *in vitro*. *Chem Biol Interact* 2002;142 (1–2):119–34.

77 Kapetanovic IM, Torchin CD, Thompson CD, Miller TA, McNeilly PJ, Macdonald TL, et al. Potentially reactive cyclic carbamate metabolite of the antiepileptic drug felbamate produced by human liver tissue *in vitro*. *Drug Metab Dispos* 1998;26 (11):1089–95.

78 Popovic M, Nierkens S, Pieters R, Uetrecht J. Investigating the role of 2-phenylpropenal in felbamate-induced idiosyncratic drug reactions. *Chem Res Toxicol* 2004;17 (12):1568–76.

79 Roller SG, Dieckhaus CM, Santos WL, Sofia RD, Macdonald TL. Interaction between human serum albumin and the felbamate metabolites 4-Hydroxy-5-phenyl-[1,3]oxazinan-2-one and 2-phenylpropenal. *Chem Res Toxicol* 2002;15(6): 815–24.

80 Thompson CD, Kinter MT, Macdonald TL. Synthesis and *in vitro* reactivity of 3-carbamoyl-2-phenylpropionaldehyde and 2-phenylpropenal: putative reactive metabolites of felbamate. *Chem Res Toxicol* 1996;9(8):1225–9.

81 Leone AM, Kao LM, McMillian MK, Nie AY, Parker JB, Kelley MF, et al. Evaluation of felbamate and other antiepileptic drug toxicity potential based on hepatic protein covalent binding and gene expression. *Chem Res Toxicol* 2007;20(4):600–8.

82 Potschka H, Fedrowitz M, Loscher W. P-Glycoprotein-mediated efflux of phenobarbital, lamotrigine, and felbamate at the blood-brain barrier: evidence from microdialysis experiments in rats. *Neurosci Lett* 2002;327(3):173–6.

83 Potschka H, Fedrowitz M, Loscher W. Brain access and anticonvulsant efficacy of carbamazepine, lamotrigine, and felbamate in ABCC2/MRP2-deficient TR- rats. *Epilepsia* 2003;44 (12):1479–86.

84 Amery KV. Clinical evaluation of the effects of flucloxacillin in skin and soft tissue infections in the Ivory Coast. *Pharmatherapeutica* 1988;5(3):193–7.

85 Harding JW, Knudsen ET. General practitioners' forum. Flucloxacillin in the treatment of skin and soft-tissue infections. *Practitioner* 1970;205 (230):801–6.

86 Lacey RW, Lewis EL. Further evolution of a strain of *Staphylococcus aureus in vivo*: evidence for significant inactivation of flucloxacillin by penicillinase. *J Med Microbiol* 1975;8(2):337–47.

87 Brogi G. Experience in pediatrics with a combination of ampicillin and flucloxacillin drops. *Minerva Pediatr* 1979;31 (10):813–8.

88 Chew R, Woods ML. Flucloxacillin does not achieve therapeutic cerebrospinal fluid levels against meticillin-sensitive *Staphylococcus aureus* in adults: a case report and review of the literature. *Int J Antimicrob Agents* 2016;47(3):229–31.

89 Ritchie SR, Rupali P, Roberts SA, Thomas MG. Flucloxacillin treatment of *Staphylococcus aureus* meningitis. *Eur J Clin Microbiol Infect Dis* 2007;26(7):501–4.

90 Moghissi K, Lutley C, Green J, Moghissi AJ. A trial comparing the use of penicillin and streptomycin, and flucloxacillin and ampicillin prophylactically in patients undergoing major thoracic surgery. *Br J Clin Pract* 1981;35 (7–8):250–3.

91 Wilson AP, Gruneberg RN, Treasure T, Sturridge MF. A clinical trial of teicoplanin compared with a combination of flucloxacillin and tobramycin as antibiotic prophylaxis for cardiac surgery: the use of a scoring method to assess the incidence of wound infection. *J Hosp Infect.* 1986;7 Suppl A:105–12.

92 Holm S, Larsson SE. The penetration of flucloxacillin into cortical and cancellous bone during arthroplasty of the knee. *Int Orthop* 1982;6(4):243–7.

93 Koek GH, Stricker BH, Blok AP, Schalm SW, Desmet VJ. Flucloxacillin-associated hepatic injury. *Liver* 1994;14(5):225–9.

94 Turner IB, Eckstein RP, Riley JW, Lunzer MR. Prolonged hepatic cholestasis after flucloxacillin therapy. *Med J Aust* 1989;151 (11–12):701–5.

95 Teixeira M, Macedo S, Batista T, Martins S, Correia A, Matos LC. Flucloxacillin-induced hepatotoxicity-association with HLA-B*5701. *Rev Assoc Med Bras (1992).* 2020;66(1):12–7.

96 Nicoletti P, Aithal GP, Chamberlain TC, Coulthard S, Alshabeeb M, Grove JI, et al. Drug-induced liver injury due to flucloxacillin: relevance of multiple human leukocyte antigen alleles. *Clin Pharmacol Ther* 2019;106(1):245–53.

97 Daly AK, Donaldson PT, Bhatnagar P, Shen Y, Pe'er I, Floratos A, et al. HLA-B*5701 genotype is a major determinant of drug-induced liver injury due to flucloxacillin. *Nat Genet* 2009;41(7):816–9.

98 Gath J, Charles B, Sampson J, Smithurst B. Pharmacokinetics and bioavailability of flucloxacillin in elderly hospitalized patients. *J Clin Pharmacol* 1995;35(1):31–6.

99 Thijssen HH, Wolters J. The metabolic disposition of flucloxacillin in patients with impaired kidney function. *Eur J Clin Pharmacol* 1982;22(5):429–34.

100 Dekker SJ, Dohmen F, Vermeulen NPE, Commandeur JNM. Characterization of kinetics of human cytochrome P450s involved in bioactivation of flucloxacillin: inhibition of CYP3A-catalysed hydroxylation by sulfaphenazole. *Br J Pharmacol.* 2019;176(3):466–77.

101 Lakehal F, Dansette PM, Becquemont L, Lasnier E, Delelo R, Balladur P, et al. Indirect cytotoxicity of flucloxacillin toward human biliary epithelium via metabolite formation in hepatocytes. *Chem Res Toxicol* 2001;14(6):694–701.

102 Huwyler J, Wright MB, Gutmann H, Drewe J. Induction of cytochrome P450 3A4 and P-glycoprotein by the isoxazolyl-penicillin antibiotic flucloxacillin. *Curr Drug Metab* 2006;7(2):119–26.

103 Jenkins RE, Meng X, Elliott VL, Kitteringham NR, Pirmohamed M, Park BK. Characterisation of flucloxacillin and 5-hydroxymethyl flucloxacillin haptenated HSA *in vitro* and *in vivo*. *Proteomics Clin Appl* 2009;3(6):720–9.

104 Waddington JC, Ali SE, Penman SL, Whitaker P, Hamlett J, Chadwick A, et al. Cell membrane transporters facilitate the accumulation of hepatocellular flucloxacillin protein adducts: implication in flucloxacillin-induced liver injury. *Chem Res Toxicol* 2020, 33(12):2939–2943.

105 Cyr M, Brown CS. Nefazodone: its place among antidepressants. *Ann Pharmacother* 1996;30(9):1006–12.

106 DeVane CL, Grothe DR, Smith SL. Pharmacology of antidepressants: focus on nefazodone. *J Clin Psychiatry* 2002;63 Suppl 1:10–7.

107 Eison AS, Eison MS, Torrente JR, Wright RN, Yocca FD. Nefazodone: preclinical pharmacology of a new antidepressant. *Psychopharmacol Bull* 1990;26(3):311–5.

108 Ellingrod VL, Perry PJ. Nefazodone: a new antidepressant. *Am J Health Syst Pharm* 1995;52 (24):2799–812.

109 Goldberg RJ. Nefazodone: a novel antidepressant. *Psychiatr Serv* 1995;46 (11):1113–4.

110 Aranda-Michel J, Koehler A, Bejarano PA, Poulos JE, Luxon BA, Khan CM, et al. Nefazodone-induced liver failure: report of three cases. *Ann Intern Med* 1999;130 (4 Pt 1):285–8.

111 Conway CR, McGuire JM, Baram VY. Nefazodone-induced liver failure. *J Clin Psychopharmacol* 2004;24(3):353–4.

112 Eloubeidi MA, Gaede JT, Swaim MW. Reversible nefazodone-induced liver failure. *Dig Dis Sci* 2000;45(5):1036–8.

113 Lucena MI, Andrade RJ, Gomez-Outes A, Rubio M, Cabello MR. Acute liver failure after treatment with nefazodone. *Dig Dis Sci* 1999;44 (12):2577–9.

114 Schirren CA, Baretton G. Nefazodone-induced acute liver failure. *Am J Gastroenterol* 2000;95(6):1596–7.

115 van Battum PL, van de Vrie W, Metselaar HJ, Verstappen VM, Zondervan PE, de Man RA. Acute liver failure ascribed to nefazodone: importance of 'postmarketing surveillance' for recently introduced drugs. *Ned Tijdschr Geneeskd* 2000;144 (41):1964–7.

116 Choi S. Nefazodone (Serzone) withdrawn because of hepatotoxicity. *CMAJ* 2003;169 (11):1187.

117 Edwards IR. Withdrawing drugs: nefazodone, the start of the latest saga. *Lancet* 2003;361 (9365):1240.

118 Stewart DE. Hepatic adverse reactions associated with nefazodone. *Can J Psychiatry* 2002;47(4):375–7.

119 Kalgutkar AS, Vaz AD, Lame ME, Henne KR, Soglia J, Zhao SX, et al. Bioactivation of the nontricyclic antidepressant nefazodone to a reactive quinone-imine species in human liver microsomes and recombinant cytochrome P450 3A4. *Drug Metab Dispos* 2005;33(2):243–53.

120 Dykens JA, Jamieson JD, Marroquin LD, Nadanaciva S, Xu JJ, Dunn MC, et al. *in vitro* assessment of mitochondrial dysfunction and cytotoxicity of nefazodone, trazodone, and buspirone. *Toxicol Sci* 2008;103(2):335–45.

121 Zhang J, Doshi U, Suzuki A, Chang CW, Borlak J, Li AP, et al. Evaluation of multiple mechanism-based toxicity endpoints in primary cultured human hepatocytes for the identification of drugs with clinical hepatotoxicity: results from 152 marketed drugs with known liver injury profiles. *Chem Biol Interact* 2016;255:3–11.

122 Kostrubsky SE, Strom SC, Kalgutkar AS, Kulkarni S, Atherton J, Mireles R, et al. Inhibition of hepatobiliary transport as a predictive method for clinical hepatotoxicity of nefazodone. *Toxicol Sci* 2006;90(2):451–9.

123 Oorts M, Baze A, Bachellier P, Heyd B, Zacharias T, Annaert P, et al. Drug-induced cholestasis risk assessment in sandwich-cultured human hepatocytes. *Toxicol in vitro* 2016;34:179–86.

124 Saab L, Peluso J, Muller CD, Ubeaud-Sequier G. Implication of hepatic transporters (MDR1 and MRP2) in inflammation-associated idiosyncratic drug-induced hepatotoxicity investigated by microvolume cytometry. *Cytometry A* 2013;83(4):403–8.

125 Markham A, Keam SJ. Obeticholic acid: first global approval. *Drugs* 2016;76 (12):1221–6.

126 Jhaveri MA, Kowdley KV. New developments in the treatment of primary biliary cholangitis - role of obeticholic acid. *Ther Clin Risk Manag* 2017;13: 1053–60.

127 Jindal A, Gupta A, Sarin S. Obeticholic acid in primary biliary cholangitis. *N Engl J Med* 2016;375 (20):e41.

128 Jones DE. Obeticholic acid for the treatment of primary biliary cirrhosis. *Expert Rev Gastroenterol Hepatol* 2016;10(10):1091–1099.

129 Silveira MG, Lindor KD. Obeticholic acid and budesonide for the treatment of primary biliary cirrhosis. *Expert Opin Pharmacother* 2014;15(3):365–72.

130 Forman BM, Goode E, Chen J, Oro AE, Bradley DJ, Perlmann T, et al. Identification of a nuclear receptor that is activated by farnesol metabolites. *Cell* 1995;81(5):687–93.

131 Parks DJ, Blanchard SG, Bledsoe RK, Chandra G, Consler TG, Kliewer SA, et al. Bile acids: natural ligands for an orphan nuclear receptor. *Science* 1999;284 (5418):1365–8.

132 Wang H, Chen J, Hollister K, Sowers LC, Forman BM. Endogenous bile acids are ligands for the nuclear receptor FXR/BAR. *Mol Cell* 1999;3(5):543–53.

133 Lu TT, Makishima M, Repa JJ, Schoonjans K, Kerr TA, Auwerx J, et al. Molecular basis for feedback regulation of bile acid synthesis by nuclear receptors. *Mol Cell* 2000;6(3):507–15.

134 Ananthanarayanan M, Balasubramanian N, Makishima M, Mangelsdorf DJ, Suchy FJ. Human bile salt export pump promoter is transactivated by the farnesoid X receptor/bile acid receptor. *J Biol Chem* 2001;276 (31):28857–65.

135 Frankenberg T, Rao A, Chen F, Haywood J, Shneider BL, Dawson PA. Regulation of the mouse organic solute transporter alpha-beta, Ostalpha-Ostbeta, by bile acids. *Am J Physiol Gastrointest Liver Physiol* 2006;290(5): G912–22.

136 Landrier JF, Eloranta JJ, Vavricka SR, Kullak-Ublick GA. The nuclear receptor for bile acids, FXR, transactivates human organic solute transporter-alpha and -beta genes. *Am J Physiol Gastrointest Liver Physiol* 2006;290(3):G476–85.

137 Hirschfield GM, Mason A, Luketic V, Lindor K, Gordon SC, Mayo M, et al. Efficacy of obeticholic acid in patients with primary biliary cirrhosis and inadequate response to ursodeoxycholic acid. *Gastroenterology* 2015;148(4): 751–61 e8.

138 Ali AH, Lindor KD. Obeticholic acid for the treatment of primary biliary cholangitis. *Expert Opin Pharmacother* 2016;17 (13):1809–15.

139 Erlinger S. Obeticholic acid in primary biliary cholangitis. *Clin Res Hepatol Gastroenterol* 2017;41(1):3–5.

140 Nevens F, Lindor KD, Jones DE. Obeticholic acid in primary biliary cholangitis. *N Engl J Med* 2016;375 (20):e41.

141 Spacek LA, Solga SF. Obeticholic acid in primary biliary cholangitis. *N Engl J Med.* 2016;375 (20):e41.

142 van Golen RF. Obeticholic acid in primary biliary cholangitis. *N Engl J Med* 2016;375 (20):e41.

143 Quigley G, Al Ani M, Nadir A. Occurrence of jaundice following simultaneous ursodeoxycholic acid cessation and obeticholic acid initiation. *Dig Dis Sci* 2018;63(2):529–32.

144 Chiang JY. Bile acid metabolism and signaling. *Compr Physiol* 2013;3(3):1191–212.

145 Li T, Chiang JY. Nuclear receptors in bile acid metabolism. *Drug Metab Rev* 2013;45(1):145–55.

146 Li T, Chiang JY. Bile acid signaling in liver metabolism and diseases. *J Lipids* 2012;2012:754067.

147 Edwards JE, Eliot L, Parkinson A, Karan S, MacConell L. Assessment of pharmacokinetic interactions between obeticholic acid and caffeine, midazolam, warfarin, dextromethorphan, omeprazole, rosuvastatin, and digoxin in phase 1 studies in healthy subjects. *Adv Ther* 2017;34(9):2120–38.

148 Zhang Y, Jackson JP, St Claire RL, 3rd, Freeman K, Brouwer KR, Edwards JE. Obeticholic acid, a selective farnesoid X receptor agonist, regulates bile acid homeostasis in sandwich-cultured human hepatocytes. *Pharmacol Res Perspect.* 2017;5(4), e00329.

149 Guo C, LaCerte C, Edwards JE, Brouwer KR, Brouwer KLR. Farnesoid X receptor agonists obeticholic acid and chenodeoxycholic acid increase bile acid efflux in sandwich-cultured human hepatocytes: functional evidence and mechanisms. *J Pharmacol Exp Ther* 2018;365(2):413–21.

150 Kahler CM, Graziadei I, Vogelsinger H, Desole S, Cima K, Vogel W. Successful treatment of portopulmonary hypertension with the selective endothelin receptor antagonist Sitaxentan. *Wien Klin Wochenschr* 2011;123 (7–8):248–52.

151 Scott LJ. Sitaxentan: in pulmonary arterial hypertension. *Drugs* 2007;67(5): 761–70; discussion 71-2.

152 Gholam P, Sehr T, Enk A, Hartmann M. Successful treatment of systemic-sclerosis-related digital ulcers with a selective endothelin type A receptor antagonist (sitaxentan). *Dermatology* 2009;219(2):171–3.

153 Zhang J, Kong W, Wang C. Mechanism of plasma endothelin-1 level elevation and its relation with pulmonary hypertension in chronic cor pulmonale. *Zhonghua Nei Ke Za Zhi* 1996;35(2):110–3.

154 Galie N, Hoeper MM, Simon J, Gibbs R, Simonneau G, Task Force for the D, et al. Liver toxicity of sitaxentan in pulmonary arterial hypertension. *Eur Heart J* 2011;32(4):386–7.

155 Lavelle A, Sugrue R, Lawler G, Mulligan N, Kelleher B, Murphy DM, et al. Sitaxentan-induced hepatic failure in two patients with pulmonary arterial hypertension. *Eur Respir J* 2009;34(3):770–1.

156 Lee WT, Kirkham N, Johnson MK, Lordan JL, Fisher AJ, Peacock AJ. Sitaxentan-related acute liver failure in a patient with pulmonary arterial hypertension. *Eur Respir J* 2011;37(2):472–4.

157 Hoeper MM, Olsson KM, Schneider A, Golpon H. Severe hepatitis associated with sitaxentan and response to glucocorticoid therapy. *Eur Respir J* 2009;33(6): 1518–9.

158 Owen K, Cross DM, Derzi M, Horsley E, Stavros FL. An overview of the preclinical toxicity and potential carcinogenicity of sitaxentan (Thelin(R)), a potent endothelin receptor antagonist developed for pulmonary arterial hypertension. *Regul Toxicol Pharmacol* 2012;64(1):95–103.

159 Erve JC, Gauby S, Maynard JW, Jr., Svensson MA, Tonn G, Quinn KP. Bioactivation of sitaxentan in liver microsomes, hepatocytes, and expressed human P450s with characterization of the glutathione conjugate by liquid chromatography tandem mass spectrometry. *Chem Res Toxicol* 2013;26(6): 926–36.

160 Whitley RJ. Sorivudine: a potent inhibitor of varicella zoster virus replication. *Adv Exp Med Biol* 1996;394:41–4.

161 Whitley RJ. Sorivudine: a promising drug for the treatment of varicella-zoster virus infection. *Neurology* 1995;45 (12 Suppl 8):S73–5.

162 Burdge DR, Voigt R, Lindley JI, Gage L, Sacks SL. Sorivudine (BV-ara-U) for the treatment of complicated refractory varicella zoster virus infection in HIV-infected patients. *AIDS* 1995;9(7):810–2.

163 Pinnolis MK, Foxworthy D, Kemp B. Treatment of progressive outer retinal necrosis with sorivudine. *Am J Ophthalmol* 1995;119(4):516–7.

164 Wunderli W, Miner R, Wintsch J, von Gunten S, Hirsch HH, Hirschel B. Outer retinal necrosis due to a strain of varicella-zoster virus resistant to acyclovir, ganciclovir, and sorivudine. *Clin Infect Dis* 1996;22(5):864–5.

165 Bodsworth NJ, Boag F, Burdge D, Genereux M, Borleffs JC, Evans BA, et al. Evaluation of sorivudine (BV-araU) versus acyclovir in the treatment of acute localized herpes zoster in human immunodeficiency virus-infected adults. The Multinational Sorivudine Study Group. *J Infect Dis* 1997;176(1):103–11.

166 Okuda H, Nishiyama T, Ogura K, Nagayama S, Ikeda K, Yamaguchi S, et al. Lethal drug interactions of sorivudine, a new antiviral drug, with oral 5-fluorouracil prodrugs. *Drug Metab Dispos* 1997;25(5):270–3.

167 Okuda H, Ogura K, Kato A, Takubo H, Watabe T. A possible mechanism of eighteen patient deaths caused by interactions of sorivudine, a new antiviral drug, with oral 5-fluorouracil prodrugs. *J Pharmacol Exp Ther* 1998;287(2):791–9.

168 Diasio RB. Sorivudine and 5-fluorouracil; a clinically significant drug-drug interaction due to inhibition of dihydropyrimidine dehydrogenase. *Br J Clin Pharmacol* 1998;46(1):1–4.

169 Dawson RM. Reversibility of the inhibition of acetylcholinesterase by tacrine. *Neurosci Lett* 1990;118(1):85–7.

170 Summers WK, Koehler AL, Marsh GM, Tachiki K, Kling A. Long-term hepatotoxicity of tacrine. *Lancet* 1989;1(8640):729.

171 Watkins PB, Zimmerman HJ, Knapp MJ, Gracon SI, Lewis KW. Hepatotoxic effects of tacrine administration in patients with Alzheimer's disease. *JAMA* 1994;271 (13):992–8.

172 Fredj G, Dietlin F, Barbier G, Jasmin C, Bonhomme L, Esctein S, et al. Comparison of tacrine hepatotoxicity in patients with Alzheimer disease or AIDS. *Therapie* 1992;47(3):245–7.

173 Becquemont L, Le Bot MA, Riche C, Beaune P. Influence of fluvoxamine on tacrine metabolism *in vitro*: potential implication for the hepatotoxicity *in vivo*. *Fundam Clin Pharmacol* 1996;10(2):156–7.

174 Madden S, Woolf TF, Pool WF, Park BK. An investigation into the formation of stable, protein-reactive and cytotoxic metabolites from tacrine *in vitro*. Studies with human and rat liver microsomes. *Biochem Pharmacol* 1993;46(1):13–20.

175 Hendrickson HP, Scott DO, Lunte CE. Identification of 9-hydroxylamine-1,2,3,4-tetrahydroacridine as a hepatic microsomal metabolite of tacrine by high-performance liquid chromatography and electrochemistry. *J Chromatogr* 1989;487(2):401–8.

176 Becquemont L, Ragueneau I, Le Bot MA, Riche C, Funck-Brentano C, Jaillon P. Influence of the CYP1A2 inhibitor fluvoxamine on tacrine pharmacokinetics in humans. *Clin Pharmacol Ther* 1997;61(6):619–27.

177 Pool WF, Reily MD, Bjorge SM, Woolf TF. Metabolic disposition of the cognition activator tacrine in rats, dogs, and humans. Species comparisons. *Drug Metab Dispos* 1997;25(5):590–7.

178 Madden S, Spaldin V, Hayes RN, Woolf TF, Pool WF, Park BK. Species variation in the bioactivation of tacrine by hepatic microsomes. *Xenobiotica* 1995;25(1):103–16.

179 Sung JH, Yu KH, Park JS, Tsuruo T, Kim DD, Shim CK, et al. Saturable distribution of tacrine into the striatal extracellular fluid of the rat: evidence of involvement of multiple organic cation transporters in the transport. *Drug Metab Dispos* 2005;33(3):440–8.

180 Fontana RJ, Turgeon DK, Woolf TF, Knapp MJ, Foster NL, Watkins PB. The caffeine breath test does not identify patients susceptible to tacrine hepatotoxicity. *Hepatology* 1996;23(6):1429–35.

181 Simon T, Becquemont L, Mary-Krause M, de Waziers I, Beaune P, Funck-Brentano C, et al. Combined glutathione-S-transferase M1 and T1 genetic polymorphism and tacrine hepatotoxicity. *Clin Pharmacol Ther* 2000;67(4):432–7.

182 Becquemont L, Lecoeur S, Simon T, Beaune P, Funck-Brentano C, Jaillon P. Glutathione S-transferase theta genetic polymorphism might influence tacrine hepatotoxicity in Alzheimer's patients. *Pharmacogenetics* 1997;7(3):251–3.

183 Biglin KE, Faraon MS, Constance TD, Lieh-Lai M. Drug-induced torsades de pointes: a possible interaction of terfenadine and erythromycin. *Ann Pharmacother* 1994;28(2):282.

184 Wynn RL. Erythromycin and ketoconazole (Nizoral) associated with terfenadine (Seldane)-induced ventricular arrhythmias. *Gen Dent* 1993;41(1):27–9.

185 Honig PK, Woosley RL, Zamani K, Conner DP, Cantilena LR, Jr. Changes in the pharmacokinetics and electrocardiographic pharmacodynamics of terfenadine with concomitant administration of erythromycin. *Clin Pharmacol Ther* 1992;52(3):231–8.

186 Woosley RL, Chen Y, Freiman JP, Gillis RA. Mechanism of the cardiotoxic actions of terfenadine. *JAMA* 1993;269 (12):1532–6.

187 Honig PK, Wortham DC, Zamani K, Conner DP, Mullin JC, Cantilena LR. Terfenadine-ketoconazole interaction. Pharmacokinetic and electrocardiographic consequences. *JAMA* 1993;269 (12):1513–8.

188 Zimmermann M, Duruz H, Guinand O, Broccard O, Levy P, Lacatis D, et al. Torsades de Pointes after treatment with terfenadine and ketoconazole. *Eur Heart J* 1992;13(7):1002–3.

189 Kamisako T, Adachi Y, Nakagawa H, Yamamoto T. Torsades de pointes associated with terfenadine in a case of liver cirrhosis and hepatocellular carcinoma. *Intern Med* 1995;34(2):92–5.

190 Mason J, Reynolds R, Rao N. The systemic safety of fexofenadine HCl. *Clin Exp Allergy* 1999;29 Suppl 3:163–70; discussion 71-3.

191 Balian JD, Rahman A. Metabolic drug-drug interactions: perspective from FDA medical and clinical pharmacology reviewers. *Adv Pharmacol* 1997;43:231–8.

192 Davit B, Reynolds K, Yuan R, Ajayi F, Conner D, Fadiran E, et al. FDA evaluations using *in vitro* metabolism to predict and interpret *in vivo* metabolic drug-drug interactions: impact on labeling. *J Clin Pharmacol* 1999;39(9):899–910.

193 Huang SM, Strong JM, Zhang L, Reynolds KS, Nallani S, Temple R, et al. New era in drug interaction evaluation: US Food and Drug Administration update on CYP enzymes, transporters, and the guidance process. *J Clin Pharmacol* 2008;48(6):662–70.

194 Yu J, Petrie ID, Levy RH, Ragueneau-Majlessi I. Mechanisms and clinical significance of pharmacokinetic-based drug-drug interactions with drugs approved by the U.S. Food and Drug Administration in 2017. *Drug Metab Dispos* 2019;47(2):135–44.

195 Aslanian R, Piwinski JJ, Zhu X, Priestley T, Sorota S, Du XY, et al. Structural determinants for histamine H(1) affinity, hERG affinity and QTc prolongation in a series of terfenadine analogs. *Bioorg Med Chem Lett* 2009;19 (17):5043–7.

196 Kamiya K, Niwa R, Morishima M, Honjo H, Sanguinetti MC. Molecular determinants of hERG channel block by terfenadine and cisapride. *J Pharmacol Sci* 2008;108(3):301–7.

197 Roy M, Dumaine R, Brown AM. HERG, a primary human ventricular target of the nonsedating antihistamine terfenadine. *Circulation* 1996;94(4):817–23.

198 Suessbrich H, Waldegger S, Lang F, Busch AE. Blockade of HERG channels expressed in *Xenopus* oocytes by the histamine receptor antagonists terfenadine and astemizole. *FEBS Lett* 1996;385 (1–2):77–80.

199 Raeissi SD, Hidalgo IJ, Segura-Aguilar J, Artursson P. Interplay between CYP3A-mediated metabolism and polarized efflux of terfenadine and its metabolites in intestinal epithelial Caco-2 (TC7) cell monolayers. *Pharm Res* 1999;16(5):625–32.

200 Jurima-Romet M, Crawford K, Cyr T, Inaba T. Terfenadine metabolism in human liver. *in vitro* inhibition by macrolide antibiotics and azole antifungals. *Drug Metab Dispos* 1994;22(6):849–57.

201 Herings RM, Stricker BH, Leufkens HG, Bakker A, Sturmans F, Urquhart J. Public health problems and the rapid estimation of the size of the population at risk. Torsades de pointes and the use of terfenadine and astemizole in The Netherlands. *Pharm World Sci* 1993;15(5):212–8.

202 Pohjola-Sintonen S, Viitasalo M, Toivonen L, Neuvonen P. Itraconazole prevents terfenadine metabolism and increases risk of torsades de pointes ventricular tachycardia. *Eur J Clin Pharmacol* 1993;45(2):191–3.

203 Paris DG, Parente TF, Bruschetta HR, Guzman E, Niarchos AP. Torsades de pointes induced by erythromycin and terfenadine. *Am J Emerg Med* 1994;12(6):636–8.

204 Pohjola-Sintonen S. Treatment with terfenadine and ketoconazole or itraconazole can cause torsades de pointes ventricular tachycardia. *Duodecim* 1993;109(2):164–6.

205 Benton RE, Honig PK, Zamani K, Cantilena LR, Woosley RL. Grapefruit juice alters terfenadine pharmacokinetics, resulting in prolongation of repolarization on the electrocardiogram. *Clin Pharmacol Ther* 1996;59(4):383–8.

206 Clifford CP, Adams DA, Murray S, Taylor GW, Wilkins MR, Boobis AR, et al. The cardiac effects of terfenadine after inhibition of its metabolism by grapefruit juice. *Eur J Clin Pharmacol* 1997;52(4):311–5.

207 Honig PK, Wortham DC, Lazarev A, Cantilena LR. Grapefruit juice alters the systemic bioavailability and cardiac repolarization of terfenadine in poor metabolizers of terfenadine. *J Clin Pharmacol* 1996;36(4):345–51.

208 Rau SE, Bend JR, Arnold MO, Tran LT, Spence JD, Bailey DG. Grapefruit juice-terfenadine single-dose interaction: magnitude, mechanism, and relevance. *Clin Pharmacol Ther* 1997;61(4):401–9.

209 Kohlroser J, Mathai J, Reichheld J, Banner BF, Bonkovsky HL. Hepatotoxicity due to troglitazone: report of two cases and review of adverse events reported to the United States Food and Drug Administration. *Am J Gastroenterol* 2000;95(1):272–6.

210 Gitlin N, Julie NL, Spurr CL, Lim KN, Juarbe HM. Two cases of severe clinical and histologic hepatotoxicity associated with troglitazone. *Ann Intern Med* 1998;129(1):36–8.

211 Herrine SK, Choudhary C. Severe hepatotoxicity associated with troglitazone. *Ann Intern Med* 1999;130(2):163–4.

212 Kaplowitz N. Idiosyncratic drug hepatotoxicity. *Nat Rev Drug Discov* 2005;4(6):489–99.

213 Meier Y, Cavallaro M, Roos M, Pauli-Magnus C, Folkers G, Meier PJ, et al. Incidence of drug-induced liver injury in medical inpatients. *Eur J Clin Pharmacol* 2005;61(2):135–43.

214 Bjornsson ES, Bergmann OM, Bjornsson HK, Kvaran RB, Olafsson S. Incidence, presentation, and outcomes in patients with drug-induced liver injury in the general population of Iceland. *Gastroenterology*. 2013;144(7):1419–25, 25 e1-3; quiz e19-20.

215 Faich GA, Moseley RH. Troglitazone (Rezulin) and hepatic injury. *Pharmacoepidemiol Drug Saf* 2001;10(6):537–47.

216 Dixit VA, Bharatam PV. Toxic metabolite formation from Troglitazone (TGZ): new insights from a DFT study. *Chem Res Toxicol* 2011;24(7):1113–22.

217 Prabhu S, Fackett A, Lloyd S, McClellan HA, Terrell CM, Silber PM, et al. Identification of glutathione conjugates of troglitazone in human hepatocytes. *Chem Biol Interact* 2002;142 (1–2):83–97.

218 Honma W, Shimada M, Sasano H, Ozawa S, Miyata M, Nagata K, et al. Phenol sulfotransferase, ST1A3, as the main enzyme catalyzing sulfation of troglitazone in human liver. *Drug Metab Dispos* 2002;30(8):944–9.

219 Kostrubsky VE, Sinclair JF, Ramachandran V, Venkataramanan R, Wen YH, Kindt E, et al. The role of conjugation in hepatotoxicity of troglitazone in human and porcine hepatocyte cultures. *Drug Metab Dispos* 2000;28 (10):1192–7.

220 Masubuchi Y. Metabolic and non-metabolic factors determining troglitazone hepatotoxicity: a review. *Drug Metab Pharmacokinet* 2006;21(5):347–56.

221 Smith MT. Mechanisms of troglitazone hepatotoxicity. *Chem Res Toxicol* 2003;16(6):679–87.

222 Tafazoli S, Spehar DD, O'Brien PJ. Oxidative stress mediated idiosyncratic drug toxicity. *Drug Metab Rev* 2005;37(2):311–25.

223 Galati G, Tafazoli S, Sabzevari O, Chan TS, O'Brien PJ. Idiosyncratic NSAID drug induced oxidative stress. *Chem Biol Interact* 2002;142 (1–2):25–41.

224 Shibuya A, Watanabe M, Fujita Y, Saigenji K, Kuwao S, Takahashi H, et al. An autopsy case of troglitazone-induced fulminant hepatitis. *Diabetes Care* 1998;21 (12):2140–3.

225 Uetrecht J. Drug metabolism by leukocytes and its role in drug-induced lupus and other idiosyncratic drug reactions. *Crit Rev Toxicol* 1990;20(4):213–35.

226 Funk C, Pantze M, Jehle L, Ponelle C, Scheuermann G, Lazendic M, et al. Troglitazone-induced intrahepatic cholestasis by an interference with the hepatobiliary export of bile acids in male and female rats. Correlation with the gender difference in troglitazone sulfate formation and the inhibition of the canalicular bile salt export pump (BSEP) by troglitazone and troglitazone sulfate. *Toxicology* 2001;167(1):83–98.

227 Funk C, Ponelle C, Scheuermann G, Pantze M. Cholestatic potential of troglitazone as a possible factor contributing to troglitazone-induced hepatotoxicity: *in vivo* and *in vitro* interaction at the canalicular bile salt export pump (BSEP) in the rat. *Mol Pharmacol* 2001;59(3):627–35.

228 Okada R, Maeda K, Nishiyama T, Aoyama S, Tozuka Z, Hiratsuka A, et al. Involvement of different human glutathione transferase isoforms in the glutathione conjugation of reactive metabolites of troglitazone. *Drug Metab Dispos* 2011;39 (12):2290–7.

229 Hewitt NJ, Lloyd S, Hayden M, Butler R, Sakai Y, Springer R, et al. Correlation between troglitazone cytotoxicity and drug metabolic enzyme activities in cryopreserved human hepatocytes. *Chem Biol Interact* 2002;142 (1–2):73–82.

230 Stieger B. Role of the bile salt export pump, BSEP, in acquired forms of cholestasis. *Drug Metab Rev* 2010;42(3):437–45.

231 Shaw PJ, Ganey PE, Roth RA. Idiosyncratic drug-induced liver injury and the role of inflammatory stress with an emphasis on an animal model of trovafloxacin hepatotoxicity. *Toxicol Sci* 2010;118(1):7–18.

232 Li AP. A review of the common properties of drugs with idiosyncratic hepatotoxicity and the "multiple determinant hypothesis" for the manifestation of idiosyncratic drug toxicity. *Chem Biol Interact* 2002;142 (1–2):7–23.

233 Bachoual R, Dubreuil L, Soussy CJ, Tankovic J. Roles of gyrA mutations in resistance of clinical isolates and *in vitro* mutants of *Bacteroides fragilis* to the new fluoroquinolone trovafloxacin. *Antimicrob Agents Chemother* 2000;44(7):1842–5.

234 Gootz TD, Zaniewski RP, Haskell SL, Kaczmarek FS, Maurice AE. Activities of trovafloxacin compared with those of other fluoroquinolones against purified topoisomerases and gyrA and grlA mutants of *Staphylococcus aureus*. *Antimicrob Agents Chemother* 1999;43(8):1845–55.

235 Lazarczyk DA, Goldstein NS, Gordon SC. Trovafloxacin hepatotoxicity. *Dig Dis Sci* 2001;46(4):925–6.

236 Lucena MI, Andrade RJ, Rodrigo L, Salmeron J, Alvarez A, Lopez-Garrido MJ, et al. Trovafloxacin-induced acute hepatitis. *Clin Infect Dis* 2000;30(2):400–1

237 Chen HJ, Bloch KJ, Maclean JA. Acute eosinophilic hepatitis from trovafloxacin. *N Engl J Med* 2000;342(5):359–60.

238 Sun Q, Zhu R, Foss FW, Jr., Macdonald TL. Mechanisms of trovafloxacin hepatotoxicity:studies of a model cyclopropylamine-containing system. *Bioorg Med Chem Lett* 2007;17 (24):6682–6.

239 Sun Q, Zhu R, Foss FW, Jr., Macdonald TL. *in vitro* metabolism of a model cyclopropylamine to reactive intermediate: insights into trovafloxacin-induced hepatotoxicity. *Chem Res Toxicol* 2008;21(3):711–9.

240 Shaw PJ, Ditewig AC, Waring JF, Liguori MJ, Blomme EA, Ganey PE, et al. Coexposure of mice to trovafloxacin and lipopolysaccharide, a model of idiosyncratic hepatotoxicity, results in a unique gene expression profile and interferon gamma-dependent liver injury. *Toxicol Sci* 2009;107(1):270–80.

241 Shaw PJ, Ganey PE, Roth RA. Tumor necrosis factor alpha is a proximal mediator of synergistic hepatotoxicity from trovafloxacin/lipopolysaccharide coexposure. *J Pharmacol Exp Ther* 2009;328(1):62–8.

242 Shaw PJ, Hopfensperger MJ, Ganey PE, Roth RA. Lipopolysaccharide and trovafloxacin coexposure in mice causes idiosyncrasy-like liver injury dependent on tumor necrosis factor-alpha. *Toxicol Sci* 2007;100(1):259–66.

243 Pascual A, Garcia I, Ballesta S, Perea EJ. Uptake and intracellular activity of trovafloxacin in human phagocytes and tissue-cultured epithelial cells. *Antimicrob Agents Chemother* 1997;41(2):274–7.

Part II

Drug Metabolizing Enzymes and Drug Toxicity

4

Drug-Metabolizing Enzymes and Drug Toxicity

Albert P. Li

In vitro ADMET Laboratories, Inc., 9221 Rumsey Road, Suite 8, Columbia, MD, USA

4.1 Introduction

Metabolic activation, where a relatively nontoxic parent drug is metabolized to reactive and/or toxic metabolites, and metabolic detoxification, where a toxic parent drug or metabolite is metabolized to less toxic metabolites, are critical determinants of drug toxicity. More importantly, these two events can contribute toward an individual's susceptibility to drug toxicity. The Multiple Determinant Hypothesis [1] for idiosyncratic Drug Toxicity states that idiosyncratic drug toxicity occurs due to a confluence of multiple discrete events resulting in the manifestation of serious drug toxicity. A likely confluence of events resulting in severe drug-induced liver injuries (sDILI) is the simultaneous occurrence of enhanced activating and reduced detoxifying drug metabolism enzyme activities due to environmental and/or genetic factors. A drug that undergoes metabolic activation and detoxification thereby are likely to cause sDILI. In this chapter, the involvement of key drug-metabolizing enzymes in metabolic activation and detoxification are reviewed. Examples for the identification of potential risk factors based on metabolic activation and detoxification are provided for selected marketed drugs where drug metabolism plays key roles in the observed toxicity.

4.2 Drug-Metabolism Enzymes Involved in Metabolic Activation and Detoxification

Drug-metabolizing enzymes were developed during evolution, presumably as a mean to convert toxic, hydrophobic toxicants to less toxic, hydrophilic molecules than can be excreted. The major pathways are phase 1 oxidation, the addition of oxygen molecules, and phase 2 conjugation, the conjugation of the parent or

Transporters and Drug-Metabolizing Enzymes in Drug Toxicity, First Edition. Edited by Albert P. Li.

oxygenated metabolites with highly hydrophilic molecules such as glucose an sulfate. Drug metabolism is critical to drug toxicity. A parent drug can be metabolized to toxic/reactive metabolites, a process known as metabolic activation. A toxic drug or metabolite can be metabolized to less toxic metabolites, a process known as metabolic detoxification. P450 isoforms are generally involved in metabolic activation, while conjugating pathways, especially glutathione (GSH) transferase-mediated conjugation, are responsible for metabolic detoxification. Various aspects of drug-metabolizing enzymes and their relationship to drug metabolism and toxicity have been extensively reviewed [2–7]. Here an overview is provided for the involvement of specific drug-metabolizing enzymes in metabolic activation and detoxification and the associated risk factors for drug toxicity.

4.3 Cytochrome P450 Monooxygenase (CYP)

CYP represents the major drug metabolizing enzyme superfamily involved in the metabolic activation of xenobiotics, mainly via the transfer of one molecule of oxygen to the substrate, a reaction known as monooxygenation. The oxygenated metabolite can be highly reactive, resulting in toxicity via interaction with cellular macromolecules. The cytochrome P450 (CYP) genes are grouped into 18 families and 44 subfamilies based on sequence similarity. As of this writing, over 60 human CYP isoforms have been identified, with the major ones generally accepted to be involved in human drug metabolism being CYP1A1, CYP1A2, CYP2A6, CYP2B6, CYP2C8, CYP2C9, CYP2C19, CYP2D6, CYP2E1, and CYP3A4/5/7 [8]. Based on protein contents, approximately 70% of liver P-450 could be accounted for by CYPs 1A2, 2A6, 2B6, 2C, 2D6, 2E1, and 3A. The approximate composition for the individual isoforms in the human liver has been reported to be the following: CYP3A (30%), CYP2C (20%), CYP1A2 (13%), CYP2E1 (7%), CYP2A6 (4%), CYP2D6 (2%), and CYP2B6 (<1%) [9].

4.3.1 CYP1A

The major isoforms of CYP1A is CYP1A1 and CYP1A2. CYP1A1 is the isoform present in extrahepatic tissues including lung and pulmonary tissues, and CYP1A2 as the major isoform in the liver.

4.3.1.1 Metabolic Activation

CYP1A was discovered as the P450 isoform that metabolically activates polycyclic arylhydrocarbons (PAH) and polycyclic halogenated hydrocarbons (PCH) to their mutagenic and carcinogenic metabolites. CYP1A2 is the major pathway for the metabolism [10] and one of the P450 isoforms responsible for the formation of protein-reactive metabolites [11] of the anticonvulsant carbamazepine which is associated with rare cases of liver failures [12, 13].

4.3.1.2 Drug Substrates

CYP1A2 is reported to metabolize 15% of marketed drugs including alosetron, caffeine, clozapine, duloxetine, leflunomide, melatonin, ramelteon, tacrine, tasimelteon, theophylline, and tizanidine [14]. Endogenous substrates including retinols, melatonin, steroids, uroporphyrinogen, and arachidonic acids [15].

4.3.1.3 Inducers

CYP1A is highly inducible by ligands of the aryl hydrocarbon receptor (AHR). Upon binding, the ligand–receptor complex translocates to the nucleus and interacts with the CYP1A promoter region, leading to enhanced gene expression and the subsequent protein synthesis, resulting in the induction of activity [16, 17]. Drugs that cause clinically significant CYP1A induction include carbamazepine [18], omeprazole [19], and fluvoxamine [20]. Omeprazole has been reported to be likely to enhance the toxicity of acetaminophen via CYP1A2 induction, resulting in a higher rate of formation of *N*-acetyl-*p*-benzoquinone imine (NAPQI), the toxic metabolite, but mainly in CYP2C19 poor metabolizers [19]. Smoking and exposure to PAH and PCH are major environmental factors leading to CYP1A1 and CYP1A2 induction [21].

4.3.1.4 Inhibitors

Ciprofloxacin, enoxacin, fluvoxamine, methoxsalen, mexiletine, and oral contraceptives are known inhibitors of CYP1A2 [14].

4.3.1.5 Individual Variations

It has been reported that CYP1A2 mRNA content varied >40-fold among individuals while CYP1A1 content varied >20-fold [22]. The interindividual difference of CYP1A2 are generally attributed to environmental inducers such as cigarette smoke [23] and foods such as cruciferous vegetables [24]. There are both *in vitro* [25] and *in vivo* [26] evidence suggesting gender difference in the expression of human CYP1A, with males having higher activities than females [26, 27]. Drug toxicity influenced by CYP1A2 genotype has been reported for leflunomide where patients with *CYP1A2*1F CC* genotype had a 9.7-fold higher risk for overall leflunomide-induced toxicity than did the carriers of *CYP1A2*1F* A allele [28].

4.3.1.6 Involvement in Drug Toxicity

CYP1A metabolism of the environmental pollutant benzo(a)pyrene [29], resulting in the formation of the highly reactive diol epoxide diol-epoxide, 9,10-dihydroxy-11,12-epoxy-9,10,11,-12-tetrahydro-B[e]P [30] is one of the very first observations introducing the concept of metabolic activation [31]. CYP1A2 is now known to bioactivate a number of procarcinogens including polycyclic aromatic hydrocarbons, heterocyclic aromatic amines/amides, mycotoxins, and phytotoxins. Drugs metabolically activated by CYP1A2 include tacrine, leflunomide, teriflunomide, clozapine. CYP1A2 bioactivation of tacrine involving an initial 7-hydroxylation followed by a postulated 2-electron oxidation to yield a reactive quinone methide [32–34], suggesting that patients with elevated CYP1A2 activity may be susceptible to tacrine hepatotoxicity. However, CYP1A2 activity phenotyping of patients based on caffeine breath test was not successful in the identification of susceptible individuals to its hepatotoxicity [35]. The major metabolite of leflunomide, teriflunomide, is reported to significantly decrease the expression of Na+-taurocholate cotransporting polypeptide (NTCP) in human hepatocytes and has been postulated to be one of the mechanisms of its hepatotoxicity [36]. Metabolic clearance of clozapine by CYP1A2 is a major determinant of clozapine, with a correlation of toxic side effects associated with decreased CYP1A2 activity due to genetic polymorphism or concurrent exposure to CYP1A2 inhibitors.

4.3.2 CYP2A6

CYP2A6 is found mainly in the liver and is approximately 5–10% of the total CYP450 content [37], and is expressed also in the nasal mucosa, respiratory tracts [38], and brain [39], but not in the intestinal mucosa [40].

4.3.2.1 Substrates

When compared with isoforms such as CYP3A and CYP2C, CYP2A6 has substantially fewer substrates. Drugs that are selectively metabolized by CYP2A6 are coumarin, nicotine, and novel platelet-activating factor receptor antagonist SM-1250 [41]. Drugs that are partially metabolized by CYP2A6 include losigamone, valproic acid, letrozole, and disulfiram [41].

4.3.2.2 Inducers

CYP2A6 is an inducible CYP isoform is the principal enzyme, with its induction reported to include a variety of nuclear receptors including constitutive androstane receptor (CAR), pregnane X receptor (PXR) peroxisome proliferator-activated receptor-γ coactivator 1α (PGC-1α), hepatocyte nuclear factor 4 alpha (HNF4α), and the glucocorticoid receptor (GR) [42–44].

4.3.2.3 Inhibitors

Drugs that are CYP2A6 inhibitors include the metabolism-based inhibitors 8-methoxypsoralen and selegiline, and the competitive inhibitor tranylcypromine and ketoconazole [45].

4.3.2.4 Individual Variations

Genetic variation is the primary contributor to variation in CYP2A6 enzyme activity, ranging from higher than wild type activity in individuals with gene duplications and no activity in individuals homozygous for genetically null *CYP2A6* alleles [46]. *in vivo* measure of CYP2A6 activity based on the ratio of nicotine metabolite 3′-hydroxycotinin to parent nicotine (nicotine metabolite ratio, NMR), show that 60–80% of the variation in the CYP2A6 activity can be attributed to genetic influences and lack of correlation with age, and alcohol consumption [47]. Asian and African American populations have been reported to exhibit lower overall activity due to higher frequencies of CYP2A6 decrease- or loss-of-function genetic variants. An interesting observation is that slow metabolizers were found to smoke fewer cigarettes per day and to initiate smoking at an earlier age [48].

4.3.2.5 Involvement in Drug Toxicity

Tobacco nitrosamines need to be mentioned although they are not drug substances as CYP2A6 plays a key role in tobacco carcinogenesis. CYP2A6 is involved in the metabolic activation of tobacco-specific nitrosamines to their ultimate carcinogenic forms and metabolism of nicotine. Male smokers possessing the wildtype *1A/*1A genotype have been reported to be associated with a higher risk for tobacco-induced lung cancers than individuals with the homozygous deletion (*4/*4) genotype [49]. Coumarin 7-hydroxylation is specifically catalyzed by CYP2A6 and is identified as a major pathway for coumarin detoxification [50]. CYP2A6 has been reported to be one of the P450 isoforms that catalyzes the formation of the toxic metabolite NAPQI from acetaminophen [51]. Acetaminophen (APAP)-associated liver failure [52–54] is a well-documented phenomenon, and NAPQI is generally considered to be the metabolite responsible for acetaminophen hepatotoxicity [55–58]. Tegafur is an anticancer prodrug that is metabolized primarily by CYP2A6 to the active metabolite 5-fluorouracil which is also responsible for its toxicity [59].

4.3.3 CYP2B6

CYP2B6 is generally considered a minor P450 isoform mainly expressed in the human liver but is also present in extrahepatic tissues such as the small intestines [60–64], brain, kidney, endometrium, bronchoalveolar macrophages, peripheral blood lymphocytes [64, 65], and skin [66]. It was one of the first P450 isoforms found to be highly inducible by phenobarbital (PB) in animals and in cultured rodent hepatocytes [67–70].

4.3.3.1 Substrates

CYP2B6 has been reported to metabolize 7.6% of marketed drugs, including a number of CYP3A4 substrates [71]. Bupropion hydroxylation is generally considered a selective marker for the quantification of CYO2B6 activity [72, 73]. Other drugs with CYP2B6 as the major contributor for their metabolism include propofol, mephenytoin, mephobarbital, artemether, artemisinin, efavirenz, cyclophosphamide, isofosamide, selegiline, and pethidine [74].

4.3.3.2 Inducers

CYP2B6 is a highly inducible P450 isoform, with the CAR responsible for its induction. Binding of CAR by its ligand such as PB leads to its nuclear translocation and binding to the phenobarbital-responsive enhancer module (PBREM), resulting in enhanced CYP2B6 gene expression, a process observed both *in vivo* in animals [75] and in cultured human hepatocytes [76]. Besides CAR, GR, and PXR have also been found to be also involved in CYP2B6 induction [77]. CYP2B6 inducers are generally also CYP3A4 inducers, including PB, phenytoin, rifampin, clotrimazole, carbamazepine, efavirenz, and artemisinin [74].

4.3.3.3 Inhibitors

Clopidogrel and ticlopidine are potent and specific inhibitors have been approved as clinical tools for the evaluation of the role of CYP2B6 in drug metabolism in human subjects [78]. 2-phenyl-2-(1-piperidinyl)propane (PPP) is also a potent and specific CYP2B6 inhibitor for *in vitro* studies [79, 80]. An interesting inhibitor is pyrrolo[2,1-c][1,4]benzodiazepine dimer (PBD), a DNA-minor groove alkylating agent being investigated as a cytotoxic antibody-drug conjugate warhead for oncology targets which has been found to inactivate CYP2B6 via direct covalent binding in the absence of NADPH [81].

4.3.3.4 Individual Variations

CYP2B6 polymorphism are associated with both enhanced and loss of activity [82]. Environment inducers and inhibitors are expected to alter an individual's CYP2B6 activity.

4.3.3.5 Involvement in Drug Toxicity

Drugs that are metabolized by CYP2B6, resulting in the formation of reactive metabolites include cyclophosphamide [83], tamoxifen [84], and ifosfamide [85]. Other toxicants metabolically activated by CYP2B6 include the environmental pollutants polychlorinated biphenyl [86] and polybrominated biphenyl ether [87], the organophosphorus pesticide chlorpyrifos [88]. Patients with CYP2B6 slow metabolizer genotypes have been found to have high plasma concentrations and the associated unwarranted side effects and toxicity for efavirenz [89], and methadone [90].

4.3.4 CYP2C8

CYP2C8 is one of the major hepatic P450 isoforms, comprising 7% of the total hepatic CYP content, and is also present in extrahepatic sites such as the kidney, heart, adrenal gland, brain, uterus, mammary gland, ovary, and intestines. A notable CYP2C8-related drug toxicity is the toxic drug interaction between cerivastatin, a CYP2C8 substrate, with gemfibrozil, a CYP2C8 inhibitor. Co-administration of cerivastatin and gemfibrozil resulted in a higher incidence of rhabdomyolysis due to cerivastatin toxicity. The observation ultimately led to the withdrawal of cerivastatin from the worldwide market [91].

4.3.4.1 Substrates

Drug substrates of CYP2C8 include amiodarone, amodiaquine, cerivastatin, chloroquine, paclitaxel, pioglitazone, repaglinide, and the thiazolidinediones rosiglitazone, pioglitazone and troglitazone [92].

4.3.4.2 Inducers

As the PXR, CAR, and GR are involved in the regulation of CYP2C8, their ligands have the potential to induce CYP2C8. *in vitro* inducers include dexamethasone, PB and rifampin. Clinically observed CYP2C8 induction has been observed for rifampin [93] for enterocytes after oral administration.

4.3.4.3 Inhibitors

An *in vitro* study evaluating 209 drugs identified leukotriene receptor antagonist, montelukast, to be the most potent CYP2C8 inhibitor, with an IC50 of 19.6 nM, with other potent inhibitors (IC50 $<1\,\mu M$) being candesartan cilexetil, zafirlukast, clotrimazole, felodipine, and mometasone furoatethe, and moderate inhibitors (IC50 >1 and $<10\,\mu M$) being salmeterol, raloxifene, fenofibrate, ritonavir, levothyroxine, tamoxifen, loratadine, quercetin, oxybutynin, medroxyprogesterone, simvastatin, ketoconazole, ethinyl estradiol, spironolactone, lovastatin, nifedipine, and irbesartan [94]. It is interesting that this study did not identify gemfibrozil which has caused clinically significant CYP2C8 inhibition, leading to fatal interaction with cerivastatin. It is now recognized that the metabolite gemfibrozil 1-*O*-beta-glucuronide is responsible for the inhibitory effects [95] which is not formed by the human microsomes under the experimental conditions used for the study. This discrepancy between *in vitro* and *in vivo* findings illustrates the importance of using an *in vitro* system with complete drug metabolizing enzyme pathways for the evaluation of drug–drug interaction potential of drug candidates.

4.3.4.4 Individual Variations

CYP2C8*3 is the most frequent variant allele in the Caucasian population, with an allele frequency of about 10–20%. While *in vitro* studies have suggested that CYP2C8*3 has reduced activity *in vitro* [96], evidence for enhanced metabolism was observed *in vivo* for CYP2C8 substrates thiazolidinediones [97, 98], and repaglinide [99].

4.3.4.5 Involvement in Drug Toxicity

An increased risk for paclitaxel neuropathy has been reported for patients with the CYP2C8*3 genotype [100, 101] which has been attributed due to the lower rate of its metabolic clearance [102] as a result of the lower enzyme activity accompanied with the genotype.

4.3.5 CYP2C9

CYP2C9, originally discovered as the P450 isoform for *s*-warfarin metabolism, is an inducible CYP isoform. CYP2C9 accounts for approximately 20% of total hepatic P450 protein in human liver microsomes [103], thereby ranking second to CYP3A as the most abundant P450 isoform. CYP2C9 is also highly active in the human intestinal mucosa [63, 104].

4.3.5.1 Substrates

CYP2C9 is responsible for the metabolic clearance of up to 15–20% of all drugs undergoing phase I metabolism [105]. Diclofenac 4′-hydroxylation, *S*-flurbiprofen 4′-hydroxylation, losartan carboxylation, phenytoin 4′-hydroxylation, tolbutamide and torsemide tolylmethyl hydroxylation, and *S*-warfarin 7-hydroxylation activities have been used to assess CYP2C9 activity both *in vitro* and *in vivo* [106]. The crystal structure of CYP2C9 shows a relatively large active site that may simultaneously accommodate multiple ligands which may be one of the reasons for the vast structural varieties of its substrates [107, 108].

4.3.5.2 Inducers

While PXR is the best characterized nuclear receptor for CYP2C9 induction, CAR, GR, estrogen receptor and vitamin D receptor may also be involved. CYP2C9 inducers are generally also CYP3A4 inducers. There is clear evidence that CYP2C9 is inducible *in vivo* [109] which may have toxicological consequences for protoxicants that are metabolically activated by this P450 isoform.

4.3.5.3 Inhibitors

Sulfaphenazole is a specific CYP2C9 inhibitor used routinely for both *in vitro* and *in vivo* studies for the evaluation of the role of this isoform in drug metabolism.

Other inhibitors include fluconazole, voriconazole, and amiodarone. Coadministration of amiodarone and warfarin in patients has led to unwarranted decreases in warfarin clearances [110–112].

4.3.5.4 Individual Variations

CYP2C9 has more than 60 polymorphisms, the most clinically important of which are CYP2C9*2 and CYP2C9*3, both associated with decreased enzymatic activity. Individual variations may also result from concomitant exposure to inducers and inhibitors.

4.3.5.5 Involvement in Drug Toxicity

CYP2C9 has been implicated in the metabolic activation of hepatotoxic drugs including benzbromarone [113], diclofenac [114], tienilic acid [115], and losartan [116]. No evidence was found for CYP2C9 polymorphism to be a determinant of diclofenac-induced hepatotoxicity [117].

4.3.6 CYP2C19

CYP2C19, together with CYP2C8 and CYP2C19, belongs to the CYP2C P450 family. Besides the liver, it is also expressed in extrahepatic tissues such as the human small intestines. It is one of the first P450 isoforms discovered to have large individual differences due to genetic polymorphism, with subpopulations of poor metabolizers, intermediate metabolizers, extensive metabolizers, and ultrarapid metabolizers. Its drug substrates include proton pump inhibitors, antiepileptic agents, antiplatelet drugs and antidepressants.

4.3.6.1 Substrates

CYP2C19 has been estimated to metabolize 6.8% of drugs that are P450 substrates. Clinically defined CYP2C19 substrates include citalopram [118], clopidogrel [119], escitalopram [120], nelfinavir [121], mephenytoin [122], omeprazole [123], and lansoprazole [119].

4.3.6.2 Inducers

In general, CYP2C19 is inducible by ligands of PXR and CAR, but to a lesser extent than CYP3A4 and CYP2C9. Clinically observed inducers include rifampin [124], PB, dicloxacillin [125], efavirenz [126], leading to a lower plasma concentration of substrate drugs such as the proton pump omeprazole.

4.3.6.3 Inhibitors

Clinically significant inhibition of CYP2C19 has been reported for esomeprazole [127], fluconazole [128], and fluoxetine [123]. Ticlodipine and clopidogrel are mechanism-based inhibitors of CYP2C19 [129].

4.3.6.4 Individual Variations

The *CYP2C19*2A* allele causes aberrant mRNA splicing, resulting in no detectable hepatic protein and a poor metabolizer phenotype. The *CYP2C19*17* allele is associated with ultrarapid metabolism of CYP2C19 substrates [130]. CYP2C19 rapid metabolizer genotypes have been reported to be predictors of loss of response to proton pump inhibitors (PPIs), requiring dose escalation [131].

4.3.6.5 Involvement in Drug Toxicity

CYP2C19 is reported to be one of the major P450 isoform responsible for the metabolic activation of hepatotoxicant/immunotoxicant cyclophosphamide [132], and the hepatoxican clopidogrel [133].

4.3.7 CYP2D6

CYP2D6 is expressed mainly in the liver and also in the human small intestines. CYP2D6 mediates hepatic metabolism of approximately 20% of marketed drugs. It is well studied for its genetic polymorphism and is one of the first P450 isoforms with gene sequence correlated with functional activity [134, 135].

4.3.7.1 Substrates

CYP2D6 substrates are mainly psychotropic drugs including the tricyclic antidepressant amitriptyline [136, 137], the antipsychotic aripiprazole [138], the analgesic and antitussive codeine [139], the *in vitro* and *in vivo* probe substrate dextromethorphan [140], the substrate and inhibitor antidepressant fluoxetine [141, 142], the selective serotonin reuptake inhibitor (SSRI) fluvoxamine [143], (*R*)-methadone [144], the SSRI paroxetine [145], and the antipsychotic risperidone [146].

4.3.7.2 Inducers

Nuclear factors found to be responsible for transcriptional regulation of CYP2D6 include hepatocyte nuclear factor (HNF) 4α and CCAAT/enhancer-binding protein (C/EBP) *α* [147]. As of this writing, there are no known CYP2D6 inducers.

4.3.7.3 Inhibitors

Potent, clinically significant CYP2D6 inhibitors including fluoxetine, quinidine, sertraline, and paroxetine [134, 135].

4.3.7.4 Individual Variations

Patients groups have been defined as extensive metabolizers, intermediate metabolizers, poor metabolizers as well as ultra-rapid metabolizers based on genetic polymorphism of CYP2D6.

4.3.7.5 Involvement in Drug Toxicity

CYP2D6 is involved in the formation of reactive metabolites. An interesting observation demethylenation of paroxetine by CYP2D6 to a highly reactive catechol intermediate, leading to mechanism-based inactivation of the isoform [148]. CYP2D6 is one of the isoforms involved in the metabolic activation of benzbromarone [149], ochratoxin A [150], trazodone [151], and the tobacco-specific nitrosamine such as NNK and nicotinamide [152].

4.3.8 CYP2E1

Based on proteomic analysis, CYP2E1 represents approximately 21% of the total P450 protein in the human liver [153]. CYP2E1 is also expressed in the human small intestines [63, 104, 154, 155].

4.3.8.1 Substrates

CYP2E1 has broad substrate specificity toward typically small (molecular weight <100) molecules such as ethanol, including short chain alcohols, ketones, aldehydes, nitrosamines, alkanes, and halogenated alkanes [156]. Drugs known to be metabolized by CYP2E1 include acetaminophen, chlorzoxazone, cisplatin, isoniazid, lidocaine, and prilocaine [157]. Chlorzoxazone and *p*-nitrophenol are routinely employed as probe substrates for CYP2E1 activity.

4.3.8.2 Inducers

Although CYP2E1 is generally believed to be inducible, demonstration of its induction in primary cultured human hepatocytes has been difficult to achieve. Ethanol induction of CYP2E1 is mainly extrapolated from rodent studies both *in vivo* [158, 159]. The herbal supplement St. John's wort [160] has been reported to induce CYP2E1 in cultured human hepatocytes. Protein stabilization rather than enhanced expression has been proposed to be a mechanism of ethanol induction of CYP2E1 protein and activity [161].

4.3.8.3 Inhibitors

Inhibitors of CYP2E1 include disulfiram [162], diethyldithiocarbamate [163], clomethiazole [164], and diallylsulfide [165].

4.3.8.4 Involvement in Drug Toxicity

CYP2E1 is an inducible P450 isoform found be responsible for alcohol metabolism and related toxicity and carcinogenicity [166, 167], and is now known to be involved in the metabolic activation of myriad drugs and xenobiotics including benzbromarone [149], furan [168, 169], and isoniazid [170].

4.3.9 CYP2J2

CYP2J2, although expressed in the liver, is generally considered an extrahepatic P450 isoform. Recently, CYP2J2 activity has been quantified in the human small intestine showing extensive interindividual and minimal inter-regional differences [63, 104, 155]. Other organs expressing CYP2J2 include heart [171], kidney [172, 173], and lung [174]. Cytochrome P450 2J2 has attracted particular attention for its ability to epoxidize arachidonic acid regioselectively to 5,6-, 8,9-, 11,12-, or 14,15-epoxyeicosatrienoic acids (EETs) which has a protective role in drug and disease-induced tissue injuries [175–177].

4.3.9.1 Substrates

Arachidonic acid is the endogenous substrate of CYP2J2. Drug substrate includes, amiodarone, astemizole diclofenac, bufurarol, dasatinib, ebastine, nilotinib, ritonavir, sorafenib, and terfenadine [178, 179].

4.3.9.2 Inhibitors

in vitro inhibitors of CYP2J2 in human liver microsomes include amiodarone danazol [180], terfenadone [180], phytocannabinoids [181], angiotensin II receptor blockers (in ascending order of *in vitro* IC50: telmisartan < irbesartan < olmesartan < losartan < candesartan/valsartan [182]. Rosiglitazone and ritonavir were fund to inhibit CYP2J2 activity in primary cultured human cardiomyocytes [175].

4.3.9.3 Inducers

There is a scarcity of information on CYP2J2 induction. Rosiglitazone was reported to increase CYP2J2 mRNA expression by 1.8-fold in primary cultured human cardiomyocytes [175].

4.3.9.4 Individual Variations

CYP2J2 polymorphism, especially CYP2J2*7 which is associated with decrease in gene expression, has been reported to be associated with ischemic stroke risk of coronary heart disease in African-Americans [183], Taiwanese [184], and Chinese [185, 186] populations. The population with decreased CYP2J2 may be more susceptible to drug toxicity resulting from inflammation and oxidative stress.

4.3.9.5 Involvement in Drug Toxicity

CYP2J2 has an interesting role of being a protective P450 isoform. CYP2J2 synthesis of tissue protective EETs from arachidonic acids has been attributed to the attenuation of inflammation, oxidative stress and the subsequent toxic events including ischemia, alcohol, and drug-induced injuries in kidneys [173, 187, 188], and heart [183, 189–195]. CYP2J2 inhibitor drugs have been reported to enhance drug toxicity via suppression of EET production for drug-induced oxidative stress. Human intestinal CYP2J2 has been reported to play a significant role in astemizole *O*-demethylation [196] and thereby contributes to astemizole drug interactions with its inhibitors.

4.3.10 CYP3A

CYP3A is arguably the most important P450 isoform found responsible for the metabolism of >50% of marketed drugs [197] and is the target for a majority of clinically observed drug–drug interactions. The major adult isoforms are CYP3A4 and CYP3A5, with CYP3A7 as the fetal isoform. It is one of the most abundant P450 isoforms in the human liver and intestine.

4.3.10.1 Substrates

A large variety of marketed drugs of diverse chemical structures have been found to be CYP3A substrates [197–199]. The promiscuity of CYP3A has been attributed to the large substrate pocket which has been further defined to contain two substrate binding sites [200, 201]. Midazolam is the preferred substrate used in clinical studies for the evaluation of *in vivo* CYP3A4 drug–drug interactions [202].

4.3.10.2 Inducers

Rifampin is one of the first clinically significant CYP3A inducer, resulting in loss of efficacy of drugs with narrow therapeutic index drugs including oral anticoagulants, oral contraceptives, glucocorticoids, digitoxin, quinidine, methadone, hypoglycemics, and barbiturates [203]. Other clinically significant CYP3A inducers include PB [204], phenytoin [205], efavirenz [206], carbamazepine and oxcarbazepine [207], rifabutin [208], and troglitazone [209], and the herbal supplement St. John's wort [210].

4.3.10.3 Inhibitors

CYP3A inhibition is a major cause of drug–drug interactions. Inhibitors with clinically significant drug interactions include diltiazem, erythromycin, fluconazole, itraconazole, ketoconazole, and ritonavir, [211, 212].

4.3.10.4 Individual Variations

There is an extremely high (>100-fold) individual variability in human liver [213] and various regions of the human intestine [63, 104]. The large interindividual variations are likely due to environmental factors such as exposure to inducers and inhibitors in the diet, herbal supplements, and drugs. Endogenous factors controlling including CYP3A expression by hormones and inflammatory cytokines also contribute to interindividual variations. Genetic polymorphism of CYP3A4 has not been found to affect its activity. CYP3A5 is polymorphically expressed in adults with readily detectable expression in about 10–20% in Caucasians, 33% in Japanese, and 55% in African-Americans [214], therefore can have a significant impact on drugs specifically metabolized by this CYP3A isoform.

4.3.10.5 Involvement in Drug Toxicity

Numerous hepatotoxic drugs throughout the history of drug development are metabolically activated by CYP3A. These drugs include the tyrosine kinase inhibitors lapatinib [215] sunitinib [216], the selective phosphatidylinositol-3-kinase delta inhibitor idelalisib [217], and a variety of drug molecules associated with severe hepatotoxicity including benzbromarone [218], dauricine [219], 4-ipomeanol [220], nevirapine [221], raloxifene [222], trazodone [151], and troglitazone [223–226]. Torsades de pointes resulting from coadministration of terfenadine with CYP3A inhibitors erythromycin and clarithromycin is pivotal in the development of regulatory requirements for the evaluation of drug–drug interaction potential of drug candidates [227, 228], resulting in the adoption of drug–drug interaction as a routine practice in drug development and the rapid advances of *in vitro* experimental approaches [229–231].

4.4 Non-P450 Drug-Metabolizing Enzymes

4.4.1 Flavin-Containing Monooxygenases (FMOs)

FMOs are universally present in virtually all plant and animal species. FMOs are known to bind NAD(P)H and reduced FAD in the absence of substrate ("cocked and loaded") in preparation for substrate metabolism [6]. The five FMO isoforms (FMO1 to 5) have differential distribution in adult human tissues. The major isoforms in each tissue are FMO3 and 5 in the liver, FMO1 and 2 in the kidney, FMO2 in lung, and FMO5 in the small intestine [232]. FMO activity is the highest of all drug-metabolizing enzymes in the human small intestine [104, 154].

4.4.1.1 Substrates

FMOs catalyze the oxygenation of a wide variety of chemicals with nucleophilic nitrogen, sulfur, and phosphorous heteroatoms. Drugs metabolized by FMO include benzydamine, cimetidine, clozapine, ethylenethiourea, fenthion, imipramine, methimazole, moclobemide, trimethylamine, ranitidine, tamoxifen, tyramine, and xanomeline [233–235]. Benzydamine N-oxidation is generally used as an *in vitro* probe for the quantification of FMO activity [236–238].

FMOs catalyze the oxygenation of a wide variety of chemicals with nucleophilic nitrogen, sulfur, and phosphorous heteroatoms. Drugs metabolized by include benzydamine, cimetidine, clozapine, ethylenethiourea, fenthion, imipramine, methimazole, moclobemide, trimethylamine, ranitidine, tamoxifen, tyramine, and xanomeline [233–235]. Benzydamine N-oxidation is generally used as an *in vitro* probe for the quantification of FMO activity [236-238].

4.4.1.2 Inducers

Although generally believed not to be inducible [233, 234], mRNAs for FMO1, FMO2, and FMO3 have been reported to be highly induced in livers of male mice treated with 2,3,7,8-tetraclorodibenzo-*p*-dioxin [239] and rapamycin [240].

4.4.1.3 Inhibitors

FMO inhibitors include the methimazole (MMI), nicotine and 5,6-dimethylxanthenone-4-acetic acid, with MMI as the most commonly used inhibitor for experimental investigations [235].

4.4.1.4 Individual Variations

FMO expression is affected by polymorphism [241], age [242] as well as various environmental factors including nutrition and disease [243, 244]. FMO activity has been reported to vary from 2- to 20-fold among individuals, depending on the age bracket [242], thereby showing less inter-individual variations than most P450 isoforms.

4.4.1.5 Involvement in Drug Toxicity

Drugs known to be activated by FMO include ketoconazole, with deacetyl ketoconazole as a primary toxic metabolite which is converted to a secondary hydroxylamine metabolite that is proposed to be further oxidized to reactive nitrone and dialdehyde reactive metabolites [245]. Other drugs known to be bioactivated by FMO include ethionamide [246], thiacetazone [247], and albendazole [248].

4.4.2 Monoamine Oxidase (MAO)

Monoamine oxidase (MAO) catalyzes the oxidative deamination of monoamines. Major physiological functions of MAO include catabolism of monamines in foods in the small intestine and the inactivation of neurotransmitters in the brain. MAO-A and MAO-B are the two major forms of MAO that are differentially expressed in human tissues, with both present in the brain and neuronal tissues, and only MAO-B in the liver and small intestine [104, 154]. MAO inhibitor drugs have been developed for the treatment of various psychological disorders, with MAO-A inhibitors for depression and anxiety, and MAO-B inhibitors for Alzheimer's disease and Parkinson's disease.

4.4.2.1 Substrates

MAO substrates are monoamines include adrenaline, benzylamine, dopamine, epinephrine, melatonin, metanephrine, 3-methoxytyramine, noradrenaline, norepinephrine, normetanephrine, phenethylamine, serotonin, tyramine, and tryptamine. Kynuramine is routinely used as a nonselective substrate for experimental evaluation of MAO activity [249].

4.4.2.2 Inducers

There is no information on MAO induction in humans *in vivo* nor in human hepatocytes. Results with nonhepatic systems show that MAO-A and MAO-B expression are differentially upregulated. Cytokines IL-4 and IL-13 upregulate expression of monoamine oxidase A (MAO-A) but not MAO-B in human peripheral monocytes [250]. Phorbol 12-myristate 13-acetate) induces MAO-B but not MAO-A gene expression during the differentiation of Caco-2 cells [251]. Retinoic acid induces MAO B but not MAO-A activity in cultured chick embryonic hepatocytes [252].

4.4.2.3 Inhibitors

FDA-approved MAO inhibitors for the treatment of depression include isocarboxazid, phenelzine, selegiline, and tranylcypromine, and rasagiline. Other inhibitors include procaine [253], clorgyline [254], substituted benzthiazoles [255], 6-hydroxydopamine [256], substituted thiosemicarbazides [257], and endocannabinoids [258]. MAO inhibitory activity of tobacco product has been postulated as one of the determinants of tobacco dependence [259]. Several herbal supplements are reported to contain MAO inhibitory ingredients [260].

4.4.2.4 Individual Variations

Inter-individual variations in MAO-A is a well-established phenomenon. For instance, MAOA levels vary widely between healthy non-smoking men but with no correlation with genetic polymorphism [261]. Two common alleles in the promoter region that are referred to as the “high” and “low” genotypes (reflecting the corresponding behavioral phenotypes) represent the known polymorphism of MAO-A which, however, has no effects on activity [262]. There is evidence suggesting that inter-individual variations may be a result of epigenetic modifications of the genome [263].

4.4.2.5 Involvement in Drug Toxicity

MAO is known to catalyze metabolic activation of several drugs with severe toxicity. A classical example is the metabolism of the opioid impurity 1-methyl-4-phenyl-1,2,3,6-tetrahydropyridine (MPTP) to a potent neurotoxin, 1-methyl-4-phenylpyridinium, via a 1-methyl-4-phenyl-2,3-dihydropyridinium intermediate, resulting in Parkinson's disease [264]. Another example is the bioactivation of the vascular toxin allylamine [265].

4.4.3 Alcohol Dehydrogenase (ADH) and Aldehyde Dehydrogenase (ALDH)

Both alcohol dehydrogenase (ADH) and aldehyde dehydrogenase (ALDH) are the major enzymes for alcohol metabolism, with ADH metabolizing alcohol to aldehyde, and ALDH metabolizing the aldehyde to acetic acid. Both enzymes are present in high quantities in the liver, with ADH also found in the stomach and may be responsible for the bioavailability of alcohol [266]. ADH is considered a toxifying enzyme, catalyzing the formation of the toxic aldehyde metabolites, while ALDH is considered a detoxifying enzyme, metabolizing the aldehyde metabolites to the nontoxic acids.

4.4.3.1 Substrates

Alcohols and aldehydes are the substrates of ADH and ALDH, respectively.

4.4.3.2 Inducers

There is only limited information on the inducibility of ADH and ALDH. Ethanol and methanol are considered inducers of ADH [267, 268]. 20-methylcholanthrene (20-MC), beta-naphthoflavone (beta-NF) and benzo[alpha]pyrene (B[alpha]P) are *in vivo* inducers ALDH in rat liver [269], and arsenic for rat bladder [270].

4.4.3.3 Inhibitors

ADH inhibitors include 2(3)-tert-butyl-4-hydroxyanisole (BHA), pyrazole, chloral hydrate and trichloroethanol, chlorpromazine [271], isoniazid and disulfiram [272]. ALDH inhibitors include ampal, benomyl, citral, chloral, cyanamide, disulfiram, molinate, nitroglycerin, and pargyline [273].

4.4.3.4 Individual Variations

There are seven genes, *ADH1A*, *ADH1B*, *ADH1C*, *ADH4*, *ADH5*, *ADH6*, and *ADH7*, for human ADH. Polymorphism of the allele exists, with the ADH1B*2 allele associated with 8× higher activity and protective against alcoholism [274]. Individuals with similar ADH genotype have been reported to have fourfold differences in ADH activity, suggesting nongenetic factors controlling gene expression [275]. The human ALDH superfamily comprises of 19 known NAD(P+)-dependent enzymes [273].

4.4.3.5 Involvement in Drug Toxicity

ADH metabolism of ethanol to acetaldehyde is a well-established mechanism for the hepatotoxicity of ethanol via the formation of the hepatotoxic acetaldehyde [276, 277]. ADH catalyzed toxic aldehyde formation has been shown to be for the hepatotoxicant allyl alcohol [278], the toxicant and teratogen ethylene glycol monomethyl ether [279], carcinogens and mutagens *N*-nitrosodiethanolamine [280, 281] and methylazoxymethanol (MAM) acetate [282]. Enhancement of toxicity by ADH has been shown for abacavir, where the parent drug is metabolized to the reactive α,β-unsaturated aldehyde intermediate upon ADH-catalyzed oxidation of abacavir [283, 284]. ALDH is generally considered to be protective of drug toxicity via the oxidative detoxification of the toxic aldehyde metabolites [285]. Inhibition of ALDH by carbon tetrachloride is reported to be a reason for its potentiation of alcohol hepatotoxicity [286]. ALDH has been reported to detoxify the toxic metabolites of cyclophosphamide [287, 288].

4.4.4 Aldehyde Oxidase (AOX)

AOX is a non-NADPH dependent, molybdenum cofactor containing cytosolic enzyme present in the liver and extrahepatic tissues including small intestines [104, 154] and skin [289]. It was originally discovered to catalyze the oxidation of aldehyde to carboxylic acid, and now known responsible for the oxidation of nitrogen-containing heterocyclic compounds. The absence of AOX in human liver microsomes accounts for the under-appreciation of its involvement in drug metabolism. Recent discovery on the role of AOX on metabolic clearance has stimulated interest in this drug metabolism enzyme in drug development [290, 291].

4.4.4.1 Substrates

AOX substrates are aldehydes as well as nitrogen-containing heterocyclic compounds such as pyridines, diazines, benzimidazoles, purines [292]. Drugs that are substrates of AOX1 include brimonidine [293], carbazeran [289], *N*-[(2-dimethylamino)ethyl] acridine-4-carboxamide (DACA) [294] methotrexate [295], zaleplon [296], zoniporide [289], ziprasidone [297], and quinoline-containing c-Met kinase inhibitors [298] and metabolites of kinase inhibitors [299]. Carbazeran 4-oxidation is routinely used for the quantification of AOX activity [300].

4.4.4.2 Inducers

As of this writing, nimesulide [301, 302] is the only known inducer of AOX.

4.4.4.3 Inhibitors

AOX inhibitors include hydralazine [296], raloxifene [303], gefitinib and erlotinib [295] thioridazine [304], oxygen and sulfur azaheterocycles [305], and phenothiazine drugs trifluperazine, chlorpromazine, perphenazine, thioridazine and promethazine [306]. Raloxifene, perphenazine, menadione, maprotiline, ketoconazole, erythromycin, and estradiol have been reported to inhibit AOX in cultured human hepatocytes [307].

4.4.4.4 Individual Variations

Individual variations of AOX activity has been reported for cytosols derived from frozen human livers, with >40-fold for N1-methylnicotinamide oxidase activity and <3.6-fold for 6-methylpurine and benzaldehyde oxidase activity in one study [308], and 16.6-fold based on DACA clearance and 2.75-fold with benzaldehyde clearance in another study [309]. One potential artifact for these studies is the relative instability of AOX in frozen and thawed tissues.

4.4.4.5 Involvement in drug toxicity

Nimesulide (NIM) has been found to enhance the liver injury risk of methotrexate via induction of AOX1 activity [301], resulting in elevated plasma levels of the hepatotoxic 7OH-methrotrexate [310]. AOX has been reported to metabolize a tyrosine kinase c-met inhibitor JNJ-38877605 to insoluble metabolites, resulting in renal toxicity in a phase 1 clinical trial [311].

4.4.5 Carboxylesterases (CESs)

Carboxylesterases (CESs) are categorized as phase-I drug-metabolizing enzymes that can hydrolyze a variety of ester-containing drugs and prodrugs. In human, CES1 is predominantly expressed in the liver, and CES2 in the small intestine [104, 154]. CES is known in the activation of ester prodrugs to their active metabolites [312].

4.4.5.1 Substrates

CES substrates include angiotensin-converting enzyme inhibitors (temocapril, cilazapril, quinapril, and imidapril), anti-tumor drugs (CPT-11 and Capecitabin), and narcotics (cocaine, heroin, and meperidine). Other substrates include malathion, clofibrate, diethylsuccinate, diethylphthalate, and *p*-nitrophenylacetate [313], the protein kinase C activators 12-*O*-tetradecanoylphorbol-13-acetate and 1-oleoyl-2-acetyl-rac-glycerol [314]. Imidapril and irinotecan hydrochloride (CPT-11) are used as specific probe substrates for CES1 and CES2, respectively [315].

4.4.5.2 Inducers

CESs are found to be inducible in rodents *in vivo*. Inducers include peroxisome proliferators such as clofibrate, perfluoro-*n*-decanoic acid, perfluoro-*n*-octanoic acid, perfluorooctane sulphonic acid and 1-*H*,1-*H*-pentadecafluoro-*n*-octanol which induces CES activity in rat livers [313, 316], pyridine in rat nasal epithelium [317]. Dexamethasone has been reported to induce CES in both rats *in vivo* [315, 318, 319] and in cultured human hepatocytes [320].

4.4.5.3 Inhibitors

CES inhibitors include bis-(*p*-nitrophenyl) [321], substituted trifluoroketones [313].

4.4.5.4 Individual Variations

CES1 and CES2 are polymorphic enzymes. CES1 A(-816)C polymorphism has been reported to correlate increased response to imidapril [322]. The 830C>G SNP of CES2 which occurs in 10% of the human population is reported to be unlikely to have significant functional consequences on CES2 expression, activity or function [323]. Patients heterozygous or homozygous for the polymorphism CES 2 5′UTR 823 C/G exhibited a significantly greater response rate to capecitabine [324].

4.4.5.5 Involvement in Drug Toxicity

CES mainly serves a detoxifying function, for instance, in the conversion of a naturally occurring T-2 toxin to nontoxic metabolites [325]. Human hepatic carboxylesterase 1 (hCE1) has been found to be involved in the transesterification of cocaine in the presence of ethanol to the toxic metabolite cocaethylene [326].

4.4.6 *N*-Acetyltransferase (NAT)

N-Acetylation, which is catalyzed by the enzymes *N*-acetyltransferase (NAT), is an important biotransformation pathway for the elimination of a wide variety of xenobiotics. NAT catalyzes the transfer of the acetyl group from acetyl-CoA to arylamines, arylhydroxylamines, and arylhydrazines, with NAT1 and NAT2 as the major isoforms. NAT1 is expressed in virtually all tissues, while NAT2 is mainly expressed in the liver and intestine [104, 154].

4.4.6.1 Substrates

Drugs reported to be metabolized by NAT2 include isoniazid, sulfadimidine, hydralazine, dapsone, procaine amide, sulfapyridine, nitrazepam, and some sulfa drugs [327].

4.4.6.2 Inducers

NAT inducers include androgens [328], the histone deacetylase inhibitor trichostatin A [329].

4.4.6.3 Inhibitors

NAT inhibitors include acrolein [330]. Suppression of NAT expression has been reported as a result of inflammation [331] and exposure to in botanical inhibitors such as herbal medicines [332].

4.4.6.4 Individual Variations

Both NAT1 and NAT2 genes are polymorphic. A single amino acid substitution of the wild type (fast acetylator) enzyme leads to the slow acetylator phenotype. NAT polymorphism is known to be associated with individual differences in drug toxicity [333]. Examples include with the higher sensitivity of slow acetylators toward hepatotoxicity of the tuberculosis drug isoniazid [334–337], a higher incidence of esophageal squamous cell carcinoma upon exposure to the carcinogenic arylamines present in tobacco smoke and overcooked red meat [338].

4.4.6.5 Involvement in Drug Toxicity

N-hydroxylation by P450 followed by NAT catalyzed O-acetylation has been found to be critical for the genotoxicity and carcinogenicity of various protoxicants including environmental pollutant 3,6-dinitrobenzo[e]pyrene [339], 2-phenylbenzotriazole dyes [340], and aromatic amines such as 4-aminobiphenyl (ABP) and 2-aminofluorene (AF) [341]. N-acetylation as a detoxifying pathway includes the increased reactive metabolite formation from the hepatotoxic drug flupirtine [342].

4.4.7 Glutathione Transferase (GST)

Glutathione transferases (GSTs) are key enzymes in the detoxification of reactive metabolites via catalyzing their conjugation with cellular reduced glutathione. GSTs belong to a superfamily consisting of three families based on their subcellular distribution: cytosolic GSTs (alpha, zeta, theta, mu, pi, sigma, and omega), mitochondrial GSTs (kappa class GSTs) and microsomal GSTs (membrane-associated proteins) [343]. Most extensively studied are four main families of human cytosolic GST: GSTAlpha(A), GSTMu(M), GSTPi(P), and GSTTheta(T) [344]. In human, membrane bound GSTs represent 45% of the total proteins present in endoplasmic reticulum and the outer mitochondria while the cytosolic GSTs account for 10% of total cellular proteins. The properties of GST have been extensively reviewed [345–347].

4.4.7.1 Substrates

GST catalyzes the conjugation of reduced GSH with highly reactive, electrophilic metabolites including arene oxides, α,β-unsaturated carbonyls, *ortho-* and *para-*quinones, quinone imines. 1-chloro-2,4-dinitrobenzene (CDNB) is often used as a probe substrate for *in vitro* quantification of GST activity [348]. GSTM and GSTP1 catalyze glutathione conjugation upon a large variety of carcinogenic electrophiles [349, 350].

4.4.7.2 Inducers

Planar aromatic compounds (e.g. 3-methylcholanthrene, beta-naphthoflavone, and 2,3,7,8-tetrachlorodibenzo-*p*-dioxin), electrophiles (e.g. *trans*-4-phenyl-3-buten-2-one and dimethyl fumarate) and compounds easily oxidized to electrophiles (e.g. tert-butylhydroquinone) are inducers of GST [351]. PB [352], antioxidants 3,5-di-tert-butyl-4-hydroxytoluene (BHT), BHA, ethoxyquin (EQ) and 5-(2-pyrizinyl)-4-methyl-1,2-dithiol-3-thione (oltipraz) (Kensler, 1985 #30741) [353].

4.4.7.3 Inhibitors

Bilirubin, biliverdin, and chenodeoxycholic acid are endogenous inhibitors of GST, raising the possibility cholestasis may enhance the hepatotoxicity of toxic electrophiles [354].

4.4.7.4 Individual Variations

Genetic polymorphism is a major determinant of individual variations in GST activity. About 50% of the individuals in most Caucasian populations lack GSTM1 activity which has been reported to be associated with increased susceptibility to different types of cancers [355]. A significant correlation between the absence of GSTM1-1 activity and adenocarcinoma of the lung was observed in a study with smokers in the US population [356].

4.4.7.5 Involvement in Drug Toxicity

GST-mediated conjugation is well established to be a major detoxifying mechanism for drugs with toxicity as a function of reactive metabolite formation such as the reactive metabolite NAPQI, the ultimate toxic metabolite for acetaminophen hepatotoxicity [357, 358]. Reduced GSH GSTT1 null patients are found to have a higher risk toward drug induced liver injuries upon exposure to isoniazid [359]. LC/MS–MS quantification of GSH-conjugates is a common approach for the demonstration of reactive metabolite formation [360] and to provide information allowing structural modification to reduce toxic liability of drug candidates [361]. Substrates bioactivated rather than detoxified by GST include several halogenated solvents which upon renal excretion, are metabolized in the kidneys by either cysteine conjugate β-lyase or flavin-containing monooxygenase to produce a reactive intermediate, resulting in renal toxicity [362].

4.4.8 Methyltransferase (MT)

Methyltransferase (MT) catalyzes the transfer of a methyl group from a methyl donor, generally using *S*-adenosyl methionine (SAM) as the methyl donor. MTs are named based on the substrates. For instance, thiopurine methyl transferase (TMPT) catechol *O*-methyltransferase (COMT), histamine *N*-methyltransferase (HNMT).

4.4.8.1 Substrates

N-methyltransferase substrates include primary and secondary amines including amphetamine, tetrahydroisoquinolines imidazole, histamine, thiobendazole nicotinamide, pyridine. *S*-methylation substrates include thiopurines, thiophenols, propylthiouracil, and captopril. *O*-methyltransferase substrates include L-dopa, entacopone, nitecapone, tocapone.

4.4.8.2 Inhibitors

The drugs for the treatment of Parkinson's disease, tolcapone, and entacapone, are potent inhibitors of COMT [363].

4.4.8.3 Individual Variations

Single nucleotide substitution (e.g. PMT*3A/*3A) can result in MT deficiency, resulting in heightened sensitivity towards narrow therapeutic drug that are MT substrates [364–366].

4.4.8.4 Involvement in Drug Toxicity

MT in general serves a detoxifying enzyme. Examples of the involvement of MT in drug toxicity include thiopurines (6-thioguanine, 6-mercaptopurine, and azathioprine) for the treatment of acute lymphoblastic leukemia, autoimmune diseases and inflammatory bowel disease. Patients with deficient TPMT activity due to NUDT15 genetic polymorphism or co-exposure to TPMT inhibitors may develop life-threatening bone marrow toxicity and liver toxicity due to elevated plasma drug levels, thereby requiring TPMY genotyping before the onset of therapy [364–366]. Quinolines have been reported to be bioactivated by MT to estrogenic metabolites [367].

4.4.9 Uridine Glucuronosyltransferase (UGT)

Uridine glucuronosyltransferase (UGT) catalyzes the transfer of glucuronic acid from uridine diphosphate-glucuronic acid (UDPGA) to acceptor molecules containing hydroxyl, phenol, carboxylic acid, thiol, or amine groups.

The glucuronides generated in the liver are excreted into the bile via efflux transporters MRP3 and MRP4 on the basolateral membrane of the hepatocytes. Generally considered a detoxification, UGT metabolism may also be a bioactivation pathways resulting in toxic acyl glucuronides that are covered in detail in a later chapter by Mark Grillo. UGT is highly expressed in hepatic and extrahepatic tissues, especially the small intestine. UGT consists of multiple isoforms. The superfamily UGT is divided into four subfamilies (UGT 1, 2, 3, and 4) based on amino acid sequence homology. The liver expresses UGTs 1A1, 1A3, 1A4, 1A6, 1A9, 2B4, 2B7, 2B10, 2B15, and 2B17 [368].

4.4.9.1 Substrates

Most drugs/metabolites with hydroxyl and carboxylic acid functional groups are subjected to UGT conjugation. Isoform selective UGT substrates include SN38 (UGT1A1), chenodeoxycholic acid (UGT1A3), trifluoperazine (1A4), 4-hydroxyindole (1A6), propofol (1A9), hyodeoxycholic acid (2B4), naloxone (2B7), trifluoperazine (2B10), oxazepam (2B15), and vorinostat (2B17) [369–374]. Bilirubin is an endogenous substrate for UGT.

4.4.9.2 Inducers

UGT gene expression is regulated by nuclear receptors including AhR, PXR, and CAR and thereby can be induced by ligands of these receptors as observed for P450 isoforms [375]. Rifampin induction of UGT resulting in clinically significant drug interactions has been reported for the UGT1A9 substrate mycophenolic acid [376] and the UGT1A1 substrate ezetimibe [377].

4.4.9.3 Inhibitors

UGT inhibitors include the kinase inhibitors dabrafenib, ibrutinib, nintedanib, trametinib, and BIBF 1202 [378]. Nonselective inhibitors, inhibiting both P450 isoforms and UGT include chrysin, mefenamic acid, silibinin, tangeretin, ketoconazole, itraconazole, ritonavir, and verapamil [379], PB [380].

4.4.9.4 Individual Variations

Results of a liver rank study show that individual variation in UGT activity differ for the different isoforms, with UGT 1A1 > 1A6 > 2B15 > 1A4 = 1A9 > 2B7 with coefficient of variation values ranging from 92% to 45% [381]. Various UGT genetic polymorphism results in reduced activities, resulting in interindividual differences in drug clearance by glucuronidation.

4.4.9.5 Involvement in Drug Toxicity

UGT is generally a detoxifying enzyme, with drug toxicity often associated with genetic polymorphism associated with reduced activity. Examples include

UGT1A1*28 [382] and UGT1A1*6 associated adverse drug reactions/toxicity of atazanavir, belinostat, irinotecan pegvisomant, and methimazole [383]. Crigler-Najjar (CN) patients suffer brain damage due to bilirubin toxicity due to UGT 1A1 deficiency which may be overcome with gene therapy [384]. Drugs carboxylic functional groups can be bioactivated by UGT to form reactive/toxic acylgluronides which may result in idiosyncratic drug toxicity. Drugs and xenobiotics bioactivated by UGT include zomepirac, diclofenac, furosemide, ibuprofen, *S*-naproxen, probenecid, and tolmetin [385] and the tobacco carcinogen 2-amino-9H-pyrido[2,3-b]indole [386].

4.4.10 Sulfotransferase (SULT)

Sulfotransferase (SULT) transfers the sulfonate group of 3'-phosphoadenosine 5'-phosphosulfate (PAPS) to a hydroxyl or amino-group. The cytosolic SULTs are responsible for the metabolism of xenobiotics and small endogenous substrates. SULT is generally considered a detoxifying enzyme, playing important roles in the detoxification of drugs and toxic metabolites. SULT conjugation of bile acids protects liver damage avoid pathological conditions, such as cholestasis, liver damage, and colon cancer. SULT consists of multiple isoforms, with the major isoforms responsible for xenobiotic metabolism in the human liver being SULT1A1, SULT1A2, and SULT1A3, and SULT2A1. SULT has been extensively reviewed [387].

4.4.10.1 Substrates

Drugs and their metabolites with hydroxyl and amine functional groups are subjected to SULT metabolism. APAP, a SULT1A1 specific substrate, is extensively sulfonated in humans. Substrates of SULT include phenols (2-naphthol), primary (ethanol) and secondary alcohols (2-butanol), *N*-hydroxy arylamines (2-acetylaminofluorene; 2-AAF), *N*-hydroxy heterocyclic amines (2-amino-1-methyl-6-phenylimidazo[4,5-b]pyridine; PhIP), benzylic alcohols of polycyclic aromatic hydrocarbons (1-hydroxymethylpyrene; 1-HMP), phenolic (17β-estradiol; E2) and alicyclic hydroxysteroids (dehydroepiandrosterone; DHEA), and iodothyronines [387].

4.4.10.2 Inducers

SULT expression is regulated by nuclear receptors which also regulate P450 and UGT expression, including AhR, PXR, liver X receptor (LXR), farnesoid X receptor (FXR), peroxisome proliferator-activated receptors (PPARs), and vitamin D receptor (VDR), the expression of SULT thereby can be modified by ligands of these receptors. Induction of SULT expression has been reported by carbamazepine (SULT1A1) [18] in human *in vivo* by rifampin (SULTA1) [388], ciprofibrate (SULT2A1) [389], and dexamethasone (SULT1A3 expression and dopamine sulfation) [390], the phytoestrogen biochanin A (SULT1A2), in primary cultured human hepatocytes; by pyrene (SULT1A1) [391], octachlorostyrene (SULT1A1), diallyl sulfide (SULT1E1) [392] in mice.

4.4.10.3 Inhibitors

Clomiphene, danazol, meclofenamate, mefenamic acid, NIM, salicylic acid, and salicylic acid have been reported to be competitive inhibitors of SULT1A1 [393–395].

4.4.10.4 Individual Variations

Large interindividual variability of SULT has been reported to be a result of genetic variation and non-genetic, epigenetic, and environmental factors [396].

4.4.10.5 Involvement in Drug Toxicity

While generally considered a detoxifying enzyme, SULT also participates in bioactivation of protoxicants. Drugs bioactivated by SULT, resulting in hepatotoxicity include the non-nucleoside reverse transcriptase inhibitor nevirapine [397], the vasopressin receptor 2 antagonist tolvaptan [398]. SULT bioactivation of aromatic amines, nitroarenes, alkenylbenzenes to mutagenic metabolites has been reported [399].

4.5 Conclusions

The involvement of drug-metabolizing enzymes in drug toxicity is reviewed in this chapter. The most important conclusion is that virtually all drug-metabolizing enzymes reviewed play important roles in drug toxicity, either via metabolic clearance to remove systemic burden and organ exposure, metabolic activation of the less toxic parent drugs to more toxic metabolites, and metabolic detoxification of the toxic parent and/or metabolites. An implication of the significant role of drug metabolism in drug toxicity is that species and individuation variation in drug toxicity may be a result of interspecies and interindividual differences in drug metabolizing enzyme activities, respectively.

The rare incidence of idiosyncratic drug toxicity may be a result of variations in drug metabolizing enzyme activity between individuals, and at various times of drug administration for an individual. Drug-metabolizing enzyme activities for the formation of reactive metabolites, as well as key detoxification pathways including metabolic clearance and transporter-mediated efflux of toxic parent drugs and GSH conjugation of reactive metabolites, can be influenced by environmental (e.g. exposure to inducers and inhibitors of enzyme pathways and GSH depleting agents) and genetic (e.g. polymorphism leading to enhanced or reduced activities) factors, there will be individual differences in drug toxicity occurring among the human population, as well as differences within an individual due to the particular environment at the time of drug administration. Genetic risk factors can be minimized via genotyping, but not environmental risk factors. The Multiple Determinant Hypothesis [1] of idiosyncratic drug toxicity which proposes that toxicity is a result of a confluence of risk factors – environmental and genetic factors leading to the cooccurrence of drug metabolism and transporter activities favoring the manifestation of toxicity, may explain the difficulty in the identification of at-risk patient populations. A thorough understanding of the roles of drug metabolism and transport and the corresponding risk factors in drug toxicity is of utmost importance in the development of safe drugs.

References

1. Li AP. A review of the common properties of drugs with idiosyncratic hepatotoxicity and the "multiple determinant hypothesis" for the manifestation of idiosyncratic drug toxicity. *Chem Biol Interact* 2002;142(1–2):7–23.
2. Sim SC, Kacevska M, Ingelman-Sundberg M. Pharmacogenomics of drug-metabolizing enzymes: a recent update on clinical implications and endogenous effects. *Pharmacogenomics J* 2013;13(1):1–11.
3. Conney AH. Induction of drug-metabolizing enzymes: a path to the discovery of multiple cytochromes P450. *Annu Rev Pharmacol Toxicol* 2003;43:1–30.
4. Zanger UM, Schwab M. Cytochrome P450 enzymes in drug metabolism: regulation of gene expression, enzyme activities, and impact of genetic variation. *Pharmacol Ther* 2013;138(1):103–41.
5. Coon MJ. Cytochrome P450: nature's most versatile biological catalyst. *Annu Rev Pharmacol Toxicol* 2005;45:1–25.
6. Guengerich FP. Mechanisms of Cytochrome P450-Catalyzed Oxidations. *ACS Catal* 2018;8(12):10964–76.
7. Omiecinski CJ, Remmel RP, Hosagrahara VP. Concise review of the cytochrome P450s and their roles in toxicology. *Toxicol Sci* 1999;48(2):151–56.
8. Wrighton SA, Stevens JC. The human hepatic cytochromes P450 involved in drug metabolism. *Crit Rev Toxicol* 1992;22(1):1–21.
9. Shimada T, Yamazaki H, Mimura M, Inui Y, Guengerich FP. Interindividual variations in human liver cytochrome P-450 enzymes involved in the oxidation of drugs, carcinogens and toxic chemicals: studies with liver microsomes of 30 Japanese and 30 Caucasians. *J Pharmacol Exp Ther* 1994;270(1):414–23.
10. Djordjevic N, Milovanovic DD, Radovanovic M, Radosavljevic I, Obradovic S, Jakovljevic M, et al. CYP1A2 genotype affects carbamazepine pharmacokinetics in children with epilepsy. *Eur J Clin Pharmacol* 2016;72(4):439–45.

11 Pearce RE, Lu W, Wang Y, Uetrecht JP, Correia MA, Leeder JS. Pathways of carbamazepine bioactivation in vitro. III. The role of human cytochrome P450 enzymes in the formation of 2,3-dihydroxycarbamazepine. *Drug Metab Dispos* 2008;36(8):1637–49.

12 Young CR, Mazure CM. Fulminant hepatic failure from acetaminophen in an anorexic patient treated with carbamazepine. *J Clin Psychiatry* 1998; 59(11):622.

13 Berkowitz FE, Henderson SL, Fajman N, Schoen B, Naughton M. Acute liver failure caused by isoniazid in a child receiving carbamazepine. *Int J Tuberc Lung Dis* 1998;2(7):603–06.

14 Zhou SF, Yang LP, Zhou ZW, Liu YH, Chan E. Insights into the substrate specificity, inhibitors, regulation, and polymorphisms and the clinical impact of human cytochrome P450 1A2. *AAPS J* 2009;11(3):481–94.

15 Lu J, Shang X, Zhong W, Xu Y, Shi R, Wang X. New insights of CYP1A in endogenous metabolism: a focus on single nucleotide polymorphisms and diseases. *Acta Pharm Sin B* 2020;10(1):91–104.

16 Tukey RH, Hannah RR, Negishi M, Nebert DW, Eisen HJ. The Ah locus: correlation of intranuclear appearance of inducer-receptor complex with induction of cytochrome P1-450 mRNA. *Cell* 1982;31(1):275–84.

17 Nebert DW, Dalton TP, Okey AB, Gonzalez FJ. Role of aryl hydrocarbon receptor-mediated induction of the CYP1 enzymes in environmental toxicity and cancer. *J Biol Chem* 2004;279(23):23847–50.

18 Oscarson M, Zanger UM, Rifki OF, Klein K, Eichelbaum M, Meyer UA. Transcriptional profiling of genes induced in the livers of patients treated with carbamazepine. *Clin Pharmacol Ther* 2006;80(5):440–56.

19 Sarich T, Kalhorn T, Magee S, al-Sayegh F, Adams S, Slattery J, et al. The effect of omeprazole pretreatment on acetaminophen metabolism in rapid and slow metabolizers of S-mephenytoin. *Clin Pharmacol Ther* 1997;62(1):21–28.

20 Rasmussen BB, Jeppesen U, Gaist D, Brosen K. Griseofulvin and fluvoxamine interactions with the metabolism of theophylline. *Ther Drug Monit* 1997;19(1):56–62.

21 Koonrungsesomboon N, Khatsri R, Wongchompoo P, Teekachunhatean S. The impact of genetic polymorphisms on CYP1A2 activity in humans: a systematic review and meta-analysis. *Pharmacogenomics J.* 2018;18(6):760–68.

22 Schweikl H, Taylor JA, Kitareewan S, Linko P, Nagorney D, Goldstein JA. Expression of CYP1A1 and CYP1A2 genes in human liver. *Pharmacogenetics* 1993;3(5):239–49.

23 Xie C, Pogribna M, Word B, Lyn-Cook L, Jr., Lyn-Cook BD, Hammons GJ. in vitro analysis of factors influencing CYP1A2 expression as potential determinants of interindividual variation. *Pharmacol Res Perspect* 2017;5(2):e00299.

24 Reed GA, Peterson KS, Smith HJ, Gray JC, Sullivan DK, Mayo MS, et al. A phase I study of indole-3-carbinol in women: tolerability and effects. *Cancer Epidemiol Biomark Prev* 2005;14(8):1953–60.

25 Parkinson A, Mudra DR, Johnson C, Dwyer A, Carroll KM. The effects of gender, age, ethnicity, and liver cirrhosis on cytochrome P450 enzyme activity in human liver microsomes and inducibility in cultured human hepatocytes. *Toxicol Appl Pharmacol* 2004;199(3):193–209.

26 Carrillo JA, Benitez J. CYP1A2 activity, gender and smoking, as variables influencing the toxicity of caffeine. *Br J Clin Pharmacol* 1996;41(6):605–08.

27 Relling MV, Lin JS, Ayers GD, Evans WE. Racial and gender differences in N-acetyltransferase, xanthine oxidase, and CYP1A2 activities. *Clin Pharmacol Ther* 1992;52(6):643–58.

28 Bohanec Grabar P, Rozman B, Tomsic M, Suput D, Logar D, Dolzan V. Genetic polymorphism of CYP1A2 and the toxicity of leflunomide treatment in rheumatoid arthritis patients. *Eur J Clin Pharmacol* 2008;64(9):871–76.

29 Lewis DF, Lake BG. Molecular modelling of CYP1A subfamily members based on an alignment with CYP102: rationalization of CYP1A substrate specificity in terms of active site amino acid residues. *Xenobiotica* 1996;26(7):723–53.

30 MacLeod MC, Levin W, Conney AH, Lehr RE, Mansfield BK, Jerina DM, et al. Metabolism of benzo(e)pyrene by rat liver microsomal enzymes. *Carcinogenesis* 1980;1(2):165–73.

31 Ma Q, Lu AY. CYP1A induction and human risk assessment: an evolving tale of in vitro and in vivo studies. *Drug Metab Dispos* 2007;35(7):1009–16.

32 Spaldin V, Madden S, Pool WF, Woolf TF, Park BK. The effect of enzyme inhibition on the metabolism and activation of tacrine by human liver microsomes. *Br J Clin Pharmacol* 1994;38(1):15–22.

33 Meng Q, Ru J, Zhang G, Shen C, Schmitmeier S, Bader A. Re-evaluation of tacrine hepatotoxicity using gel entrapped hepatocytes. *Toxicol Lett* 2007;168(2):140–47.

34 Madden S, Spaldin V, Hayes RN, Woolf TF, Pool WF, Park BK. Species variation in the bioactivation of tacrine by hepatic microsomes. *Xenobiotica* 1995;25(1):103–16.

35 Fontana RJ, Turgeon DK, Woolf TF, Knapp MJ, Foster NL, Watkins PB. The caffeine breath test does not identify patients susceptible to tacrine hepatotoxicity. *Hepatology* 1996;23(6):1429–35.

36 Ma LL, Wu ZT, Wang L, Zhang XF, Wang J, Chen C, et al. Inhibition of hepatic cytochrome P450 enzymes and sodium/bile acid cotransporter exacerbates leflunomide-induced hepatotoxicity. *Acta Pharmacol Sin* 2016;37(3):415–24.

37 Shimada T, Yamazaki H, Guengerich FP. Ethnic-related differences in coumarin 7-hydroxylation activities catalyzed by cytochrome P4502A6 in liver microsomes of Japanese and Caucasian populations. *Xenobiotica* 1996;26(4):395–403.

38 Chiang HC, Wang CK, Tsou TC. Differential distribution of CYP2A6 and CYP2A13 in the human respiratory tract. *Respiration* 2012;84(4):319–26.

39 Bhagwat SV, Boyd MR, Ravindranath V. Multiple forms of cytochrome P450 and associated monooxygenase activities in human brain mitochondria. *Biochem Pharmacol* 2000;59(5):573–82.

40 Su T, Bao Z, Zhang QY, Smith TJ, Hong JY, Ding X. Human cytochrome P450 CYP2A13: predominant expression in the respiratory tract and its high efficiency metabolic activation of a tobacco-specific carcinogen, 4-(methylnitrosamino)-1-(3-pyridyl)-1-butanone. *Cancer Res* 2000;60(18):5074–79.

41 Raunio H, Rautio A, Gullsten H, Pelkonen O. Polymorphisms of CYP2A6 and its practical consequences. *Br J Clin Pharmacol* 2001;52(4):357–63.

42 Moore LB, Parks DJ, Jones SA, Bledsoe RK, Consler TG, Stimmel JB, et al. Orphan nuclear receptors constitutive androstane receptor and pregnane X receptor share xenobiotic and steroid ligands. *J Biol Chem* 2000;275(20): 15122–27.

43 Itoh M, Nakajima M, Higashi E, Yoshida R, Nagata K, Yamazoe Y, et al. Induction of human CYP2A6 is mediated by the pregnane X receptor with peroxisome proliferator-activated receptor-gamma coactivator 1alpha. *J Pharmacol Exp Ther* 2006;319(2):693–702.

44 Onica T, Nichols K, Larin M, Ng L, Maslen A, Dvorak Z, et al. Dexamethasone-mediated up-regulation of human CYP2A6 involves the glucocorticoid receptor and increased binding of hepatic nuclear factor 4 alpha to the proximal promoter. *Mol Pharmacol* 2008;73(2):451–60.

45 Koenigs LL, Peter RM, Thompson SJ, Rettie AE, Trager WF. Mechanism-based inactivation of human liver cytochrome P450 2A6 by 8-methoxypsoralen. *Drug Metab Dispos* 1997;25(12):1407–15.

46 Tanner JA, Tyndale RF. Variation in CYP2A6 activity and personalized medicine. *J Pers Med* 2017;7(4).

47 Ueda R, Muranushi Y, Yoshimoto H. Considerations of the early diagnosis of an acoustic neurinoma. *Jibiinkoka* 1968;40(4):275–81.

48 Schoedel KA, Hoffmann EB, Rao Y, Sellers EM, Tyndale RF. Ethnic variation in CYP2A6 and association of genetically slow nicotine metabolism and smoking in adult Caucasians. *Pharmacogenetics* 2004;14(9):615–26.

49 Ariyoshi N, Miyamoto M, Umetsu Y, Kunitoh H, Dosaka-Akita H, Sawamura Y, et al. Genetic polymorphism of CYP2A6 gene and tobacco-induced lung cancer risk in male smokers. *Cancer Epidemiol Biomark Prev* 2002;11(9):890–94.

50 Rietjens IM, Boersma MG, Zaleska M, Punt A. Differences in simulated liver concentrations of toxic coumarin metabolites in rats and different human populations evaluated through physiologically based biokinetic (PBBK) modeling. *Toxicol in vitro* 2008;22(8):1890–901.

51 Chen W, Koenigs LL, Thompson SJ, Peter RM, Rettie AE, Trager WF, et al. Oxidation of acetaminophen to its toxic quinone imine and nontoxic catechol metabolites by baculovirus-expressed and purified human cytochromes P450 2E1 and 2A6. *Chem Res Toxicol* 1998;11(4):295–301.

52 Dinakaran D, Sergi CM. Co-ingestion of aspirin and acetaminophen promoting fulminant liver failure: a critical review of Reye syndrome in the current perspective at the dawn of the 21st century. *Clin Exp Pharmacol Physiol* 2018;45(2):117–21.

53 Ozkaya O, Genc G, Bek K, Sullu Y. A case of acetaminophen (paracetamol) causing renal failure without liver damage in a child and review of literature. *Ren Fail* 2010;32(9):1125–27.

54 Bailey B, Amre DK, Gaudreault P. Fulminant hepatic failure secondary to acetaminophen poisoning: a systematic review and meta-analysis of prognostic criteria determining the need for liver transplantation. *Crit Care Med* 2003;31(1):299–305.

55 Mian P, van den Anker JN, van Calsteren K, Annaert P, Tibboel D, Pfister M, et al. Physiologically based pharmacokinetic modeling to characterize acetaminophen pharmacokinetics and N-acetyl-p-benzoquinone imine (NAPQI) formation in non-pregnant and pregnant women. *Clin Pharmacokinet* 2020;59(1):97–110.

56 Walker V, Mills GA, Anderson ME, Ingle BL, Jackson JM, Moss CL, et al. The acetaminophen metabolite N-acetyl-p-benzoquinone imine (NAPQI) inhibits glutathione synthetase in vitro a clue to the mechanism of 5-oxoprolinuric acidosis? *Xenobiotica* 2017;47(2):164–75.

57 LeBlanc A, Shiao TC, Roy R, Sleno L. Absolute quantitation of NAPQI-modified rat serum albumin by LC-MS/MS: monitoring acetaminophen covalent binding in vivo. *Chem Res Toxicol* 2014;27(9):1632–39.

58 Moyer AM, Fridley BL, Jenkins GD, Batzler AJ, Pelleymounter LL, Kalari KR, et al. Acetaminophen-NAPQI hepatotoxicity: a cell line model system genome-wide association study. *Toxicol Sci* 2011;120(1):33–41.

59 Komatsu T, Yamazaki H, Shimada N, Nakajima M, Yokoi T. Roles of cytochromes P450 1A2, 2A6, and 2C8 in 5-fluorouracil formation from tegafur, an anticancer prodrug, in human liver microsomes. *Drug Metab Dispos* 2000;28(12):1457–63.

60 Krogstad V, Peric A, Robertsen I, Kringen MK, Wegler C, Angeles PC, et al. A comparative analysis of cytochrome P450 activities in paired liver and small intestinal samples from patients with obesity. *Drug Metab Dispos* 2020;48(1):8–17.

61 Clermont V, Grangeon A, Barama A, Turgeon J, Lallier M, Malaise J, et al. Activity and mRNA expression levels of selected cytochromes P450 in various sections of the human small intestine. *Br J Clin Pharmacol* 2019;85(6):1367–77.

62 Drozdzik M, Busch D, Lapczuk J, Muller J, Ostrowski M, Kurzawski M, et al. Protein abundance of clinically relevant drug-metabolizing enzymes in the human liver and intestine: a comparative analysis in paired tissue specimens. *Clin Pharmacol Ther* 2018;104(3):515–24.

63 Ho MD, Ring N, Amaral K, Doshi U, Li AP. Human enterocytes as an in vitro model for the evaluation of intestinal drug metabolism: characterization of drug-metabolizing enzyme activities of cryopreserved human enterocytes from twenty-four donors. *Drug Metab Dispos* 2017;45(6):686–91.

64 Ding X, Kaminsky LS. Human extrahepatic cytochromes P450: function in xenobiotic metabolism and tissue-selective chemical toxicity in the respiratory and gastrointestinal tracts. *Annu Rev Pharmacol Toxicol* 2003;43:149–73.

65 Gervot L, Rochat B, Gautier JC, Bohnenstengel F, Kroemer H, de Berardinis V, et al. Human CYP2B6: expression, inducibility and catalytic activities. *Pharmacogenetics* 1999;9(3):295–306.

66 Janmohamed A, Dolphin CT, Phillips IR, Shephard EA. Quantification and cellular localization of expression in human skin of genes encoding flavin-containing monooxygenases and cytochromes P450. *Biochem Pharmacol* 2001;62(6):777–86.

67 Sapeika N, Kaplan ER. Effect of the antiepileptic drug sodium valproate on induction of hepatic microsomal P450. *Res Commun Chem Pathol Pharmacol* 1975;10(4):767–68.

68 Wright MC, Paine AJ. Induction of the cytochrome P450 3A subfamily in rat liver correlates with the binding of inducers to a microsomal protein. *Biochem Biophys Res Commun* 1994;201(2):973–79.

69 Madra S, Smith AG. Induction of cytochrome P450 activities by polychlorinated biphenyls in isolated mouse hepatocytes. Influence of Ah-phenotype and iron. *Biochem Pharmacol* 1992;44(3):455–64.

70 Hammond AH, Fry JR. The in vivo induction of rat hepatic cytochrome P450-dependent enzyme activities and their maintenance in culture. *Biochem Pharmacol* 1990;40(3):637–42.

71 Rendic S. Summary of information on human CYP enzymes: human P450 metabolism data. *Drug Metab Rev* 2002;34(1–2):383–448.

72 Hesse LM, Venkatakrishnan K, Court MH, von Moltke LL, Duan SX, Shader RI, et al. CYP2B6 mediates the in vitro hydroxylation of bupropion: potential drug interactions with other antidepressants. *Drug Metab Dispos* 2000;28(10):1176–83.

73 Faucette SR, Hawke RL, Lecluyse EL, Shord SS, Yan B, Laethem RM, et al. Validation of bupropion hydroxylation as a selective marker of human cytochrome P450 2B6 catalytic activity. *Drug Metab Dispos* 2000;28(10):1222–30.

74 Hedrich WD, Hassan HE, Wang H. Insights into CYP2B6-mediated drug-drug interactions. *Acta Pharm Sin B* 2016;6(5):413–25.

75 Kawamoto T, Sueyoshi T, Zelko I, Moore R, Washburn K, Negishi M. Phenobarbital-responsive nuclear translocation of the receptor CAR in induction of the CYP2B gene. *Mol Cell Biol* 1999;19(9):6318–22.

76 Li H, Chen T, Cottrell J, Wang H. Nuclear translocation of adenoviral-enhanced yellow fluorescent protein-tagged-human constitutive androstane receptor (hCAR): a novel tool for screening hCAR activators in human primary hepatocytes. *Drug Metab Dispos* 2009;37(5):1098–106.

77 Gerbal-Chaloin S, Pascussi JM, Pichard-Garcia L, Daujat M, Waechter F, Fabre JM, et al. Induction of CYP2C genes in human hepatocytes in primary culture. *Drug Metab Dispos* 2001;29(3):242–51.

78 Turpeinen M, Tolonen A, Uusitalo J, Jalonen J, Pelkonen O, Laine K. Effect of clopidogrel and ticlopidine on cytochrome P450 2B6 activity as measured by bupropion hydroxylation. *Clin Pharmacol Ther* 2005;77(6):553–59.

79 Walsky RL, Obach RS. A comparison of 2-phenyl-2-(1-piperidinyl)propane (ppp), 1,1′,1″-phosphinothioylidynetrisaziridine (thioTEPA), clopidogrel, and ticlopidine as selective inactivators of human cytochrome P450 2B6. *Drug Metab Dispos* 2007;35(11):2053–59.

80 Chun J, Kent UM, Moss RM, Sayre LM, Hollenberg PF. Mechanism-based inactivation of cytochromes P450 2B1 and P450 2B6 by 2-phenyl-2-(1-piperidinyl) propane. *Drug Metab Dispos* 2000;28(8):905–11.

81 Kosaka M, Zhang D, Wong S, Yan Z. NADPH-independent inactivation of CYP2B6 and NADPH-dependent inactivation of CYP3A4/5 by PBD: potential implication for assessing covalent modulators for time-dependent inhibition. *Drug Metab Dispos* 2020;48(8):655–61.

82 Wang PF, Neiner A, Kharasch ED. Efavirenz metabolism: influence of polymorphic CYP2B6 variants and stereochemistry. *Drug Metab Dispos* 2019;47(10):1195–205.

83 El-Serafi I, Afsharian P, Moshfegh A, Hassan M, Terelius Y. Cytochrome P450 oxidoreductase influences CYP2B6 activity in cyclophosphamide bioactivation. *PLoS One* 2015;10(11):e0141979.

84 Sridar C, D'Agostino J, Hollenberg PF. Bioactivation of the cancer chemopreventive agent tamoxifen to quinone methides by cytochrome P4502B6 and identification of the modified residue on the apoprotein. *Drug Metab Dispos* 2012;40(12):2280–88.

85 Huang Z, Roy P, Waxman DJ. Role of human liver microsomal CYP3A4 and CYP2B6 in catalyzing N-dechloroethylation of cyclophosphamide and ifosfamide. *Biochem Pharmacol* 2000;59(8):961–72.

86 Uwimana E, Ruiz P, Li X, Lehmler HJ. Human CYP2A6, CYP2B6, and CYP2E1 atropselectively metabolize polychlorinated biphenyls to hydroxylated metabolites. *Environ Sci Technol* 2019;53(4):2114–23.

87 Feo ML, Gross MS, McGarrigle BP, Eljarrat E, Barcelo D, Aga DS, et al. Biotransformation of BDE-47 to potentially toxic metabolites is predominantly mediated by human CYP2B6. *Environ Health Perspect* 2013;121(4):440–46.

88 Crane AL, Klein K, Olson JR. Bioactivation of chlorpyrifos by CYP2B6 variants. *Xenobiotica* 2012;42(12):1255–62.

89 Haas DW, Ribaudo HJ, Kim RB, Tierney C, Wilkinson GR, Gulick RM, et al. Pharmacogenetics of efavirenz and central nervous system side effects: an Adult AIDS Clinical Trials Group study. *AIDS* 2004;18(18):2391–400.

90 Kharasch ED, Regina KJ, Blood J, Friedel C. Methadone pharmacogenetics: CYP2B6 polymorphisms determine plasma concentrations, clearance, and metabolism. *Anesthesiology* 2015;123(5):1142–53.

91 Griffin JP. The withdrawal of Baycol (cerivastatin). *Adverse Drug React Toxicol Rev* 2001;20(4):177–80.

92 Van Rysselberge M, Puissant F, Barlow P, Lejeune B, Delvigne A, Leroy F. Fertility prognosis in IVF treatment of patients with cancelled cycles. *Hum Reprod* 1989;4(6):663–66.

93 Glaeser H, Drescher S, Eichelbaum M, Fromm MF. Influence of rifampicin on the expression and function of human intestinal cytochrome P450 enzymes. *Br J Clin Pharmacol* 2005;59(2):199–206.

94 Walsky RL, Gaman EA, Obach RS. Examination of 209 drugs for inhibition of cytochrome P450 2C8. *J Clin Pharmacol* 2005;45(1):68–78.

95 Ogilvie BW, Zhang D, Li W, Rodrigues AD, Gipson AE, Holsapple J, et al. Glucuronidation converts gemfibrozil to a potent, metabolism-dependent inhibitor of CYP2C8: implications for drug-drug interactions. *Drug Metab Dispos* 2006;34(1):191–97.

96 Yu L, Shi D, Ma L, Zhou Q, Zeng S. Influence of CYP2C8 polymorphisms on the hydroxylation metabolism of paclitaxel, repaglinide and ibuprofen enantiomers in vitro. *Biopharm Drug Dispos* 2013;34(5):278–87.

97 Dawed AY, Donnelly L, Tavendale R, Carr F, Leese G, Palmer CN, et al. CYP2C8 and SLCO1B1 variants and therapeutic response to thiazolidinediones in patients with Type 2 diabetes. *Diabetes Care* 2016;39(11):1902–08.

98 Daily EB, Aquilante CL. Cytochrome P450 2C8 pharmacogenetics: a review of clinical studies. *Pharmacogenomics* 2009;10(9):1489–510.

99 Rodriguez-Antona C, Niemi M, Backman JT, Kajosaari LI, Neuvonen PJ, Robledo M, et al. Characterization of novel CYP2C8 haplotypes and their contribution to paclitaxel and repaglinide metabolism. *Pharmacogenomics J.* 2008;8(4):268–77.

100 Hertz DL, Roy S, Motsinger-Reif AA, Drobish A, Clark LS, McLeod HL, et al. CYP2C8*3 increases risk of neuropathy in breast cancer patients treated with paclitaxel. *Ann Oncol* 2013;24(6):1472–78.

101 Hertz DL, Motsinger-Reif AA, Drobish A, Winham SJ, McLeod HL, Carey LA, et al. CYP2C8*3 predicts benefit/risk profile in breast cancer patients receiving neoadjuvant paclitaxel. *Breast Cancer Res Treat* 2012;134(1):401–10.

102 Bergmann TK, Brasch-Andersen C, Green H, Mirza M, Pedersen RS, Nielsen F, et al. Impact of CYP2C8*3 on paclitaxel clearance: a population pharmacokinetic and pharmacogenomic study in 93 patients with ovarian cancer. *Pharmacogenomics J.* 2011;11(2):113–20.

103 Zhang HF, Wang HH, Gao N, Wei JY, Tian X, Zhao Y, et al. Physiological content and intrinsic activities of 10 cytochrome P450 isoforms in human normal liver microsomes. *J Pharmacol Exp Ther* 2016;358(1):83–93.

104 Li AP, Alam N, Amaral K, Ho MD, Loretz C, Mitchell W, et al. Cryopreserved human intestinal mucosal epithelium: a novel in vitro experimental system for the evaluation of enteric drug metabolism, cytochrome P450 induction, and enterotoxicity. *Drug Metab Dispos* 2018;46(11):1562–71.

105 Van Booven D, Marsh S, McLeod H, Carrillo MW, Sangkuhl K, Klein TE, et al. Cytochrome P450 2C9-CYP2C9. *Pharmacogenet Genomics* 2010;20(4):277–81.

106 Daly AK, Rettie AE, Fowler DM, Miners JO. Pharmacogenomics of CYP2C9: functional and clinical considerations. *J Pers Med.* 2017;8(1).

107 Williams PA, Cosme J, Ward A, Angove HC, Matak Vinkovic D, Jhoti H. Crystal structure of human cytochrome P450 2C9 with bound warfarin. *Nature* 2003;424(6947):464–68.

108 Wester MR, Yano JK, Schoch GA, Yang C, Griffin KJ, Stout CD, et al. The structure of human cytochrome P450 2C9 complexed with flurbiprofen at 2.0-A resolution. *J Biol Chem* 2004;279(34):35630–37.

109 Shibata S, Takahashi H, Baba A, Takeshita K, Atsuda K, Matsubara H, et al. Delayed de-induction of CYP2C9 compared to CYP3A after discontinuation of rifampicin: report of two cases. *Int J Clin Pharmacol Ther* 2017;55(5):449–52.

110 Almog S, Shafran N, Halkin H, Weiss P, Farfel Z, Martinowitz U, et al. Mechanism of warfarin potentiation by amiodarone: dose--and concentration--dependent inhibition of warfarin elimination. *Eur J Clin Pharmacol* 1985;28(3):257–61.

111 McDonald MG, Au NT, Wittkowsky AK, Rettie AE. Warfarin-amiodarone drug-drug interactions: determination of [I](u)/K(I,u) for amiodarone and its plasma metabolites. *Clin Pharmacol Ther* 2012;91(4):709–17.

112 O'Reilly RA, Trager WF, Rettie AE, Goulart DA. Interaction of amiodarone with racemic warfarin and its separated enantiomorphs in humans. *Clin Pharmacol Ther* 1987;42(3):290–94.

113 McDonald MG, Rettie AE. Sequential metabolism and bioactivation of the hepatotoxin benzbromarone: formation of glutathione adducts from a catechol intermediate. *Chem Res Toxicol* 2007;20(12):1833–42.

114 Yan Z, Li J, Huebert N, Caldwell GW, Du Y, Zhong H. Detection of a novel reactive metabolite of diclofenac: evidence for CYP2C9-mediated bioactivation via arene oxides. *Drug Metab Dispos* 2005;33(6):706–13.

115 Koenigs LL, Peter RM, Hunter AP, Haining RL, Rettie AE, Friedberg T, et al. Electrospray ionization mass spectrometric analysis of intact cytochrome P450: identification of tienilic acid adducts to P450 2C9. *Biochemistry* 1999;38(8):2312–19

116 Iwamura A, Fukami T, Hosomi H, Nakajima M, Yokoi T. CYP2C9-mediated metabolic activation of losartan detected by a highly sensitive cell-based screening assay. *Drug Metab Dispos* 2011;39(5):838–46.

117 Aithal GP, Day CP, Leathart JB, Daly AK. Relationship of polymorphism in CYP2C9 to genetic susceptibility to diclofenac-induced hepatitis. *Pharmacogenetics* 2000;10(6):511–18.

118 Mrazek DA, Biernacka JM, O'Kane DJ, Black JL, Cunningham JM, Drews MS, et al. CYP2C19 variation and citalopram response. *Pharmacogenet Genomics* 2011;21(1):1–9.

119 Kobayashi M, Kajiwara M, Hasegawa S. A randomized study of the safety, tolerability, pharmacodynamics, and pharmacokinetics of clopidogrel in three different CYP2C19 genotype groups of healthy Japanese subjects. *J Atheroscler Thromb* 2015;22(11):1186–96.

120 Ohlsson Rosenborg S, Mwinyi J, Andersson M, Baldwin RM, Pedersen RS, Sim SC, et al. Kinetics of omeprazole and escitalopram in relation to the CYP2C19*17 allele in healthy subjects. *Eur J Clin Pharmacol* 2008;64(12): 1175–79.

121 Kattel K, Evande R, Tan C, Mondal G, Grem JL, Mahato RI. Impact of CYP2C19 polymorphism on the pharmacokinetics of nelfinavir in patients with pancreatic cancer. *Br J Clin Pharmacol* 2015;80(2):267–75.

122 Tamminga WJ, Wemer J, Oosterhuis B, Wieling J, Touw DJ, de Zeeuw RA, et al. Mephenytoin as a probe for CYP2C19 phenotyping:effect of sample storage, intra-individual reproducibility and occurrence of adverse events. *Br J Clin Pharmacol* 2001;51(5):471–74.

123 Sager JE, Lutz JD, Foti RS, Davis C, Kunze KL, Isoherranen N. Fluoxetine- and norfluoxetine-mediated complex drug-drug interactions: in vitro to in vivo correlation of effects on CYP2D6, CYP2C19, and CYP3A4. *Clin Pharmacol Ther* 2014;95(6):653–62.

124 Park GJ, Bae SH, Park WS, Han S, Park MH, Shin SH, et al. Drug-drug interaction of microdose and regular-dose omeprazole with a CYP2C19 inhibitor and inducer. *Drug Des Devel Ther* 2017;11:1043–53.

125 Stage TB, Graff M, Wong S, Rasmussen LL, Nielsen F, Pottegard A, et al. Dicloxacillin induces CYP2C19, CYP2C9 and CYP3A4 in vivo and in vitro. *Br J Clin Pharmacol* 2018;84(3):510–19.

126 Michaud V, Kreutz Y, Skaar T, Ogburn E, Thong N, Flockhart DA, et al. Efavirenz-mediated induction of omeprazole metabolism is CYP2C19 genotype dependent. *Pharmacogenomics J.* 2014;14(2):151–59.

127 Kaartinen TJK, Tornio A, Tapaninen T, Launiainen T, Isoherranen N, Niemi M, et al. Effect of high-dose esomeprazole on CYP1A2, CYP2C19, and CYP3A4 activities in humans: evidence for substantial and long-lasting inhibition of CYP2C19. *Clin Pharmacol Ther* 2020.

128 Dymond AW, So K, Martin P, Huang Y, Severin P, Mathews D, et al. Effects of cytochrome P450 (CYP3A4 and CYP2C19) inhibition and induction on the exposure of selumetinib, a MEK1/2 inhibitor, in healthy subjects: results from two clinical trials. *Eur J Clin Pharmacol* 2017;73(2):175–84.

129 Nishiya Y, Hagihara K, Kurihara A, Okudaira N, Farid NA, Okazaki O, et al. Comparison of mechanism-based inhibition of human cytochrome P450 2C19 by ticlopidine, clopidogrel, and prasugrel. *Xenobiotica* 2009;39(11):836–43.

130 Shirasaka Y, Chaudhry AS, McDonald M, Prasad B, Wong T, Calamia JC, et al. Interindividual variability of CYP2C19-catalyzed drug metabolism due to differences in gene diplotypes and cytochrome P450 oxidoreductase content. *Pharmacogenomics J.* 2016;16(4):375–87.

131 Molina-Infante J, Rodriguez-Sanchez J, Martinek J, van Rhijn BD, Krajciova J, Rivas MD, et al. Long-term loss of response in proton pump inhibitor-responsive esophageal eosinophilia is uncommon and influenced by CYP2C19 genotype and rhinoconjunctivitis. *Am J Gastroenterol* 2015;110(11):1567–75.

132 Helsby NA, Hui CY, Goldthorpe MA, Coller JK, Soh MC, Gow PJ, et al. The combined impact of CYP2C19 and CYP2B6 pharmacogenetics on cyclophosphamide bioactivation. *Br J Clin Pharmacol* 2010;70(6):844–53.

133 Zhai Y, Wang L, Yang F, Feng G, Feng S, Cui T, et al. The mechanism and risk factors of clopidogrel-induced liver injury. *Drug Chem Toxicol* 2016;39(4): 367–74.

134 Bahar MA, Setiawan D, Hak E, Wilffert B. Pharmacogenetics of drug-drug interaction and drug-drug-gene interaction: a systematic review on CYP2C9, CYP2C19 and CYP2D6. *Pharmacogenomics* 2017;18(7):701–39.

135 Ingelman-Sundberg M. Genetic polymorphisms of cytochrome P450 2D6 (CYP2D6): clinical consequences, evolutionary aspects and functional diversity. *Pharmacogenomics J.* 2005;5(1):6–13.

136 Steimer W, Zopf K, von Amelunxen S, Pfeiffer H, Bachofer J, Popp J, et al. Allele-specific change of concentration and functional gene dose for the prediction of steady-state serum concentrations of amitriptyline and nortriptyline in CYP2C19 and CYP2D6 extensive and intermediate metabolizers. *Clin Chem* 2004;50(9):1623–33.

137 Coutts RT, Bach MV, Baker GB. Metabolism of amitriptyline with CYP2D6 expressed in a human cell line. *Xenobiotica* 1997;27(1):33–47.

138 Kubo M, Koue T, Inaba A, Takeda H, Maune H, Fukuda T, et al. Influence of itraconazole co-administration and CYP2D6 genotype on the pharmacokinetics of the new antipsychotic ARIPIPRAZOLE. *Drug Metab Pharmacokinet* 2005;20(1):55–64.

139 Williams IS, Gatchie L, Bharate SB, Chaudhuri B. Biotransformation, using recombinant CYP450-expressing Baker's yeast cells, identifies a novel CYP2D6.10(A122V) variant which is a superior metabolizer of codeine to morphine than the wild-type enzyme. *ACS Omega* 2018;3(8):8903–12.

140 Jacqz-Aigrain E, Funck-Brentano C, Cresteil T. CYP2D6- and CYP3A-dependent metabolism of dextromethorphan in humans. *Pharmacogenetics* 1993;3(4):197–204.

141 Asensi-Bernardi L, Martin-Biosca Y, Escuder-Gilabert L, Sagrado S, Medina-Hernandez MJ. Fast evaluation of enantioselective drug metabolism by electrophoretically mediated microanalysis: application to fluoxetine metabolism by CYP2D6. *Electrophoresis* 2013;34(22–23):3214–20.

142 Amchin J, Ereshefsky L, Zarycranski W, Taylor K, Albano D, Klockowski PM. Effect of venlafaxine versus fluoxetine on metabolism of dextromethorphan, a CYP2D6 probe. *J Clin Pharmacol* 2001;41(4):443–51.

143 Carrillo JA, Dahl ML, Svensson JO, Alm C, Rodriguez I, Bertilsson L. Disposition of fluvoxamine in humans is determined by the polymorphic CYP2D6 and also by the CYP1A2 activity. *Clin Pharmacol Ther* 1996;60(2):183–90.

144 Begre S, von Bardeleben U, Ladewig D, Jaquet-Rochat S, Cosendai-Savary L, Golay KP, et al. Paroxetine increases steady-state concentrations of (R)-methadone in CYP2D6 extensive but not poor metabolizers. *J Clin Psychopharmacol* 2002;22(2):211–15.

145 Ozdemir V, Tyndale RF, Reed K, Herrmann N, Sellers EM, Kalow W, et al. Paroxetine steady-state plasma concentration in relation to CYP2D6 genotype in extensive metabolizers. *J Clin Psychopharmacol* 1999;19(5):472–75.

146 Roh HK, Kim CE, Chung WG, Park CS, Svensson JO, Bertilsson L. Risperidone metabolism in relation to CYP2D6*10 allele in Korean schizophrenic patients. *Eur J Clin Pharmacol* 2001;57(9):671–75.

147 Pan X, Ning M, Jeong H. Transcriptional regulation of CYP2D6 expression. *Drug Metab Dispos* 2017;45(1):42–48.

148 Zhao SX, Dalvie DK, Kelly JM, Soglia JR, Frederick KS, Smith EB, et al. NADPH-dependent covalent binding of [3H]paroxetine to human liver microsomes and S-9 fractions: identification of an electrophilic quinone metabolite of paroxetine. *Chem Res Toxicol* 2007;20(11):1649–57.

149 Kitagawara Y, Ohe T, Tachibana K, Takahashi K, Nakamura S, Mashino T. Novel bioactivation pathway of benzbromarone mediated by cytochrome P450. *Drug Metab Dispos* 2015;43(9):1303–06.

150 Pfohl-Leszkowicz A. [Ochratoxin A, ubiquitous mycotoxin contaminating human food]. *C R Seances Soc Biol Fil* 1994;188(4):335–53.

151 Najibi A, Heidari R, Zarifi J, Jamshidzadeh A, Firoozabadi N, Niknahad H. Evaluating the role of drug metabolism and reactive intermediates in trazodone-induced cytotoxicity toward freshly-isolated rat hepatocytes. *Drug Res (Stuttg)* 2016;66(11):592–96.

152 Yan Z, Wu Y, Wu Y. [Relation between cytochrome P450 2D6 and lung cancer susceptibility caused by smoking]. *Wei Sheng Yan Jiu* 2007;36(1):112–13, 116.

153 Couto N, Al-Majdoub ZM, Achour B, Wright PC, Rostami-Hodjegan A, Barber J. Quantification of proteins involved in drug metabolism and disposition in the human liver using label-free global proteomics. *Mol Pharm* 2019;16(2):632–47.

154 Li AP. in vitro human cell-based experimental models for the evaluation of enteric metabolism and drug interaction potential of drugs and natural products. *Drug Metab Dispos* 2020.

155 Li AP, Amaral K, Ho MD. A novel in vitro experimental system for the evaluation of enteric drug metabolism: cofactor-supplemented permeabilized cryopreserved human enterocytes (MetMax Cryopreserved Human Enterocytes). *Drug Metab Lett* 2018;12(2):132–37.

156 Collom SL, Laddusaw RM, Burch AM, Kuzmic P, Perry MD, Jr., Miller GP. CYP2E1 substrate inhibition. Mechanistic interpretation through an effector site for monocyclic compounds. *J Biol Chem* 2008;283(6):3487–96.

157 Chen J, Jiang S, Wang J, Renukuntla J, Sirimulla S, Chen J. A comprehensive review of cytochrome P450 2E1 for xenobiotic metabolism. *Drug Metab Rev* 2019;51(2):178–95.

158 Ki SH, Choi JH, Kim CW, Kim SG. Combined metadoxine and garlic oil treatment efficaciously abrogates alcoholic steatosis and CYP2E1 induction in rat liver with restoration of AMPK activity. *Chem Biol Interact* 2007;169(2):80–90.

159 Daiker DH, Shipp BK, Schoenfeld HA, Klimpel GR, Witz G, Moslen MT, et al. Effect of CYP2E1 induction by ethanol on the immunotoxicity and genotoxicity of extended low-level benzene exposure. *J Toxicol Environ Health A* 2000;59(3):181–96.

160 Hellum BH, Hu Z, Nilsen OG. Trade herbal products and induction of CYP2C19 and CYP2E1 in cultured human hepatocytes. *Basic Clin Pharmacol Toxicol.* 2009;105(1):58–63.

161 Roberts BJ, Song BJ, Soh Y, Park SS, Shoaf SE. Ethanol induces CYP2E1 by protein stabilization. Role of ubiquitin conjugation in the rapid degradation of CYP2E1. *J Biol Chem* 1995;270(50):29632–35.

162 Emery MG, Jubert C, Thummel KE, Kharasch ED. Duration of cytochrome P-450 2E1 (CYP2E1) inhibition and estimation of functional CYP2E1 enzyme half-life after single-dose disulfiram administration in humans. *J Pharmacol Exp Ther* 1999;291(1):213–19.

163 Tassaneeyakul W, Veronese ME, Birkett DJ, Gonzalez FJ, Miners JO. Validation of 4-nitrophenol as an in vitro substrate probe for human liver CYP2E1 using cDNA expression and microsomal kinetic techniques. *Biochem Pharmacol* 1993;46(11):1975–81.

164 Stresser DM, Perloff ES, Mason AK, Blanchard AP, Dehal SS, Creegan TP, et al. Selective time- and NADPH-dependent inhibition of human CYP2E1 by clomethiazole. *Drug Metab Dispos* 2016;44(8):1424–30.

165 Zhai Q, Lu SR, Lin Y, Yang QL, Yu B. Oxidative stress potentiated by diallylsulfide, a selective CYP2E1 inhibitor, in isoniazid toxic effect on rat primary hepatocytes. *Toxicol Lett* 2008;183(1-3):95–98.

166 Kaphalia L, Calhoun WJ. Alcoholic lung injury: metabolic, biochemical and immunological aspects. *Toxicol Lett* 2013;222(2):171–79.

167 Seitz HK, Wang XD. The role of cytochrome P450 2E1 in ethanol-mediated carcinogenesis. *Subcell Biochem* 2013;67:131–43.

168 Gates LA, Phillips MB, Matter BA, Peterson LA. Comparative metabolism of furan in rodent and human cryopreserved hepatocytes. *Drug Metab Dispos* 2014;42(7):1132–36.

169 Moro S, Chipman JK, Antczak P, Turan N, Dekant W, Falciani F, et al. Identification and pathway mapping of furan target proteins reveal mitochondrial energy production and redox regulation as critical targets of furan toxicity. *Toxicol Sci* 2012;126(2):336–52.

170 Huang YS, Chern HD, Su WJ, Wu JC, Chang SC, Chiang CH, et al. Cytochrome P450 2E1 genotype and the susceptibility to antituberculosis drug-induced hepatitis. *Hepatology* 2003;37(4):924–30.

171 Michaud V, Frappier M, Dumas MC, Turgeon J. Metabolic activity and mRNA levels of human cardiac CYP450s involved in drug metabolism. *PLoS One* 2010;5(12):e15666.

172 Enayetallah AE, French RA, Thibodeau MS, Grant DF. Distribution of soluble epoxide hydrolase and of cytochrome P450 2C8, 2C9, and 2J2 in human tissues. *J Histochem Cytochem* 2004;52(4):447–54.

173 Zhu Y, Ding A, Yang D, Cui T, Yang H, Zhang H, et al. CYP2J2-produced epoxyeicosatrienoic acids attenuate ischemia/reperfusion-induced acute kidney injury by activating the SIRT1-FoxO3a pathway. *Life Sci* 2020;246: 117327.

174 Zeldin DC, Foley J, Ma J, Boyle JE, Pascual JM, Moomaw CR, et al. CYP2J subfamily P450s in the lung: expression, localization, and potential functional significance. *Mol Pharmacol* 1996;50(5):1111–17.

175 Evangelista EA, Kaspera R, Mokadam NA, Jones JP, 3rd, Totah RA. Activity, inhibition, and induction of cytochrome P450 2J2 in adult human primary cardiomyocytes. *Drug Metab Dispos* 2013;41(12):2087–94.

176 Pfister SL, Gauthier KM, Campbell WB. Vascular pharmacology of epoxyeicosatrienoic acids. *Adv Pharmacol* 2010;60:27–59.

177 Lee CA, Neul D, Clouser-Roche A, Dalvie D, Wester MR, Jiang Y, et al. Identification of novel substrates for human cytochrome P450 2J2. *Drug Metab Dispos* 2010;38(2):347–56.

178 Lu J, Qin X, Liu M, Wang X. A note on CYP2J2-mediated terfenadine hydroxylation in human liver microsomes. *Food Chem Toxicol* 2014; 71:284–85.

179 Matsumoto S, Yamazoe Y. Involvement of multiple human cytochromes P450 in the liver microsomal metabolism of astemizole and a comparison with terfenadine. *Br J Clin Pharmacol* 2001;51(2):133–42.

180 Lee E, Kim JH, Shon JC, Wu Z, Kim HJ, Gim M, et al. Terfenadone is a strong inhibitor of CYP2J2 present in the human liver and intestinal microsomes. *Drug Metab Pharmacokinet* 2018;33(3):159–63.

181 Arnold WR, Weigle AT, Das A. Cross-talk of cannabinoid and endocannabinoid metabolism is mediated via human cardiac CYP2J2. *J Inorg Biochem* 2018;184:88–99.

182 Senda A, Mukai Y, Hayakawa T, Kato Y, Eliasson E, Rane A, et al. Angiotensin II receptor blockers inhibit the generation of epoxyeicosatrienoic acid from arachidonic acid in recombinant CYP2C9, CYP2J2 and human liver microsomes. *Basic Clin Pharmacol Toxicol* 2017;121(4):239–45.

183 Lee CR, North KE, Bray MS, Couper DJ, Heiss G, Zeldin DC. CYP2J2 and CYP2C8 polymorphisms and coronary heart disease risk: the Atherosclerosis Risk in Communities (ARIC) study. *Pharmacogenet Genomics* 2007;17(5):349–58.

184 Liu PY, Li YH, Chao TH, Wu HL, Lin LJ, Tsai LM, et al. Synergistic effect of cytochrome P450 epoxygenase CYP2J2*7 polymorphism with smoking on the onset of premature myocardial infarction. *Atherosclerosis* 2007;195(1):199–206.

185 Wang SY, Xing PF, Zhang CY, Deng BQ. Association of CYP2J2 gene polymorphisms with ischemic stroke and stroke subtypes in Chinese population. *Medicine (Baltimore)* 2017;96(10):e6266.

186 Yan H, Kong Y, He B, Huang M, Li J, Zheng J, et al. CYP2J2 rs890293 polymorphism is associated with susceptibility to Alzheimer's disease in the Chinese Han population. *Neurosci Lett* 2015;593:56–60.

187 Genvigir FDV, Nishikawa AM, Felipe CR, Tedesco-Silva H, Jr., Oliveira N, Salazar ABC, et al. Influence of ABCC2, CYP2C8, and CYP2J2 polymorphisms on tacrolimus and mycophenolate sodium-based treatment in Brazilian kidney transplant recipients. *Pharmacotherapy* 2017;37(5):535–45.

188 Zhao G, Tu L, Li X, Yang S, Chen C, Xu X, et al. Delivery of AAV2-CYP2J2 protects remnant kidney in the 5/6-nephrectomized rat via inhibition of apoptosis and fibrosis. *Hum Gene Ther* 2012;23(7):688–99.

189 Zhao G, Wang X, Edwards S, Dai M, Li J, Wu L, et al. NLRX1 knockout aggravates lipopolysaccharide (LPS)-induced heart injury and attenuates the anti-LPS cardioprotective effect of CYP2J2/11,12-EET by enhancing activation of NF-kappaB and NLRP3 inflammasome. *Eur J Pharmacol* 2020;881:173276.

190 Wang X, Ni L, Yang L, Duan Q, Chen C, Edin ML, et al. CYP2J2-derived epoxyeicosatrienoic acids suppress endoplasmic reticulum stress in heart failure. *Mol Pharmacol* 2014;85(1):105–15.

191 Hoffmann MM, Bugert P, Seelhorst U, Wellnitz B, Winkelmann BR, Boehm BO, et al. The -50G>T polymorphism in the promoter of the CYP2J2 gene in coronary heart disease: the Ludwigshafen Risk and Cardiovascular Health study. *Clin Chem* 2007;53(3):539–40.

192 Zhou C, Huang J, Li Q, Zhan C, Xu X, Zhang X, et al. CYP2J2-derived EETs attenuated ethanol-induced myocardial dysfunction through inducing autophagy and reducing apoptosis. *Free Radic Biol Med* 2018;117:168–79.

193 Zhang Y, El-Sikhry H, Chaudhary KR, Batchu SN, Shayeganpour A, Jukar TO, et al. Overexpression of CYP2J2 provides protection against doxorubicin-induced cardiotoxicity. *Am J Physiol Heart Circ Physiol* 2009;297(1):H37–46.

194 Evangelista EA, Aliwarga T, Sotoodehnia N, Jensen PN, McKnight B, Lemaitre RN, et al. CYP2J2 modulates diverse transcriptional programs in adult human cardiomyocytes. *Sci Rep* 2020;10(1):5329.

195 Evangelista EA, Lemaitre RN, Sotoodehnia N, Gharib SA, Totah RA. CYP2J2 expression in adult ventricular myocytes protects against reactive oxygen species toxicity. *Drug Metab Dispos* 2018;46(4):380–86.

196 Matsumoto S, Hirama T, Kim HJ, Nagata K, Yamazoe Y. in vitro inhibition of human small intestinal and liver microsomal astemizole O-demethylation: different contribution of CYP2J2 in the small intestine and liver. *Xenobiotica* 2003;33(6):615–23.

197 Li AP, Kaminski DL, Rasmussen A. Substrates of human hepatic cytochrome P450 3A4. *Toxicology* 1995;104(1–3):1–8.

198 Evans WE, Relling MV. Pharmacogenomics: translating functional genomics into rational therapeutics. *Science* 1999;286(5439):487–91.

199 Rendic S, Di Carlo FJ. Human cytochrome P450 enzymes: a status report summarizing their reactions, substrates, inducers, and inhibitors. *Drug Metab Rev* 1997;29(1–2):413–580.

200 He YA, Roussel F, Halpert JR. Analysis of homotropic and heterotropic cooperativity of diazepam oxidation by CYP3A4 using site-directed mutagenesis and kinetic modeling. *Arch Biochem Biophys* 2003;409(1): 92–101.

201 Ekroos M, Sjogren T. Structural basis for ligand promiscuity in cytochrome P450 3A4. *Proc Natl Acad Sci USA* 2006;103(37):13682–87.

202 Sabarathinam S, Vijayakumar TM. A short exploration of selected sensitive CYP3A4 substrates (Probe Drug). *Drug Metab Lett* 2020.

203 Baciewicz AM, Self TH. Rifampin drug interactions. *Arch Intern Med* 1984;144(8):1667–71.

204 Goldberg MR, Lo MW, Deutsch PJ, Wilson SE, McWilliams EJ, McCrea JB. Phenobarbital minimally alters plasma concentrations of losartan and its active metabolite E-3174. *Clin Pharmacol Ther* 1996;59(3):268–74.

205 Fleishaker JC, Pearson LK, Peters GR. Phenytoin causes a rapid increase in 6 beta-hydroxycortisol urinary excretion in humans--a putative measure of CYP3A induction. *J Pharm Sci* 1995;84(3):292–94.

206 Mouly S, Lown KS, Kornhauser D, Joseph JL, Fiske WD, Benedek IH, et al. Hepatic but not intestinal CYP3A4 displays dose-dependent induction by efavirenz in humans. *Clin Pharmacol Ther* 2002;72(1):1–9.

207 Andreasen AH, Brosen K, Damkier P. A comparative pharmacokinetic study in healthy volunteers of the effect of carbamazepine and oxcarbazepine on cyp3a4. *Epilepsia* 2007;48(3):490–96.

208 Khalilieh SG, Yee KL, Sanchez RI, Liu R, Fan L, Martell M, et al. Multiple doses of Rifabutin reduce exposure of Doravirine in healthy subjects. *J Clin Pharmacol* 2018.

209 Dimaraki EV, Jaffe CA. Troglitazone induces CYP3A4 activity leading to falsely abnormal dexamethasone suppression test. *J Clin Endocrinol Metab* 2003;88(7):3113–16.

210 Wang Z, Gorski JC, Hamman MA, Huang SM, Lesko LJ, Hall SD. The effects of St John's wort (*Hypericum perforatum*) on human cytochrome P450 activity. *Clin Pharmacol Ther* 2001;70(4):317–26.

211 Vereerstraeten P, Thiry P, Kinnaert P, Toussaint C. Influence of erythromycin on cyclosporine pharmacokinetics. *Transplantation* 1987;44(1):155–56.

212 Niwa T, Imagawa Y, Yamazaki H. Drug interactions between nine antifungal agents and drugs metabolized by human cytochromes P450. *Curr Drug Metab* 2014;15(7):651–79.

213 Ohtsuki S, Schaefer O, Kawakami H, Inoue T, Liehner S, Saito A, et al. Simultaneous absolute protein quantification of transporters, cytochromes P450, and UDP-glucuronosyltransferases as a novel approach for the characterization of individual human liver: comparison with mRNA levels and activities. *Drug Metab Dispos* 2012;40(1):83–92.

214 Lamba JK, Lin YS, Schuetz EG, Thummel KE. Genetic contribution to variable human CYP3A-mediated metabolism. *Adv Drug Deliv Rev* 2002;54(10):1271–94.

215 Bissada JE, Truong V, Abouda AA, Wines KJ, Crouch RD, Jackson KD. Interindividual variation in CYP3A activity influences Lapatinib bioactivation. *Drug Metab Dispos* 2019;47(11):1257–69.

216 Amaya GM, Durandis R, Bourgeois DS, Perkins JA, Abouda AA, Wines KJ, et al. Cytochromes P450 1A2 and 3A4 catalyze the metabolic activation of sunitinib. *Chem Res Toxicol* 2018;31(7):570–84.

217 Zhu J, Wang P, Shehu AI, Lu J, Bi H, Ma X. Identification of novel pathways in idelalisib metabolism and bioactivation. *Chem Res Toxicol* 2018;31(7):548–55.

218 Masubuchi Y, Kondo S. Inactivation of CYP3A4 by benzbromarone in human liver microsomes. *Drug Metab Lett* 2016;10(1):16–21.

219 Wang Y, Zhong D, Chen X, Zheng J. Identification of quinone methide metabolites of dauricine in human liver microsomes and in rat bile. *Chem Res Toxicol* 2009;22(5):824–34.

220 Alvarez-Diez TM, Zheng J. Mechanism-based inactivation of cytochrome P450 3A4 by 4-ipomeanol. *Chem Res Toxicol* 2004;17(2):150–57.

221 Dekker SJ, Zhang Y, Vos JC, Vermeulen NP, Commandeur JN. Different reactive metabolites of nevirapine require distinct glutathione S-transferase isoforms for bioinactivation. *Chem Res Toxicol* 2016;29(12):2136–44.

222 Pearson JT, Wahlstrom JL, Dickmann LJ, Kumar S, Halpert JR, Wienkers LC, et al. Differential time-dependent inactivation of P450 3A4 and P450 3A5 by raloxifene: a key role for C239 in quenching reactive intermediates. *Chem Res Toxicol* 2007;20(12):1778–86.

223 Ikeda T. Drug-induced idiosyncratic hepatotoxicity: prevention strategy developed after the troglitazone case. *Drug Metab Pharmacokinet* 2011;26(1): 60–70.

224 Tettey JN, Maggs JL, Rapeport WG, Pirmohamed M, Park BK. Enzyme-induction dependent bioactivation of troglitazone and troglitazone quinone in vivo. *Chem Res Toxicol* 2001;14(8):965–74.

225 Yoshikawa Y, Hosomi H, Fukami T, Nakajima M, Yokoi T. Establishment of knockdown of superoxide dismutase 2 and expression of CYP3A4 cell system to evaluate drug-induced cytotoxicity. *Toxicol in vitro* 2009;23(6):1179–87.

226 Yamamoto Y, Yamazaki H, Ikeda T, Watanabe T, Iwabuchi H, Nakajima M, et al. Formation of a novel quinone epoxide metabolite of troglitazone with cytotoxicity to HepG2 cells. *Drug Metab Dispos* 2002;30(2):155–60.

227 Hakim A, Stahl A. The FDA and terfenadine. *West J Med* 1994;161(6): 619–20.

228 Zhang L, Zhang YD, Zhao P, Huang SM. Predicting drug-drug interactions: an FDA perspective. *AAPS J* 2009;11(2):300–06.

229 Li AP. in vitro human hepatocyte-based experimental systems for the evaluation of human drug metabolism, drug-drug interactions, and drug toxicity in drug development. *Curr Top Med Chem* 2014;14(11):1325–38.

230 Li AP. in vitro evaluation of metabolic drug-drug interactions: a descriptive and critical commentary. *Curr Protoc Toxicol.* 2007 4 (25):1–11.

231 Li AP. Primary hepatocyte cultures as an in vitro experimental model for the evaluation of pharmacokinetic drug-drug interactions. *Adv Pharmacol* 1997;43:103–30.

232 Zhang J, Cashman JR. Quantitative analysis of FMO gene mRNA levels in human tissues. *Drug Metab Dispos* 2006;34(1):19–26.

233 Krueger SK, Williams DE. Mammalian flavin-containing monooxygenases: structure/function, genetic polymorphisms and role in drug metabolism. *Pharmacol Ther* 2005;106(3):357–87.

234 Hines RN, Koukouritaki SB, Poch MT, Stephens MC. Regulatory polymorphisms and their contribution to interindividual differences in the expression of enzymes influencing drug and toxicant disposition. *Drug Metab Rev* 2008;40(2): 263–301.

235 Yeniceli D, Deng X, Adams E, Dogrukol-Ak D, Van Schepdael A. Development of a CD-MEKC method for investigating the metabolism of tamoxifen by flavin-containing monooxygenases and the inhibitory effects of methimazole, nicotine and DMXAA. *Electrophoresis* 2013;34(3):463–70.

236 Stormer E, Roots I, Brockmoller J. Benzydamine N-oxidation as an index reaction reflecting FMO activity in human liver microsomes and impact of FMO3 polymorphisms on enzyme activity. *Br J Clin Pharmacol* 2000;50(6):553–61.

237 Ubeaud G, Schiller CD, Hurbin F, Jaeck D, Coassolo P. Estimation of flavin-containing monooxygenase activity in intact hepatocyte monolayers of rat, hamster, rabbit, dog and human by using N-oxidation of benzydamine. *Eur J Pharm Sci* 1999;8(4):255–60.

238 Kawaji A, Ohara K, Takabatake E. An assay of flavin-containing monooxygenase activity with benzydamine N-oxidation. *Anal Biochem* 1993;214(2):409–12.

239 Celius T, Roblin S, Harper PA, Matthews J, Boutros PC, Pohjanvirta R, et al. Aryl hydrocarbon receptor-dependent induction of flavin-containing monooxygenase mRNAs in mouse liver. *Drug Metab Dispos* 2008;36(12):2499–505.

240 Miller RA, Harrison DE, Astle CM, Fernandez E, Flurkey K, Han M, et al. Rapamycin-mediated lifespan increase in mice is dose and sex dependent and metabolically distinct from dietary restriction. *Aging Cell* 2014;13(3):468–77.

241 Koukouritaki SB, Poch MT, Cabacungan ET, McCarver DG, Hines RN. Discovery of novel flavin-containing monooxygenase 3 (FMO3) single nucleotide polymorphisms and functional analysis of upstream haplotype variants. *Mol Pharmacol* 2005;68(2):383–92.

242 Koukouritaki SB, Simpson P, Yeung CK, Rettie AE, Hines RN. Human hepatic flavin-containing monooxygenases 1 (FMO1) and 3 (FMO3) developmental expression. *Pediatr Res* 2002;51(2):236–43.

243 Motika MS, Zhang J, Cashman JR. Flavin-containing monooxygenase 3 and human disease. *Expert Opin Drug Metab Toxicol* 2007;3(6):831–45.

244 Takamura T, Sakurai M, Ota T, Ando H, Honda M, Kaneko S. Genes for systemic vascular complications are differentially expressed in the livers of type 2 diabetic patients. *Diabetologia* 2004;47(4):638–47.

245 Rodriguez RJ, Buckholz CJ. Hepatotoxicity of ketoconazole in Sprague-Dawley rats: glutathione depletion, flavin-containing monooxygenases-mediated bioactivation and hepatic covalent binding. *Xenobiotica* 2003;33(4):429–41.

246 Vannelli TA, Dykman A, Ortiz de Montellano PR The antituberculosis drug ethionamide is activated by a flavoprotein monooxygenase. *J Biol Chem* 2002;277(15):12824–29.

247 Qian L, Ortiz de Montellano PR. Oxidative activation of thiacetazone by the *Mycobacterium tuberculosis* flavin monooxygenase EtaA and human FMO1 and FMO3. *Chem Res Toxicol* 2006;19(3):443–49.

248 Rawden HC, Kokwaro GO, Ward SA, Edwards G. Relative contribution of cytochromes P-450 and flavin-containing monoxygenases to the metabolism of albendazole by human liver microsomes. *Br J Clin Pharmacol* 2000;49(4): 313–22.

249 Mills J, Kattau R, Slater IH, Fuller RW. N-substituted cyclopropylamines as monoamine oxidase inhibitors. Structure-activity relationships. Dopa potentiation in mice and in vitro inhibition of kynuramine oxidation. *J Med Chem* 1968;11(1):95–97.

250 Chaitidis P, Billett EE, O'Donnell VB, Fajardo AB, Fitzgerald J, Kuban RJ, et al. Th2 response of human peripheral monocytes involves isoform-specific induction of monoamine oxidase-A. *J Immunol* 2004;173(8):4821–27.

251 Shih JC, Chen K. Regulation of MAO-A and MAO-B gene expression. *Curr Med Chem* 2004;11(15):1995–2005.

252 Nicotra A, Falasca L, Senatori O, Conti Devirgiliis L. Monoamine oxidase A and B activities in embryonic chick hepatocytes: differential regulation by retinoic acid. *Cell Biochem Funct* 2002;20(2):87–94.

253 MacFarlane MD. Procaine HCl (Gerovital H3): a weak, reversible, fully competitive inhibitor of monoamine oxidase. *Fed Proc* 1975;34(1):108–10.

254 Youdim MB, Holman B. The nature of inhibition of cat brain mitochondrial monoamine oxidase by clorgyline. *J Neural Transm* 1975;37(1):11–24.

255 Singh SP, Misra RS, Parmar SS, Brumleve SJ. Synthesis of 2-(4-arylthiosemicarbazidocarbonylthio)benzthiazoles and their monoamine oxidase inhibitory and anticonvulsant properties. *J Pharm Sci* 1975;64(7):1245–47.

256 Youdim MB. in vitro inhibition of brain mitochondrial monoamine oxidase by 6-hydroxydopamine. *Br J Pharmacol* 1975;54(3):403–05.

257 Ali B, Kumar R, Parmar SS, Dwivedi C, Harbison RD. Antihemolytic and anticonvulsant activities of 1-(2,4-dichloro/2,4,5-trichlorophenoxyacetyl)-4-alkyl/arylthiosemicarbazides and their inhibition of NAD-dependent oxidations and monoamine oxidase. *J Pharm Sci* 1975;64(8):1329–33.

258 Pandey P, Chaurasiya ND, Tekwani BL, Doerksen RJ. Interactions of endocannabinoid virodhamine and related analogs with human monoamine oxidase-A and -B. *Biochem Pharmacol* 2018;155:82–91.

259 Truman P, Stanfill S, Heydari A, Silver E, Fowles J. Monoamine oxidase inhibitory activity of flavoured e-cigarette liquids. *Neurotoxicology* 2019;75:123–28.

260 Chaurasiya ND, Zhao J, Pandey P, Doerksen RJ, Muhammad I, Tekwani BL. Selective inhibition of human monoamine oxidase B by acacetin 7-methyl ether isolated from *Turnera diffusa* (Damiana). *Molecules.* 2019;24(4).

261 Fowler JS, Alia-Klein N, Kriplani A, Logan J, Williams B, Zhu W, et al. Evidence that brain MAO A activity does not correspond to MAO A genotype in healthy male subjects. *Biol Psychiatry* 2007;62(4):355–58.

262 Pal B, Gibson C, Passmore J, Griffiths ID, Dick WC. A study of headaches and migraine in Sjogren's syndrome and other rheumatic disorders. *Ann Rheum Dis* 1989;48(4):312–16.

263 Shumay E, Logan J, Volkow ND, Fowler JS. Evidence that the methylation state of the monoamine oxidase A (MAOA) gene predicts brain activity of MAO A enzyme in healthy men. *Epigenetics* 2012;7(10):1151–60.

264 Glover V, Gibb C, Sandler M. The role of MAO in MPTP toxicity – a review. *J Neural Transm Suppl* 1986;20:65–76.

265 Ramos K, Grossman SL, Cox LR. Allylamine-induced vascular toxicity in vitro: prevention by semicarbazide-sensitive amine oxidase inhibitors. *Toxicol Appl Pharmacol* 1988;95(1):61–71.

266 Hernandez-Munoz R, Caballeria J, Baraona E, Uppal R, Greenstein R, Lieber CS. Human gastric alcohol dehydrogenase: its inhibition by H2-receptor antagonists, and its effect on the bioavailability of ethanol. *Alcohol Clin Exp Res* 1990;14(6):946–50.

267 Wang CH, Singh SM. Genetic considerations in the effects of ethanol in mice. II. A trans-acting inducibility regulator(s) affecting alcohol dehydrogenase (ADH) activity. *Biochem Genet* 1984;22(7–8):597–609.

268 Chrostek L, Szmitkowski M, Jelski W. Alcohol and aldehyde dehydrogenase activity measured with fluorogenic substrates in the liver of rats poisoned with methanol. *Exp Toxicol Pathol* 2001;53(1):77–80.

269 Ritter E, Eriksson LC. Kinetics of induction of cytosolic benzaldehyde: NADP and propionaldehyde: NAD aldehyde dehydrogenase activities in rat livers from male Wistar rats. *Carcinogenesis* 1991;12(5):751–58.

270 Huang YC, Yu HS, Chai CY. Aldehyde dehydrogenase induction in arsenic-exposed rat bladder epithelium. *Exp Toxicol Pathol* 2016;68(1):61–68.

271 Messiha FS. Cerebral and peripheral neurotoxicity of chlorpromazine and ethanol interaction: implications for alcohol and aldehyde dehydrogenase. *Neurotoxicology* 1991;12(3):559–70.

272 Karamanakos PN, Pappas P, Boumba V, Vougiouklakis T, Marselos M. The alcohol intolerance produced by isoniazid is not due to a disulfiram-like reaction despite aldehyde dehydrogenase inhibition. *Pharmacology* 2016;98(5–6):267–71.

273 Koppaka V, Thompson DC, Chen Y, Ellermann M, Nicolaou KC, Juvonen RO, et al. Aldehyde dehydrogenase inhibitors: a comprehensive review of the pharmacology, mechanism of action, substrate specificity, and clinical application. *Pharmacol Rev* 2012;64(3):520–39.

274 Chen CC, Lu RB, Chen YC, Wang MF, Chang YC, Li TK, et al. Interaction between the functional polymorphisms of the alcohol-metabolism genes in protection against alcoholism. *Am J Hum Genet* 1999;65(3):795–807.

275 O'Connor S, Morzorati S, Christian J, Li TK. Clamping breath alcohol concentration reduces experimental variance: application to the study of acute tolerance to alcohol and alcohol elimination rate. *Alcohol Clin Exp Res* 1998;22(1):202–10.

276 Ranganathan S, Davis DG, Leeper JD, Hood RD. Effects of differential alcohol dehydrogenase activity on the developmental toxicity of ethanol in *Drosophila melanogaster*. *Teratology* 1987;36(3):329–34.

277 Mitchell DY, Petersen DR. Inhibition of rat hepatic mitochondrial aldehyde dehydrogenase-mediated acetaldehyde oxidation by trans-4-hydroxy-2-nonenal. *Hepatology* 1991;13(4):728–34.

278 Rikans LE, Moore DR. Effect of age and sex on allyl alcohol hepatotoxicity in rats: role of liver alcohol and aldehyde dehydrogenase activities. *J Pharmacol Exp Ther* 1987;243(1):20–26.

279 Eisses KT. Teratogenicity and toxicity of ethylene glycol monomethyl ether (2-methoxyethanol) in *Drosophila melanogaster*: involvement of alcohol dehydrogenase activity. *Teratog Carcinog Mutagen* 1989;9(5):315–25.

280 Denkel E, Pool BL, Schlehofer JR, Eisenbrand G. Biological activity of N-nitrosodiethanolamine and of potential metabolites which may arise after activation by alcohol dehydrogenase in Salmonella typhimurium, in mammalian cells, and in vivo. *J Cancer Res Clin Oncol* 1986;111(2):149–53.

281 Henn I, Eisenbrand G, Zankl H. Increased mutagenicity of N-nitrosodiethanolamine in human lymphocyte cultures after activation by alcohol dehydrogenase. *J Cancer Res Clin Oncol* 1989;115(5):445–48.

282 Wattenberg LW, Sparnins VL. Inhibitory effects of butylated hydroxyanisole on methylazoxymethanol acetate-induced neoplasia of the large intestine and on nicotinamide adenine dinucleotide-dependent alcohol dehydrogenase activity in mice. *J Natl Cancer Inst* 1979;63(1):219–22.

283 Walsh JS, Reese MJ, Thurmond LM. The metabolic activation of abacavir by human liver cytosol and expressed human alcohol dehydrogenase isozymes. *Chem Biol Interact* 2002;142(1–2):135–54.

284 Charneira C, Godinho AL, Oliveira MC, Pereira SA, Monteiro EC, Marques MM, et al. Reactive aldehyde metabolites from the anti-HIV drug abacavir: amino acid adducts as possible factors in abacavir toxicity. *Chem Res Toxicol* 2011;24(12):2129–41.

285 Usatenko MS, Petrova MA, Sokolovskaia NE, Anakina RP, Borodkin Iu S. [The harmful action of ethanol on the liver: the role of alcohol dehydrogenase and aldehyde dehydrogenase]. *Tsitologiia* 1989;31(5):604–07.

286 Kenel MF, Kulkarni AP. Ethanol potentiation of carbon tetrachloride hepatotoxicity: possible role for the in vivo inhibition of aldehyde dehydrogenase. *Gen Pharmacol* 1985;16(4):355–60.

287 Misra RR, Lorr NA, Bloom SE. Cyclophosphamide metabolism in the primary immune organs of the chick: assays of drug activation, P450 expression, and aldehyde dehydrogenase. *Arch Toxicol* 1991;65(1):32–38.

288 Magni M, Shammah S, Schiro R, Mellado W, Dalla-Favera R, Gianni AM. Induction of cyclophosphamide-resistance by aldehyde-dehydrogenase gene transfer. *Blood* 1996;87(3):1097–103.

289 Manevski N, Balavenkatraman KK, Bertschi B, Swart P, Walles M, Camenisch G, et al. Aldehyde oxidase activity in fresh human skin. *Drug Metab Dispos* 2014;42(12):2049–57.

290 Hutzler JM, Obach RS, Dalvie D, Zientek MA. Strategies for a comprehensive understanding of metabolism by aldehyde oxidase. *Expert Opin Drug Metab Toxicol* 2013;9(2):153–68.

291 Sanoh S, Tayama Y, Sugihara K, Kitamura S, Ohta S. Significance of aldehyde oxidase during drug development: effects on drug metabolism, pharmacokinetics, toxicity, and efficacy. *Drug Metab Pharmacokinet* 2015;30(1):52–63.

292 Dalvie D, Zientek M. Metabolism of xenobiotics by aldehyde oxidase. *Curr Protoc Toxicol* 2015;63:4 41 1–4 13.

293 Acheampong AA, Chien DS, Lam S, Vekich S, Breau A, Usansky J, et al. Characterization of brimonidine metabolism with rat, rabbit, dog, monkey and human liver fractions and rabbit liver aldehyde oxidase. *Xenobiotica* 1996;26(10):1035–55.

294 Schofield PC, Robertson IG, Paxton JW. Inter-species variation in the metabolism and inhibition of N-[(2'-dimethylamino)ethyl]acridine-4-carboxamide (DACA) by aldehyde oxidase. *Biochem Pharmacol* 2000;59(2):161–65.

295 Tan WK, Tan ARY, Sivanandam P, Goh EJH, Yap ZP, Saburulla NF, et al. in vitro inhibition of human aldehyde oxidase activity by clinically relevant concentrations of gefitinib and erlotinib: comparison with select metabolites, molecular docking analysis, and impact on hepatic metabolism of zaleplon and methotrexate. *J Pharmacol Exp Ther* 2020;374(2):295–307.

296 Strelevitz TJ, Orozco CC, Obach RS. Hydralazine as a selective probe inactivator of aldehyde oxidase in human hepatocytes: estimation of the contribution of aldehyde oxidase to metabolic clearance. *Drug Metab Dispos* 2012;40(7):1441–48.

297 Obach RS, Prakash C, Kamel AM. Reduction and methylation of ziprasidone by glutathione, aldehyde oxidase, and thiol S-methyltransferase in humans: an in vitro study. *Xenobiotica* 2012;42(11):1049–57.

298 Zhang JW, Xiao W, Gao ZT, Yu ZT, Zhang JYJ. Metabolism of c-Met kinase inhibitors containing quinoline by aldehyde oxidase, electron donating, and steric hindrance effect. *Drug Metab Dispos* 2018;46(12):1847–55.

299 Dick RA. Refinement of in vitro methods for identification of aldehyde oxidase substrates reveals metabolites of kinase inhibitors. *Drug Metab Dispos* 2018;46(6):846–59.

300 Chen S, Austin-Muttitt K, Zhang LH, Mullins JGL, Lau AJ. in vitro and in silico analyses of the inhibition of human aldehyde oxidase by bazedoxifene, lasofoxifene, and structural analogues. *J Pharmacol Exp Ther* 2019;371(1):75–86.

301 Zhou L, Pang XY, Hou XY, Liu L, Guo ZT, Chen XY. Nimesulide increases the aldehyde oxidase activity of humans and rats. *Acta Pharmacol Sin* 2020;41(6):843–51

302 Beedham C, Miceli JJ, Obach RS. Ziprasidone metabolism, aldehyde oxidase, and clinical implications. *J Clin Psychopharmacol* 2003;23(3):229–32.

303 Obach RS. Potent inhibition of human liver aldehyde oxidase by raloxifene. *Drug Metab Dispos* 2004;32(1):89–97.

304 Coelho C, Foti A, Hartmann T, Santos-Silva T, Leimkuhler S, Romao MJ. Structural insights into xenobiotic and inhibitor binding to human aldehyde oxidase. *Nat Chem Biol* 2015;11(10):779–83.

305 Gristwood W, Wilson K. Kinetics of some benzothiazoles, benzoxazoles, and quinolines as substrates and inhibitors of rabbit liver aldehyde oxidase. *Xenobiotica* 1988;18(8):949–54.

306 Deris-Abdolahpour F, Abdolalipouran-Sadegh L, Dastmalchi S, Hamzeh-Mivehroud M, Zarei O, Dehgan G, et al. Effects of phenothiazines on aldehyde oxidase activity towards aldehydes and N-heterocycles: an in vitro and in silico study. *Eur J Drug Metab Pharmacokinet* 2019;44(2):275–86.

307 Sahi J, Khan KK, Black CB. Aldehyde oxidase activity and inhibition in hepatocytes and cytosolic fractions from mouse, rat, monkey and human. *Drug Metab Lett* 2008;2(3):176–83.

308 Rodrigues AD. Comparison of levels of aldehyde oxidase with cytochrome P450 activities in human liver in vitro. *Biochem Pharmacol* 1994;48(1):197–200.

309 Al-Salmy HS. Individual variation in hepatic aldehyde oxidase activity. *IUBMB Life* 2001;51(4):249–53.

310 Morgan SL, Baggott JE. The importance of inhibition of a catabolic pathway of methotrexate metabolism in its efficacy for rheumatoid arthritis. *Med Hypotheses* 2019;122:10–15.

311 Lolkema MP, Bohets HH, Arkenau HT, Lampo A, Barale E, de Jonge MJA, et al. The c-Met tyrosine kinase inhibitor JNJ-38877605 causes renal toxicity through species-specific insoluble metabolite formation. *Clin Cancer Res* 2015;21(10): 2297–304.

312 Hosokawa M. Structure and catalytic properties of carboxylesterase isozymes involved in metabolic activation of prodrugs. *Molecules*. 2008;13(2):412–31.

313 Ashour MB, Moody DE, Hammock BD. Apparent induction of microsomal carboxylesterase activities in tissues of clofibrate-fed mice and rats. *Toxicol Appl Pharmacol* 1987;89(3):361–69.

314 Maki T, Hosokawa M, Satoh T, Sato K. Changes in carboxylesterase isoenzymes of rat liver microsomes during hepatocarcinogenesis. *Jpn J Cancer Res* 1991;82(7):800–06.

315 Zhang C, Gao P, Yin W, Xu Y, Xiang D, Liu D. Dexamethasone regulates differential expression of carboxylesterase 1 and carboxylesterase 2 through activation of nuclear receptors. *J Huazhong Univ Sci Technol Med Sci* 2012;32(6):798–805.

316 Hosokawa M, Satoh T. Differences in the induction of carboxylesterase isozymes in rat liver microsomes by perfluorinated fatty acids. *Xenobiotica* 1993;23(10): 1125–33.

317 Nikula KJ, Novak RF, Chang IY, Dahl AR, Kracko DA, Zangar RC, et al. Induction of nasal carboxylesterase in F344 rats following inhalation exposure to pyridine. *Drug Metab Dispos* 1995;23(5):529–35.

318 Furihata T, Hosokawa M, Fujii A, Derbel M, Satoh T, Chiba K. Dexamethasone-induced methylprednisolone hemisuccinate hydrolase: its identification as a member of the rat carboxylesterase 2 family and its unique existence in plasma. *Biochem Pharmacol* 2005;69(8):1287–97.

319 Hori T, Jin L, Fujii A, Furihata T, Nagahara Y, Chiba K, et al. Dexamethasone-mediated transcriptional regulation of rat carboxylesterase 2 gene. *Xenobiotica* 2012;42(7):614–23.

320 Zhu W, Song L, Zhang H, Matoney L, LeCluyse E, Yan B. Dexamethasone differentially regulates expression of carboxylesterase genes in humans and rats. *Drug Metab Dispos* 2000;28(2):186–91.

321 Kuykendall JR, Taylor ML, Bogdanffy MS. Cytotoxicity and DNA-protein crosslink formation in rat nasal tissues exposed to vinyl acetate are carboxylesterase-mediated. *Toxicol Appl Pharmacol* 1993;123(2):283–92.

322 Geshi E, Kimura T, Yoshimura M, Suzuki H, Koba S, Sakai T, et al. A single nucleotide polymorphism in the carboxylesterase gene is associated with the responsiveness to imidapril medication and the promoter activity. *Hypertens Res* 2005;28(9):719–25.

323 Bellott R, Le Morvan V, Charasson V, Laurand A, Colotte M, Zanger UM, et al. Functional study of the 830C>G polymorphism of the human carboxylesterase 2 gene. *Cancer Chemother Pharmacol* 2008;61(3):481–88.

324 Ribelles N, Lopez-Siles J, Sanchez A, Gonzalez E, Sanchez MJ, Carabantes F, et al. A carboxylesterase 2 gene polymorphism as predictor of capecitabine on response and time to progression. *Curr Drug Metab* 2008;9(4):336–43.

325 Lin NN, Chen J, Xu B, Wei X, Guo L, Xie JW. The roles of carboxylesterase and CYP isozymes on the in vitro metabolism of T-2 toxin. *Mil Med Res* 2015;2:13.

326 Brzezinski MR, Spink BJ, Dean RA, Berkman CE, Cashman JR, Bosron WF. Human liver carboxylesterase hCE-1: binding specificity for cocaine, heroin, and their metabolites and analogs. *Drug Metab Dispos* 1997;25(9):1089–96.

327 Harmer D, Evans DA, Eze LC, Jolly M, Whibley EJ. The relationship between the acetylator and the sparteine hydroxylation polymorphisms. *J Med Genet* 1986;23(2):155–56.

328 Butcher NJ, Tetlow NL, Cheung C, Broadhurst GM, Minchin RF. Induction of human arylamine N-acetyltransferase type I by androgens in human prostate cancer cells. *Cancer Res* 2007;67(1):85–92.

329 Paterson S, Sin KL, Tiang JM, Minchin RF, Butcher NJ. Histone deacetylase inhibitors increase human arylamine N-acetyltransferase-1 expression in human tumor cells. *Drug Metab Dispos* 2011;39(1):77–82.

330 Bui LC, Manaa A, Xu X, Duval R, Busi F, Dupret JM, et al. Acrolein, an alpha,beta-unsaturated aldehyde, irreversibly inhibits the acetylation of aromatic amine xenobiotics by human arylamine N-acetyltransferase 1. *Drug Metab Dispos* 2013;41(7):1300–05.

331 Ramirez-Alcantara V, Montrose MH. Acute murine colitis reduces colonic 5-aminosalicylic acid metabolism by regulation of N-acetyltransferase-2. *Am J Physiol Gastrointest Liver Physiol* 2014;306(11):G1002–10.

332 Chen Y, Kang Z, Yan J, Yang GP, Tan ZR, Zhou G, et al. Liu wei di huang wan, a well-known traditional Chinese medicine, induces CYP1A2 while suppressing CYP2A6 and N-acetyltransferase 2 activities in man. *J Ethnopharmacol* 2010;132(1):213–18.

333 Mitchell SC. N-acetyltransferase: the practical consequences of polymorphic activity in man. *Xenobiotica* 2020;50(1):77–91.

334 Khan S, Mandal RK, Elasbali AM, Dar SA, Jawed A, Wahid M, et al. Pharmacogenetic association between NAT2 gene polymorphisms and isoniazid induced hepatotoxicity: trial sequence meta-analysis as evidence. *Biosci Rep* 2019;39(1) 1–15.

335 Ben Fredj N, Gam R, Kerkni E, Chaabane A, Chadly Z, Boughattas N, et al. Risk factors of isoniazid-induced hepatotoxicity in Tunisian tuberculosis patients. *Pharmacogenomics J.* 2017;17(4):372–77.

336 Huang YS, Chern HD, Su WJ, Wu JC, Lai SL, Yang SY, et al. Polymorphism of the N-acetyltransferase 2 gene as a susceptibility risk factor for antituberculosis drug-induced hepatitis. *Hepatology* 2002;35(4):883–89.

337 Chamorro JG, Castagnino JP, Musella RM, Nogueras M, Aranda FM, Frias A, et al. Sex, ethnicity, and slow acetylator profile are the major causes of hepatotoxicity induced by antituberculosis drugs. *J Gastroenterol Hepatol* 2013;28(2):323–28.

338 Matejcic M, Vogelsang M, Wang Y, Iqbal Parker M. NAT1 and NAT2 genetic polymorphisms and environmental exposure as risk factors for oesophageal squamous cell carcinoma: a case-control study. *BMC Cancer* 2015;15:150.

339 Oda Y, Hirayama T, Watanabe T. Genotoxic activation of the environmental pollutant 3,6-dinitrobenzo[e]pyrene in *Salmonella typhimurium* umu strains expressing human cytochrome P450 and N-acetyltransferase. *Toxicol Lett* 2009;188(3):258–62.

340 Oda Y, Watanabe T, Terao Y, Nukaya H, Wakabayashi K. Genotoxic activation of 2-phenylbenzotriazole-type compounds by human cytochrome P4501A1 and N-acetyltransferase expressed in *Salmonella typhimurium* umu strains. *Mutat Res* 2008;654(1):52–57.

341 Baldauf KJ, Salazar-Gonzalez RA, Doll MA, Pierce WM, Jr., States JC, Hein DW. Role of human N-acetyltransferase 2 genetic polymorphism on aromatic amine carcinogen-induced DNA damage and mutagenicity in a Chinese hamster ovary cell mutation assay. *Environ Mol Mutagen* 2020;61(2):235–45.

342 Konishi K, Fukami T, Ogiso T, Nakajima M. in vitro approach to elucidate the relevance of carboxylesterase 2 and N-acetyltransferase 2 to flupirtine-induced liver injury. *Biochem Pharmacol* 2018;155:242–51.

343 Allocati N, Masulli M, Di Ilio C, Federici L. Glutathione transferases: substrates, inihibitors and pro-drugs in cancer and neurodegenerative diseases. *Oncogenesis* 2018;7(1):8.

344 Verlaan M, te Morsche RH, Roelofs HM, Laheij RJ, Jansen JB, Peters WH, et al. Glutathione S-transferase Mu null genotype affords protection against alcohol induced chronic pancreatitis. *Am J Med Genet A* 2003;120A(1):34–39.

345 Josephy PD. Genetic variations in human glutathione transferase enzymes: significance for pharmacology and toxicology. *Hum Genomics Proteomics* 2010;2010:876940.

346 Vaish S, Gupta D, Mehrotra R, Mehrotra S, Basantani MK. Glutathione S-transferase: a versatile protein family. *3 Biotech*. 2020;10(7):321.

347 Strange RC, Jones PW, Fryer AA. Glutathione S-transferase: genetics and role in toxicology. *Toxicol Lett*. 2000;112–113:357–63.

348 Lei SB, Peng RX. [Subcellular distribution of glutathione S-transferase in Chinese fetal liver]. *Zhongguo Yao Li Xue Bao* 1990;11(5):389–91.

349 Sophonnithiprasert T, Saelee P, Pongtheerat T. Glutathione S-transferase P1 polymorphism on exon 6 and risk of hepatocellular carcinoma in thai male patients. *Oncology* 2020;98(4):243–47.

350 Fontana X, Peyrotte I, Valente E, Rossi C, Ettore F, Namer M, et al. [Glutathione S-transferase mu 1 (GSTM1): susceptibility gene of breast cancer]. *Bull Cancer* 1997;84(1):35–40.

351 Friling RS, Bensimon A, Tichauer Y, Daniel V. Xenobiotic-inducible expression of murine glutathione S-transferase Ya subunit gene is controlled by an electrophile-responsive element. *Proc Natl Acad Sci USA* 1990;87(16):6258–62.

352 Crawford MJ, Hutson DH, King PA. Metabolic demethylation of the insecticide dimethylvinphos in rats, in dogs, and in vitro. *Xenobiotica* 1976;6(12):745–62.

353 Reddy BS, Maeura Y. Dose-response studies of the effect of dietary butylated hydroxyanisole on colon carcinogenesis induced by methylazoxymethanol acetate in female CF1 mice. *J Natl Cancer Inst* 1984;72(5):1181–87.

354 Boyer TD, Vessey DA. Inhibition of human cationic glutathione S-transferase by nonsubstrate ligands. *Hepatology* 1987;7(5):843–48.

355 Lanciotti M, Coco S, Michele PD, Haupt R, Boni L, Pigullo S, et al. Glutathione S-transferase polymorphisms and susceptibility to neuroblastoma. *Pharmacogenet Genomics* 2005;15(6):423–26.

356 Ketterer B, Harris JM, Talaska G, Meyer DJ, Pemble SE, Taylor JB, et al. The human glutathione S-transferase supergene family, its polymorphism, and its effects on susceptibility to lung cancer. *Environ Health Perspect* 1992;98:87–94.

357 Moles A, Torres S, Baulies A, Garcia-Ruiz C, Fernandez-Checa JC. Mitochondrial-lysosomal axis in acetaminophen hepatotoxicity. *Front Pharmacol* 2018;9:453.

358 Coen M. Metabolic phenotyping applied to pre-clinical and clinical studies of acetaminophen metabolism and hepatotoxicity. *Drug Metab Rev* 2015;47(1):29–44.

359 Santos EA, Goncalves JCS, Fleury MK, Kritski AL, Oliveira MM, Velasque LS, et al. Relationship of anti-tuberculosis drug-induced liver injury and genetic polymorphisms in CYP2E1 and GST. *Braz J Infect Dis* 2019;23(6):381–87.

360 Gan J, Harper TW, Hsueh MM, Qu Q, Humphreys WG. Dansyl glutathione as a trapping agent for the quantitative estimation and identification of reactive metabolites. *Chem Res Toxicol* 2005;18(5):896–903.

361 Humphreys WG. Overview of strategies for addressing BRIs in drug discovery: impact on optimization and design. *Chem Biol Interact* 2011;192(1–2):56–59.

362 Lash LH. Glutathione-dependent bioactivation. *Curr Protoc Toxicol* 2007;Chapter 6:Unit6 12.

363 von Kleist L, Michaelis S, Bartho K, Graebner O, Schlief M, Dreger M, et al. Identification of potential off-target toxicity liabilities of catechol-O-methyltransferase inhibitors by differential competition capture compound mass spectrometry. *J Med Chem* 2016;59(10):4664–75.

364 Marinaki AM, Arenas-Hernandez M. Reducing risk in thiopurine therapy. *Xenobiotica* 2020;50(1):101–09.

365 Chang JY, Cheon JH. Thiopurine therapy in patients with inflammatory bowel disease: a focus on metabolism and pharmacogenetics. *Dig Dis Sci* 2019;64(9): 2395–403.

366 Dean L. Azathioprine therapy and TPMT genotype. In: Pratt VM, McLeod HL, Rubinstein WS, Scott SA, Dean LC, Kattman BL, et al. *Medical Genetics Summaries*. Bethesda (MD): National Center for Biotechnology Information (US); 2012 1–19.

367 Brinkmann M, Barz B, Carriere D, Velki M, Smith K, Meyer-Alert H, et al. Bioactivation of quinolines in a recombinant estrogen receptor transactivation assay is catalyzed by N-methyltransferases. *Chem Res Toxicol* 2019;32(4):698–707.

368 Court MH. Interindividual variability in hepatic drug glucuronidation: studies into the role of age, sex, enzyme inducers, and genetic polymorphism using the human liver bank as a model system. *Drug Metab Rev* 2010;42(1): 209–24.

369 Barre L, Fournel-Gigleux S, Finel M, Netter P, Magdalou J, Ouzzine M. Substrate specificity of the human UDP-glucuronosyltransferase UGT2B4 and UGT2B7. Identification of a critical aromatic amino acid residue at position 33. *FEBS J* 2007;274(5):1256–64.

370 Kato Y, Izukawa T, Oda S, Fukami T, Finel M, Yokoi T, et al. Human UDP-glucuronosyltransferase (UGT) 2B10 in drug N-glucuronidation: substrate screening and comparison with UGT1A3 and UGT1A4. *Drug Metab Dispos* 2013;41(7):1389–97.

371 Court MH, Hao Q, Krishnaswamy S, Bekaii-Saab T, Al-Rohaimi A, von Moltke LL, et al. UDP-glucuronosyltransferase (UGT) 2B15 pharmacogenetics: UGT2B15 D85Y genotype and gender are major determinants of oxazepam glucuronidation by human liver. *J Pharmacol Exp Ther* 2004;310(2):656–65.

372 Ando Y, Saka H, Asai G, Sugiura S, Shimokata K, Kamataki T. UGT1A1 genotypes and glucuronidation of SN-38, the active metabolite of irinotecan. *Ann Oncol* 1998;9(8):845–47.

373 Seo KA, Kim HJ, Jeong ES, Abdalla N, Choi CS, Kim DH, et al. in vitro assay of six UDP-glucuronosyltransferase isoforms in human liver microsomes, using cocktails of probe substrates and liquid chromatography-tandem mass spectrometry. *Drug Metab Dispos* 2014;42(11):1803–10.

374 Kang SP, Ramirez J, House L, Zhang W, Mirkov S, Liu W, et al. A pharmacogenetic study of vorinostat glucuronidation. *Pharmacogenet Genomics* 2010;20(10): 638–41.

375 Soars MG, Petullo DM, Eckstein JA, Kasper SC, Wrighton SA. An assessment of udp-glucuronosyltransferase induction using primary human hepatocytes. *Drug Metab Dispos* 2004;32(1):140–48.

376 Kuypers DR, Verleden G, Naesens M, Vanrenterghem Y. Drug interaction between mycophenolate mofetil and rifampin: possible induction of uridine diphosphate-glucuronosyltransferase. *Clin Pharmacol Ther* 2005;78(1):81–88.

377 Oswald S, Haenisch S, Fricke C, Sudhop T, Remmler C, Giessmann T, et al. Intestinal expression of P-glycoprotein (ABCB1), multidrug resistance associated protein 2 (ABCC2), and uridine diphosphate-glucuronosyltransferase 1A1 predicts the disposition and modulates the effects of the cholesterol absorption inhibitor ezetimibe in humans. *Clin Pharmacol Ther* 2006;79(3): 206–17.

378 Korprasertthaworn P, Chau N, Nair PC, Rowland A, Miners JO. Inhibition of human UDP-glucuronosyltransferase (UGT) enzymes by kinase inhibitors: effects of dabrafenib, ibrutinib, nintedanib, trametinib and BIBF 1202. *Biochem Pharmacol* 2019;169:113616.

379 Walsky RL, Bauman JN, Bourcier K, Giddens G, Lapham K, Negahban A, et al. Optimized assays for human UDP-glucuronosyltransferase (UGT) activities: altered alamethicin concentration and utility to screen for UGT inhibitors. *Drug Metab Dispos* 2012;40(5):1051–65.

380 Mutlib AE, Goosen TC, Bauman JN, Williams JA, Kulkarni S, Kostrubsky S. Kinetics of acetaminophen glucuronidation by UDP-glucuronosyltransferases 1A1, 1A6, 1A9 and 2B15. Potential implications in acetaminophen-induced hepatotoxicity. *Chem Res Toxicol* 2006;19(5):701–09.

381 Stopinski M. Studies on psychrophilic bacteria in two lakes of different trophy. *Acta Microbiol Pol* 1981;30(3):283–94.

382 Steventon G. Uridine diphosphate glucuronosyltransferase 1A1. *Xenobiotica* 2020;50(1):64–76.

383 Li X, Yang J, Jin S, Dai Y, Fan Y, Fan X, et al. Mechanistic examination of methimazole-induced hepatotoxicity in patients with Grave's disease: a metabolomic approach. *Arch Toxicol* 2020;94(1):231–44.

384 Seppen J, Bakker C, de Jong B, Kunne C, van den Oever K, Vandenberghe K, et al. Adeno-associated virus vector serotypes mediate sustained correction of bilirubin UDP glucuronosyltransferase deficiency in rats. *Mol Ther* 2006;13(6): 1085–92.

385 Jinno N, Ohashi S, Tagashira M, Kohira T, Yamada S. A simple method to evaluate reactivity of acylglucuronides optimized for early stage drug discovery. *Biol Pharm Bull* 2013;36(9):1509–13.

386 Tang Y, LeMaster DM, Nauwelaers G, Gu D, Langouet S, Turesky RJ. UDP-glucuronosyltransferase-mediated metabolic activation of the tobacco carcinogen 2-amino-9H-pyrido[2,3-b]indole. *J Biol Chem* 2012;287(18): 14960–72.

387 Gamage N, Barnett A, Hempel N, Duggleby RG, Windmill KF, Martin JL, et al. Human sulfotransferases and their role in chemical metabolism. *Toxicol Sci* 2006;90(1):5–22.

388 Fang HL, Strom SC, Ellis E, Duanmu Z, Fu J, Duniec-Dmuchowski Z, et al. Positive and negative regulation of human hepatic hydroxysteroid sulfotransferase (SULT2A1) gene transcription by rifampicin: roles of hepatocyte nuclear factor 4alpha and pregnane X receptor. *J Pharmacol Exp Ther* 2007;323(2):586–98.

389 Fang HL, Strom SC, Cai H, Falany CN, Kocarek TA, Runge-Morris M. Regulation of human hepatic hydroxysteroid sulfotransferase gene expression by the peroxisome proliferator-activated receptor alpha transcription factor. *Mol Pharmacol* 2005;67(4):1257–67.

390 Bian HS, Ngo SY, Tan W, Wong CH, Boelsterli UA, Tan TM. Induction of human sulfotransferase 1A3 (SULT1A3) by glucocorticoids. *Life Sci* 2007;81(25–26):1659–67.

391 Lee CH, Ito Y, Yanagiba Y, Yamanoshita O, Kim H, Zhang SY, et al. Pyrene-induced CYP1A2 and SULT1A1 may be regulated by CAR and not by AhR. *Toxicology* 2007;238(2–3):147–56.

392 Sueyoshi T, Green WD, Vinal K, Woodrum TS, Moore R, Negishi M. Garlic extract diallyl sulfide (DAS) activates nuclear receptor CAR to induce the Sult1e1 gene in mouse liver. *PLoS One* 2011;6(6):e21229.

393 King RS, Ghosh AA, Wu J. Inhibition of human phenol and estrogen sulfotransferase by certain non-steroidal anti-inflammatory agents. *Curr Drug Metab* 2006;7(7):745–53.

394 Cook I, Wang T, Leyh TS. Isoform-specific therapeutic control of sulfonation in humans. *Biochem Pharmacol* 2019;159:25–31.

395 Wang LQ, James MO. Inhibition of sulfotransferases by xenobiotics. *Curr Drug Metab* 2006;7(1):83–104.

396 Marto N, Morello J, Monteiro EC, Pereira SA. Implications of sulfotransferase activity in interindividual variability in drug response: clinical perspective on current knowledge. *Drug Metab Rev* 2017;49(3):357–71.

397 Fang JL, Loukotkova L, Chitranshi P, Gamboa da Costa G, Beland FA. Effects of human sulfotransferases on the cytotoxicity of 12-hydroxynevirapine. *Biochem Pharmacol* 2018;155:455–67.

398 Fang JL, Wu Y, Gamboa da Costa G, Chen S, Chitranshi P, Beland FA Human sulfotransferases enhance the cytotoxicity of tolvaptan. *Toxicol Sci* 2016;150(1):27–39.

399 Oda Y, Zhang Y, Buchinger S, Reifferscheid G, Yang M. Roles of human sulfotransferases in genotoxicity of carcinogens using genetically engineered umu test strains. *Environ Mol Mutagen* 2012;53(2):152–64.

5

Genetic Polymorphism of Drug-Metabolizing Enzymes and Drug Transporters in Drug Toxicity

Ann K. Daly

Translational and Clinical Research Institute, Newcastle University, Newcastle upon Tyne, UK

5.1 Introduction

Drug toxicity remains an important problem worldwide for those developing drugs, for prescribers, and for patients. Toxicity is a potential problem not only at all phases of drug development, where it may lead to discontinuation of the development project or drug withdrawal if the drug has been already licensed, but is also an issue even for well established, effective, and commonly prescribed drugs. There is increasing evidence that genetic risk factors contribute to susceptibility to adverse drug reactions involving toxicity. As discussed elsewhere (Chapter 1 and [1]), such reactions are often divided into concentration-dependent reactions (type A) and rarer idiosyncratic reactions (type B). Idiosyncratic reactions are generally regarded as not dependent on drug concentration, though it is increasingly clear that drug concentration may also sometimes contribute to type B in addition to type A even if it is not the main underlying factor.

The emphasis in this chapter is on genetic factors affecting idiosyncratic adverse drug reactions. As mentioned above, idiosyncratic reactions are rare and, while traditionally considered not to be concentration-dependent, genetic polymorphisms which affect genes concerned with drug metabolism and transport will often result in effects on drug concentration which are likely to contribute at least in part to these reactions. There are a wide range of different types of idiosyncratic reactions which can affect many tissues and organs in humans [2]. This chapter will emphasize those affecting liver, skin, and muscle since there are good examples of genes relevant to drug metabolism and transport which affect risk of developing such reactions. Additional gene classes, particularly human leukocyte antigen (HLA) and other immune genes, are also important in risk of idiosyncratic reactions, with these associations described in detail elsewhere [3]; they will not be considered further here.

Transporters and Drug-Metabolizing Enzymes in Drug Toxicity, First Edition. Edited by Albert P. Li.

5.2 Drug-Induced Liver Injury

5.2.1 Background

Drug-induced liver injury (DILI) is one of the most common reasons for drug withdrawal [4]. DILI is considered to be a major cause of serious liver disease, though the majority of DILI cases that lead to acute liver failure are due to acetaminophen overdose rather than idiosyncratic DILI where the disease develops after the causative drug is used at the recommended dose [5]. Genetic susceptibility appears to be more important in idiosyncratic compared with concentration-dependent DILI. Metabolism of the drug causing DILI to a reactive intermediate is believed to be an important step in the disease process [6]. Impaired membrane transport in the hepatocyte, affecting either drug transport directly or bile acid transport, is also considered to contribute to idiosyncratic DILI, possibly by promoting accumulation of reactive intermediates from the drug or of toxic bile acids in hepatocytes [7]. The availability of extensive data on the functional effect of many polymorphisms in genes coding for drug metabolizing enzymes and transporters has prompted a large number of studies on relationships between genotype and risk of DILI development to be performed. This has provided a few interesting observations which provide insights into the underlying mechanism, but overall the area is still not very clear with most effects being small and not well replicated. This is in contrast to data on HLA genes as DILI risk factors where reported effects of certain HLA genotypes on susceptibility to a number of forms of DILI are strong. However importantly, not all DILI shows an association with a particular HLA genotype. In particular, risk of DILI due to commonly prescribed drugs such as diclofenac, statins, and isoniazid appears HLA-independent [8, 9]. For DILI due to exposure to these drugs, formation of reactive intermediates may be an important contributor. This may lead to an inappropriate immune response independent of HLA genotype as well as causing other effects leading to direct cellular damage.

5.2.2 Polymorphisms Affecting Drug Metabolism and DILI

Table 5.1 summarizes a range of possible associations between DILI and genes coding for enzymes of phase I and phase II drug metabolism. In view of the large number of studies, this section will focus on well-accepted associations where there has been some element of replication with a number of specific drug causes considered in turn.

5.2.2.1 Isoniazid

The most extensively studied association relates to an association between DILI due to anti-tuberculosis (TB) drug regimens containing isoniazid and NAT2, the

Table 5.1 Associations between polymorphisms affecting drug metabolism and DILI susceptibility from candidate gene studies.

Gene	Substrate	Summary of effect	Reference
CYP2B6	Ticlopidine	rs7254579 (−2320T>C) is more common in cases (Allelic odds ratio 2.09, $p = 0.04$)	[10]
CYP2B6	Efavirenz	rs3745274 (516G>T) is more common in cases ($p = 0.04$)	[11]
CYP2C8	Diclofenac	*CYP2C8*4* allele more common in cases (Odds ratio for carriage 3.5, $p = 0.09$)	[12]
CYP2C9	Bosentan	*CYP2C9*2* allele more common in cases (Odds ratio 2.29, $p = 0.003$)	[13]
CYP2D6	Perhexiline	75% of cases were CYP2D6 phenotypic poor metabolizers	[14]
CYP2E1	Isoniazid	*CYP2E1*5* allele more common in cases (OR, 2.38; $p = 0.017$)	[15]
UGT1A6	Tolcapone	Sequencing study; $p = 0.0027$ for haplotype more common in cases than controls	[16]
UGT2B7	Diclofenac	*UGT2B7*2* is more common in cases (Odds ratio 8.5, $p = 0.03$)	[12]
NAT2	Isoniazid	See Table 5.2 for more detail	—
GSTM1/T1	Troglitazone	Increased risk in cases homozygous null for both alleles (Odds ratio 3.69, $p = 0.008$)	[46]
GSTM1/T1	Amoxicillin-clavulanate	Case homozygous for both null alleles had a 2.81-fold increased risk ($p = 0.037$)	[47]
GSTM1	Isoniazid	Increased risk in cases homozygous for GSTM1 null (Odd ratio 2.13, $p < 0.05$)	[48]

Data provided is based on original report only. Several associations have not been confirmed by subsequent studies or are based on a single study only.

gene encoding the phase II metabolizing enzyme *N*-acetyltransferase 2. There are several distinctive features: (i) NAT2 is of limited interest in relation to currently prescribed drugs with the important exception of its key role in detoxication of isoniazid and (ii) isoniazid is one of the best established causes of DILI with case reports dating back to 1950s, soon after extensive prescription of isoniazid had been initiated [49]. NAT2 is subject to a well-studied genetic polymorphism which results in some individuals having an absence of enzyme activity, with this deficiency seen in up to 50% of individuals in Europe and North America. Those with no NAT2 activity are usually termed slow acetylators and those with detectable activity fast acetylators. In the 1970s, the possibility that the NAT2 polymorphism might be a predictor of isoniazid DILI was investigated in detail for the first time, though these early studies did not yield clear-cut findings [50, 51]. Subsequently, there have been a large number of further studies on this, especially because the molecular basis for the NAT2 deficiency is now well understood, facilitating genotyping to predict NAT2 status. As summarized in Table 5.2, candidate gene studies on NAT2 where this is the main or only gene studied have been somewhat contradictory on whether acetylation status is a risk factor for DILI due to isoniazid. However, since approximately 2012, most studies have reported an increased risk of DILI in NAT2 slow acetylators. This reflects the fact that the recent studies tend to be on larger numbers of cases than the older studies. Another factor is that most recent studies have been based in East Asia where slow acetylators typically account for approximately 10% of the population. Seeing a significant difference between slow and fast acetylators in these Asian populations is easier to detect statistically than in those elsewhere. There are some issues with definition of DILI which may also affect study outcome. In general, some studies may include mostly cases with mild liver enzyme elevation where drug treatment can be maintained. Whether NAT2 genotype is most important in determining risk of mild elevation of liver enzymes but is also a risk factor for serious DILI which may result in liver failure is still unclear. At least four different meta-analyses with large numbers of isoniazid DILI cases of varying severity have concluded that slow acetylators have an increased risk of isoniazid-related DILI with an overall odds ratio (OR) varying from 1.6 to 6, although the risk varies depending on the precise genotypic definition of slow acetylation and the population studied [52–55].

A genome-wide association study (GWAS) involving patients from Thailand recently found a genome-wide significant signal for NAT2, suggesting a stronger risk for DILI development in those positive for slow acetylator alleles, in line with many of the earlier candidate gene studies [56]. However, two other GWAS which included small numbers of European cases with anti-TB DILI did not detect genome-wide associations with either NAT2 or any other gene [8, 9]. The apparent discrepancy seen between East Asians and other populations, which also extends to some candidate gene genotyping studies (for example Refs [19, 24] from Table 5.2), may relate to differences in worldwide distributions of NAT2 variant alleles. The NAT2*5 allele, which is more common in Europe, Africa, and North America than the other loss of function alleles NAT2*6 and *7, is now suggested to be associated with decreased but not complete absence of NAT2 activity [57, 58]. This means that it is possible that the risk of developing isoniazid DILI is less in those positive for NAT2*5, though this needs further investigation. It has been suggested that knowledge of NAT2 genotype may be helpful in optimizing isoniazid treatment and avoiding DILI from a study based in Japan where NAT2*6 and NAT2*7 are the common "slow acetylator" alleles [59]. Whether this extends to other populations worldwide needs further investigation.

A possible association between CYP2E1 and DILI due to isoniazid has been quite extensively investigated but studies have not shown consistent associations

Table 5.2 Summary of selected NAT2 candidate gene studies concerned with isoniazid-related DILI.

No of cases	Effect	Ethnicity	Odds ratio (unless stated otherwise for slow acetylators against rapid acetylators)	Reference	Year published
14	Increased risk of DILI in slow acetylators	Japanese	10.00 (95% CI 2.713–36.86) $p = 0.0008$	[17]	2000
33	Increased risk of DILI in slow acetylators	Taiwanese	3.66 (95% CI 1.58–8.49) $p = 0.003$	[18]	2002
34	No increased risk of DILI in slow acetylators	Mixed	Not significant	[19]	2006
18	Increased risk of DILI in slow acetylators	Korean	5.4 (95% CI, 1.8–16.6) $p = 0.005$	[20]	2007
30	Increased incidence of slow acetylators	Turkish	8.8 (95% CI 3.3–23.9) $p < 0.0001$	[21]	2008
14	Increased risk of DILI in slow acetylators	Brazilian	5.4 (95% CI 1.57–19.4), $p = 0.003$	[22]	2008
67	Increased risk of DILI in slow acetylators	Korean	2.33 (95% CI 1.45–3.78) $p = 0.0049$ for *NAT2*6* possession	[23]	2009
23	No increased risk of DILI in slow acetylators	Mixed	Not significant	[24]	2009
45	Increased risk of DILI in slow acetylators	Taiwanese	OR 3.28 (95%CI 1.53–7.06) $p = 0.0019$	[25]	2010
50	Increased risk of DILI in slow acetylators compared with Spanish population controls but not with TB patients without DILI	Spanish	OR 2.07 (95% CI 1.12–2.79) ($p = 0.016$)	[26]	2011

(*Continued*)

Table 5.2 (Continued)

No of cases	Effect	Ethnicity	Odds ratio (unless stated otherwise for slow acetylators against rapid acetylators)	Reference	Year published
41	Increased risk of DILI in slow acetylators	Indian	OR 2.99 (95% CI 1.4–6.2) ($p = 0.0045$)	[27]	2011
89	No increased risk of DILI in slow acetylators	Chinese	Not significant	[28]	2012
50	Increased risk of DILI in slow acetylators	Indian	OR 3.5 (95% CI 1.8–7.0) ($p = 0.0006$)	[29]	2012
14	Increased risk of DILI in slow acetylators	Tunisian	OR 4.3 (95% CI 1.5–18.0) ($p = 0.04$)	[30]	2012
101	Increased risk of DILI in slow acetylators	Chinese	OR 4.74 (95%CI 2.35–9.59) ($p < 0.0001$)	[31]	2012
47	Increased risk of DILI in slow acetylators	Argentinian and Bolivian	OR 2.62 (95%CI 1.26–5.41) ($p = 0.01$)	[32]	2013
18	Increased risk of DILI in slow acetylators	Brazilian	OR 3.05 (95%CI 1.07–8.64) ($p = 0.036$)	[33]	2013
59	Trend toward increased risk of mild DILI in slow acetylators	Brazilian	$p = 0.061$	[34]	2013
50	Increased risk of DILI in slow acetylators	Indian	OR 2.3 (95% CI 1.16–4.62) ($p = 0.02$)	[35]	2013
20	Increased risk of DILI in slow acetylators	Chinese	OR 6.71 (95% CI 2.56–17.86) ($p = 0.0001$)	[36]	2013

89	Trend toward increased risk of DILI in slow acetylators	Chinese Uyghur	OR 2.0 (95% CI 0.95–4.20) ($p = 0.06$)	[37]	2014
17	Increased risk of DILI in slow acetylators	Indian	OR 6.27 (95% CI 1.35–40.22) ($p = 0.01$)	[38]	2014
52	Increased risk of DILI in slow acetylators	Brazilian	OR 2.94 (95% CI 1.40–6.25) ($p < 0.01$)	[39]	2014
26	Increased risk of DILI in slow acetylators	British and South Asian	OR 4.25 (95% CI 1.36–13.22) ($p = 0.012$)	[40]	2014
50	Increased risk of DILI in slow acetylators	Indonesian	OR 3.45 (95% CI 1.79–6.67) ($p = 0.00017$)	[41]	2016
73	Increased risk of DILI in slow acetylators	Japanese	OR 4.32 (95% CI 1.93–9.66) ($p = 0.0006$)	[42]	2016
53	Increased risk of DILI in slow acetylators	Thai	OR 8.80 (95%CI 4.01–19.31) ($p = 1.53 \times 10^{-8}$)	[43]	2016
24	Increased risk of DILI in slow acetylators	Mixed Singaporean (50% Chinese)	OR 9.98 (95% CI 3.32–33.80) ($p = 8.36 \times 10^{-5}$)	[44]	2017
100	Increased risk of DILI in slow acetylators	Indonesian	OR 3.37 (95% CI 2.00–5.68) ($p = 4.3 \times 10^{-6}$)	[45]	2019

following the first positive report in a candidate gene study [15]. CYP2E1 contributes to isoniazid metabolism by activating acetylhydrazine [60] so is a good candidate biologically. Most studies concerned the CYP2E1*5 variant (rs2031920) (an allele often detected using the restriction enzyme *Rsa*I) for which the biological significance remains unclear. Importantly, none of the published GWAS to date found a signal in the area of CYP2E1 [8, 9, 56], though a small effect cannot be ruled out entirely.

A few studies on the possibility that genotype for the glutathione *S*-transferases GSTM1 and GSTT1 might also modulate risk of isoniazid DILI have also been performed. Data are contradictory with two studies reporting an increased frequency of GSTM1 homozygous null genotypes only [48, 61] and a third increased GSTT1 homozygous null only [62]. Similar to the CYP2E1 reports, there is no GWAS replication of these associations.

5.2.2.2 Diclofenac

Diclofenac is a widely prescribed non-steroidal anti-inflammatory which is a relatively common cause of DILI worldwide [63, 64]. This compound undergoes phase I metabolism by CYP2C9 and some other P450s including CYP2C8 and phase II metabolism by the UDP-glucuronosyltransferase UGT2B7. There is no evidence that the common CYP2C9 variant alleles are relevant to risk of DILI [8, 12, 65] but some evidence that CYP2C8 genotype might contribute [12, 66]. In a study on susceptibility to diclofenac-related DILI, possession of UGT2B7*2 was associated with an increased risk of toxicity, possibly due to increased formation of reactive diclofenac acylglucuronide metabolites [12]. A follow-up study involved a GWAS approach and found that single nucleotide polymorphisms (SNPs) in the UGT2B region which were in linkage disequilibrium with the UGT2B7*2-related SNPs were more common in the diclofenac DILI cases [8]. As discussed in more detail in Section 5.2.3, there is also evidence that genotype for the transporter ABCC2 is relevant to risk of diclofenac DILI.

5.2.2.3 Tolcapone

Another reported DILI association with UGT genotype relates to tolcapone, a catechol-*O*-methyltransferase inhibitor used in treatment of Parkinson's disease. Several polymorphisms in UGT1A6 were found to be significantly associated with elevated transaminase levels [16]. Because these polymorphisms appear to be associated with decreased enzyme levels, it was suggested that the toxicity might be linked to slow metabolism of the parent drug. It was suggested subsequently that promoter region polymorphisms in the nearby UGT1A9 may also be relevant to DILI from this drug [67].

5.2.2.4 Ticlopidine

For DILI due to ticlopidine, an upstream SNP (rs7254579 which results in −2320T>C) which has been reported to be associated with increased CYP2B6 expression [68] was significantly more common among cases [10]. It is possible that this variant could be associated with higher reactive intermediate levels leading to increased risk of DILI. In addition to the association with CYP2B6, there are replicated data showing that individuals positive for HLA-A*33 alleles show an increased risk of DILI [9, 69]. Ticlopidine is one of only a few drugs associated with DILI where both HLA genotype and a genotype for a drug-metabolizing enzyme may contribute jointly to risk.

5.2.2.5 Efavirenz

In DILI cases due to efavirenz, the CYP2B6 variant rs3745274 (516G>T) appears to be more common [11]. This variant is associated with lower than normal CYP2B6 activity suggesting that for efavirenz, accumulation of the parent drug might be an important factor in the toxicity. Some limited independent replication of this association has been reported [70, 71].

5.2.2.6 Troglitazone

Trogilitazone is an important example of a drug withdrawn from the market soon after approval due to an association with serious DILI. While only limited DNA samples are available to investigate genetic susceptibility to this form of DILI directly, individual homozygous null for two glutathione *S*-transferase genes, GSTM1 and GSTT1, have been reported to be increased risk of DILI due to troglitazone [46]. This finding is consistent with the finding that GSTM1 but not GSTT1 has a role in detoxication of a quinone metabolite of this drug [72]

5.2.3 Polymorphisms Affecting Transporters and DILI

There have been a relatively large number of studies on genetic polymorphisms in ATP-binding cassette (ABC) transporters for which the biological plausibility of an association appears strong. This is especially the case in relation to cholestatic forms of DILI where bile acids accumulate within hepatocytes, resulting in local toxicity with a likely contribution by the bile acid transporter BSEP encoded by ABCB11. Other ABC transporters, including MDR1 and MRP2 which both contribute to biliary excretion of a range of drugs, are also plausible candidates as DILI risk factors [7]. Both genetic variation and inhibition of normal ABC transporter function by certain drugs may contribute to risk of DILI [73]. Inhibition may be more problematic if transporter expression or activity is lower than normal due to the presence of genetic polymorphisms. To date, GWAS on DILI have failed to demonstrate any statistically significant signals from ABC transporters, suggesting that the role of these transporters as genetic risk factors is relatively small (compared with, for example, HLA genotype) or is specific to only a few drugs [8]. Some more positive findings have emerged from candidate gene studies.

In a study on a small group of patients who had suffered DILI relating to a range of different drugs, an association between cholestatic injury and a polymorphism in exon 13 of ABCB11 was reported [74]. The risk from this common polymorphism was small and it therefore seems unlikely to be a major risk factor for DILI. In a more detailed study involving genotyping for a larger number of polymorphisms including the exon 13 variant, no association with DILI was detected [75]. In contrast to the original finding about ABCB11 and cholestatic DILI, a further study on the exon 13 variant suggested that possession of this allele was associated with increased risk of hepatocellular disease, especially with non-steroidal anti-inflammatory drugs [76].

MRP2 which is encoded by ABCC2 has a major role in the biliary excretion of many glucuronide conjugates. In the study on diclofenac DILI described above (Section 5.2.2.2), an upstream polymorphism in ABCC2 (rs717620; -24C>T) was found to be significantly more common among DILI cases [12]. This finding is consistent with increased levels of the reactive diclofenac acyl glucuronide being linked to toxicity on the basis of findings that -24C>T results in lower expression of the MRP2 protein which would favor cellular accumulation of the glucuronide [77]. In a study on DILI caused by a range of drugs, a polymorphism in linkage disequilibrium with -24C>T was a significant risk factor for the development of hepatocellular toxicity whereas a second promoter region polymorphism was a risk factor for cholestatic or mixed disease [78]. These findings from Korea regarding ABCC2-24C>T are in broad agreement with the data on diclofenac toxicity. In a sub-study concerned only with drug disposition genes which formed part of a larger GWAS on DILI, the lowest p values for drug disposition-related genes were seen for a series of ABCC2 polymorphisms [8]. More recently, a study based in Spain on ABC transporter polymorphisms as risk factors for DILI found some associations with a rare ABCC2 genotype and also reported an association between an extended ABCC2 haplotype and disease severity [79].

Genotypes for another transporter gene, ABCB1 which codes for MDR1, have been studied in patients with DILI due to nevirapine. In African DILI cases, there was a significantly decreased frequency of the ABCB1 SNP 3435C>T (rs1045642) [80] and a similar association was also reported in a US patient group [81]. However, this was not confirmed in a study on a separate group of European nevirapine DILI patients [82]. No association between ABCB1 polymorphisms and DILI more generally has been found to date.

Two other ABC transporters, MRP3 and MRP4 (encoded by the ABCC3 and ABCC4 genes) are found on the basolateral membrane of the hepatocyte and are also of potential relevance to DILI susceptibility. MRP4 is expressed at higher levels than MRP3 in human hepatocytes and also appears more important in efflux of bile acids, especially in cholestasis [83, 84]. It has been proposed that MRP4 inhibition by some compounds is an important underlying mechanism for bile acid accumulation in cholestatic DILI [85]. For example, troglitazone and its sulfate metabolite inhibit both BSEP and MRP4 [86]. In the case of MRP3, inhibition of transport by drugs appears less relevant to DILI generally but there is evidence that in an Mrp3 mouse knockout model, basolateral efflux of diclofenac acylglucuronide is lost and that the animals show increased gastrointestinal ulceration [87]. Hepatic accumulation of diclofenac acylglucuronide when both MRP2 and MRP3 levels are low due to genetics or drug inhibition could be a factor in the human liver toxicity [88]. However, as with the other candidate transporter polymorphisms, to date no association involving ABCC3 (MRP3) has been detected in a GWAS [8].

5.3 Drug-Induced Skin Injury and Related Hypersensitivity Reactions

Skin reactions following drug exposure may also involve additional organs including liver, lungs, and kidney. A variety of drugs cause these rare reactions which vary considerably in severity ranging from severe blistering skin reactions such as Stevens–Johnson syndrome to a mild skin rash. The best studied causes from a pharmacogenomic point of view are those induced by anticonvulsants including carbamazepine and phenytoin, by beta lactam antibiotics (penicillins and cephalosporins) and by anti-HIV drugs including nevirapine and abacavir. Similar to DILI (see Section 5.2), HLA genes are the important risk factors for these reactions (for review see [89]). However, the overall contribution by HLA is stronger for skin rash than for DILI. There is an interesting overlap between reactions affecting liver and skin with some DILI reactions also involving a skin rash. Unlike DILI, data on non-HLA genetic risk factors, especially polymorphisms affecting drug disposition, are limited with a few exceptions.

The cytochrome P450 CYP2C9 has a major role in phase I metabolism of phenytoin and common variants, especially CYP2C9*2 and CYP2C9*3, have been shown to be associated with decreased rates of oxidative metabolism of this drug [90]. Phenytoin is a well-established cause of skin injury and, is similar to carbamazepine-induced injury; susceptibility to this type of toxicity has been shown to be due to association with HLA B*15:02 [91]. However, when the effect size for B*15:02 as a risk factor for carbamazepine versus phenytoin-induced toxicity is compared, the HLA-associated risk is much lower for phenytoin, suggesting that other genetic factors might have contributed. Using a GWAS approach, genetic factors for risk of skin injury with phenytoin were investigated in 168 cases of skin injury due to this drug recruited from a number of East Asian centers [92]. The phenotypes for the skin injury were variable but a clear history of phenytoin use was confirmed. GWAS analysis gave a signal on chromosome 10 in the CYP2C gene cluster with the strongest effect ($p = 10^{-12}$) seen for rs1057910 which codes for the non-synonymous polymorphism associated with the CYP2C9*3 allele. This finding was then replicated in additional cases from the same region. Subsequent studies from East Asia are confirmatory of this association [93]. The reason for the observed association with CYP2C9*3 only relates to studies so far being based in East Asia where this allele is quite common but other variants such as CYP2C9*2 are rare. It is likely that CYP2C9 variants common elsewhere in the world such as rs1799853 (CYP2C9*2) may also be relevant to risk of this toxicity. A recent small study based in the USA suggests that homozygosity for CYP2C9*2 or any carriage of CYP2C9*3 increases risk of skin reaction [94], but further larger studies are needed to fully evaluate effects of individual alleles.

As mentioned above, carbamazepine is another important cause of adverse drug reactions affecting the skin. Similar to reactions involving phenytoin, HLA-B*15:02 is an important risk factor in East Asians [89]. However, unlike phenytoin, there appears to be no additional risk of developing this reaction from genes affecting drug metabolism based on GWAS analysis [95, 96], even though carbamazepine is subject to a complex metabolic pathway involving CYP3A4 and epoxide hydrolase (EPHX1) [97].

The anti-HIV drug nevirapine causes skin rash and related hypersensitivity reactions including DILI quite commonly [98]. This drug is subject to cytochrome P450-mediated metabolism by several isoforms with CYP2B6 showing major involvement [99]. CYP2B6 is subject to a number of common polymorphisms with the relationship between the reported wide interindividual variation in expression of this isoform in human liver and the effect of individual polymorphisms now well understood [100]. The relationship between selected CYP2B6 variants and nevirapine hypersensitivity including both skin rash and liver toxicity has been investigated in several studies. The earliest of these studies was worldwide and genotyped for rs3745274 (516G>T) in CYP2B6. As discussed in Section 5.2.2.5, this decreased activity variant may also be associated with DILI linked to efavirenz. A significantly increased risk for development of nevirapine skin reactions with carriage of one or more of these variant alleles, especially in cases of African ethnicity was found [82]. In a separate smaller study based in Mozambique which was concerned with Stevens–Johnson syndrome and toxic epidermal necrolysis only, the same variant was investigated together with rs28399499 (983T>C), a variant associated with absence of activity [101]. The significant association of 516G>T could not be confirmed though a trend toward significance in the same direction as previously was found. For 983T>C, a highly significant association in this small group of 27 cases was found ($p = 0.0047$). In a larger study based in Malawi [102], 516G>T and 983T>C were again studied in a group of 207 hypersensitivity cases and 425 tolerant controls. It was found that carriage of 983T>C was significantly more common in the cases with Stevens–Johnson syndrome or toxic epidermal necrolysis only (total of 70). Attempts to replicate this directly were not successful, possibly due to a small replication cohort. However, no effect was seen for 516G>T either in the entire cohort or more restricted subgroups. A recent study based in Cameroon was unable to confirm associations with either CYP2B6 variant but was limited by a small number of cases and a broad definition of adverse reaction [103].

Nevirapine hypersensitivity also shows HLA associations (for review see [89]). Attempts have been made to assess risk for combined HLA-CYP2B6 genotypes in several of the studies discussed above [82, 102] without clear associations being seen. Assessing joint HLA-metabolic risks reliably is likely to require large numbers of cases with a well-defined phenotype.

5.4 Statin-Induced Myopathy

5.4.1 Background

A number of drugs from different therapeutic classes can give rise to myopathy. In most cases this involves subacute myopathic symptoms such as muscle weakness, myalgia, and creatine phosphokinase elevation but occasionally a very severe condition known as rhabdomyolysis can be induced (see [104]). Because of the widespread use of statins worldwide and the fact that myopathy of varying severity is one of the more commonly reported adverse reactions to these drugs, the pharmacogenomic basis for susceptibility to this reaction has been studied quite widely. To date, the most positive findings for statin-induced myopathy relate to associations with genes relevant to drug metabolism and transport though there may also be additional associations with various cellular functions and, possibly for an immune-related subgroup of cases, HLA associations. A detailed recent account of all aspects of statin-induced myopathy has appeared recently [105]. The current section will consider associations with cytochrome P450 genes and transporter genes in detail. These represent the best studied group of genes in relation to statin-induced myopathy but also those genes for which strong associations have been reported and, at least in some cases, these associations have been well replicated.

5.4.2 Cytochromes P450

The seven different statins currently licensed worldwide show some differences with respect to P450-mediated metabolism with simvastatin, atorvastatin, fluvastatin, and lovastatin undergoing substantial P450-mediated metabolism whereas pitavastatin, pravastatin, and rosuvastatin undergo more limited P450 metabolism [106]. It is well established that CYP3A4 is the main statin-metabolizing cytochrome P450 but CYP2C9 has a major role in fluvastatin, pitavastatin and rosuvastatin oxidation [107]. Cerivastatin which is no longer licensed due to its association with adverse reactions is mainly metabolized by CYP2C8. The clear understanding of a major role for cytochromes P450 in statin metabolism prompted a number of early candidate gene studies to consider whether P450 genotype was a risk factor for myopathy. Overall the results obtained in relation to P450 genotype as a predictor of statin-induced myopathy are inconclusive. The most commonly studied variants are CYP3A4*22, CYP3A5*3, and the two CYP2C9 variants CYP2C9*2 and *3. Despite the key role of CYP3A4 and related enzymes in statin metabolism, no significant associations have been reported between the functionally significant CYP3A4*22 and CYP3A5*3 alleles and statin-induced myopathy in candidate gene studies [108–111]. GWAS and related open approaches have also failed to show a significant signal in this region [106, 112–114]. Two studies on CYP2C9 variants reported no associations [110, 115] but, slightly paradoxically since CYP2D6 does not contribute to statin oxidation, associations of myopathy with the common CYP2D6 loss of function variant CYP2D6*4 have been reported in studies on several different statins, predominantly simvastatin [116, 117] but these were not confirmed in other candidate gene studies [110, 115] or in the various GWAS analyses.

5.4.3 Transporters

A role for the anionic uptake transporter OATP1B1 which is encoded by SLCO1B1 was suggested in pharmacokinetic studies on several different statins including pravastatin, simvastatin, pitavastatin, and atorvastatin [118–121]. These studies showed that polymorphisms in SLCO1B1 which decrease transport activity, particularly 521T>C (rs4149056) which is most commonly seen as part of the SLCO1B1*15 allele, were associated with higher plasma concentrations of the various statins. This finding was extended in an independent manner by a GWAS involving 80 cases of simvastatin-associated myopathy from a UK-based population which found that only one variant (rs4363657 on chromosome 12) was genome-wide significant [112]. The variant was known to be in strong linkage disequilibrium with rs4149056 and so this study confirmed that a variant associated with a higher than normal circulating level of simvastatin was a risk factor for myopathy. The finding for simvastatin has been well replicated in independent cohorts which either included simvastatin cases predominantly or involved a mix of different statins especially atorvastatin [111, 113, 122–124]. Whether the association with the SLCO1B1 variant extends to all statins remains controversial.

There is still uncertainty in relation to rosuvastatin [125, 126] and based on *in vitro* data, there appears to be no effect in relation to fluvastatin [120].

ABC transporters also contribute to statin disposition. Breast cancer resistance protein (BCRP), the transporter encoded by ABCG2, contributes to statin absorption in its location on the mucosal membrane of enterocytes together with biliary excretion via the hepatocyte canalicular membrane and renal excretion. A relatively common nonsynonymous polymorphism in ABCG2 (rs2231143) is associated with decreased transport activity which appears to correlate with increased circulating levels of statin among those positive for this variant [127]. This appears to be particularly relevant for rosuvastatin and also fluvastatin and atorvastatin [128]. A number of candidate gene studies on this gene as a risk factor for statin myopathy have reported positive though not strong findings [129, 130] but a significant signal for this gene has not been by GWAS to date.

The ABCB1 gene has also received attention as a risk factor for statin myopathy, Most statins are substrates for MDR1 which is encoded by ABCB1 [131]. This transporter is expressed in a number of cell types relevant to drug disposition including enterocytes, hepatocytes, and renal tubule cells. ABCB1 is subject to three relatively common genetic polymorphisms which result in three common haplotypes relating to this gene. The functional significance of these haplotypes and their relevance to drug disposition generally is still a controversial area [132]. It has been reported that genotype for the common variants in ABCB1 is relevant to simvastatin and atorvastatin disposition but not to other statins including pravastatin, fluvastatin, and rosuvastatin [133, 134]. A few candidate gene studies report increased risks of statin myopathy for particular ABCB1 genotypes [109, 135]. However, there are no GWAS data to confirm the relevance of this gene to statin myopathy.

Overall, there is strong data from both candidate gene and GWAS linking SLCO1B1 genotype to susceptibility to myopathy from some statins particularly simvastatin. The overall observed effect sizes do not fully explain susceptibility and to date there is only limited evidence that genes which are unrelated to statin disposition or relevant to statin disposition locally within muscle tissue make any additional contribution see [105]. There is a recommendation from the USA-based Clinical Pharmacogenetics Implementation Consortium (CPIC) that patients positive for one or two copies of the rs4149056 SLCO1B1 variant should be prescribed a low dose of simvastatin (e.g. 20 mg) or an alternative statin to protect against myopathy [136]. A similar group based in the Netherlands (Dutch Pharmacogenetics Working Group) recommend an alternative statin to simvastatin in all patients positive for the rs4149056 SLCO1B1 variant and also recommend that these patients be observed carefully if prescribed atorvastatin [137]. Implementation of these guidelines worldwide is still limited but seems likely to increase over time as genetic data becomes more widely available in patient medical records. At present, SLCO1B1 rs4149056 is the best example of a polymorphism relevant to drug disposition that can be genotyped to decrease risk of an idiosyncratic adverse reaction.

5.5 Conclusions

It is clear that for idiosyncratic adverse drug reactions genetic polymorphisms affecting drug metabolism and drug transporters make a contribution to susceptibility. While the widespread use of GWAS has helped confirm the importance of genes such as NAT2 and SLCO1B1 to a subgroup of these reactions, it has also shown that the overall importance of this gene class is less than that envisaged originally when common polymorphisms in metabolic genes such as the cytochromes P450 were described and later characterized in detail. The importance of the immune system, especially HLA genes, to idiosyncratic reactions was not appreciated fully until quite recently also. For some forms of toxicity, there may be a dual role for both HLA and metabolic genes. This aspect needs further investigation though is complicated by the need to study large numbers of cases relating to specific drugs and reactions which is difficult for these relatively rare events.

This chapter has focused on idiosyncratic reactions affecting three specific organs mainly because genetic studies on other types of reaction, especially with respect to the involvement of metabolic and transporter genes, have been very limited to date. Understanding genetic risk factors for other types of organ toxicity, including those affecting the kidney, pancreas, brain, and vascular tissue, remains important. It remains possible that polymorphisms in drug metabolism or transporter genes could be relevant to these currently less well studied toxicities.

References

1 Uetrecht J, Naisbitt DJ (2013) Idiosyncratic adverse drug reactions: current concepts. *Pharmacol Rev* 65 (2): 779–808

2 Uetrecht J (2007) Idiosyncratic drug reactions: current understanding. *Annu Rev Pharmacol Toxicol* 47: 513–539

3 Usui T, Naisbitt DJ (2017) Human leukocyte antigen and idiosyncratic adverse drug reactions. *Drug Metab Pharmacokinet* 32 (1): 21–30

4 Wilke RA, Lin DW, Roden DM et al. (2007) Identifying genetic risk factors for serious adverse drug reactions: current progress and challenges. *Nat Rev Drug Discov* 6 (11): 904–916

5 Ostapowicz G, Fontana RJ, Schiodt FV et al. (2002) Results of a prospective study of acute liver failure at 17 tertiary care centers in the United States. *Ann Intern Med* 137 (12): 947–954

6 Park BK, Kitteringham NR, Maggs JL, Pirmohamed M, Williams DP (2005) The role of metabolic activation in drug-induced hepatotoxicity. *Annu Rev Pharmacol Toxicol* 45: 177–202

7 Russmann S, Jetter A, Kullak-Ublick GA Pharmacogenetics of drug-induced liver injury. *Hepatology* 52 (2): 748–761.

8 Urban TJ, Shen Y, Stolz A et al. (2012) Limited contribution of common genetic variants to risk for liver injury due to a variety of drugs. *Pharmacogenet Genomics* 22 (11): 784–795

9 Nicoletti P, Aithal GP, Bjornsson ES et al. (2017) Association of liver injury from specific drugs, or groups of drugs, with polymorphisms in HLA and other genes in a genome-wide association study. *Gastroenterology* 152 (5): 1078–1089

10 Ariyoshi N, Iga Y, Hirata K et al. (2010) Enhanced susceptibility of HLA-mediated ticlopidine-induced idiosyncratic hepatotoxicity by CYP2B6 polymorphism in Japanese. *Drug Metab Pharmacokinet* 25 (3): 298–306

11 Yimer G, Amogne W, Habtewold A et al. (2012) High plasma efavirenz level and CYP2B6*6 are associated with efavirenz-based HAART-induced liver injury in the treatment of naive HIV patients from Ethiopia: a prospective cohort study. *Pharmacogenomics J* 12 (6): 499–506

12 Daly AK, Aithal GP, Leathart JB et al. (2007) Genetic susceptibility to diclofenac-induced hepatotoxicity: contribution of UGT2B7, CYP2C8, and ABCC2 genotypes. *Gastroenterology* 132 (1): 272–281

13 Markova SM, De Marco T, Bendjilali N et al. (2013) Association of CYP2C9*2 with bosentan-induced liver injury. *Clin Pharmacol Ther* 94 (6): 678–686

14 Morgan MY, Reshef R, Shah RR et al. (1984) Impaired oxidation of debrisoquine in patients with perhexiline liver injury. *Gut* 25 (10): 1057–1064

15 Huang YS, Chern HD, Su WJ et al. (2003) Cytochrome p450 2E1 genotype and the susceptibility to antituberculosis drug-induced hepatitis. *Hepatology* 37 (4): 924–930

16 Acuna G, Foernzler D, Leong D et al. (2002) Pharmacogenetic analysis of adverse drug effect reveals genetic variant for susceptibility to liver toxicity. *Pharmacogenomics J* 2 (5): 327–334

17 Ohno M, Yamaguchi I, Yamamoto I et al. (2000) Slow *N*-acetyltransferase 2 genotype affects the incidence of isoniazid and rifampicin-induced hepatotoxicity. *Int J Tuberc Lung Dis* 4 (3): 256–261

18 Huang YS, Chern HD, Su WJ et al. (2002) Polymorphism of the *N*-acetyltransferase 2 gene as a susceptibility risk factor for antituberculosis drug-induced hepatitis. *Hepatology* 35 (4): 883–889

19 Vuilleumier N, Rossier MF, Chiappe A et al. (2006) CYP2E1 genotype and isoniazid-induced hepatotoxicity in patients treated for latent tuberculosis. *Eur J Clin Pharmacol* 62 (6): 423–429

20 Cho HJ, Koh WJ, Ryu YJ et al. (2007) Genetic polymorphisms of NAT2 and CYP2E1 associated with antituberculosis drug-induced hepatotoxicity in Korean patients with pulmonary tuberculosis. *Tuberculosis (Edinburgh, Scotland)* 87 (6): 551–556

21 Bozok Cetintas V, Erer OF, Kosova B et al. (2008) Determining the relation between *N*-acetyltransferase-2 acetylator phenotype and antituberculosis drug induced hepatitis by molecular biologic tests. *Tuberk Toraks* 56 (1): 81–86

22 Possuelo LG, Castelan JA, de Brito TC et al. (2008) Association of slow *N*-acetyltransferase 2 profile and anti-TB drug-induced hepatotoxicity in patients from Southern Brazil. *Eur J Clin Pharmacol* 64 (7): 673–681

23 Kim SH, Kim SH, Bahn JW et al. (2009) Genetic polymorphisms of drug-metabolizing enzymes and anti-TB drug-induced hepatitis. *Pharmacogenomics* 10 (11): 1767–1779

24 Yamada S, Tang M, Richardson K et al. (2009) Genetic variations of NAT2 and CYP2E1 and isoniazid hepatotoxicity in a diverse population. *Pharmacogenomics* 10 (9): 1433–1445

25 Lee SW, Chung LS, Huang HH et al. (2010) NAT2 and CYP2E1 polymorphisms and susceptibility to first-line anti-tuberculosis drug-induced hepatitis. *Int J Tuberc Lung Dis* 14 (5): 622–626

26 Leiro-Fernandez V, Valverde D, Vazquez-Gallardo R et al. (2011) *N*-acetyltransferase 2 polymorphisms and risk of anti-tuberculosis drug-induced hepatotoxicity in Caucasians. *Int J Tuberc Lung Dis* 15 (10): 1403–1408

27 Bose PD, Sarma MP, Medhi S et al. (2011) Role of polymorphic *N*-acetyl transferase2 and cytochrome P4502E1 gene in antituberculosis treatment-induced hepatitis. *J Gastroenterol Hepatol* 26 (2): 312–318

28 Lv X, Tang S, Xia Y et al. (2012) NAT2 genetic polymorphisms and anti-tuberculosis drug-induced hepatotoxicity in Chinese community population. *Ann Hepatol* 11 (5): 700–707

29 Rana SV, Ola RP, Sharma SK et al. (2012) Comparison between acetylator phenotype and genotype polymorphism of *N*-acetyltransferase-2 in tuberculosis patients. *Hepatol Int* 6: 397–402

30 Ben Mahmoud L, Ghozzi H, Kamoun A et al. (2012) Polymorphism of the *N*-acetyltransferase 2 gene as a susceptibility risk factor for antituberculosis drug-induced hepatotoxicity in Tunisian patients with tuberculosis. *Pathol Biol (Paris)* 60 (5): 324–330

31 An HR, Wu XQ, Wang ZY, Zhang JX, Liang Y (2012) NAT2 and CYP2E1 polymorphisms associated with antituberculosis drug-induced hepatotoxicity in Chinese patients. *Clin Exp Pharmacol Physiol* 39 (6): 535–543

32 Chamorro JG, Castagnino JP, Musella RM et al. (2013) Sex, ethnicity, and slow acetylator profile are the major causes of hepatotoxicity induced by antituberculosis drugs. *J Gastroenterol Hepatol* 28 (2): 323–328

33 Santos NP, Callegari-Jacques SM, Ribeiro Dos Santos AK et al. (2013) *N*-acetyl transferase 2 and cytochrome P450 2E1 genes and isoniazid-induced hepatotoxicity in Brazilian patients. *Int J Tuberc Lung Dis* 17 (4): 499–504

34 Forestiero FJ, Cecon L, Hirata MH et al. (2013) Relationship of NAT2, CYP2E1 and GSTM1/GSTT1 polymorphisms with mild elevation of liver enzymes in Brazilian individuals under anti-tuberculosis drug therapy. *Clin Chim Acta* 415: 215–219

35 Gupta VH, Amarapurkar DN, Singh M et al. (2013) Association of *N*-acetyltransferase 2 and cytochrome P450 2E1 gene polymorphisms with antituberculosis drug-induced hepatotoxicity in Western India. *J Gastroenterol Hepatol* 28 (8): 1368–1374

36 Ho HT, Wang TH, Hsiong CH et al. (2013) The NAT2 tag SNP rs1495741 correlates with the susceptibility of antituberculosis drug-induced hepatotoxicity. *Pharmacogenet Genomics* 23 (4): 200–207

37 Xiang Y, Ma L, Wu W et al. (2014) The incidence of liver injury in Uyghur patients treated for TB in Xinjiang Uyghur Autonomous Region, China, and its association with hepatic enzyme polymorphisms NAT2, CYP2E1, GSTM1 and GSTT1. *PLoS ONE* 9 (1): e85905

38 Singla N, Gupta D, Birbian N, Singh J (2014) Association of NAT2, GST and CYP2E1 polymorphisms and anti-tuberculosis drug-induced hepatotoxicity. *Tuberculosis (Edinburgh, Scotland)* 94 (3): 293–298

39 Zaverucha-do-Valle C, Monteiro SP, El-Jaick KB et al. (2014) The role of cigarette smoking and liver enzymes polymorphisms in anti-tuberculosis drug-induced hepatotoxicity in Brazilian patients. *Tuberculosis (Edinburgh, Scotland)* 94 (3): 299–305

40 Ng CS, Hasnat A, Al Maruf A et al. (2014) *N*-acetyltransferase 2 (NAT2) genotype as a risk factor for development of drug-induced liver injury relating to antituberculosis drug treatment in a mixed-ethnicity patient group. *Eur J Clin Pharmacol* 70 (9): 1079–1086

41 Yuliwulandari R, Susilowati RW, Wicaksono BD et al. (2016) NAT2 variants are associated with drug-induced liver injury caused by anti-tuberculosis drugs in Indonesian patients with tuberculosis. *J Hum Genet* 61 (6): 533–537

42 Mushiroda T, Yanai H, Yoshiyama T et al. (2016) Development of a prediction system for anti-tuberculosis drug-induced liver injury in Japanese patients. *Hum Genome Var* 3: 16014

43 Wattanapokayakit S, Mushiroda T, Yanai H et al. (2016) NAT2 slow acetylator associated with anti-tuberculosis drug-induced liver injury in Thai patients. *Int J Tuberc Lung Dis* 20 (10): 1364–1369

44 Chan SL, Chua APG, Aminkeng F et al. (2017) Association and clinical utility of NAT2 in the prediction of isoniazid-induced liver injury in Singaporean patients. *PLoS ONE* 12 (10): e0186200

45 Yuliwulandari R, Prayuni K, Susilowati RW et al. (2019) NAT2 slow acetylator is associated with anti-tuberculosis drug-induced liver injury severity in indonesian population. *Pharmacogenomics* 20 (18): 1303–1311

46 Watanabe I, Tomita A, Shimizu M et al. (2003) A study to survey susceptible genetic factors responsible for troglitazone-associated hepatotoxicity in Japanese patients with type 2 diabetes mellitus. *Clin Pharmacol Ther* 73 (5): 435–455

47 Lucena MI, Andrade RJ, Martinez C et al. (2008) Glutathione *S*-transferase m1 and t1 null genotypes increase susceptibility to idiosyncratic drug-induced liver injury. *Hepatology* 48 (2): 588–596

48 Roy B, Chowdhury A, Kundu S et al. (2001) Increased risk of antituberculosis drug-induced hepatotoxicity in individuals with glutathione *S*-transferase M1 'null' mutation. *J Gastroenterol Hepatol* 16 (9): 1033–1037

49 Randolph H, Joseph S (1953) Toxic hepatitis with jaundice occuring in a patient treated with isoniazid. *J Am Med Assoc* 152 (1): 38–40

50 Mitchell JR, Thorgeirsson UP, Black M et al. (1975) Increased incidence of isoniazid hepatitis in rapid acetylators: possible relation to hydrazine metabolites. *Clin Pharmacol Ther* 18 (1): 70–79

51 Gurumurthy P, Krishnamurthy MS, Nazareth O et al. (1984) Lack of relationship between hepatic toxicity and acetylator phenotype in three thousand South Indian patients during treatment with isoniazid for tuberculosis. *Am Rev Respir Dis* 129 (1): 58–61

52 Zhang M, Wang S, Wilffert B et al. (2018) The association between the NAT2 genetic polymorphisms and risk of DILI during anti-TB treatment: a systematic review and meta-analysis. *Br J Clin Pharmacol* 84 (12): 2747–2760

53 Khan S, Mandal RK, Elasbali AM et al. (2019) Pharmacogenetic association between NAT2 gene polymorphisms and isoniazid induced hepatotoxicity: trial sequence meta-analysis as evidence. *Biosci Rep* 39 (1), BSR20180845.

54 Richardson M, Kirkham J, Dwan K et al. (2019) NAT2 variants and toxicity related to anti-tuberculosis agents: a systematic review and meta-analysis. *Int J Tuberc Lung Dis* 23 (3): 293–305

55 Cai Y, Yi J, Zhou C, Shen X (2012) Pharmacogenetic study of drug-metabolising enzyme polymorphisms on the risk of anti-tuberculosis drug-induced liver injury: a meta-analysis. *PLoS ONE* 7 (10): e47769

56 Suvichapanich S, Wattanapokayakit S, Mushiroda T et al. (2019) Genomewide association study confirming the association of NAT2 with susceptibility to antituberculosis drug-induced liver injury in Thai patients. *Antimicrob Agents Chemother* 63 (8), e02692-18.

57 Selinski S, Blaszkewicz M, Ickstadt K, Hengstler JG, Golka K (2013) Refinement of the prediction of *N*-acetyltransferase 2 (NAT2) phenotypes with respect to enzyme activity and urinary bladder cancer risk. *Arch Toxicol* 87 (12): 2129–2139

58 Doll MA, Hein DW (2017) Genetic heterogeneity among slow acetylator *N*-acetyltransferase 2 phenotypes in cryopreserved human hepatocytes. *Arch Toxicol* 91 (7): 2655–2661

59 Azuma J, Ohno M, Kubota R et al. (2013) NAT2 genotype guided regimen reduces isoniazid-induced liver injury and early treatment failure in the 6-month four-drug standard treatment of tuberculosis: a randomized controlled trial for pharmacogenetics-based therapy. *Eur J Clin Pharmacol* 69 (5): 1091–1101

60 Wang P, Pradhan K, Zhong XB, Ma X (2016) Isoniazid metabolism and hepatotoxicity. *Acta Pharm Sin B* 6 (5): 384–392

61 Huang YS, Su WJ, Huang YH et al. (2007) Genetic polymorphisms of manganese superoxide dismutase, NAD(P)H:quinone oxidoreductase, glutathione *S*-transferase M1 and T1, and the susceptibility to drug-induced liver injury. *J Hepatol* 47 (1): 128–134

62 Leiro V, Fernandez-Villar A, Valverde D et al. (2008) Influence of glutathione *S*-transferase M1 and T1 homozygous null mutations on the risk of antituberculosis drug-induced hepatotoxicity in a Caucasian population. *Liver Int* 28 (6): 835–839

63 Banks AT, Zimmerman HJ, Ishak KG, Harter JG (1995) Diclofenac-associated hepatotoxicity - analysis of 180 cases reported to the Food-and-Drug-Administration as adverse reactions. *Hepatology* 22 (3): 820–827

64 Bjornsson ES, Bergmann OM, Bjornsson HK, Kvaran RB, Olafsson S (2013) Incidence, presentation, and outcomes in patients with drug-induced liver injury in the general population of Iceland. *Gastroenterology* 144 (7): 1419–1425, 1425 e1411-1413; quiz e1419-1420.

65 Aithal GP, Day CP, Leathart JBS, Daly AK (2000) Relationship of polymorphism in CYP2C9 to genetic susceptibility to diclofenac-induced hepatitis. *Pharmacogenetics* 10: 511–518

66 Lazarska KE, Dekker SJ, Vermeulen NPE, Commandeur JNM (2018) Effect of UGT2B7*2 and CYP2C8*4 polymorphisms on diclofenac metabolism. *Toxicol Lett* 284: 70–78

67 Martignoni E, Cosentino M, Ferrari M et al. (2005) Two patients with COMT inhibitor-induced hepatic dysfunction and UGT1A9 genetic polymorphism. *Neurology* 65 (11): 1820–1822

68 Hesse LM, He P, Krishnaswamy S et al. (2004) Pharmacogenetic determinants of interindividual variability in bupropion hydroxylation by cytochrome P450 2B6 in human liver microsomes. *Pharmacogenetics* 14 (4): 225–238

69 Hirata K, Takagi H, Yamamoto M et al. (2008) Ticlopidine-induced hepatotoxicity is associated with specific human leukocyte antigen genomic subtypes in

Japanese patients: a preliminary case-control study. *Pharmacogenomics J* 8 (1): 29–33

70 Elsharkawy AM, Schwab U, McCarron B et al. (2013) Efavirenz induced acute liver failure requiring liver transplantation in a slow drug metaboliser. *J Clin Virol* 58 (1): 331–333

71 Manosuthi W, Sukasem C, Lueangniyomkul A et al. (2014) CYP2B6 haplotype and biological factors responsible for hepatotoxicity in HIV-infected patients receiving efavirenz-based antiretroviral therapy. *Int J Antimicrob Agents* 43 (3): 292–296

72 Okada R, Maeda K, Nishiyama T et al. (2011) Involvement of different human glutathione *S*-transferase isoforms in the glutathione conjugation of reactive metabolites of troglitazone. *Drug Metab Dispos*, 39(12):2290–7.

73 Morgan RE, Trauner M, van Staden CJ et al. (2010) Interference with bile salt export pump function is a susceptibility factor for human liver injury in drug development. *Toxicol Sci* 118 (2): 485–500

74 Lang C, Meier Y, Stieger B et al. (2007) Mutations and polymorphisms in the bile salt export pump and the multidrug resistance protein 3 associated with drug-induced liver injury. *Pharmacogenet Genomics* 17 (1): 47–60

75 Bhatnagar P, Day CP, Aithal G et al. (2008) Genetic variants of hepatic transporters and susceptibility to drug induced liver injury. *Toxicology* 253 (1–3): 10–10

76 Ulzurrun E, Stephens C, Crespo E et al. (2013) Role of chemical structures and the 1331T>C bile salt export pump polymorphism in idiosyncratic drug-induced liver injury. *Liver Int* 33 (9): 1378–1385

77 Haenisch S, Zimmermann U, Dazert E et al. (2007) Influence of polymorphisms of ABCB1 and ABCC2 on mRNA and protein expression in normal and cancerous kidney cortex. *Pharmacogenomics J* 7 (1): 56–65

78 Choi JH, Ahn BM, Yi J et al. (2007) MRP2 haplotypes confer differential susceptibility to toxic liver injury. *Pharmacogenet Genomics* 17 (6): 403–415

79 Ulzurrun E, Stephens C, Ruiz-Cabello F et al. (2014) Selected ABCB1, ABCB4 and ABCC2 polymorphisms do not enhance the risk of drug-induced hepatotoxicity in a Spanish cohort. *PLoS ONE* 9 (4): e94675.

80 Haas DW, Bartlett JA, Andersen JW et al. (2006) Pharmacogenetics of nevirapine-associated hepatotoxicity: an Adult AIDS Clinical Trials Group collaboration. *Clin Infect Dis* 43 (6): 783–786

81 Ritchie MD, Haas DW, Motsinger AA et al. (2006) Drug transporter and metabolizing enzyme gene variants and nonnucleoside reverse-transcriptase inhibitor hepatotoxicity. *Clin Infect Dis* 43 (6): 779–782

82 Yuan J, Guo S, Hall D et al. (2011) Toxicogenomics of nevirapine-associated cutaneous and hepatic adverse events among populations of African, Asian, and European descent. *AIDS* 25 (10): 1271–1280

83 Rius M, Nies AT, Hummel-Eisenbeiss J, Jedlitschky G, Keppler D (2003) Cotransport of reduced glutathione with bile salts by MRP4 (ABCC4) localized to the basolateral hepatocyte membrane. *Hepatology* 38 (2): 374–384

84 Slot AJ, Molinski SV, Cole SP (2011) Mammalian multidrug-resistance proteins (MRPs). *Essays Biochem* 50 (1): 179–207

85 Kock K, Ferslew BC, Netterberg I et al. (2014) Risk factors for development of cholestatic drug-induced liver injury: inhibition of hepatic basolateral bile acid transporters multidrug resistance-associated proteins 3 and 4. *Drug Metab Dispos* 42 (4): 665–674

86 Yang K, Woodhead JL, Watkins PB, Howell BA, Brouwer KL (2014) Systems pharmacology modeling predicts delayed presentation and species differences in bile acid-mediated troglitazone hepatotoxicity. *Clin Pharmacol Ther* 96 (5): 589–598

87 Scialis RJ, Csanaky IL, Goedken MJ, Manautou JE (2015) Multidrug resistance-associated protein 3 plays an important role in protection against acute toxicity of diclofenac. *Drug Metab Dispos* 43 (7): 944–950

88 Lagas JS, Sparidans RW, Wagenaar E, Beijnen JH, Schinkel AH (2010) Hepatic clearance of reactive glucuronide metabolites of diclofenac in the mouse is dependent on multiple ATP-binding cassette efflux transporters. *Mol Pharmacol* 77 (4): 687–694

89 Oussalah A, Yip V, Mayorga C et al. (2020) Genetic variants associated with T cell-mediated cutaneous adverse drug reactions: a PRISMA-compliant systematic review-An EAACI position paper. *Allergy* 75 (5): 1069–1098

90 Aynacioglu AS, Brockmoller J, Bauer S et al. (1999) Frequency of cytochrome P450 CYP2C9 variants in a Turkish population and functional relevance for phenytoin. *Br J Clin Pharmacol* 48 (3): 409–415

91 Hung SI, Chung WH, Liu ZS et al. (2010) Common risk allele in aromatic antiepileptic-drug induced Stevens–Johnson syndrome and toxic epidermal necrolysis in Han Chinese. *Pharmacogenomics* 11 (3): 349–356

92 Chung WH, Chang WC, Lee YS et al. (2014) Genetic variants associated with phenytoin-related severe cutaneous adverse reactions. *JAMA* 312 (5): 525–534

93 Hikino K, Ozeki T, Koido M et al. (2020) HLA-B*51:01 and CYP2C9*3 are risk factors for phenytoin-induced eruption in the Japanese population: analysis of data from the Biobank Japan project. *Clin Pharmacol Ther* 107 (5): 1170–1178

94 Fohner AE, Rettie AE, Thai KK et al. (2020) Associations of CYP2C9 and CYP2C19 pharmacogenetic variation with phenytoin-induced cutaneous adverse drug reactions. *Clin Transl Sci*, 13(5):1004–1009

95 McCormack M, Alfirevic A, Bourgeois S et al. (2011) HLA-A*3101 and carbamazepine-induced hypersensitivity reactions in Europeans. *N Engl J Med* 364 (12): 1134–1143

96 Nicoletti P, Barrett S, McEvoy L et al. (2019) Shared genetic risk factors across carbamazepine-induced hypersensitivity reactions. *Clin Pharmacol Ther* 106 (5): 1028–1036

97 Klotz U (2007) The role of pharmacogenetics in the metabolism of antiepileptic drugs: pharmacokinetic and therapeutic implications. *Clin Pharmacokinet* 46 (4): 271–279

98 Phillips E, Gutiérrez S, Jahnke N et al. (2007) Determinants of nevirapine hypersensitivity and its effect on the association between hepatitis C status and mortality in antiretroviral drug-naive HIV-positive patients. *AIDS* 21 (12): 1561–1568

99 Riska P, Lamson M, MacGregor T et al. (1999) Disposition and biotransformation of the antiretroviral drug nevirapine in humans. *Drug Metab Dispos* 27 (8): 895–901

100 Zanger UM, Klein K (2013) Pharmacogenetics of cytochrome P450 2B6 (CYP2B6): advances on polymorphisms, mechanisms, and clinical relevance. *Front Genet* 4: 24

101 Ciccacci C, Di Fusco D, Marazzi MC et al. (2013) Association between CYP2B6 polymorphisms and Nevirapine-induced SJS/TEN: a pharmacogenetics study. *Eur J Clin Pharmacol* 69 (11): 1909–1916

102 Carr DF, Chaponda M, Cornejo Castro EM et al. (2014) CYP2B6 c.983T>C polymorphism is associated with nevirapine hypersensitivity in Malawian and Ugandan HIV populations. *J Antimicrob Chemother* 69 (12): 3329–3334

103 Nguefeu Nkenfou C, Atogho Tiedeu B, Nguefeu Nkenfou C et al. (2019) Adverse drug reactions associated with CYP 2B6 polymorphisms in HIV/AIDS-treated patients in Yaoundé, Cameroon. *Appl Clin Genet* 12: 261–268

104 Dalakas MC (2009) Toxic and drug-induced myopathies. *J Neurol Neurosurg Psychiatry* 80 (8): 832–838

105 Turner RM, Pirmohamed M (2019) Statin-related myotoxicity: a comprehensive review of pharmacokinetic, pharmacogenomic and muscle components. *J Clin Med* 9 (1): 22

106 Turner RM, Fontana V, Zhang JE et al. (2020) A genome-wide association study of circulating levels of atorvastatin and its major metabolites. *Clin Pharmacol Ther*, 108 (2), 287–297.

107 Canestaro WJ, Austin MA, Thummel KE (2014) Genetic factors affecting statin concentrations and subsequent myopathy: a HuGENet systematic review. *Genet Med* 16 (11): 810–819

108 Wilke RA, Moore JH, Burmester JK (2005) Relative impact of CYP3A genotype and concomitant medication on the severity of atorvastatin-induced muscle damage. *Pharmacogenet Genomics* 15 (6): 415–421

109 Fiegenbaum M, da Silveira FR, Van der Sand CR et al. (2005) The role of common variants of ABCB1, CYP3A4, and CYP3A5 genes in lipid-lowering efficacy and safety of simvastatin treatment. *Clin Pharmacol Ther* 78 (5): 551–558

110 Zuccaro P, Mombelli G, Calabresi L et al. (2007) Tolerability of statins is not linked to CYP450 polymorphisms, but reduced CYP2D6 metabolism improves cholesteraemic response to simvastatin and fluvastatin. *Pharmacol Res* 55 (4): 310–317

111 Bakar NS, Neely D, Avery P et al. (2018) Genetic and clinical factors are associated with statin-related myotoxicity of moderate severity: a case-control study. *Clin Pharmacol Ther* 104 (1): 178–187

112 Link E, Parish S, Armitage J et al. (2008) SLCO1B1 variants and statin-induced myopathy – a genomewide study. *N Engl J Med* 359 (8): 789–799

113 Carr DF, Francis B, Jorgensen AL et al. (2019) Genomewide association study of statin-induced myopathy in patients recruited using the UK Clinical Practice Research Datalink. *Clin Pharmacol Ther* 106 (6): 1353–1361

114 Floyd JS, Bloch KM, Brody JA et al. (2019) Pharmacogenomics of statin-related myopathy: meta-analysis of rare variants from whole-exome sequencing. *PLoS ONE* 14 (6): e0218115

115 Voora D, Shah SH, Spasojevic I et al. (2009) The SLCO1B1*5 genetic variant is associated with statin-induced side effects. *J Am Coll Cardiol* 54 (17): 1609–1616

116 Mulder AB, van Lijf HJ, Bon MA et al. (2001) Association of polymorphism in the cytochrome CYP2D6 and the efficacy and tolerability of simvastatin. *Clin Pharmacol Ther* 70 (6): 546–551

117 Frudakis TN, Thomas MJ, Ginjupalli SN et al. (2007) CYP2D6*4 polymorphism is associated with statin-induced muscle effects. *Pharmacogenet Genomics* 17 (9): 695–707

118 Nishizato Y, Ieiri I, Suzuki H et al. (2003) Polymorphisms of OATP-C (SLC21A6) and OAT3 (SLC22A8) genes: consequences for pravastatin pharmacokinetics. *Clin Pharmacol Ther* 73 (6): 554–565

119 Pasanen MK, Neuvonen M, Neuvonen PJ, Niemi M (2006) SLCO1B1 polymorphism markedly affects the pharmacokinetics of simvastatin acid. *Pharmacogenet Genomics* 16 (12): 873–879

120 Niemi M, Pasanen MK, Neuvonen PJ (2006) SLCO1B1 polymorphism and sex affect the pharmacokinetics of pravastatin but not fluvastatin. *Clin Pharmacol Ther* 80 (4): 356–366

121 Pasanen MK, Fredrikson H, Neuvonen PJ, Niemi M (2007) Different effects of SLCO1B1 polymorphism on the pharmacokinetics of atorvastatin and rosuvastatin. *Clin Pharmacol Ther* 82 (6): 726–733

122 Donnelly LA, Doney AS, Tavendale R et al. (2011) Common nonsynonymous substitutions in SLCO1B1 predispose to statin intolerance in routinely treated individuals with type 2 diabetes: a go-DARTS study. *Clin Pharmacol Ther* 89 (2): 210–216

123 Carr DF, O'Meara H, Jorgensen AL et al. (2013) SLCO1B1 genetic variant associated with statin-induced myopathy: a proof-of-concept study using the clinical practice research datalink. *Clin Pharmacol Ther* 94 (6): 695–701

124 de Keyser CE, Peters BJ, Becker ML et al. (2014) The SLCO1B1 c.521T>C polymorphism is associated with dose decrease or switching during statin therapy in the Rotterdam Study. *Pharmacogenet Genomics* 24 (1): 43–51

125 Danik JS, Chasman DI, MacFadyen JG et al. (2013) Lack of association between SLCO1B1 polymorphisms and clinical myalgia following rosuvastatin therapy. *Am Heart J* 165 (6): 1008–1014

126 Xiang Q, Chen SQ, Ma LY et al. (2018) Association between SLCO1B1 T521C polymorphism and risk of statin-induced myopathy: a meta-analysis. *Pharmacogenomics J* 18 (6): 721–729

127 Niemi M (2010) Transporter pharmacogenetics and statin toxicity. *Clin Pharmacol Ther* 87 (1): 130–133

128 Keskitalo JE, Zolk O, Fromm MF et al. (2009) ABCG2 polymorphism markedly affects the pharmacokinetics of atorvastatin and rosuvastatin. *Clin Pharmacol Ther* 86 (2): 197–203

129 Mirošević Skvrce N, Macolić Šarinić V, Šimić I et al. (2015) ABCG2 gene polymorphisms as risk factors for atorvastatin adverse reactions: a case-control study. *Pharmacogenomics* 16 (8): 803–815

130 Miroševic Skvrce N, Božina N, Zibar L et al. (2013) CYP2C9 and ABCG2 polymorphisms as risk factors for developing adverse drug reactions in renal transplant patients taking fluvastatin: a case-control study. *Pharmacogenomics* 14 (12): 1419–1431

131 Chen C, Mireles RJ, Campbell SD et al. (2005) Differential interaction of 3-hydroxy-3-methylglutaryl-coa reductase inhibitors with ABCB1, ABCC2, and OATP1B1. *Drug Metab Dispos* 33 (4): 537–546

132 Wolking S, Schaeffeler E, Lerche H, Schwab M, Nies AT (2015) Impact of genetic polymorphisms of ABCB1 (MDR1, *P*-glycoprotein) on drug disposition and potential clinical implications: update of the literature. *Clin Pharmacokinet* 54 (7): 709–735

133 Keskitalo JE, Kurkinen KJ, Neuvoneni PJ, Niemi M (2008) ABCB1 haplotypes differentially affect the pharmacokinetics of the acid and lactone forms of simvastatin and atorvastatin. *Clin Pharmacol Ther* 84 (4): 457–461

134 Keskitalo JE, Kurkinen KJ, Neuvonen M et al. (2009) No significant effect of ABCB1 haplotypes on the pharmacokinetics of fluvastatin, pravastatin, lovastatin, and rosuvastatin. *Br J Clin Pharmacol* 68 (2): 207–213

135 Ferrari M, Guasti L, Maresca A et al. (2014) Association between statin-induced creatine kinase elevation and genetic polymorphisms in SLCO1B1, ABCB1 and ABCG2. *Eur J Clin Pharmacol* 70 (5): 539–547

136 Ramsey LB, Johnson SG, Caudle KE et al. (2014) The clinical pharmacogenetics implementation consortium guideline for SLCO1B1 and simvastatin-induced myopathy: 2014 update. *Clin Pharmacol Ther* 96 (4): 423–428

137 Bank PCD, Swen JJ, Guchelaar HJ (2019) Estimated nationwide impact of implementing a preemptive pharmacogenetic panel approach to guide drug prescribing in primary care in The Netherlands. *BMC Med* 17 (1): 110

6

Acyl Glucuronidation and Acyl-CoA Formation Mechanisms Mediating the Bioactivation and Potential Toxicity of Carboxylic Acid-containing Drugs

Mark P. Grillo

MyoKardia, Inc., 1000 Sierra Point Parkway, Brisbane, CA, 94005, USA

6.1 Introduction

This chapter deals with the mechanisms involved in the metabolism of carboxylic acid-containing drugs to chemically reactive, and hence potentially toxic, metabolites by a general process known as bioactivation or metabolic activation. Many of the drugs withdrawn from clinical use or clinical trials over the past few decades worldwide have been compounds containing the carboxylic acid-moiety [1–3]. Prominent examples include the nonsteroidal inflammatory drugs (NSAIDs) such as benoxaprofen [4], bromfenac [5], ibufenac [6], fenclozic acid [7], suprofen [8], and zomepirac [9], and other carboxylic acid drugs such as the hypolipidemic agent nafenopin [10], the diuretic tienilic acid [11], and the antibiotic trovafloxacin [12]. Examples of other carboxylic acid drugs currently used clinically that are also known to cause idiosyncratic drug toxicity include the hypolipidemic drug clofibric acid [13], the NSAIDs diclofenac [14] and ibuprofen [15], and the anticonvulsant valproic acid [16]. The chemical structures of these carboxylic acid-containing drugs and other compounds cited in the text are shown in Figure 6.1. Adverse effects observed for these and other carboxylic acid drugs leading to warnings or for their withdrawal from clinical use have ranged from mild liver enzyme elevations, jaundice, skin rash and eczema, to serious and sometimes fatal hepatotoxicity, and anaphylaxis [3, 17]. The primary hypothesis focusing research toward an understanding of the mechanism for carboxylic acid drug-mediated toxicity is that these drugs are metabolized to chemically reactive metabolites that bind irreversibly to tissue proteins in the body leading to the formation of drug-protein adducts that are recognized by the immune system as foreign and thus eliciting an immune response and subsequent organ failure [3]. This is termed idiosyncratic hypersensitivity or allergic toxicity and has been observed for many carboxylic acid-containing drugs [2].

Transporters and Drug-Metabolizing Enzymes in Drug Toxicity, First Edition. Edited by Albert P. Li.

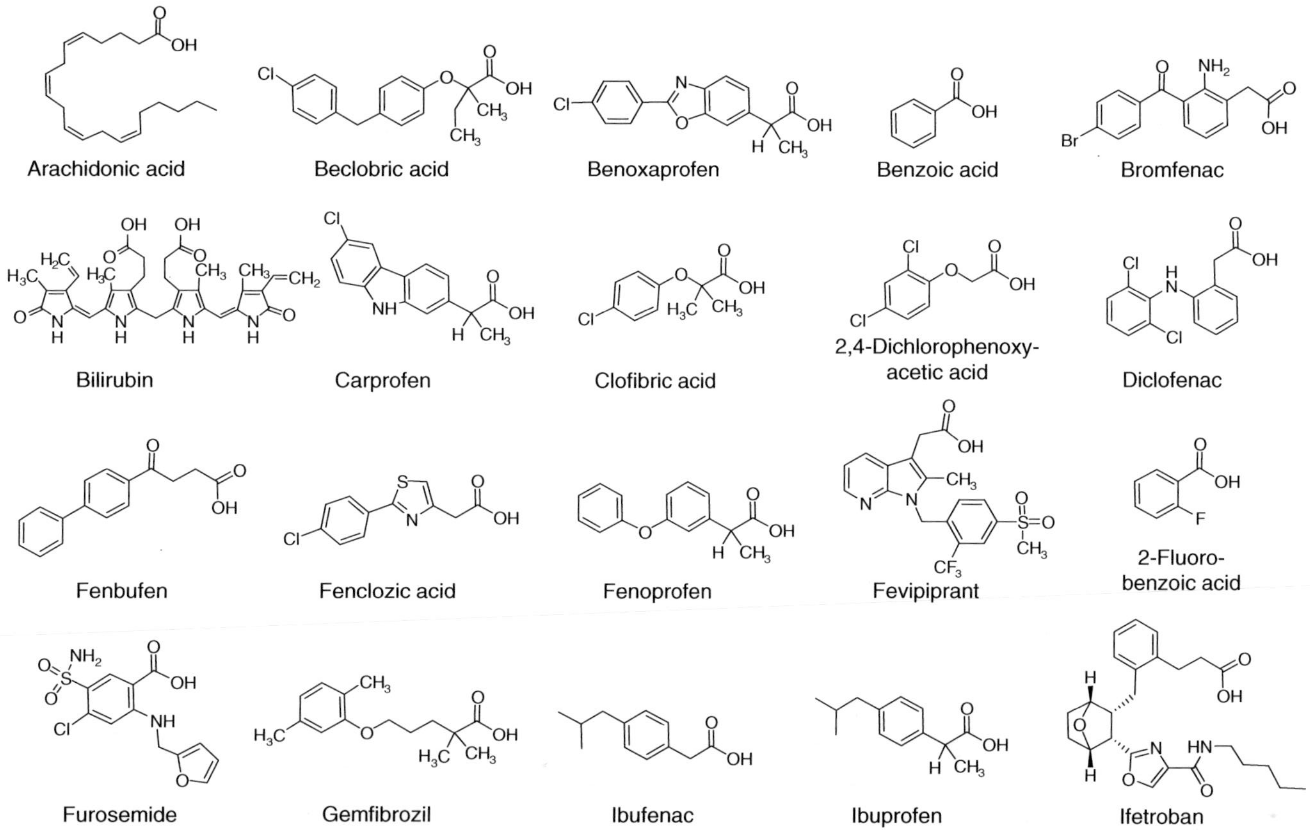

Figure 6.1 Chemical structures of carboxylic acid-containing drugs and compounds cited in the text.

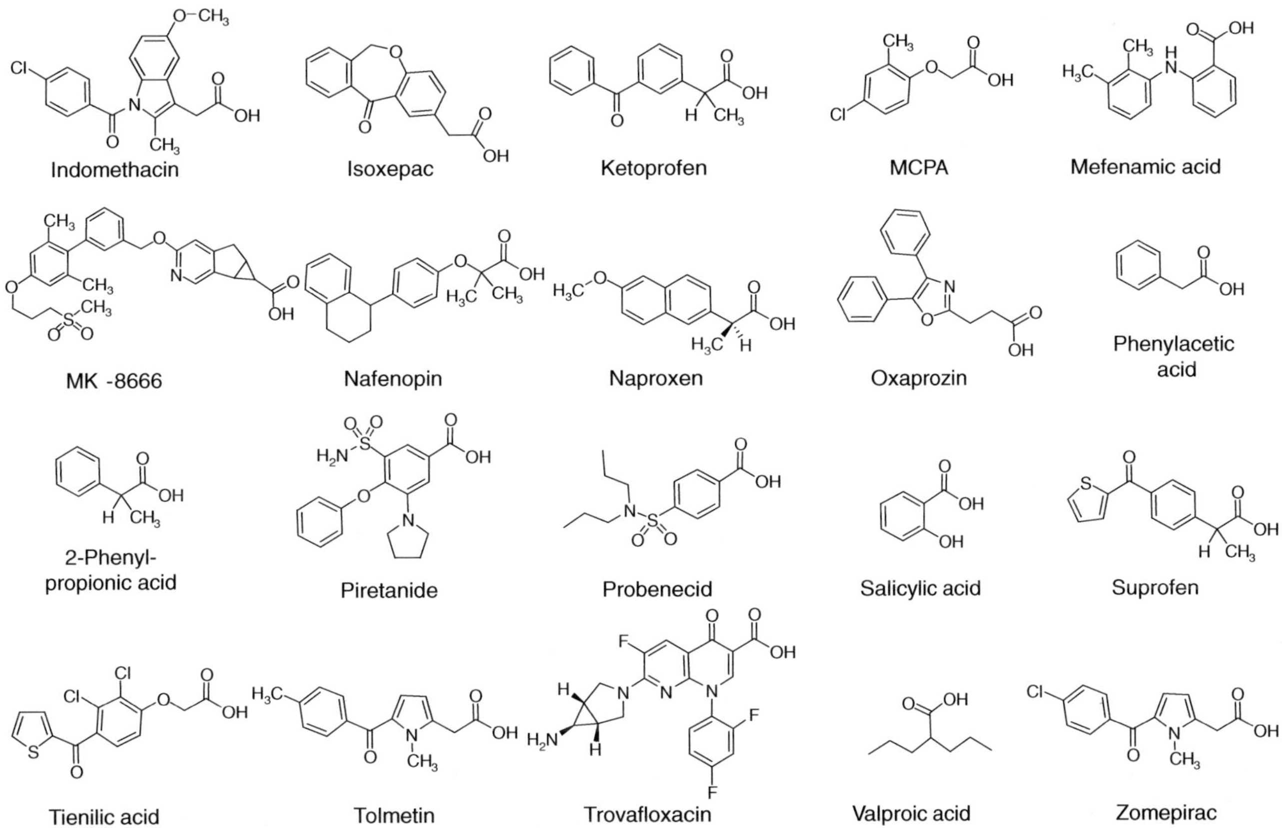

Figure 6.1 (Continued)

Carboxylic acid drugs can undergo both phase I- and phase II-mediated metabolism. Phase I metabolism includes enzyme-mediated transformation of drugs by processes such as oxidation by cytochrome P450, reduction by various reductases, and hydrolysis by amidase and esterase enzymes. Phase I-mediated transformation of drug molecules usually leads to drug metabolites with increased water solubility and often leading to more efficient excretion from the body. However, in some cases phase I-mediated metabolism leads to the formation of chemically reactive metabolites containing reactive electrophilic groups such as quinones, iminoquinones, epoxides, and α,β-unsaturated ketones that can bind to nucleophilic groups on proteins and potentially leading to organ toxicity [18]. Examples of withdrawn carboxylic acid-containing drugs known to form reactive metabolites mediated by phase-I-mediated oxidative metabolism include the diuretic tienilic acid [19], the NSAID suprofen [20], and the fluoroquinolone antibiotic trovafloxacin [12]. The chemical structures of both tienilic acid and suprofen contain the structural alert thiophene moiety [18] which can undergo P450-mediated *S*-oxidation or thiophene ring epoxidation leading to chemically reactive metabolites and the subsequent formation of protein adducts. Trovafloxacin contains a cyclopropylamine substructure that can be metabolized to chemically reactive ring-opened derivatives including a carbon-centered radical and an α,β-unsaturated aldehyde capable of forming covalent tissue protein adducts potentially resulting in liver damage. The currently used NSAID diclofenac has been shown to form P450-medieated benzoquinone imine-type reactive metabolites that are capable of binding covalently to protein [21]. For each of these carboxylic acid-containing drugs, it is possible that the reactive metabolites formed by phase-I-mediated bioactivation in human, separate from, or in combination with, phase II-mediated metabolic activation of the carboxylic acid moiety, contribute to the respective observed drug-mediated hepatic injury.

Phase II-mediated metabolism involves the conjugation of a functional group of a drug or a drug phase I metabolite with endogenous substances such as glucuronic acid, sulfate, acetate, glutathione, or an amino acid such as glycine. In most cases the phase II-mediated conjugation of drugs or drug metabolites leads to conjugates with increased water solubility, as well as providing for increased substrate affinity for efflux transporters such as multidrug resistance protein 2 (MRP2), leading potentially to more efficient elimination [22]. The most extensively studied mechanism for carboxylic acid drug bioactivation is by the phase II process of glucuronidation leading to acyl glucuronide-linked metabolites. A second important mechanism of phase II bioactivation is by acyl-CoA formation leading to acyl-CoA thioester-linked chemically reactive intermediates. Both of these phase-II bioactivation mechanisms are discussed below regarding the enzymatic formation, chemical stability, and chemical reactivity of acyl glucuronide metabolites compared to acyl-CoA derivatives toward an increased understanding of relative toxicological importance.

6.2 Phase II Metabolism

6.2.1 Glucuronidation

The biotransformation of drugs by phase II-mediated glucuronidation is an important route of metabolic clearance of drugs and drug metabolites from the body [23, 24]. The glucuronidation of drugs occurs primarily in the liver but can also occur in other tissues including kidney, heart, intestine, and brain [25, 26]. Glucuronidation leads to the formation of glucuronide metabolites and is catalyzed by two families of uridine 5′-diphospho-glucuronosyltransferase (UDP-glucuronosyltransferase, UGT, EC 2.4.1.17) enzymes UGT1 and UGT2, that are subdivided into three subfamilies UGT1A, UGT2A, and UGT2B. UGT enzymes are located intracellularly in the smooth endoplasmic reticulum and are membrane bound [23]. These enzymes utilize the cofactor UDP-α-D-glucuronic acid (UDPGA). Substrates for UGTs include compounds containing nucleophilic hydroxyl, amine, sulfhydryl, and carboxylic acid functional groups which lead to the generation of ether-, amino-, thioether-, and acyl-linked glucuronides, respectively. The UGT enzyme mechanism occurs where the substrate nucleophile reacts with the glucuronic acid moiety of the UDPGA cofactor at the C-1 position in a nucleophilic displacement reaction with loss of UDP and the formation of the β-D-glucuronide conjugate (Figure 6.2a). For example, shown is the glucuronidation of benzoic acid forming benzoic acid 1-β-D-*O*-acyl glucuronide.

6.2.2 Acyl-CoA Thioester Formation

The metabolism of carboxylic acid-containing drugs by phase II-mediated metabolism to amino acid conjugates such as glycine and taurine amides is facilitated by first undergoing acyl-CoA thioester formation. Acyl-CoA thioester formation is catalyzed by adenosine triphosphate (ATP)-dependent acyl-CoA synthetases (EC 6.2.1.1–EC2.1.3) that are present in highest activity in liver and adipose tissues and concentrated in different intracellular locations including cytosol, smooth endoplasmic reticulum, mitochondria, and peroxisomes [27]. Acyl-CoA synthetase-mediated formation of acyl-CoA thioesters requires two cofactors, ATP and coenzyme A (CoASH). The mechanism is of acyl-CoA formation occurs by a process whereby the carboxylic acid-containing substrate is first converted to a chemically reactive transacylating acyl-adenylate monophosphate (acyl-AMP) intermediate followed by reaction with CoASH leading to formation of the acyl-CoA thioester conjugate and elimination of the AMP moiety (Figure 6.2b; [28]). For example, shown is the acyl-CoA synthetase-mediated metabolism of benzoic acid forming benzoic acid acyl-CoA thioester. Acyl-CoA thioesters cannot leave the cell due to their amphipathic property and large size, but instead are substrates for phase II reactions including conjugation with glycine or taurine prior to elimination from the body [29].

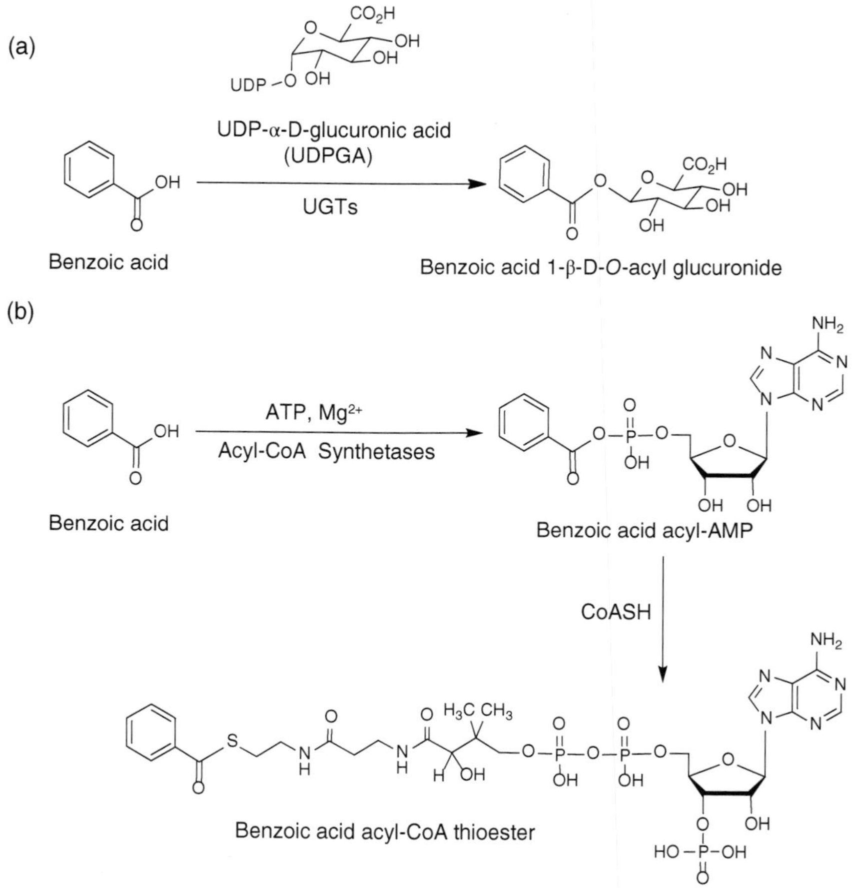

Figure 6.2 (a) Benzoic acid β-D-acyl glucuronide formation catalyzed by UGT enzymes, and (b) benzoic acid acyl-CoA thioester formation catalyzed by acyl-CoA synthetases.

6.3 Chemical Stability of Phase II Metabolites

6.3.1 Acyl Glucuronide Instability

Acyl glucuronide conjugates are known to be inherently chemically unstable in aqueous solution. Carboxylic acid drug 1-β-*O*-acyl glucuronides undergo both hydrolysis- and intramolecular acyl migration-mediated degradation *in vitro* and *in vivo*. Studies that have been performed with a range of biosynthetic 1-β-*O*-acyl glucuronides of carboxylic acid drugs incubated in phosphate buffered solution at pH 7.4 and at 37 °C have demonstrated a range of degradation rates with varying degradation half-lives from as low as 0.3 hours for tolmetin-1-β-*O*-acyl glucuronide and as high as 79 hours for valproic acid-1-β-*O*-acyl glucuronide [30–34]. The rate of degradation is determined by measuring the rate of loss of parent 1-β-*O*-acyl glucuronide *in vitro*. This can be performed by obtaining chromatographic separation of the 1-β-*O*-acyl glucuronide from the corresponding acyl migration isomers (discussed below) and with varied detection methods including ultraviolet (UV) detection [35] and liquid chromatography mass spectrometry (LC-MS; [36]) and by measuring the decreasing chromatographic peak area of the parent 1-β-*O*-acyl glucuronide over incubation time relative to the zero incubation timepoint. A valuable LC-MS tandem mass spectrometry method for measuring the 1-β-*O*-acyl glucuronide isomer from a complex mixture of corresponding isobaric acyl glucuronide migration isomers, without the need for extensive chromatographic separation, was developed by Xue et al. [37] where it was shown that by using a selected reaction monitoring (SRM) neutral loss of 176 Da transition, the sensitive and specific detection of the 1-β-*O*-acyl glucuronide isomer can be achieved.

The nonenzymatic degradation of carboxylic acid drug 1-β-*O*-acyl glucuronides in buffer occurs by two different processes which include both base-catalyzed chemical hydrolysis and intramolecular acyl migration (Figure 6.3; [38]). Hydrolysis occurs when water or hydroxide ion reacts with the carbonyl carbon of the 1-β-*O*-acyl glucuronide linkage leading to the loss of the glucuronic acid moiety and the formation of the corresponding carboxylic acid drug (aglycone). The degradation of carboxylic acid drug 1-β-*O*-acyl glucuronides also occurs by a process known as acyl migration whereby the aglycone migrates around the glucuronic acid ring to the 2-, 3-, and 4-position hydroxyl groups by reacting in an intramolecular fashion with the carbonyl-carbon of the adjacent acyl-linkage. The rearranged acyl glucuronide isomers, unlike the parent 1-β-*O*-acyl glucuronide, are resistant to enzymatic hydrolysis by treatment with β-glucuronidase [39]. Acyl migration-mediated degradation of 1-β-*O*-acyl glucuronides predominates and can be 10–20-fold more rapid than the rate of hydrolysis [40].

Acyl glucuronide stability in buffer is pH-dependent. For example, the intramolecular rearrangement of clofibric acid acyl glucuronide was studied over a pH range 5.2–8.6 and where the acyl glucuronide was observed to be completely stable at pH values up to 7, but increasingly unstable above pH 7 [41]. After three hours of incubation at pH 8.6 only 15% of parent clofibric acid 1-β-*O*-acyl glucuronide remained. Similar results have been observed for other carboxylic acid drugs tested for pH-dependent chemical instability such as isoxepac [42], zomepirac [35], furosemide [43], indomethacin [44], ifetroban [45], naproxen [46], and valproic acid [47]. Acyl glucuronide hydrolysis and acyl migration occur more readily at elevated temperatures [35] and where the general recommendation for the treatment of biological samples containing acyl glucuronide metabolites is to keep samples cool on ice, treated with phosphoric acid, and stored frozen at −20 °C or colder until bioanalysis [48].

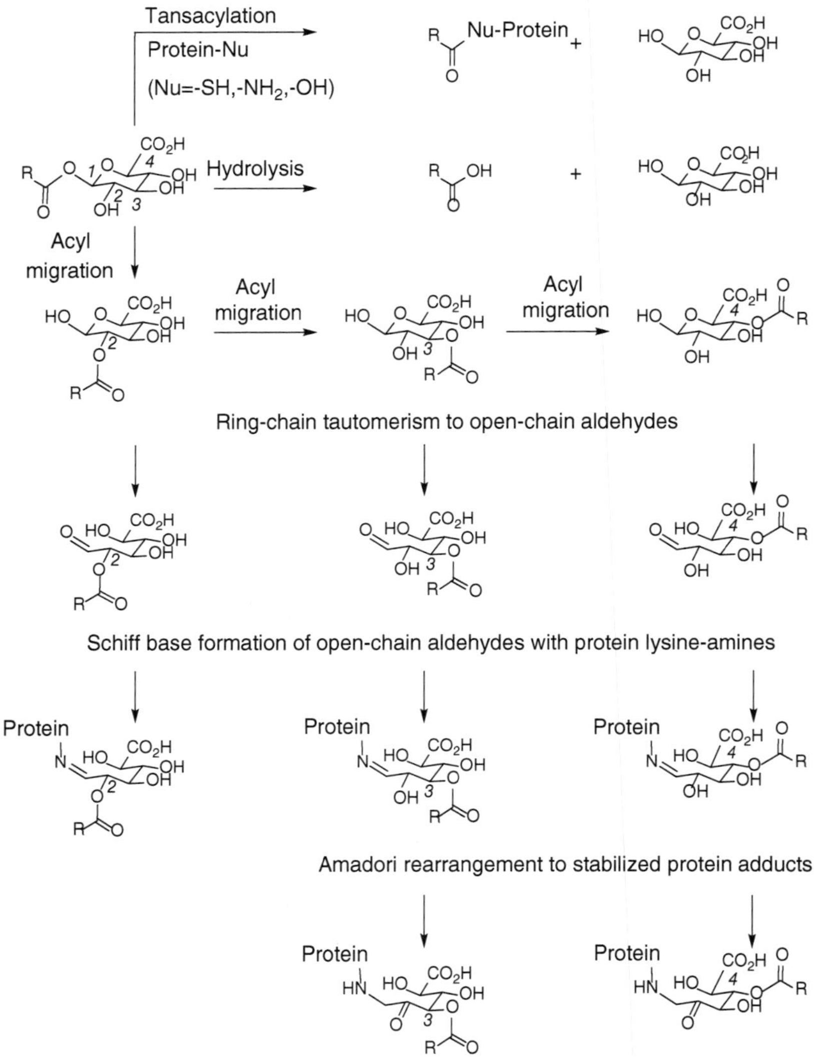

Figure 6.3 Instability of carboxylic acid-containing drug acyl glucuronide by hydrolysis and acyl migration. Reactivity with protein nucleophiles by transacylation and Schiff base formation. ("R" denotes carboxylic acid drug or compound.)

The rate of acyl migration importantly depends on the chemical structure of the carboxylic acid drug where both steric and electronic differences have been observed to lead to differences in degradation rates *in vitro* in phosphate buffer under physiological conditions [49]. In general, at sites near the carbonyl-carbon of the 1-β-*O*-acyl glucuronide linkage, sterically bulky moieties lead to decreased degradation rates and electron-withdrawing groups lead to increased degradation rates. An example of steric differences in carboxylic acid 1-β-*O*-acyl glucuronide structure leading to instability differences was observed when the *in vitro* degradation half-life values were compared between the arylacetic acid NSAID, ibufenac (0.8 hours), which contains no methyl group on the acetic acid carbon adjacent to the carboxylic acid, and the 2-phenylpropionic acid (2-PPA) NSAID, ibuprofen (2.7 hours), where there is one methyl group leading to steric hindrance at this position [34]. Interestingly, the degradation of 2-PPA 1-β-*O*-acyl glucuronides was reported to be enantioselective where the observed rate of degradation of the resolved (*R*)- and (*S*)-isomers was 28% (half-life 1.8 hours) and 15% (half-life 3.3 hours) per hour, respectively, and again where acyl migration predominated over acyl glucuronide hydrolysis [50].

An example of electronic effects on a carboxylic acid 1-β-*O*-acyl glucuronide are apparent when comparing the 1-β-*O*-acyl glucuronide stability of diclofenac (0.7 hours), which contains two electron withdrawing chlorine atoms leading to increased electrophilicity of the acyl-linkage, compared to the more stable 1-β-*O*-acyl glucuronide of mefenamic acid (17 hours), where there are two electron donating methyl groups decreasing the electrophilicity of the acyl-linkage [34]. As will be discussed below, these same steric and electronic structure activity relationships are also important for the reactivity of acyl glucuronides with protein leading to covalent adducts [31].

6.3.2 Acyl-CoA Thioester Stability

Acyl-CoA thioesters derivatives of carboxylic acid drugs have also been characterized for chemical stability toward hydrolysis. Early studies assessing the chemical stability of acyl-CoA derivatives in buffer were done with salicylic acid acyl-CoA thioester [51]. In these studies, salicylic acid acyl-CoA was incubated in buffer at 30 °C and at pH 7.5 and 9.0 and demonstrated pH-dependent hydrolysis at 0.5 and 1.6% per minute, respectively. In these same studies, hydrolysis rates of benzoic acid and 2-fluoro-benzoic acid acyl-CoA thioester analogs were also examined. Results showed less than 0.03% hydrolysis per minute for both of these acyl-CoA thioesters. These data indicated the reduced stability of salicylic acid acyl-CoA to be due presumably to the presence of the ortho-hydroxyl group providing hydrogen bonding between the hydroxyl-proton and the carbonyl-oxygen atom leading to increased electrophilicity of the carbonyl-carbon of the acyl-linkage. In addition, the results indicated the lack of an electron withdrawing effect of the ortho-fluorine atom on the stability of the 2-fluoro-benzoic acid acyl-CoA thioester derivative.

Analysis of the chemical stability of the acyl-CoA thioester of the hypolipidemic drug clofibric acid in buffer at 37 °C and pH 7.5 showed the acyl-CoA derivative to be chemically stable with a long incubation half-life of 21 days (0.002% hydrolyzed per minute; [52]). Studies of the stability of this thioester in buffer performed at 37 °C and varying pH demonstrated it to be stable in acidic and basic aqueous solutions from pH 2 to pH 11 where under these conditions, and after a one hour incubation, the incubation mixtures showed minimal hydrolysis (<3%). Approximately 50% hydrolysis occurred after one hour of incubation at pH 12 and complete hydrolysis after one hour of incubation at pH 12.5. Presumably the relatively high chemical stability toward hydrolysis is due to the presence of the two sterically hindering methyl groups located next to the carbonyl-carbon of the thioester-linkage of clofibric acid acyl-CoA.

Studies directly comparing the chemical instability of 1-β-*O*-acyl glucuronides and the corresponding acyl-CoA thioesters in buffer were performed for 2-PPA [53] and mefenamic acid [54]. Experiments conducted with (*R*)- and (*S*)-2-PPA showed the chemical stability observed in phosphate buffer pH 7.4 and 37 °C for the acyl glucuronide and acyl-CoA thioester derivatives to be very different. The acyl-CoA thioesters were very stable with slow, nonenantioselective hydrolysis over 24 hours of incubation (0.18% hydrolyzed per hour). By contrast, the 1-β-*O*-acyl glucuronide conjugates of the (*R*)- and (*S*)-2-PPA showed rapid enantioselective degradation at 38% (half-life 1.3 hours) and 21% (half-life 2.4 hours) per hour, respectively. In studies conducted with the NSAID mefenamic acid for comparing the chemical stability of the acyl-CoA thioester with that of the 1-β-*O*-acyl glucuronide in buffer, results showed the acyl-CoA thioester to be completely stable after 15 hours of incubation in phosphate buffer at pH 7.4 and 37 °C. By contrast, the 1-β-*O*-acyl glucuronide was unstable and degraded at a rate of 3% per hour (half-life 16.5 hours).

An example of a relatively unstable acyl-CoA thioester was shown in studies with the carboxylic acid-containing herbicide, 2,4-dichlorophenoxyacetic acid (2,4-D; [55]). Stability studies conducted in phosphate buffer at pH 7.4 and 37 °C showed 2,4-D-acyl-CoA thioester to have a degradation rate of 9% per hour (half-life 5.6 hours). The increased instability of 2,4-D-acyl-CoA thioester relative to the observed stability of phenylacetic acid (PAA) acyl-CoA thioester, where no hydrolysis to PAA was observed over a corresponding one hour incubation [56], is proposed to be due to both the electronic effects of the β-position oxygen atom and the ortho- and para-chlorine atoms of 2,4-D making the carbonyl-carbon of the thioester linkage more electrophilic, and hence more chemically unstable toward hydrolysis.

6.4 Phase II Metabolite Chemical Reactivity

6.4.1 Acyl Glucuronide Reactivity with Nucleophiles *In vitro*

Acyl glucuronides metabolites of carboxylic acid-containing drugs are chemically reactive with nucleophiles. Carboxylic acid 1-β-*O*-acyl glucuronides can form covalent adducts with proteins by two known routes (Figure 6.3). The first mechanism occurs where the electrophilic carbonyl-carbon of the 1-β-*O*-acyl linkage reacts with protein nucleophiles (including lysinyl-NH_2, tyrosine-OH and serine-OH, and cysteinyl-SH) by nucleophilic displacement of the glucuronic acid moiety to form corresponding amide-, ester-, and thioester-linked protein adducts, respectively. The second mechanism occurs whereby the 1-β-*O*-acyl glucuronide first undergoes intramolecular acyl migration of the aglycone to the adjacent hydroxyl groups around the glucuronic acid ring forming 2-, 3-, and 4-*O*-acyl migration isomers [35]. Different from the initial 1-β-*O*-acyl-linked glucuronide, acyl migration isomers are resistant to hydrolysis by the enzyme β-glucuronidase [42]. The migrated acyl glucuronide isomers in the intact glucuronic acid ring form are in equilibrium with their corresponding open-chain isomer forms which have a C-1 position aldehyde-moiety that reacts with amino groups of protein-lysine residues forming Schiff base-linked protein adducts. The 3-and 4-*O*-acyl glucuronide Schiff base-linked adducts can be stabilized through an Amadori rearrangement reaction forming 1-amino-2-keto-linked acyl glucuronide protein adducts [57, 58].

One of the first reports on the reactivity of carboxylic acid drug 1-β-*O*-acyl glucuronide was on the reactivity of clofibric acid 1-β-*O*-acyl glucuronide *in vitro* in buffer with a model nucleophile ethanethiol [59]. Biosynthetic clofibric acid 1-β-*O*-acyl glucuronide was reacted with ethanethiol in phosphate buffer at pH 7.4 and 37 °C for one hour where approximately 10% of the acyl glucuronide was converted to the clofibric acid ethanethiol thioester. In this same report, the detection of the *N*-acetylcysteine (NAC) (mercapturate) thioester of clofibric acid in urine from clofibric acid-dosed patients was observed and demonstrated the first drug acyl-linked mercapturate found in man. The mercapturate *in vivo* was initially conjugated with glutathione but was excreted in urine as the mercapturate after degradation of the acyl-glutathione adduct by the mercapturic acid pathway [60]. Whether the formation of the clofibric acid acyl-glutathione thioester was formed from the reaction of glutathione with clofibric acid 1-β-*O*-acyl glucuronide or by another mechanism, such as reaction with clofibric acid acyl-CoA in human, was not known. Based on these data alone, Stogniew and Fenselau [59] proposed that the electrophilic acyl glucuronide may play an important role in the cytotoxicity of clofibric acid, and where it was the first example of metabolism by glucuronidation leading to an activated electrophile.

Further studies assessing the chemical reactivity of clofibric acid 1-β-*O*-acyl glucuronide was performed in testing the acylation of the standard chemical nucleophile 4-(*p*-nitrobenzyl)-pyridine (NBP; [44]) with direct comparison to the chemical reactivity of the powerful alkylating agent chlorambucil. Results showed chlorambucil to react 127 times faster with NBP than clofibric acid 1-β-*O*-acyl glucuronide, thus indicating that acyl glucuronides are considerably weaker electrophiles than toxic nitrogen mustards.

Later studies with clofibric acid showed that glutathione (5 mM) reacts nonenzymatically in phosphate buffer (pH 7.5, 37 °C) in a nucleophilic displacement reaction with the clofibric acid 1-β-*O*-acyl glucuronide (1 mM), but not with the acyl migration acyl glucuronide isomers, to form clofibric acid acyl-glutathione thioester adduct at a rate of 0.103 nmol min^{-1} [61]. After five hours of incubation, the glutathione-adduct formed amounted to only ~0.6% of the 1-β-*O*-acyl glucuronide at the start of the incubation. Glutathione *S*-transferases from rat liver was found to catalyze the formation of the thioester conjugate eightfold and demonstrated acyl glucuronides as a new class of substrates for the enzymes. In these same studies, it was shown that the thioester was detected in bile collected from clofibric acid-dosed rats. Subsequent studies in rat indicated that clofibric acid acyl-glutathione thioester undergoes degradation by the mercapturic acid degradation pathway *in vivo* leading primarily to the clofibric acid *N*-acyl-cysteine adducts rather than the clofibric acid NAC conjugate excreted in urine [62]. From these results, it was proposed that drug-*N*-acyl-cysteine derivatives be used as markers of acyl-glutathione conjugate formation occurring *in vivo* when examining urine extracts for evidence of carboxylic acid drug bioactivation by acyl glucuronidation or acyl-CoA formation.

An important early report on the chemical reaction of an acyl glucuronide with protein comes from a study concluding that the formation of biliprotein (bilirubin acylated albumin) *in vivo* occurs from the nonenzymatic transacylation of a nucleophilic site on albumin by bilirubin 1-β-*O*-acyl glucuronide [63].

Studies with the withdrawn hepatotoxic NSAID, zomepirac were performed to determine the mechanism of covalent binding (CVB) of zomepirac 1-β-*O*-acyl glucuronide to protein *in vitro* [57]. Prior to these studies, the mechanism of CVB to protein was uncertain, where it was proposed initially to involve only the nucleophilic displacement of the glucuronic acid moiety by cysteine thiols, tyrosine hydroxyl, or lysine amine residues of proteins. Results from studies with zomepirac indicated that transacylation of cysteine thiols was actually not occurring, but more importantly, that zomepirac 1-β-*O*-acyl glucuronide reacted after intramolecular rearrangement with human serum albumin (HSA) leading to covalent adducts, and where the binding increased 8.3- and 5.5-fold in incubations containing the imine trapping reagent, sodium cyanide, or the imine reducing reagent, sodium cyanoborohydride. These results provided strong support for the mechanism of CVB of acyl glucuronides to protein where acyl migration of the glucuronide followed by ring-chain tautomerism to the open-chain aldehyde with subsequent Schiff base-adduct formation with protein-lysine groups.

Clofibric acid 1-β-*O*-acyl glucuronide has also been shown to react covalently with HSA *in vitro* in buffer [64]. These studies compared the CVB to HSA in incubations with radiolabeled clofibric acid 1-β-*O*-acyl glucuronide using two different [^{14}C]-labeled derivatives with one labeled on clofibric acid and the other on the glucuronic acid moiety. Results showed greater CVB to HSA in reactions with clofibric acid 1-β-*O*-acyl glucuronide where the [^{14}C]-label was located on the glucuronic acid moiety. This result provided strong evidence for CVB to protein through acyl migration and Schiff base formation followed by subsequent Amadori rearrangement.

Studies with the 1-β-*O*-acyl glucuronide of the NSAID tolmetin in incubations with HSA *in vitro* used mass spectrometry to show the formation and location of the CVB sites toward elucidating the mechanism of covalent adduct formation [58]. Results from these studies were the first to show unambiguously that the glucuronic acid moiety of acyl glucuronides is reactive with protein lysine groups by undergoing acyl migration followed by Schiff base-adduct formation as a mechanism for covalent adduct formation *in vivo*. In subsequent studies the chemical structures of adducts formed from the *in vitro* incubation of tolmetin 1-β-*O*-acyl glucuronide with HSA were determined by mass spectrometry [65]. Data showed CVB via the glucuronic acid moiety to lysine 199, 195 and 525, and by transacylation of lysines 199 and 541, serines 229, 232, and 480, and arginine 222. These results established that CVB of acyl glucuronides to nucleophilic sites on proteins occurs by the two different mechanisms involving Schiff base formation and transacylation involving nucleophilic displacement of the glucuronic acid moiety.

Similar mechanistic mass spectrometric studies with benoxaprofen 1-β-*O*-acyl glucuronide, the acyl glucuronide conjugate of the hepatotoxic NSAID benoxaprofen, examining tryptic peptides from extracts of *in vitro* reactions with HSA established that the acyl glucuronide formed covalent adducts with protein nucleophiles on HSA by both nucleophilic displacement and Schiff base reactions with Lys-159 and Lys-199 ε-amino-lysine residues, where Lys-159 was the major amino acid involved in protein adduct formation [66]. Other CVB sites of transacylation by benoxaprofen-1-β-*O*-acyl glucuronide included Ser-312, Ser-480, and Arg-222.

More recently, studies with the NSAID diclofenac were performed to study the covalent adducts formed from the reaction of diclofenac 1-β-*O*-acyl glucuronide with HSA by mass spectrometric assessment [67]. Diclofenac 1-β-*O*-acyl glucuronide was incubated with HSA in phosphate buffer at pH 7.4 and 37 °C and covalent modifications were investigated by mass spectrometric analysis of modified tryptic peptides. Results showed diclofenac 1-β-*O*-acyl glucuronide incubated with HSA under conditions favoring intramolecular acyl migration produced predominantly the same types of *N*-acylation and acyl migration isomer-mediated glycation adducts obtained from corresponding studies with tolmetin 1-β-*O*-acyl glucuronide [58, 65] and benoxaprofen 1-β-*O*-acyl glucuronide [66]. In these same studies, targeted mass spectrometric analyzes of HSA isolated from the plasma of diclofenac-treated patients determined the pathways of adduct formation from diclofenac acyl glucuronide and its isomeric rearrangement products *in vivo*. HSA from six patients without drug-related hypersensitivities had either a single drug-derived adduct or one of five combinations of two to eight adducts from among seven diclofenac *N*-acylations and three rearranged acyl glucuronide glycations on seven of the HSA 59 lysines. However, the authors of the study proposed that CVB acyl glucuronide to HSA is not the inevitable cause of hypersensitivity reactions to carboxylic acid-containing drugs nor is at a coincidental association.

In a more recent study, the reactivity of the 1-β-*O*-acyl glucuronide metabolite of the prostaglandin D2 receptor 2 antagonist fevipiprant with HSA *in vitro* in buffer was examined [68]. Fevipiprant is a carboxylic acid-containing drug that was in development for the treatment of asthma and atopic dermatitis. Metabolism of fevipiprant was observed to be predominant by acyl glucuronidation to fevipiprant1-β-*O*-acyl glucuronide detected in plasma and represented 28% of excreted drug-related material. During these studies, a long retention of radioactivity in plasma in addition to a low [^{14}C]-drug-related material extractability from plasma was observed. The authors proposed that the acyl glucuronide metabolite, the major metabolite detected in plasma, could have undergone CVB to plasma proteins leading to the observed long-lived radioactivity in plasma in the form of [^{14}C]-fevipiprant1-β-*O*-acyl glucuronide-mediated protein adducts. The *in vitro* reactivity of radiolabeled [^{14}C]-fevipiprant-1-β-*O*-acyl glucuronide with human plasma (1 : 1 in phosphate buffered saline at 37 °C for 24 hours) showed CVB to protein where only one protein (~68 kDa) identified as HSA was conjugated to the acyl glucuronide metabolite. Fevipiprant contains a sterically unhindered arylacetic acid moiety which, like zomepirac and diclofenac as described above, should be able to form a chemically reactive acyl glucuronide that forms covalent adducts with protein nucleophiles. The observed long terminal half-life of total radioactivity in plasma was 10.6 days, which was in the range of the reported 19 day half-life of HSA *in vivo*. Even with the ability of the drug glucuronide to form covalent adducts with HSA *in vitro* and *in vivo*, no adverse events of idiopathic drug reactions of hepatotoxicity were observed in development and in phase III clinical studies. CVB studies *in vitro* with human liver microsomes containing cofactors for P450 and UGT enzymes, and with human hepatocytes showed only negligible CVB to protein. These results suggested that acyl glucuronidation-mediated CVB of fevipiprant to human liver microsomes was insignificant.

6.4.2 Acyl-CoA Thioester Reactivity with Nucleophiles *In vitro*

Acyl-CoA thioesters possess an electrophilic carbonyl-carbon of the thioester-linkage and are able to transacylate biological nucleophiles nonenzymatically (Figure 6.4; [69, 70]).

To my knowledge, the first published report on the characterization of the chemical reactivity of a carboxylic acid-containing drug acyl-CoA thioester came

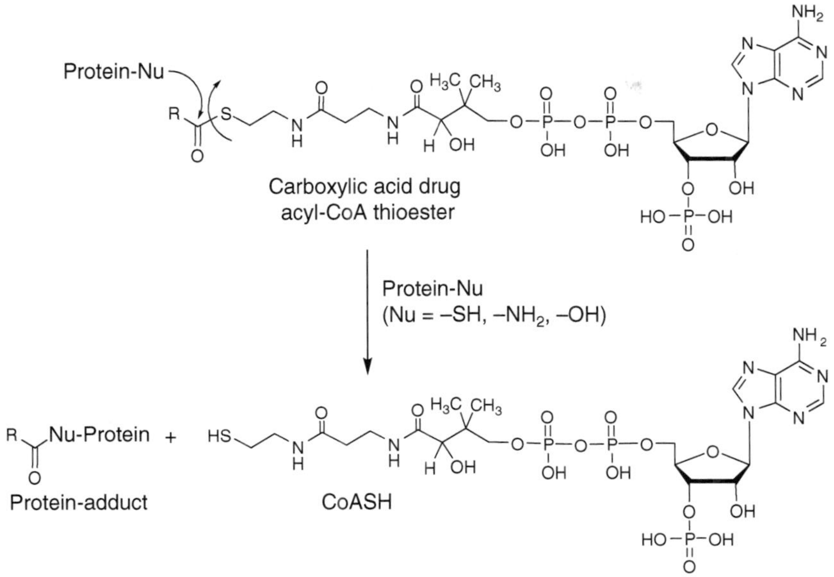

Figure 6.4 Reactivity of carboxylic acid-containing drug or compound acyl-CoA thioester with protein nucleophiles by transacylation. ("R" denotes carboxylic acid drug or compound.)

from *in vitro* studies on the nonenzymatic chemical reactivity of salicyl-CoA thioester with glycine forming the glycine amide conjugate, salicylurate [51]. The transacylation reaction of salicyl-CoA with glycine was directly compared in corresponding incubations with the acyl-CoA thioester of benzoic acid where no glycine conjugation was detected. As for the observed chemical instability described above, these results indicated that the effect of the ortho-hydroxyl group of salicylic acid was functioning to increase the chemical reactivity of the acyl-CoA thioester carbonyl-carbon through electron withdrawing and/or intramolecular hydrogen bonding interactions. The authors speculated whether such a transacylation mechanism might alter biologically important macromolecules and thus explain some of the effect of salicylic acid *in vivo*.

Other acyl-CoA thioesters have also been shown to be chemically reactive intermediates that react in a transacylation fashion with proteins. For example, the endogenous long-chain acyl-CoA thioesters arachidonoyl-acyl-CoA [71] and palmitoyl-acyl-CoA [72] were shown to react nonenzymatically with cysteine-containing proteins forming covalent thioester linkages in a time- and concentration-dependent fashion. The transacylation of UDP-glucuronosyltransferase isoforms in rat liver by arachidonoyl-acyl-CoA was shown to inhibit enzyme activity. The spontaneous autoacylation of specific cysteine residues of G protein coupled receptor alpha subunits by palmitoyl-acyl-CoA is understood as an important step in the regulation of G protein coupled receptors [73]. By a similar mechanism, protein acylation has been proposed to offer a common mode of action for some xenobiotic carboxylic acids including hypolipidemic effectors and peroxisomal proliferators, where the xenobiotic-acylated proteins might be involved in triggering the observed biological effects [74]. These results are consistent with the speculation discussed above by Tishler and Goldman [51], in that acylation of proteins by carboxylic acid drugs such as salicylic acid *in vivo* by the acyl-CoA thioester intermediate might functionally alter biologically important macromolecules.

Studies have been increasingly conducted toward providing additional information regarding the relative chemical reactivity of xenobiotic acyl-CoA thioesters with biological nucleophiles compared to acyl glucuronides. An investigation was performed on the reactivity of clofibric acid acyl-CoA thioester with glutathione *in vitro* in buffer [52] to provide a second mechanism for the formation of clofibryl-acyl-glutathione as discussed above. Incubations with glutathione in buffer were conducted as described above for clofibric acid 1-β-*O*-acyl glucuronide [61] and results showed that the rate of clofibryl-acyl-glutathione thioester adduct formation from reaction of clofibric acid acyl-CoA with glutathione was measured to be 4.2 $nmol\,min^{-1}$. This rate of reaction of glutathione with the acyl-CoA thioester was 40-fold greater than the reaction rate for the 1-β-*O*-acyl glucuronide. After five hours of incubation, the glutathione-adduct formed amounted to ~25% of the acyl-CoA thioester at the start of the incubation, which was >25-fold higher than the corresponding reaction with clofibric acid acyl glucuronide [61]. Glutathione *S*-transferase obtained from rat liver was found to catalyze the formation of the glutathione thioester conjugate by threefold and demonstrated acyl-CoA thioesters as another class of substrates for the enzyme. Results from these studies were the first to show that an acyl-CoA thioester is more reactive than the corresponding acyl glucuronide in the transacylation of glutathione *in vitro*, and therefore potentially in the transacylation of proteins *in vivo*.

Further studies performed to directly compare the chemical reactivity of acyl glucuronides with acyl-CoA thioesters were conducted with the model carboxylic acid, 2-PPA [53]. In these studies, synthetic 2-PPA-acyl-CoA and biosynthesized 2-PPA-1-β-*O*-acyl glucuronide were incubated separately in buffer at pH 7.4 and 37 °C with glutathione and the formation rates of the 2-PPA-glutathione thioester adduct compared. As in the case of the clofibric acid derivatives discussed above, the acyl-CoA thioester was shown to be 70-fold more reactive with glutathione than the 1-β-*O*-acyl glucuronide leading to acyl-glutathione thioester formation. In addition, in these same studies the reactivity of radiolabeled [^{14}C]-2-PPA-acyl-CoA thioester with bovine serum albumin indicated time- and pH-dependent CVB to protein. Later studies showed similar time-dependent CVB *in vitro* in buffer to HSA by [^{14}C]-2-PPA-acyl-CoA thioester [75]. Additional *in vitro* studies have been conducted with the acyl-CoA thioester of the herbicide 2,4-D, 2,4-D-acyl-CoA, in reactions with both glutathione and HSA [55]. Reactions of 2,4-D-acyl-CoA with glutathione (1 mM) in buffer at pH 7.4 and 37 °C showed the rapid and quantitative formation of 2,4-D-acyl-glutathione thioester after one hour of incubation. Incubation of radiolabeled [^{14}C]-2,4-D-acyl-CoA (100 μM) with HSA (30 mg ml^{-1}) also showed a rapid rate of CVB to protein leading to 440 pmol mg^{-1} protein after one hour of incubation. Together, results from *in vitro* reactivity studies such as these have provided insight into the potential importance of carboxylic acid drug acyl-CoA thioesters in the transacylation-mediated CVB to protein *in vivo*. These results have indicated the ability of acyl-CoA thioesters to transacylate biological nucleophiles *in vitro*, that the transacylation reaction rate is greater compared to the corresponding acyl glucuronides, and therefore that the transacylation of biological nucleophiles *in vivo* might be predicted to be greater mediated by acyl-CoA thioesters compared to the corresponding acyl glucuronides *in vivo*. Depending on the relative concentrations in tissues, the contribution of acyl glucuronide and acyl-CoA thioester metabolites in the CVB to tissue proteins would be predicted to vary.

6.5 Phase II Metabolite-Mediated Covalent Binding

6.5.1 Acyl Glucuronide-Mediated Covalent Binding to Protein

In this section, examples of carboxylic acid drugs that have been shown to undergo acyl glucuronidation-mediated CVB to protein *in vitro* in liver tissue preparations is presented.

In studies with the hepatotoxic NSAID zomepirac, CVB to protein in rat liver preparations was investigated [76]. Zomepirac acyl glucuronide is a major metabolite of zomepirac in rat [77], and the studies characterized the CVB to protein from incubations of zomepirac with rat hepatocytes in suspension and in culture. CVB was determined by sodium dodecyl sulphate–polyacrylamide gel electrophoresis (SDS-PAGE) and Western blotting with polyclonal zomepirac antiserum. Results implicating zomepirac acyl glucuronide mediating CVB to protein, at least in part, were observed when coincubation with (−)-borneol, an inhibitor of glucuronidation, showed decreased intensity for almost every zomepirac protein-adduct band detected.

Studies with the withdrawn NSAID benoxaprofen, where the mechanism of hepatotoxicity was proposed to be due to CVB to protein mediated by the acyl glucuronide metabolite, have been conducted in sandwich-cultured rat and human hepatocytes [78]. In these studies, benoxaprofen was incubated in the presence and absence of (−)-borneol and evaluated for effects on both acyl glucuronide formation and CVB to protein. Results showed that (−)-borneol, which is known to deplete the glucuronidation cofactor UDPGA [79], inhibited glucuronidation by >95% and led to a 46% decrease in CVB to protein. These results indicated that formation of the reactive acyl glucuronide was only partially responsible for the CVB of benoxaprofen to proteins in the hepatocyte culture incubations and that additional bioactivation mechanisms were occurring. Other studies have indicated that benoxaprofen is not bioactivated by P450-mediated metabolism leading to CVB to protein *in vitro* [80]. These results are consistent with results showing that benoxaprofen-mediated cytotoxicity in incubations with rat hepatocytes was not decreased in the presence of inhibitors of cytochrome P450 [81]. The potential role of acyl-CoA formation leading to CVB to protein has not yet been investigated for benoxaprofen.

In vitro studies performed in short term-cultured rat hepatocytes with the NSAID diclofenac indicated that CVB to hepatocellular proteins was associated with bioactivation by diclofenac acyl glucuronide formation [82]. In these studies, the role of diclofenac acyl glucuronidation on CVB to protein and cytotoxicity was explored with the use of inhibitors of acyl glucuronidation. Results showed 73% inhibition of diclofenac glucuronide formation and a corresponding ~60% decrease in CVB to protein in incubation with (−)-borneol. However, it was shown that cytotoxicity, as measured by lactate dehydrogenase release, was markedly increased. These results provided evidence that diclofenac acyl glucuronide contributed to CVB to hepatocyte protein but was not related to the observed acute cytotoxicity. This result is consistent with increase of cytotoxicity for other carboxylic acid drugs tested in rat hepatocytes in the presence of (−)-borneol [83]. In these same studies [82], inhibition of P450-mediated CYP2C metabolism of diclofenac reduced the extent of cytotoxicity but had no effect on CVB to protein. In related radiochemical and immunochemical studies with diclofenac incubated with cultured rat hepatocytes, selective protein adducts to membrane proteins were detected. In agreement with previous diclofenac CVB results presented above, the detection of protein adducts was dependent on UGT-mediated acyl glucuronidation and not on P450-mediated bioactivation [84]. Subsequent studies aimed at investigating the mechanism of CVB of diclofenac acyl glucuronide were characterized *in vitro* in incubations with rat liver microsomes [85]. Results from these studies indicated that, in incubations of [^{14}C]-diclofenac with rat liver microsomes, the drug became covalent bound to microsomal proteins when coincubated with UDPGA and CVB increased in the presence of the imine-trapping reagent sodium cyanide, a reagent known to stabilize imine bonds such as in Schiff base linkages as demonstrated by Smith et al. [57]. In corresponding incubations with [^{14}C]UDPGA and nonradiolabeled diclofenac, results showed a similar extent of CVB to microsomal proteins indicating that the glucuronic acid moiety was retained in the detected protein adducts. These results provided strong evidence for diclofenac acyl glucuronide-mediated CVB to protein occurring through Schiff base formation from the open-chain acyl migration isomers, in addition to CVB by nucleophilic displacement.

Similar *in vitro* mechanistic CVB studies using immunochemical detection were performed by Hargus et al. [86] to determine if bioactivation of diclofenac leading to CVB to hepatic proteins *in vitro* was mediated by P450, UGT, and/or acyl-CoA synthetase enzymes. Studies were conducted with diclofenac incubated with rat liver homogenate or liver microsomes in the presence of cofactors of these enzymes systems. One microsomal protein adduct (50 kDa) was detected from P450-mediated bioactivation, whereas multiple plasma protein adducts (110, 140, and 200 kDa) were detected from liver homogenate incubations including the cofactor for diclofenac acyl glucuronidation, UDPGA. No covalent adducts were detected in incubations conducted with cofactors for diclofenac-acyl-CoA formation; however, these incubations also contained excess dithiothreitol, a sulfhydryl-containing reducing reagent known to react readily with acyl-CoA thioester conjugates [52] and therefore potentially capable of competing with CVB to protein. The mechanism of adduct formation, nucleophilic displacement, and/or Schiff base formation, was not distinguishable from these studies.

The results from the studies presented above clearly indicated that acyl glucuronide metabolites of carboxylic acid-containing drugs can bind covalently to hepatic proteins and that the pathways of adduct formation involve both nucleophilic displacement transacylation and Schiff base formation mechanisms that may be important in mediating the observed hepatotoxicity.

6.5.2 Acyl-CoA Thioester-Mediated Covalent Binding to Protein

In this section, examples of carboxylic acid drugs that have been shown to undergo acyl-CoA formation-mediated CVB to protein *in vitro* in liver tissue preparations is presented. A number of carboxylic acid-containing compounds including hypolipidemic drugs, arylacetic acid- and 2-PPA-type NSAIDs, and herbicides are known to be substrates for acyl-CoA synthetase enzymes forming chemically reactive acyl-CoA thioester intermediates.

The hypolipidemic carboxylic acid agent nafenopin is a substrate for human hepatic acyl-CoA synthetase and was chosen as a model compound to test for carboxylic acid drug-mediated acylation of human liver proteins *in vitro* [87]. Radiolabeled [^{3}H]-nafenopin was incubated with human liver homogenate and the cofactors for acyl-CoA formation, ATP and CoASH, for 120 min at 37 °C with both [^{3}H]-nafenopin-acyl-CoA formation and CVB to protein measured. Results showed that protein adduct formation was directly proportional to [^{3}H]-nafenopin-acyl-CoA formation and that the protein adducts formed consisted of both thioester- (76%) and amide-linked (24%) reaction products. This was the first study to demonstrate a direct relationship between xenobiotic-CoA formation and the acylation of human liver proteins. Importantly, the authors proposed that the acylation of tissue proteins by xenobiotic carboxylic acids may be important regarding biological consequences including perturbation of endogenous regulation of protein localization and function through normal endogenous fatty acylation.

Studies with the model 2-arylpropionic acid-type NSAID substructure 2-PPA have investigated the potential involvement of the acyl-CoA as a reactive intermediary metabolite leading to protein adduct formation *in vitro* in rat liver homogenate [75]. In these experiments, radiolabeled [^{14}C]-2-PPA was incubated with rat liver homogenate in the presence of ATP and CoASH and the incubation mixture analyzed for CVB to protein and [^{14}C]-2-PPA-acyl-CoA formation. Results showed that [^{14}C]-2-PPA underwent CVB to hepatic proteins in a time- and cofactor-dependent fashion. Inhibition of [^{14}C]-2-PPA-CoA formation by acyl-CoA synthetase inhibitors palmitic acid, lauric acid, octanoic acid, and ibuprofen, markedly decreased CVB of [^{14}C]-2-PPA to liver proteins. Results from these *in vitro* studies were consistent with previous studies with nafenopin [87] and strongly suggested that acyl-CoA thioester derivatives are chemically reactive and can bind covalently to tissue proteins *in vitro*, and therefore may contribute to covalent adduct formation of carboxylic acid-containing drugs forming acyl-CoA thioester intermediates *in vivo*.

In related studies with the chlorophenoxy herbicide 2,4-D, incubations of radiolabeled [^{14}C]-2,4-D with rat hepatocytes showed a time-dependent CVB of [^{14}C]-2,4-D to hepatocyte proteins [55]. Inhibition of [^{14}C]-2,4-D-acyl-CoA formation with the acyl-CoA synthetase inhibitor trimethylacetic acid decreased the amount of CVB to protein in rat hepatocytes by 50%. These results indicated that 2,4-D-CoA thioester is a chemically reactive intermediate of 2,4-D that contributes to 2,4-D-protein adduct formation *in vitro* and potentially *in vivo* and mediating the observed hepatotoxicity.

CVB studies were also conducted with PAA which represents a model of the arylacetic acid-type NSAID substructure [56]. In rat and human, PAA is metabolized through the phenylacetyl-acyl-coenzyme A thioester (PAA-CoA) intermediary metabolite to the amino acid conjugates phenylacetylglycine amide and phenylacetylglutamine which are excreted in urine. In these studies, radiolabeled [^{14}C]-PPA was incubated with rat hepatocytes and analyzed for [^{14}C]-PPA-acyl-CoA formation and CVB to protein over time. Results showed rapid [^{14}C]-PPA-acyl-CoA formation and CVB to protein that was inhibited by coincubation with lauric acid, a substrate and direct inhibitor of acyl-CoA synthetase-mediated PPA-CoA formation. However, the CVB to protein was shown to be reversible and decreased by 72% after three hours of incubation. The major metabolite formed was phenylacetylglycine amide representing 84% of the incubation radioactivity at the three hour timepoint. No evidence for [^{14}C]-PAA-acyl glucuronide formation was observed. SDS-polyacrylamide gel electrophoresis analysis showed the selective formation of two protein adducts having molecular masses of ~29 and ~33 kDa. In summary, PAA-CoA formation in rat hepatocytes leads to the highly selective, but reversible, CVB to hepatocyte proteins. Results from these studies suggested that the CVB of arylacetic acid-containing drugs by acyl-CoA thioester-mediated transacylation of protein nucleophiles may also be reversible. In these same studies, PAA-CoA was shown to react rapidly in buffer with glutathione to form PAA-acyl-glutathione thioester; however, the glutathione adduct was not detected in incubations of [^{14}C]-PPA with rat hepatocytes. Importantly, these results indicated that a lack of detection of acyl-glutathione adducts does not preclude the formation and detection of protein adducts in hepatocytes incubations.

The CVB results from the varied reports presented above indicate that acyl-CoA thioester intermediary metabolites of carboxylic acid-containing drugs are able to mediate the CVB to hepatic proteins *in vitro* by transacylation of protein-nucleophiles and may be important in mediating carboxylic acid drug hepatotoxicity for drugs that form chemically reactive acyl-CoA thioesters *in vivo*.

6.6 Phase II Metabolite Prediction of Covalent Binding

6.6.1 Prediction of Covalent Binding to Protein by Acyl Glucuronides

From the studies described above on structurally different carboxylic-acid-containing drugs, the CVB to protein can be mediated by metabolism to chemically reactive acyl glucuronide metabolites. Results from three reports reviewed below illustrate that the instability of acyl glucuronide metabolites of structurally different acidic drugs is related to the degree of substitution at the carbon adjacent to the carbonyl-carbon of the acyl glucuronide linkage and also to the extent of substitution of electron withdrawing groups.

In early experiments, it was shown that the degradation rates of drug 1-β-*O*-acyl glucuronide derivatives of nine carboxylic acid drugs was associated with corresponding CVB rates to albumin *in vitro* and human plasma protein *in vivo* [31]. The results indicated that the chemical structure of the carboxylic acid drug can be used to predict the degree that corresponding acyl glucuronide metabolite will covalently bind to protein leading to potential toxicity. It was shown that the drugs with the arylacetic acid moiety (e.g. tolmetin and zomepirac), and having the least steric hindrance toward hydrolysis or acyl migration, had the highest degradation rates in buffer and the highest CVB to HSA *in vitro* and toward human plasma protein *in vivo*. By contrast, increasing chemical stability of the acyl glucuronide by steric hinderance due to the presence of one methyl group (e.g. the 2-arylpropionic acid-containing drugs fenoprofen and carprofen) or two methyl groups (e.g. beclobric acid) led to decreased CVB to protein. From these results, it was apparent that *in vitro* acyl glucuronide degradation rates could be a useful adjunct for potentially safer carboxylic acid drug design.

A subsequent study report, where the biosynthesis of acyl glucuronide metabolites by human liver microsomes and the subsequent *in vitro* assessment of instability and reactivity with protein were examined, showed a corresponding stability-reactivity correlation for eight carboxylic acid drug acyl glucuronides [88]. The experimental model used was to determine the drug acyl glucuronide degradation rate and the corresponding CVB to HSA from the same incubation. The model was tested with eight carboxylic acid drugs (tolmetin, zomepirac, diclofenac, fenoprofen, ketoprofen, ibuprofen, suprofen, and furosemide) known to be metabolized to acyl glucuronides. Results showed a linear relationship between the extent of acyl glucuronide degradation and the irreversible binding to HSA, just as was observed by Benet et al. [31]. Thus, from these studies, structure-reactivity observations were consistent in showing the rank order of 1-β-*O*-acyl glucuronide instability and CVB to protein to be arylacetic acid- > 2-PPA- > benzoic acid-linked acyl glucuronides.

In a later study, the reactivity of structurally varied acyl glucuronides was investigated by measuring the rate of reaction with the dipeptide lysine-phenylalanine (Lys-Phe) in buffer forming acyl glucuronide-Lys-Phe Schiff base-adducts as measured by liquid chromatography tandem mass spectrometry [89]. It was proposed that the degree of reactivity of acyl glucuronides with protein would be proportional to the amount of the corresponding Lys-Phe-adducts formed *in vitro*. In support of this proposal, results showed a structure-reactivity relationship where the rate rank order of forming the dipeptide adducts was acetic acid > propionic acid > benzoic acid derivatives, which was consistent with previous observations [31, 88]. An explanation of this rank order was proposed based on the electronic and steric properties of each carboxylic acid drug tested. In a related study, the acyl glucuronide degradation rates for a range of 21 carboxylic acid-containing drugs were analyzed in addition to the ability to form peptide adducts with dansylated Lys-Phe dipeptide [90]. Results from the assays showed consistent acyl glucuronide degradation properties related to chemical structure, but also indicated that both unstable and stable acyl glucuronides formed adducts with dansylated Lys-Phe, making it difficult to use the assay for predicting potential toxicity.

6.6.2 Prediction of Covalent Binding to Protein by Acyl-CoA Thioesters

As discussed above, a chemical structure-reactivity relationship for carboxylic acid-containing drugs exists where the extent of substitution at the alpha-carbon can be used to predict the extent to which their corresponding 1-*O*-acyl glucuronide conjugates undergo hydrolysis, acyl migration, and CVB to protein *in vitro* and *in vivo* [31, 88]. Because acyl-CoA thioester intermediates possess a similar chemically reactive carbonyl-carbon, except in this case in the form of a thioester-linkage, it makes sense that the same structure-reactivity relationship would exist for the ability of carboxylic acid drug acyl-CoA intermediates to acylate protein nucleophiles.

The investigation of a similar structure chemical reactivity relationship for a set of structurally varied carboxylic acid acyl-CoA thioester derivatives has been performed [91]. In these studies, the synthetic acyl-CoA thioesters of eight carboxylic acids (4-chloro-2-methyl-phenoxyacetic acid [MCPA], clofibric acid, fenbufen, ibuprofen, indomethacin, 2-PPA, salicylic acid, and tolmetin) were incubated in phosphate buffer at pH 7.4 and 37 °C with glutathione and examined for the rate of formation of the corresponding acyl-glutathione adducts. Results showed acyl-glutathione adduct formation occurring for each acyl-CoA thioester and where the relative reactivities of the acyl-CoA derivatives were dependent on both the extent of substitution at the carbon atom adjacent to the acyl carbonyl carbon and on the presence of an oxygen atom in a position β to the acyl carbonyl carbon as follows: phenoxyacetic acid > *o*-hydroxybenzoic acid ~ phenoxypropionic acid > arylacetic acid > 2-methyl 2-phenoxypropionic acid ~ 2-PPA. From these same incubations, the rate overall hydrolysis rates were measured and compared to the formation rate of the corresponding acyl-glutathione adduct and where a linear correlation was observed. Thus, the most reactive acyl-CoA thioester (MCPA acyl-CoA) also had the highest rate of chemical hydrolysis, whereas the least reactive acyl-CoA derivatives of clofibric acid and ibuprofen were the most stable toward chemical hydrolysis. Therefore, as in observations from previous studies on the reactivities of acyl glucuronides, in which a correlation between the degradation rate of the 1-*O*-acyl-glucuronide (hydrolysis and acyl migration) and the CVB to HSA was demonstrated, the relative rate of reactivity of acyl-CoA thioester derivatives of carboxylic acid drugs with biological nucleophiles can be predicted from the relative rate of hydrolysis of the acyl-CoA thioesters *in vitro*.

In a related report, the structure reactivity relationship of 11 structurally varied carboxylic acid acyl-glutathione derivatives with the model nucleophile NAC was characterized [52]. In these studies, it was assumed that the relative reactivity of the acyl-glutathione thioester derivative with nucleophiles *in vitro* would be essentially the same as that of the respective acyl-CoA thioesters. Incubations of acyl-glutathione derivatives with NAC resulted in the formation of the respective acyl-NAC conjugates at a rate that was dependent on the substitution at the α-carbon of the acyl-linkage. The rank order of transacylation of the acyl-glutathione thioesters with NAC was phenoxyacetic acid > arylacetic acid > 2-PPA = α,α-dimethyl-phenoxyacetic acid > α,α-dimethyl-substituted alkanoic acid, which was directly analogous to relative degradation rates observed for acyl glucuronides and acyl-CoA thioesters. The location of an oxygen atom at the β-position, rather than a methylene group as found in the phenoxyacetic acids and clofibric acid, resulted in increasingly reactive acylating species, which may be due to the electronic inductive effects of the oxygen atom. From these results, it was proposed that acyl-glutathione thioesters, in addition to acyl-CoA thioesters, are reactive in transacylation-type reactions with endogenous nucleophiles and may contribute to the overall CVB to protein *in vivo* by carboxylic acid-containing drugs.

6.7 Studies Directly Comparing Carboxylic Acid Drug Bioactivation by Acyl Glucuronidation and Acyl-CoA Formation

In this section, mechanistic studies that have been performed to compare carboxylic acid bioactivation mechanism acyl glucuronidation and acyl-CoA formation are described. These studies were conducted to determine the relative importance of the two pathways in the bioactivation of test carboxylic acids toward CVB to protein and the transacylation of glutathione *in vitro*. It was proposed that the bioactivation pathway leading to greater CVB to protein or to the transacylation of glutathione *in vitro* may be the bioactivation pathway of greater importance regarding toxicological effects *in vivo*.

The first studies to compare acyl glucuronidation and acyl-CoA formation in mediating the CVB of a carboxylic acid compound to protein were from mechanistic bioactivation experiments with the model compound 2-PPA [92]. In these studies, radiolabeled [^{14}C]-2-PPA was used in CVB studies *in vitro* in rat hepatocytes and *in vivo* in rat and where both 2-PPA-acyl glucuronide and 2-PPA-acyl-CoA were also quantified. Results from studies in rat hepatocytes showed a strong correlation between [^{14}C]-2-PPA-acyl-CoA formation, but not [^{14}C]-2-PPA-acyl glucuronide formation, in the CVB to protein. In these studies, hepatocytes were incubated with (*R*,*S*)-[^{14}C]-, (*R*)-[^{14}C]-, or (*S*)-[^{14}C]-2-PPA and analyzed for CVB, acyl glucuronidation, and acyl-CoA formation over a three hour period. The CVB of [^{14}C]-2-PPA to hepatocyte protein was time-dependent and 4.5-fold higher for the (*R*)-isomer compared to the (*S*)-isomer. The enantioselectivity of CVB was shown to correlate with a sevenfold (*R*)-/(*S*)-isomer enantioselective acyl-CoA formation, but not with the observed 0.7-fold (*R*)-/(*S*)-isomer enantioselective glucuronidation. Inhibition of (*R*,*S*)-[^{14}C]-2-PPA acyl-CoA formation by 66% by coincubation of rat hepatocytes with trimethyl acetic acid, an inhibitor of acyl-CoA formation, led to a 53% inhibition of CVB to protein, whereas a complete inhibition of acyl glucuronidation by coincubation with (−)-borneol led to only a 19% decrease in CVB to protein. These results obtained from mechanistic *in vitro* enantioselective CVB and enzyme inhibition studies clearly indicated that the 2-PPA-acyl-CoA thioester metabolite was more important than 2-PPA acyl glucuronide in the CVB of this model 2-arylpropionic acid to rat hepatocyte protein.

Further studies with [^{14}C]-2-PPA were performed to examine the importance of these two bioactivation pathways *in vivo* in rat [93]. Male Sprague–Dawley rats were pretreated with and without trimethylacetic acid or (−)-borneol before receiving [^{14}C]-2-PPA. Rat livers were collected over a two hour period and analyzed for 2-PPA acyl glucuronidation, acyl-CoA formation, and CVB to protein. Results showed that pretreatment with trimethylacetic acid led to a 49% decrease in CVB to hepatic protein and was consistent with an observed 64% decrease in the hepatic exposure of [^{14}C]-2-PPACoA. Conversely, a 95% inhibition of acyl glucuronidation by (−)-borneol only led to a 23% decrease in CVB to protein. These results are entirely consistent with corresponding results from *in vitro* mechanistic CVB studies in rat hepatocytes discussed above [92] and suggest that bioactivation through acyl-CoA formation contributes to covalent adduct formation to protein *in vivo* to a greater extent than by acyl glucuronidation.

Similar enantioselective bioactivation studies were performed with the 2-phenylproprionic class NSAID ibuprofen [94] focused on the examination of enantioselective formation of ibuprofen-acyl-glutathione thioester adducts. Enantioselective studies were conducted *in vitro* in rat hepatocytes incubations with (*R*)- and (*S*)-ibuprofen isomers and demonstrated that ibuprofen-acyl-glutathione adduct formation was highly enantioselective for the (*R*)-isomer. These results corresponded to the enantioselectivity of ibuprofen-acyl-CoA formation but not with the enantioselectivity of ibuprofen acyl glucuronidation. In addition, results from enzyme inhibition experiments showed that inhibition of ibuprofen-acyl-CoA formation by lauric acid, but not ibuprofen-1-β-*O*-G formation by coincubation with [−]-borneol, led to a corresponding reduction in ibuprofen-*S*-acyl-GSH formation. As discussed above for 2-PPA, the chemically reactive ibuprofen acyl-CoA thioester, but not ibuprofen-acyl glucuronide, mediated the transacylation of glutathione *in vitro*. Corresponding results from similar studies with (*R*)- and (*S*)-flunoxaprofen showed enantioselective acyl-glutathione thioester-adduct formation from enantioselective acyl-CoA formation in favor of the (*R*)-isomer [95]. These results provided increasing evidence for the important involvement of chemically reactive acyl-CoA thioester intermediates, rather than acyl glucuronide metabolites, in the bioactivation of carboxylic acid-containing drugs leading to the transacylation of glutathione.

Corresponding studies with the NSAIDs diclofenac [96] and mefenamic acid [54] in incubations with rat and human hepatocytes also showed that the transacylation of glutathione was not mediated by the acyl glucuronidation pathway. It was demonstrated that the coincubation of diclofenac with (−)-borneol led to a 94% decrease in diclofenac acyl glucuronide formation, although no decrease in diclofenac-acyl-glutathione adduct formation was observed. These results indicated that diclofenac was bioactivated in rat and human hepatocytes to a reactive acylating derivative that transacylated glutathione but was not dependent on transacylation by the diclofenac acyl glucuronide metabolite; however, the formation of diclofenac-acyl-CoA has not yet been reported. Studies with mefenamic acid have shown the formation of mefenamic acid-acyl-CoA and have recapitulated the results from studies with diclofenac where acyl glucuronidation did not mediate the transacylation of glutathione forming mefenamic acid-acyl-glutathione in incubations with rat hepatocytes. Instead, both mefenamic acid-acyl-CoA, and more importantly the intermediate mefenamic acid-acyl-adenylate, were implicated in mediating the formation of mefenamic acid-acyl-glutathione [97]. The mefenamic acid-adenylate was detected in higher concentrations in incubations with rat hepatocytes compared to the detection of the acyl-CoA thioester derivative. Other carboxylic acid-containing drugs and endogenous carboxylic acids have also been shown to form chemically reactive acyl-adenylate-linked intermediates [98–101]. Acyl-adenylate intermediates of endogenous bile acids, such as cholic acid, are reactive acylating derivatives that undergo transacylation-type reactions with the amino group of taurine, with peptides and proteins, and with the thiol group of GSH, leading to the formation of bile acid acyl-glutathione adducts [102].

The mechanism leading to the hepatoxicity in human of varied carboxylic acid drugs has been hypothesized to be mediated by the CVB of reactive drug metabolites to tissue proteins may be responsible or partly responsible for the hepatoxicity observed in humans. In a recent report, the CVB of a set of radiolabeled carboxylic acid-containing drugs to protein in incubations was examined to determine the bioactivation mechanisms predominating in the CVB to human liver microsome protein toward an increased understanding of the potential mechanism(s) of toxicity [103]. The bioactivation mechanisms tested were acyl glucuronidation, acyl-CoA formation, and P450-mediated metabolism of test radiolabeled drugs [^{14}C]-ibuprofen, [^{14}C]-ibufenac, [^{14}C]-fenclozic acid, [^{3}H]-zomepirac, [^{3}H]-tolmetin, [^{14}C]-tienilic acid, and [^{3}H]-suprofen. In these studies, human liver microsomes were supplemented separately with cofactors for the generation of P450-mediated reactive metabolites (NADPH), acyl glucuronides (UDPGA), and acyl-CoA conjugates (ATP and CoASH) formed from the seven test carboxylic acid drugs. The reaction mixtures were examined for metabolite formation and CVB to protein. Importantly, and consistent with mechanistic

CVB study results discussed above, results showed that all seven drugs formed acyl glucuronides, but that none of them resulted in CVB to protein. The CVB to protein via acyl-CoA formation predominated over P450-mediated reactive metabolite formation for ibuprofen, ibufenac, fenclozic, and tolmetin. The highest level of CVB to protein was observed for ibuprofen and ibufenac which was consistent with their corresponding highest level of measured acyl-CoA thioester formation. CVB of suprofen and tienilic acid to protein were mediated by NAPDPH-dependent P450-mediated bioactivation only. This experimental approach was valuable in assessing the bioactivation pathway for carboxylic acid drugs, since none of the compounds were observed to covalently bind to protein mediated by acyl glucuronidation. This result is in contrast with other observations discussed above where some of these compounds, via their acyl glucuronide conjugates, reacted covalently with plasma proteins and albumin. The importance of acyl-CoA formation in mediating the CVB to protein is consistent with observations from studies conducted in hepatocytes showing primarily acyl-CoA-mediated CVB to protein [53] and the transacylation of glutathione [94].

As a final example, a set of *in vitro* bioactivation studies were conducted recently with the carboxylic acid drug and GPR40 agonist MK-8666 toward understanding the potential bioactivation routes that may have led to observed liver safety concerns and its discontinuation from phase 1 clinical trials [104]. Results from *in vitro* studies in human liver microsomes and hepatocytes indicated that both the acyl glucuronide and acyl-CoA thioester metabolites of MK-8666 were detected and mediated the CVB to protein. It was proposed that these bioactivation pathways may represent one causative mechanism for the observed drug-induced liver injury *in vivo* in human. In accordance with results obtained by Darnell et al. [103] discussed above, CVB to protein in liver microsome incubations was higher (5.3-fold) in incubations fortified with ATP and CoASH compared to corresponding incubations with the cofactor for acyl glucuronidation, UDGPA. However, no NADPH-dependent P450-mediated CVB of MK-8666 to protein was observed in incubations with human liver microsomes. Results from these studies also showed that both MK-8666-acyl glucuronide and MK-8666-acyl-CoA thioester were able to transacylate lysine, serine, and cysteine residues via nucleophilic displacement forming acylated protein adducts. And as expected, based previous reports with varied drug acyl glucuronides, MK-8666-acyl glucuronide also degraded by acyl migration to acyl glucuronide isomers that formed Schiff base-linked protein adducts with lysine residues. From these studies it was concluded that the reactivity of MK-8666 acyl glucuronide and acyl-CoA thioester metabolites represented a probable cause for the hepatic adverse effects observed in the MK-8666 clinical studies, but it was acknowledged that the observed hepatotoxicity was likely multifaceted.

6.8 Prediction of Drug-Induced Liver Injury for Carboxylic Acid Drugs

The CVB of reactive drug metabolites has been proposed to correlate with a risk of idiosyncratic drug toxicity [105]. As discussed above, acyl glucuronides of carboxylic acid drugs are unstable under physiological conditions and undergo CVB to protein *in vitro* and *in vivo*. Sawamura et al. [34] performed studies to assess the ability to predict carboxylic acid drug toxicity risk based on the chemical stability of the corresponding acyl glucuronides. This was the first report demonstrating a relationship between risk of idiosyncratic toxicity and the degree of acyl glucuronide chemical stability. This assessment was based on the earlier work from Benet et al. [31] that demonstrated an excellent correlation between the apparent first-order degradation rate of acyl glucuronides of acidic drugs of varied structure and the extent of CVB to HSA *in vitro*. Based on this earlier work, it was proposed by Sawamura et al. [34] that the chemical instability of acyl glucuronides correlated with idiosyncratic toxicity risk in that carboxylic acid drugs with increasingly unstable acyl glucuronides were reported to induce idiosyncratic toxicity more than drugs with higher acyl glucuronide stability. In these studies, the incubation half-life values of chemically synthesized acyl glucuronides of 21 carboxylic acid drugs were tested in 0.1 M potassium phosphate buffer under standardized conditions (pH 7.4, 37 °C). The carboxylic drugs tested ranged in safety categories from "safe," "warning," and "withdrawn" in terms of their known risk and observed of idiosyncratic toxicity. Results showed that the half-life values of the safe category drug acyl glucuronides were 7.2 hours or longer, and drugs acyl glucuronides in the withdrawn category were 1.7 hours or shorter. From this analysis, the authors proposed that drug acyl glucuronide half-life value separating "safe" from "withdrawn" carboxylic acid-containing drugs was calculated to be 3.6 hours by regression analysis. Examples of "safe" drugs were montelukast and gemfibrozil with acyl glucuronide half-life values of 37.5 and 71.4 hours, respectively. Examples of withdrawn drugs were arylacetic acid compounds ibufenac and zomepirac with acyl glucuronide half-life values of 0.8 and 0.4 hours, respectively.

A related assessment of idiosyncratic toxicity based on acyl glucuronide degradation half-life values was performed by Jinno et al. [106]. In these studies, the acyl glucuronide conjugates were obtained biosynthetically from incubations of a corresponding set of 10 carboxylic acid drugs with human liver microsomes fortified with UDGPA. Protein precipitated extracts from these incubations then were added directly to incubations with potassium phosphate buffer (0.25 M, pH 7.4, 37 °C) and tested for acyl glucuronide stability. This method is useful because it does not require preliminary chemical synthesis and purification of the acyl glucuronides as was performed by Sawamura et al. [34], which can be beneficial for drug optimization at the early drug discovery stage. In the studies by Jinno et al. [106], the degradation rates of acyl glucuronides were evaluated for three "safe" drugs (telmisartan, gemfibrozil, and flufenamic acid) and seven "withdrawn" or "warning" drugs (zomepirac, diclofenac, furosemide, ibuprofen, naproxen, probenecid, and tolmetin). Results reported that the half-life values of the acyl glucuronides of the "safe" drugs were 10.6 hours or longer, and the acyl glucuronides of the "withdrawn" or "warning" drugs were 4.0 hours or shorter. For reference, the GPR40 agonist MK-8666 that was withdrawn from phase 1 clinical trials due to liver safety concerns as discussed above demonstrated an acyl glucuronide half-life in buffer under similar conditions of 4.5 hours, very close to the "withdrawn" or "warning" drugs measured half-life values proposed by Jinno et al. [106].

In a more recent study, in order to evaluate the potential toxicity risk of a set of carboxylic acid compounds, an assessment of acyl glucuronide half-life values in buffer, a test for relative reactivity of acyl glucuronides with dansylated Lys-Phe, and an examination of the effects of the acyl glucuronides in an immunostimulation assay using human peripheral blood mononuclear cells (hPBMCs) testing for their ability to induce cytokines and chemokines, was performed [90]. Acyl glucuronides of 21 carboxylic acid drugs were tested, "safe" (fourteen), "warning" (three), or "withdrawn" (four) were tested. Results showed that acyl glucuronides of each of the withdrawn drugs (e.g. zomepirac, benoxaprofen, tolmetin, and ibufenac) had short half-life values and formed peptide adducts with dansylated Lys-Phe in buffer; however, many of the safe category drugs also did (e.g. piretanide, probenecid, and oxaprozin). By contrast, acyl glucuronides of "warning" or "withdrawn" category drugs all induced the chemokine interleukin-8 in hPBMCs, whereas none of the safe category drugs did. The authors proposed that, in addition to evaluating carboxylic acid drug acyl glucuronides half-life values and reactivity with peptides in preclinical drug development, that the evaluation of immunostimulation by highly reactive acyl glucuronides using hPBMCs may contribute toward predicting acyl glucuronide-mediated risk of toxicity and to the development of safer carboxylic acid-containing drugs.

Another useful *in vitro* approach for the assessment of potential risk for idiosyncratic drug toxicity is by performing CVB studies in incubations with human hepatocytes in suspension [107]. In this approach, in part, radiolabeled drugs were incubated with cryopreserved human hepatocytes and measured for CVB to protein and metabolic stability over two to four hours. Then *in vivo* CVB burden in human was estimated from the observed CVB to protein and the fraction of drug metabolized at the end of the incubation period, and finally by factoring in the maximum prescribed daily dose. Results from statistical analysis of the CVB burden data from a set of 36 tested drugs revealed that "marked" and "severe" safety concern drugs were resolved from drugs with "low" concern at an estimated CVB burden of 0.9 mg day^{-1}. Together with other *in vitro* assays used in this approach, an integrated *in vitro* hazard matrix could discriminate drugs with a high idiosyncratic adverse drug reaction concern from those drugs with low concern.

6.9 Conclusions

Carboxylic acid-containing drugs are a widely used class of chemicals where greater than 450 of marketed drugs contain the carboxylic acid functional group [108]. However, metabolism of the carboxylic acid moiety can lead to chemically reactive metabolites that react covalently with biological macromolecules including proteins potentially leading to idiosyncratic drug toxicity. Therefore, a mechanistic understanding of the pathways of carboxylic acid drug bioactivation and the identity of the enzymes involved is important toward a better assessment and prediction of potential toxicological properties of this ubiquitous functional group.

This chapter has explored the two different phase II enzyme mechanisms of carboxylic acid drug metabolism and bioactivation namely, acyl glucuronidation and acyl-CoA formation. In both cases, the carbonyl-carbon of the acyl-linkage plays a crucial role in the chemical reactivity of the metabolites. Early studies were focused solely on the acyl glucuronide conjugate and its chemical instability and reactivity with protein nucleophiles *in vitro* and *in vivo*. Results obtained from extensive work by many investigators on varied drugs revealed that acyl glucuronide derivatives covalently bind to protein by both transacylation- and Schiff base-type mechanisms; however, whether the type of adduct formed is important toxicologically is not yet known. The prediction of potentially toxic carboxylic acid drugs metabolized to acyl glucuronide conjugates uses measured *in vitro* acyl glucuronide degradation half-life values as a tool, where shorter degradation half-life values predict increased reactivity with protein forming covalent adducts and therefore a proposed increased risk of idiosyncratic drug toxicity.

More recently, carboxylic acid drug acyl-CoA thioester metabolites have been shown to be chemically reactive intermediates also capable of transacylating protein nucleophile *in vitro* and *in vivo*. Similar to the prediction of drug acyl glucuronide potential toxicity, the prediction of potential toxicity of acyl-CoA metabolites comes from related studies showing that carboxylic acid-containing drugs that form acyl-CoA thioester metabolites with decreased stability and increased reactivity with nucleophiles are also those that are categorized as "warning" or "withdrawn" from clinical use. It is perceptible from these observations, that carboxylic acids forming unstable acyl glucuronides would also be predicted to form correspondingly unstable and chemically reactive acyl-CoA thioester intermediates. Therefore, toward the discovery of safer carboxylic acid-containing drugs, applying the recognized chemical structure-reactivity relationships would direct focus in favor of those drugs possessing steric hinderance and/or and lacking electronic effects leading to the decreased chemical reactivity of either acyl glucuronides or acyl-CoA thioester metabolites with biological nucleophiles. Of course, this concept is now well known and currently used in carboxylic acid drug design, meanwhile paying attention that the structural modifications do not lead to diminished effects on the carboxylic acid drug candidate's desired potency, selectivity, and pharmacokinetics, etc.

Mechanistic studies directly comparing the CVB to protein properties of carboxylic acid compounds that form both acyl glucuronide and acyl-CoA thioester metabolites have indicated that the acyl-CoA thioester formation pathway leading to covalent adducts to hepatic tissue *in vitro* and *in vivo* predominates. Potentially, carboxylic acid drug design might be able to determine structure–activity relationships toward blocked or minimized metabolism by acyl-CoA synthetases, for instance by the use of alpha-fluorination of carboxylic acids which have been shown to preclude metabolism by acyl-CoA formation [109, 110]. A complicating factor in delineating the importance of acyl glucuronidation versus acyl-CoA formation in mediated drug toxicity may be that many carboxylic acid drugs, for example suprofen, tienilic acid, and zomepirac, are also known to undergo bioactivation to chemically reactive metabolites by P450-mediated metabolism. Perhaps for some of these hepatotoxic carboxylic acid drugs that undergo both phase II- and phase I-mediated bioactivation, that the two events are linked and together precipitate the observed tissue toxicity. Possibly reactive phase II conjugates, acyl glucuronides and/or acyl-CoA thioesters, lead to covalent protein adducts and consequent antigen formation, but an immune response leading to idiosyncratic tissue damage may require to be mediated by an acute tissue toxicity potentially resulting from phase I-generated reactive drug metabolites and the initiation of a danger signal with subsequent activation of antigen presenting cells [111–113].

As discussed above, a significant amount of research has been conducted toward an increased understanding of carboxylic acid bioactivation pathways, but ongoing research is necessary to determine the critical pathway(s) involved in mediating carboxylic acid drug-mediated idiosyncratic toxicity.

References

1. Fung M, Thornton A, Mybeck K, et al. Evaluation of the characteristics of safety withdrawal of prescription drugs from world-wide pharmaceutical markets-1960 tc 1999. *Drug Inf J* 2001;35: 293–317.
2. Zimmerman HJ. Hepatic injury associated with nonsteroidal anti-inflammatory drugs. In: Lewis AJ, Gay GR (eds.) *Nonsteroidal Anti-Inflammatory Drugs: Mechanisms and Clinical Uses*. 2nd edition. New York: Marcel Dekker; 1994. p. 171–194.

3 Boelsterli UA, Zimmerman HJ, Kretz-Rommel A. Idiosyncratic liver toxicity of nonsteroidal anti-inflammatory drugs: molecular mechanisms and pathology. *Crit Rev Toxicol* 1995;25: 207–235.

4 Dahl SL, Ward JR. Pharmacology, clinical efficacy, and adverse effects of the nonsteroidal anti-inflammatory agent benoxaprofen. *Pharmacotherapy* 1982;2: 354–366.

5 Moses PL, Schroeder B, Alkhatib O, et al. Severe hepatoxicity associated with bromfenac sodium. *Am J Gastroenterol* 1999;94: 1393–1396.

6 Stricker BHC. *Drug-Induced Hepatic Injury*. 2nd edition. Amsterdam, Elsevier; 1992. p. 98–149.

7 Hart DF, Bain LS, Huskisson EC, et al. Hepatic effects of fenclozic acid. *Ann Rheum Dis* 1970;29: 264.

8 Henann NE, Morales JR. Suprofen-induced acute renal failure. *Drug Intell Clin Pharm* 1986;20: 860–862.

9 Richard MD, Sandler H. Anaphylactic reactions to zomepirac. *Ann Emerg Med* 1985;14: 171–174.

10 Reddy JK, Rao MS. Malignant tumors in rats fed nafenopin, a hepatic peroxisome proliferator. *J Natl Cancer Inst* 1977;59: 1645–1650.

11 Manier JW, Chang WW, Kirchner JP, Beltaos E. Hepatotoxicity associated with ticrynafen--a uricosuric diuretic. *Am J Gastroenterol*1982;77: 401–404.

12 Pannu HK, Gottlieb L, Fishman EK. Acute liver failure due to trovafloxacin: CT finding. *Emerg Radiol* 2001;8: 108–110.

13 Faed EM. Properties of acyl glucuronides: implications for studies of the pharmacokinetics and metabolism of acidic drugs. *Drug Metab Rev* 1984;15: 1213–1249.

14 Purcell P, Henry D, Melville G. Diclofenac hepatitis. *Gut* 1991;32: 1381–1385.

15 Nanau RM, Neuman MG. Ibuprofen-induced hypersensitivity syndrome. *Transl Res* 2010;155: 275–293.

16 Bota RG, Ligasan AP, Najdowski TG, Novac A. Acute hypersensitivity syndrome caused by valproic acid: a review of the literature based on a case report. *Perm J* 2011;15: 80–84.

17 Kowalski ML, Makowska JS, Blanca M, et al. Hypersensitivity to nonsteroidal anti-inflammatory drugs (NSAIDs) – classification, diagnosis and management: review of the EAACI/ENDA and GA2LEN/HANA. *Allergy* 2011;66: 818–829.

18 Nelson SD. Mechanisms of the formation and disposition of reactive metabolites that can cause acute liver injury. *Drug Metab Rev* 1995;27: 147–177.

19 Dansette PM, Bertho G, Mansuy D. First evidence that cytochrome P450 may catalyze both S-oxidation and epoxidation of thiophene derivatives. *Biochem Biophys Res Commun* 2005;338: 450–455.

20 O'Donnell JP, Dalvie DK, Kalgutkar AS, Obach RS. Mechanism-based inactivation of human recombinant P450 2C9 by the nonsteroidal anti-inflammatory drug suprofen. *Drug Metab Dispos* 2003;31: 1369–1377.

21 Tang W, Stearns RA, Bandiera SM, et al. Studies on cytochrome P-450-mediated bioactivation of diclofenac in rats and in human hepatocytes: identification of glutathione conjugated metabolites. *Drug Metab Dispos* 1999;27: 365–372.

22 Lagas JS, Sparidans RW, Wagenaar E, et al. Hepatic clearance of reactive glucuronide metabolites of diclofenac in mouse is dependent on multiple ATP-binding cassette efflux transporters. *Mol Pharmacol* 2010;77: 687–694.

23 Dutton GJ. *Glucuronidation of Drugs and Other Compounds*. Boca Raton, FL: CRC Press; 1980.

24 Miners JO, Mackenzie PI. Drug glucuronidation in humans. *Pharmacol Ther* 1991;51: 347–369.

25 Burchell B, Nebert DW, Nelson DR, et al. The UDP glucuronosyltranferase gene superfamily: suggested nomenclature based on evolutionary divergence. *DNA Cell Biol* 1991;10: 487–494.

26 Tukey RH, Strassburg CP Human UDP-glucuronosyltransferases: metabolism, expression, and disease. *Ann Rev Pharmacol Toxicol* 2000;40: 581–616.

27 Knights KM. Role of hepatic fatty acid: coenzyme A ligases in the metabolism of xenobiotic carboxylic acids. *Clin Exp Pharmacol Physiol* 1998;25: 776–782.

28 Knights KM, Sykes MJ, Miners JO. Amino acid conjugation: contribution to the metabolism and toxicity of xenobiotic carboxylic acids. *Expert Opin Drug Metab Toxicol* 2007;3: 159–168.

29 Williams K, Day R, Knihinicki R, *et al.* The stereoselective uptake of ibuprofen enantiomers into adipose tissue. *Biochem Pharmacol* 1986;35: 3403–3405.

30 Spahn-Langguth H, Benet LZ. Acyl glucuronides revisited: is the glucuronidation process a toxification as well as a detoxification mechanism? *Drug Metab Rev* 1992;24: 5–47.

31 Benet LZ, Spahn-Langguth H, Iwakawa S, et al. Predictability of covalent binding of acidic drugs in man. *Life Sci* 1993;53: (PL)141–(PL)146.

32 Boelsterli UA. Xenobiotic acyl glucuronides and acyl-CoA thioesters as protein-reactive metabolites with the potential to cause idiosyncratic drug reactions. *Curr Drug Metab* 2002;4: 3439–3450.

33 Ebner T, Heinzel G, Prox A, et al. Disposition and chemical stability of telmisartan 1-*O*-acylglucuronide. *Drug Metab Dispos* 1999;27: 1143–1149.

34 Sawamura R, Okudaira N, Watanabe K, et al. Predictability of idiosyncratic drug toxicity risk for carboxylic acid-containing drugs based on the chemical stability of acyl glucuronide. *Drug Metab Dispos* 2010;38: 1857–1864.

35 Hasegawa J, Smith PC, Benet LZ. Apparent intramolecular acyl migration of zomepirac glucuronide. *Drug Metab Dispos* 1982;10: 469–473.

36 Bolze S, Lacombe O, Durand G, et al. Standardization of a LC/MS/MS method for the determination of acyl glucuronides and their isomers. *Curr Sep* 2002; 20: 55–59.

37 Xue YJ, Akinsanya JB, Raghavan N, Zhang D. Optimization to eliminate the interference of migration isomers for measuring 1-*O*-β-acyl glucuronide without extensive chromatographic separation. *Rapid Commun Mass Spectrom* 2008;22: 109–120.

38 Camilleri P, Buch A, Soldo B, Hutt A. The influence of physicochemical properties on the reactivity and stability of acyl glucuronides. *Xenobiotica* 2018;48: 985–972.

39 Jansen PLM. β-glucuronidase-resistant bilirubin glucuronide isomers in cholestatic liver disease-determination of bilirubin metabolites in serum by means of high-pressure liquid chromatography. *Clin Chem Acta* 1981;110: 309–317.

40 Baba A, Yoshioka T. Structure-activity relationships for degradation reactions of 1-β-*O*-acyl glucuronides: kinetic description and prediction of intrinsic electrophilic reactivity under physiological conditions. *Chem Res Toxicol* 2009;22: 158–175.

41 Sinclair KA, Caldwell J. The formation of beta-glucuronide resistant glucuronides by the intramolecular rearrangement of glucuronic acid conjugates at mild alkaline pH. *Biochem Pharmacol* 1982;31: 953–957.

42 Illing HP, Wilson ID. pH dependent formation for beta-glucuronidase resistant conjugates from the biosynthetic ester glucuronide of isoxepac. *Biochem Pharmacol* 1981;30: 3381–3384.

43 Rachmel A, Hazelton GA, Yergey AL, Liberato DJ. Furosemide 1-*O*-acyl glucuronide. *In vitro* biosynthesis and pH-dependent isomerization to β-glucuronidase resistant forms. *Drug Metab Dispos* 1985;13: 705–710.

44 van Breeman RB, Fenselau CC. Reaction of 1-*O*-acyl glucuronides with 4-(p-nitrobenzyl)pyridine. *Drug Metab Dispos* 1986;14: 197–201.

45 Khan S, Teitz DS, Jemal M. Kinetic analysis by HPLC-electrospray mass spectrometry of the pH-dependent acyl migration and solvolysis as the decomposition pathways of ifetroban 1-*O*-acyl glucuronide. *Anal Chem* 1998;70: 1622–1628.

46 Mortensen RW, Sidelmann UG, Tjornelund J, Hansen SH. Stereospecific pH-dependent degradation kinetics of *R*- and *S*-naproxen-l-*O*-acyl-glucuronide. *Chirality* 2002;14: 305–312.

47 Dickinson RG, Hooper WD, Eadie MJ. pH-dependent rearrangement of the biosynthetic ester glucuronide of valproic acid to β-glucuronidase resistant forms. *Drug Metab Dispos* 1984;12: 247–252.

48 Shipkova M, Armstrong VW, Oellerich M, Wieland E. Acyl glucuronide drug metabolites: toxicological and analytical implications. *Ther Drug Monit* 2003;25: 1–16.

49 Baba A, Yoshioka T. Structure-activity relationships for the degradation reactions of 1-β-*O*-acyl glucuronides. Part 3: electronic and steric descriptors predicting the reactivity of aralkyl carboxylic acid 1-β-*O*-acyl glucuronides. *Chem Res Toxicol* 2009;22: 1998–2008.

50 Akira K, Hasegawa H, Shinohara Y, et al. Stereoselective internal acyl migration of 1β-*O*-acyl glucuronides of enantiomeric 2-phenylpropionic acids. *Biol Pharm Bull* 2000;23: 506–510.

51 Tishler SL, Goldman P. Properties and reactions of salicyl-coenzyme A. *Biochem Pharmacol* 1970;19: 143–150.

52 Grillo MP, Benet LZ. Studies on the reactivity of clofibryl-*S*-acyl-CoA thioester with glutathione *in vitro*. *Drug Metab Dispos* 2002;30: 55–62.

53 Li C, Benet LZ, Grillo MP. Studies on the chemical reactivity of 2-phenylpropionic acid 1-*O*-acyl glucuronide and *S*-acyl-CoA thioester metabolites. *Chem Res Toxicol* 2002;15: 1309–1317.

54 Grillo MP, Lohr MT, Wait JCM. Metabolic activation of mefenamic acid leading to mefenamyl-*S*-acyl-glutathione adduct formation *in vitro* and *in vivo* in rat. *Drug Metab Dispos* 2012;40: 1515–1526.

55 Li C, Grillo MP, Benet LZ. *In vitro* studies on the chemical reactivity of 2,4-dichlorophenoxyacetyl-*S*-acyl-CoA thioester. *Toxicol Appl Pharmacol* 2003;187: 101–109.

56 Grillo MP, Lohr MT. Covalent binding of phenylacetic acid to protein in incubations with freshly isolated rat hepatocytes. *Drug Metab Dispos* 2009;37: 1073–1089.

57 Smith PC, Benet LZ, McDonagh AF. Covalent binding of zomepirac glucuronide to proteins: evidence for a Schiff base mechanism. *Drug Metab Dispos* 1990;18: 639–644.

58 Ding A, Ojingwa JC, McDonagh AF, et al. Evidence for covalent binding of acyl glucuronides to serum albumin via an imine mechanism as revealed by tandem mass spectrometry. *Proc Natl Acad Sci U S A* 1993;90: 3797–3801.

59 Stogniew M, Fenselau C. Electrophilic reactions of acyl-linked glucuronides. *Drug Metab Dispos* 1982;10: 609–613.

60 Hanna PE, Anders MW. The mercapturic acid pathway. *Crit Rev Toxicol* 2019;49: 819–929.

61 Shore LJ, Fenselau C, King AR, Dickinson RG. Characterization and formation of the glutathione conjugate of clofibric acid. *Drug Metab Dispos* 1995;23: 119–123.

62 Grillo MP, Benet LZ. Interaction of γ-glutamyltranspeptidase with clofibryl-*S*-acyl-glutathione *in vitro* and *in vivo* in rat. *Chem Res Toxicol* 2001;14: 1033–1040.

63 McDonagh AF, Palma LA, Lauff JJ, Wu T-W. Origin of mammalian biliprotein and rearrangement of bilirubin glucuronides *in vivo* in the rat. *J Clin Invest* 1984;74: 763–770.

64 Grubb N, Weil A, Caldwell J. Studies on the *in vitro* reactivity of clofibryl and fenofibryl glucuronides. *Biochem Pharmacol*1993;46: 357–364.

65 Ding A, Zia-Amirhosseini P, McDonagh AF, et al. Reactivity of tolmetin glucuronide with human serum albumin. Identification of binding sites and mechanisms of reaction by tandem mass spectrometry. *Drug Metab Dispos* 1995;23: 369–376.

66 Qiu Y, Burlingame AL, Benet LZ. Mechanisms for covalent binding of benoxaprofen glucuronide to human serum albumin. Studies by tandem mass spectrometry. *Drug Metab Dispos* 1998;26: 246–256.

67 Hammond TG, Meng X, Jenkins RE, et al. Mass spectrometric characterization of circulating covalent protein adducts derived from a drug acyl glucuronide metabolite: multiple albumin adductions in diclofenac patients. *J Pharmacol Exp Ther* 2014;350: 387–402.

68 Pearson D, Weiss HM, Jin Y, et al. Absorption, distribution, metabolism, and excretion of the oral prostaglandin D2 receptor 2 antagonist fevipiprant (QAW039) in healthy volunteers and *in vitro*. *Drug Metab Dispos* 2017;45: 817–825.

69 Huxtable R. Thiols, disulfides and thioesters. In: Frieden E (ed.) *Biochemistry of Sulfur*. Plenum Press, New York: 1986. p. 230–245.

70 Brass EP. Overview of coenzyme A metabolism and its role in cellular toxicity. *Chem Biol Interact* 1994;90: 203–214.

71 Yamashita A, Watanabe M, Tonegawa T, et al. Acyl-CoA binding and acylation of UDP-glucuronosyltransferase isoforms of rat liver: their effect on enzyme activity. *Biochem J* 1995;312: 301–308.

72 Duncan JA, Gilman AG. Autoacylation of G protein alpha subunits. *J Biol Chem* 1996;271: 23594–23600.

73 Goddard AD, Watts A. Regulation of G protein-coupled receptors by palmitoylation and cholesterol. *BMC Biol* 2012;10: 27.

74 Hertz R, Bar-Tana J. The acylation of proteins by xenobiotic amphipathic carboxylic acids in cultured rat hepatocytes. *Biochem J* 1988;254: 39–44.

75 Li C, Olurinde MO, Hodges LM, et al. Covalent binding of 2-phenylpropionyl-*S*-acyl-CoA thioester to tissue proteins *in vitro*. *Drug Metab Dispos* 2003;31: 727–730.

76 Bailey MJ, Dickinson RG. Limitations of hepatocytes and liver homogenates in modelling *in vivo* formation of acyl glucuronide-derived drug-protein adducts. *J Pharmacol Toxicol* 1999;41: 27–32.

77 King AR, Dickinson RG. The utility of the bile exteriorized rat as a source of reactive acyl glucuronides-studies with zomepirac. *J Pharmacol Toxicol Methods* 1996;36: 131–136.

78 Dong JQ, Smith PC. Glucuronidation and covalent protein binding of benoxaprofen and flunoxaprofen in sandwich-cultured rat and human hepatocytes. *Drug Metab Dispos* 2009;37: 2314–2322.

79 Watkins JB, Klaassen CD. Chemically-induced alteration of UDP-glucuronic acid concentration in rat liver. *Drug Metab Dispos* 1983;11: 37–40.

80 Obach RS, Kalgutkar AS, Soglia JR, *et al.* Can *in vitro* metabolism-dependent covalent binding data in liver microsomes distinguish hepatotoxic from nonhepatotoxic drugs? An analysis of 18 drugs with consideration of intrinsic clearance and daily dose. *Chem Res Toxicol* 2008;21: 1814–1822.

81 Knights KM, Cassidy MR, Drew R. Benoxaprofen induced toxicity in rat hepatocytes. *Toxicology* 1986;40: 327–339.

82 Kretz-Rommel A, Boelsterli UA. Diclofenac covalent protein binding is dependent on acyl glucuronide formation and is inversely related to P450-mediated acute cell injury in cultured rat hepatocytes. *Toxicol Appl Pharmacol* 1993;120: 155–161.

83 Siraki AG, Chevaldina T, O'Brien PJ. Application of quantitative structure-toxicity relationships for acute NSAID cytotoxicity in rat hepatocytes. *Chem Biol Interact* 2005;151: 177–191.

84 Kretz-Rommel A, Boelsterli UA. Selective protein adducts to membrane proteins in cultured rat hepatocytes exposed to diclofenac: radiochemical and immunochemical analysis. *Mol Pharmacol* 1994;45: 237–244.

85 Kretz-Rommel A, Boelsterli UA. Mechanism of covalent adduct formation of diclofenac to rat hepatic microsomal proteins. Retention of the glucuronic acid moiety in the adduct. *Drug Metab Dispos* 1994;22: 956–961.

86 Hargus SJ, Amouzedeh HR, Pumford NR, et al. Metabolic activation and immunochemical localization of liver protein adducts of the nonsteroidal anti-inflammatory drug diclofenac. *Chem Res Toxicol* 1994;7: 575–582.

87 Sallustio BC, Nunthasomboon S, Drogemuller CJ, Knights KM. *In vitro* covalent binding of nafenopin-CoA to human liver proteins. *Toxicol Appl Pharmacol* 2000;163: 176–182.

88 Bolze S, Bromet N, Gay-Feutry C, et al. Development of an *in vitro* screening model for the biosynthesis of acyl glucuronide metabolites and the assessment of their reactivity toward human serum albumin. *Drug Metab Dispos* 2002;30: 404–413.

89 Wang J, Davis M, Li F, *et al.* A novel approach for predicting acyl glucuronide reactivity via Schiff base formation: development of rapidly formed peptide adducts for LC/MS/MS measurements. *Chem Res Toxicol* 2004;179: 1206–1216.

90 Iwamura A, Ito M, Mitsui H, et al. Toxicological evaluation of acyl glucuronides utilizing half-lives, peptide adducts, and immunostimulation assays. *Toxicol in vitro* 2015;30: 241–249.

91 Sidenius U, Skonberg C, Olsen J, et al. *in vitro* reactivity of carboxylic acid-CoA thioesters with glutathione. *Chem Res Toxicol* 2004;17: 75–81.

92 Li C, Benet LZ, Grillo MP. Enantioselective covalent binding of 2-phenylpropionic acid to protein *In vitro* in rat hepatocytes. *Chem Res Toxicol* 2003;15: 1480–1487.

93 Li C, Grillo MP, Benet LZ. *In vivo* mechanistic studies on the metabolic activation of 2-phenylpropionic acid in rat. *J Pharmacol Exp Ther* 2003;305: 250–256.

94 Grillo MP, Hua F. Enantioselective formation of ibuprofen-*S*-acyl-glutathione *in vitro* in incubations of ibuprofen with rat hepatocytes. *Chem Res Toxicol* 2008;21: 1749–1759.

95 Grillo MP, Wait JC, Tadano-Lohr M, *et al*. Stereoselective flunoxaprofen-*S*-acyl-glutathione thioester formation mediated by acyl-CoA formation in rat hepatocytes. *Drug Metab Dispos* 2010;38: 133–142.

96 Grillo MP, Hua F, Knutson CG, et al. Mechanistic studies on the bioactivation of diclofenac: identification of diclofenac-*S*-acyl-glutathione *in vitro* in incubations with rat and human hepatocytes. *Chem Res Toxicol* 2003;16: 1410–1417.

97 Horng H, Benet LZ. Characterization of the acyl-adenylate linked metabolite of mefenamic acid. *Chem Res Toxicol* 2013;26: 465–476.

98 Mao LF, Millington DS, Schulz H. Formation of a free acyl adenylate during the activation of 2-propylpentanoic acid. Valproyl-AMP: a novel cellular metabolite of valproic acid. *J Biol Chem* 1992;267: 3143–3146

99 Hall SD, Quan X. The role of coenzyme A in the biotransformation of 2-arylpropionic acids. *Chem Biol Interact* 1994;90: 235–251.

100 Menzel S, Waibel R, Brune K, Geisslinger G. Is the formation of *R*-ibuprofenyl-adenylate the first stereoselective step of chiral inversion? *Biochem Pharmacol* 1994;48: 1056–1058.

101 Ikegawa S, Ishikawa H, Oiwa H, et al. Characterization of cholyl-adenylate in rat liver microsomes by liquid chromatography/electrospray ionization-mass spectrometry. *Anal Biochem* 1999;266: 125–132.

102 Goto J, Nagata M, Mano N, et al. Bile acid acyl adenylate: a possible intermediate to produce a protein-bound bile acid. *Rapid Commun Mass Spectrom* 2001;15: 104–109.

103 Darnell M, Breitholtz K, Isin EM, et al. Significantly different covalent binding of oxidative metabolites, acyl glucuronides, and *S*-acyl CoA conjugates formed from xenobiotic carboxylic acids in human liver microsomes. *Chem Res Toxicol* 2015;28: 886–896.

104 Shang J, Tschirret-Guth R, Cancilla M, et al. Bioactivation of GPR40 agonist MK-8666: formation of protein adducts *in vitro* from reactive acyl glucuronide and acyl CoA thioester. *Chem Res Toxicol* 2020;33: 191–201.

105 Uetrecht J. Prediction of a new drug's potential to cause idiosyncratic reactions. *Curr Opin Drug Discov Devel* 2001;4: 55–59.

106 Jinno N, Ohashi S, Tagashira M, et al. A simple method to evaluate reactivity of acylglucuronides optimized for early stage drug discovery. *Biol Pharm Bull* 2013;36: 1509–1513.

107 Thompson RA, Isin EM, Yan L, et al. *In vitro* approach to assess the potential for risk of idiosyncratic adverse reactions caused by candidate drugs. *Chem Res Toxicol* 2012;25: 1616–1632.

108 Ballatore C, Huryn DM, Smith AB. Carboxylic acid (bio)isosteres in drug design. *ChemMedChem* 2013;8: 385–395.

109 Soltysiak RM, Matsuura F, Bloomer D, Sweeley CC. D,L-alpha-fluoropalmitic acid inhibits sphingosine base formation and accumulates in membrane lipids of cultured mammalian cells. *Biochim Biophys Acta* 1984;792: 214–226.

110 Grillo MP, Chiellini G, Tonelli M, Benet LZ. Effect of α-fluorination of valproic acid on valproyl-*S*-acyl-CoA formation *in vivo* in rats. *Drug Metab Dispos* 2001;29: 1210–1215.

111 Matzinger P Tolerance, danger and the extended family. *Annu Rev Immunol* 1994;12: 991–1045.

112 Pirmohamed M, Madden S, Park K. Idiosyncratic drug reactions: metabolic bioactivation as a pathogenic mechanism. *Clin Pharmacokinet* 1996;31: 215–230.

113 Dansette PM, Bonierbale E, Minoletti C, et al. Drug-induced immunotoxicity. *Eur J Drug Metab Pharmacokinet* 1998;23: 443–451.

7

Liquid Chromatography-Mass Spectrometry (LC-MS) Quantification of Reactive Metabolites

Qingping Wang and Chuang Lu

Sanofi US, Waltham, MA, USA

7.1 Introduction

It is generally understood that drug safety risk is associated with drug systemic exposure, organ-specific exposure, drug intrinsic toxicity, and patients' physiological conditions including genetic polymorphism within the drug metabolism enzymes. Drug intrinsic toxicity can be attributed to the toxicity from the parent drug molecule or from the metabolite(s) generated by **drug-metabolizing enzymes** (DMEs) in the human body after the drug has been absorbed. The latter can be further divided into (i) a metabolite(s) that contains a toxic moiety, or (ii) a metabolite(s) that is reactive, i.e. reactive metabolites that bind covalently to protein and DNA and cause irreversible toxicity or mutagenicity. This process is often referred to as bioactivation, and is considered, among several other factors, as a potential cause for drug-induced liver injury and drug-induced hypersensitivity reactions in patients. **Idiosyncratic drug reactions** (IDRs) are a specific type of drug toxicity characterized by their delayed onset, low incidence and reactive metabolite formation with little, if any, correlation between the pharmacokinetics or pharmacodynamics and general toxicological outcome [1, 2]. For example, Walgren et al. [3] reported in 2005 that five out of six drugs withdrawn from the U.S. market since 1950 showed evidence of reactive metabolite formation. Although there is no definitive proof of a relationship between reactive metabolite formation and an adverse drug effect, reactive metabolite formation is commonly considered an unwanted drug property [4, 5]. In drug discovery, bioactivation of drugs to reactive metabolites is evaluated in the early drug discovery stage to support the medicinal chemistry campaign in designing molecules with low potential of bioactivation [6–11].

Reactive metabolites, usually electrophiles that tend to bind to electron rich -S, -N, or -O groups in protein or DNA, are extremely short-lived, and so they are not

Transporters and Drug-Metabolizing Enzymes in Drug Toxicity, First Edition. Edited by Albert P. Li.

usually detectable in the systemic circulation [12]. Because the direct detection and characterization of reactive metabolites in biological systems is extremely challenging, a strategy of trapping these intermediates in situ and in *in vitro* assays is widely used in pharmaceutical industry to investigate the formation of reactive intermediates or metabolites in a biotransformation experiment. Nucleophiles classified as "Soft nucleophiles" with delocalized charges or "hard nucleophiles" with localized charges, are often used as trapping agents to react with "soft electrophiles" and "hard electrophiles" reactive intermediates, respectively. Thiols (e.g. **glutathione** (GSH) and *N*-acetylcysteine), are "soft nucleophiles" and GSH which has been shown [7] to react with alkyl or aromatic halides, epoxide, arene oxides, quinones, quinoneimines, nitrenium ions, imine methides, and Michael acceptors (Figure 7.1). Along with amines (semicarbazide and methoxylamine) and cyanide anions which have been shown to react with iminium species or aldehydes and ketones are "hard nucleophiles" [13–16]. As the parent compound undergoes biotransformation, the formed adducts can be subsequently characterized from structure elucidation and quantification of their reactive sites.

Liquid chromatography coupled with mass spectrometry (LC-MS) plays a pivotal role as the predominant analytical platform for screening and characterization the trapped reactive metabolites [17]. Starting from early triple quadrupole

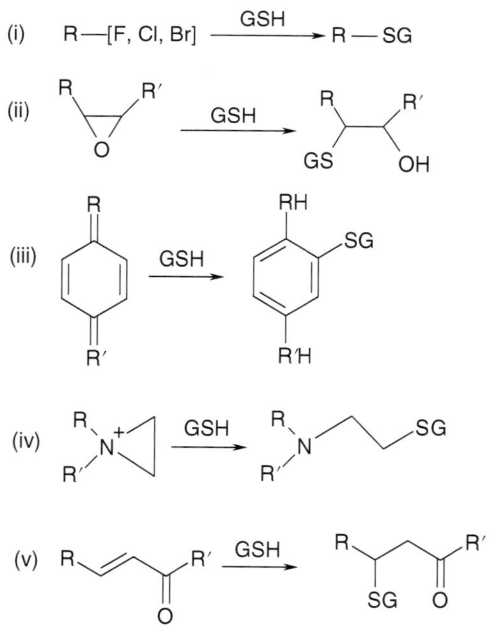

Figure 7.1 Some common conjugations with glutathione (GSH) (i) halides; (ii) epoxides; (iii) quinones; quinoneimines; (iv) nitrenium; (v) Michael acceptor.

mass spectrometry, to the most recently introduced high resolution tandem mass spectrometers, **quadrupole time-of-flight** (Q-TOF), Orbitrap mass spectrometry, etc., LC and MS technologies continue to advance. These advancements in technology have led to improvements in instrumental methods, software-aided detection and data analysis approaches, as well as workflows and assay protocols. Altogether, this has greatly enhanced our understanding and ability to assess the reactive metabolites and associated risk [18–21].

The commonly observed reactive metabolites formed from acyl-glucuronides will not be discussed in this chapter, as it is being covered in Chapter 4.

7.2 LC-MS Methods Using GSH as a Trapping Reagent

In early drug discovery, a streamlined sample preparation approach combined with a highly sensitive and highly specific LC-MS method using fast data processing algorithms are desired to support a more efficient compound design iteration cycles in drug discovery. The tripeptide GSH (γ-GluCysGly) is the most frequently used nucleophile for trapping reactive metabolites. Figure 7.1 lists some examples of GSH conjugation reactions. Typically, new chemotypes or lead compounds are subjected to reactive metabolite screening in the drug discovery stage. Test compounds are usually incubated with liver microsomes from humans and toxicology species in the presence of GSH and cofactors. The subsequent conversion of the reactive metabolic species into a stable GSH conjugate is analyzed by liquid chromatography coupled with tandem mass spectrometry (LC-MS/MS). As shown in Figure 7.2, taking advantage of characteristic fragment ions of GSH under **collision-induced dissociation** (CID) in a mass spectrometer, these GSH trapped reactive metabolites can be readily detected by a LC-MS with high sensitivity, selectivity, and fast-data processes. Table 7.1 summarized the progress of methodologies using LC-MS applications in detection of reactive metabolites and the pros and cons of these methods will be discussed in the following sections [18–39].

7.2.1 LC-MS Approaches at Positive Mode Using Constant Neutral Loss (CNL) Scan or Enhanced Product Ion (EPI) Scan

In early 1990, when triple quadrupole mass spectrometer became available, **constant neutral loss** (CNL) of 129 Da (pyroglutamic acid, Figure 7.2, fragment **d**) in a positive mode was employed for screening GSH adducts without knowing the specific structural information of an individual compound [22]. However, poor sensitivity resulting from a low abundance of **precursor ions** (PIs) and the need of a long LC run time to separate metabolites have limited its broad usage as a

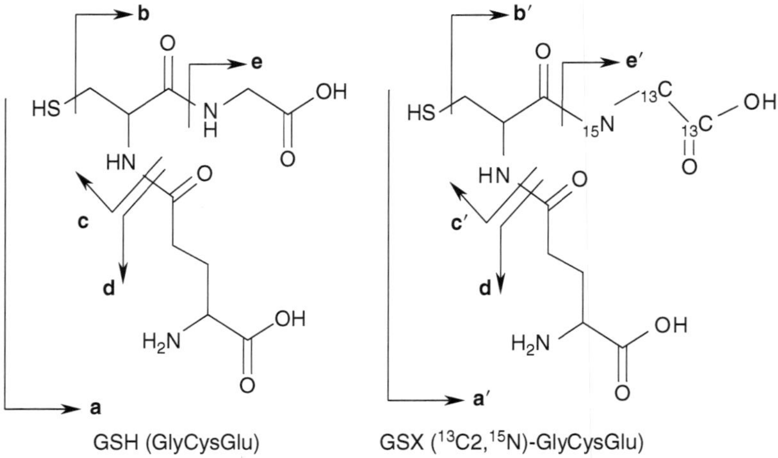

Figure 7.2 Characteristic fragment ions of glutathione used in the MS applications. *Source:* Baillie and Davis [22], Grillo et al. [23], Huang et al. [24], and Liao et al. [25].

high-throughput screen method in drug discovery. Later Yu et al. [26] applied the same CNL of 129 Da on an automated chip-based nanoelectrospray attached to a tandem mass spectrometer to demonstrate a better sensitivity that allowed for the identification of a new GSH-diclofenac adduct. Another approach to enhance sensitivity was described by Zheng et al. [27] in which **multiple reaction monitoring** (MRM) was set up as the survey scan from the protonated molecules of potential GSH adducts to their product ions derived from a **neutral loss** (NL) of 129 or 307 Da (fragment **d** or **a** in Figure 7.2), to trigger the acquisition of enhanced product ion (EPI) spectra on a hybrid triple quadrupole **linear ion trap** (LTQ) mass spectrometer. Shown in the studies with known GSH adducts of acetaminophen, diclofenac, and carbamazepine formed in **human liver microsomes** (HLM), this approach enabled the sensitive acquisition of EPI spectra with rich fragmentation in the same LC-MS run. This approach however tended to miss unexpected conjugates if they were not predicted ahead of the study. Since this approach needed a laborious prediction of the possible reactive metabolites, and consecutive generation of the MRM reactions for the predicted metabolites, its use as a high-throughput screening method was limited.

Although GSH conjugates can be readily detected by a CNL scan of 129 Da in a positive ion mode using triple quadrupole mass spectrometer, a main disadvantage of this technique is its poor selectivity and generation of false positives. For example, endogenous compounds presented in biological matrices may give rise to the loss of a neutral fragment of 129 Da (nominal mass) that is not GSH-conjugate related. As a hybrid Q-TOF high resolution mass spectrometer became

Table 7.1 MS applications in detection of GSH trapped reactive metabolites.

Mass spectrometer	[a] Reagent_Method_ions monitored in MS		References
	Positive mode	Negative mode	
Triple quadrupole	GSH_CNL_129 Da (**d**)		Baillie and Davis [22]
Nanomate-ESI Triple quadrupole	GSH_CNL_129 Da (**d**)		Yu et al. [26]
Q-Trap	GSH_MRM and NL_129 Da (**d**) and 307 Da (**a**)_EPI		Zheng et al. [27]
Q-TOF	GSH_CNL 129.0426 Da (**d**) _MS/MS		Castro-Perez et al. [28]
Triple quadrupole		GSH_PI 272 Da (**b**)	Dieckhaus et al. [29]
Triple quadrupole		GSH_PI 272 and 254 (loss H_2O) (**b**)	Mahajan and Evans [30]
Q-Trap	GSH_PI (−)_272 Da (**b**)_MS^2 (+)		Wen et al. [31]
LTQ/Orbitrap	GSH_SCID_XoPI (−) 272.0888 (**b**) and MS^n (+) and MS^n (−)		Zhu et al. [32]
Triple quadrupole	(GSH : GSX)_NL_129 Da (**d**), 75 Da (**e**)/78 Da(**e′**)		Yan and Caldwell [33]
Triple quadrupole	(GSH : GSX)_NL_129 Da (**d**), 75 Da (**e**)/78 Da (**e′**)		Yan et al. [34]
LTQ	(GSH : GSX)_ DDA(3 Da)_MS^n		Mutlib et al. [35]
LTQ XL	(GSH : GSX)_ DDA(3 Da)_MS^n (+) and MS^n (−)		Yan et al. [36]
LTQ	(GSH : GSX)_ DDA(3 Da)_MS^n		Ma et al. [37]
LTQ/Orbitrap	(GSH : GSX)_ DDA(3.003 Da)_MS^n		Lim et al. [38]
LTQ/Orbitrap	(GSH : GSX)_ DDA(3.003 Da)_MS^n		Mezine et al. [39]
Q-Trap		(GSH:GSX)_PI_272 (**b**) and 275 (**b′**)	Liao et al. [25]
Triple quadrupole		(GSH:GSX)_PI_272 (**b**) and 275 (**b′**)	Huang et al. [24]
LTQ/Orbitrap		(GSH : GSX)_SCID_ XoPI_272.0888 (**b**) and 275.0921 (**b′**) and DDA_NL 144 (**c**)	Wang et al. [21]

[a] Fragments **a, b, c, d, e** of GSH or GSX are from Figure 7.2; CNL, constant neutral loss; DDA, data-dependent scan; EPI, enhanced product ion; LTQ, linear ion trap; PI, product ion; Q-TOF, quadrupole time-of-flight; Q-Trap, quadrupole linear ion trap; SCID, in-source collision-induced dissociation; XoPI, eXtraction of Product Ion.

available, an LC-MS/MS method for screening GSH conjugates using NL of exact mass 129.0426 Da was established to eliminate the false positives [28]. The Q-TOF instrument acquired survey mass spectra sequentially at low and high energy by switching the collision energy from 5 to 20 eV. A mass difference of 129.0426 Da within a narrow mass tolerance window in the high-energy MS survey scan was examined, in which the instrument was automatically switched to the MS/MS mode to acquire a full scan MS/MS spectrum for the elucidation of the structures of detected GSH adducts. This approach allowed the selective detection and identification of GSH adducts and played a vital role in excluding false positives. Furthermore, a single injection can provide accurate mass MS and MS/MS data on each detected GSH conjugate.

7.2.2 LC-MS Approaches at Negative Mode Using Neutral Loss, Pre-Ion Scan (PIS) and XoPI (Extraction of Product Ion)

A main issue of using a CNL scan of 129 Da under CID conditions at positive mode is that this ion will not be observed for all classes of GSH adducts. For example, aliphatic and benzylic thioethers may eliminate GSH ($[M+H]^+-307$ as neutral and/or $[M+H]^+$ of 308 as protonated product ion, fragment **a** in Figure 7.2); thioester conjugates, on the other hand, may lose glutamic acid or glutamic acid followed by the loss of water ($[M+H]^+-147$, fragment **d**, and loss H_2O in Figure 7.2) [22, 23]. In addition, many GSH adducts form doubly charged $[M+2H]^{2+}$ ions, which might not generate 129 Da as NL [29], thus these GSH adducts will escape detection. To overcome this drawback, Dieckhaus et al. demonstrated that negative ion scanning for precursors of the product ion (PIS) m/z 272 (deprotonated γ-glutamyl-dehydroalanyl-glycine, fragment **b**, Figure 7.2) can successfully detect GSH conjugates of different classes of molecules, including benzylic, aromatic, aliphatic, and thioester GSH conjugates [29]. In contrast, in the same LC-MS/MS experiments, some of GSH adducts were not identified using CNL scan of 307 Da in positive mode, CNL scan of 129 Da in positive mode, or PIS of m/z 130 in the positive mode. To achieve a further increase in selectivity, Mahajan et al. [30] incorporated a simultaneous dual negative PI scan for m/z 272 and 254 (the dehydrated form of m/z 272) to extend this methodology, as both anions are abundant and originate from the glutathionyl moiety regardless of structural classes. However, since few product ions are derived from CID fragmentation of GSH conjugates in the negative mode, that can significantly hinder the completeness of structure elucidation. Therefore, the current strategy uses negative ion mode PI scanning of m/z 272 ion as a detection technique for GSH-adducts, followed by positive ion tandem mass spectrometry of the corresponding $[M+H]^+$ ion for structural elucidation. A high-throughput method was developed for screening and characterization of reactive metabolites on a QTRAP (triple quadrupole LTQ mass spectrometer) instrument using "polarity switching" with initial negative PI *m/z* 272 scanning as the trigger for the acquisition of positive ion-EPI spectra in a single LC-MS/MS run [31]. Results from the analysis of reactive metabolites of model compounds suggest that this PI-EPI approach is a feasible high-throughput method that can be utilized to screen a large number of structurally diverse compounds with high sensitivity and selectivity in a drug discovery setting [31].

Recently LTQ mass spectrometers have become a powerful MS platform widely used for the identification and structural characterization of small molecules, peptides, and proteins, largely owing to its full MS scanning speed and MS^n acquisition capability. However, since ion trap instruments were not capable of performing NL and PI scans, its utility in detecting uncommon metabolites with certain fragmentation patterns was limited. This limitation was overcome and the new methods were recently published [21, 32]. Zhu et al. [32] described the use of **high-resolution mass spectrometry** (HRMS), LTQ Orbitrap Velos, to monitor *m/z* 272.0888 (fragment **b** in Figure 7.2) through extraction of product ion (XoPI) under **in-source collision-induced dissociation** (SCID) in negative mode. This method was able to achieve high sensitivity by XoPI at exact mass 272.0888 ± 5 ppm which was generated from select ion monitoring (SIM) at narrow mass range (*m/z* 269.5–274.5 Da), as negligible interferences were observed at the same mass range with non-selective CID in the negative mode. This method was also able to achieve specificity by obtaining structural information of GSH adducts through the acquisition of accurate mass in full scan and data-dependent MS^2 in either negative or positive modes. For example, GSH-trapped reactive metabolites from incubation of HLM with amodiaquine, clozapine, diclofenac, or fipexide, in the presence of β-nicotinamide adenine dinucleotide 2′-phosphate reduced tetrasodium salt (NADPH) and GSH, were analyzed and characterized. This study was able to detect many GSH conjugates which were not previously reported.

7.2.3 LC-MS Approaches Using Stable Isotopic-GSH

The methods described above presented the pros and cons of each GSH trapping method. Some earlier methods have sensitivity issues and interference from endogenous matrix, whereas some later developed methods may still not be specific enough to detect all GSH-reactive metabolite conjugates. Recently, a novel approach using a mixture of natural and stable-isotope-labeled [^{13}C2,^{15}N-Gly]GSH or GSX (Figure 7.2) as a trapping agent for screening of reactive metabolites was developed to demonstrate a great selectivity [33–39]. In detail, incubations were performed with test compounds in HLM in the presence of NADPH and a fixed mixture of GSH and GSX. The conjugate elucidation was performed using a LTQ LC-MS/MS setting for data-dependent MS^2 triggered by isotopic pattern analysis [38]. This method focused on mass spectrometric signature with a unique characteristic pattern of "twin ions" separated by 3 Da in full scan mass spectra of GSH adducts. In such, the data-dependent neutral loss (DDA-NL) or MS^n in ion trap mass spectrometry provided an unbiased screening and/or structural information of any reactive metabolites regardless of their CID fragmentation pathways. Although, the stable-isotope trapping procedures did not enhance the signal-to-noise ratio in either the total-ion chromatogram or the neutral-loss scan, the unique isotopic doublet was easily recognized visually even at low signal-to-noise ratios. Hence, the GSH adducts can be detected reliably at low levels and thus provides a high sensitivity. For example, Yan et al. demonstrated this novel approach initially on a triple quadrupole MS for rapid detection and structural characterization of minor GSH-adducts derived from reactive metabolites of 2-acetylthiophene, clozapine, troglitazone, and 7-methylindole [33, 34]. This approach was further applied to a LTQ mass spectrometry to acquire data-dependent MS^2 triggered by isotopic pattern analysis [35–37]. In their method [36], a polarity switch between the mass tag-triggered data dependent MS^2 scan, and thus ESI- and ESI + MS^2 spectra of both labeled and nonlabeled GSH conjugates were obtained in a single LC-MS run, and provided more information for structural elucidation with unambiguous identification of GSH adducts of several known compounds. Recently, high-resolution accurate mass spectrometry such as LTQ-Orbitrap became available, Lim et al. [38] showed that the mass accuracies measured for the precursor and product ions by the Orbitrap were <2 ppm in external mass calibration mode. Therefore, the detected GSH and GSX conjugate of acetaminophen, tienilic acid, clozapine, ticlopidine, and mifepristone were within the top five hits by isotope search and easily identified with high confidence. This isotope-labeled strategy with LTQ-Orbitrap system was also applied for detection of GSH conjugates in cryopreserved human hepatocytes [39].

Furthermore, methods with high sensitivity and fast data processing were developed on a triple quadrupole instrument using characteristic fragments of GSH *m/z* 272 and 275 (fragment **b** and **b**′, Figure 7.2) in negative modes [24, 25]. Overall, this GSH/GSX isotope label approach greatly simplified data collection and processing and made it a good high-throughput assay for trapping reactive metabolites in early drug discovery.

7.2.4 LC-MS Approaches Using Combined XoPI and Stable-Isotopic GSH

Most recently, **Ultra performance liquid chromatography** (UPLC)-high-resolution accurate mass spectrometry such as LTQ-Orbitrap is a preferred LC-MS platform for biotransformation studies in the aspects of **drug**

metabolism and pharmacokinetics (DMPK) in drug discovery. Applications using this kind of instrument for high sensitivity, high selectivity, and fast data analysis of reactive metabolite is highly desired. Despite the great sensitivity of using XoPI in negative mode, it cannot detect GSH-adducts other than those that generate the fragment of 272.0888 under SCID [32]. This approach would also require more time for detailed data analysis including examination of MS^2 fragmentation spectra to avoid false positives. While the stable-isotopic method offers high selectivity [38], it lacks sensitivity due to the need for a full scan survey and associated settings on threshold and isotopic ratio for generating data-dependent MS^2. To overcome this limitation, Wang et al. [21] recently developed a **high-throughput screening** (HTS) GSH screening assay with a more streamlined sample preparation and fast data analysis approach by applying a high sensitivity and high specificity LC-MS method. As depicted in Figure 7.3, reactive intermediates (labeled as **R**) are formed from incubation of compounds with HLM, NADPH and the presence of GSH and GSX at 1 : 1 ratio, resulting in the formation of adducts of **R**-GSH : **R**-GSX at 1 : 1 ratio. Samples were prepared in a 96-well plate, after simple protein precipitation and centrifugation; the supernatants were directly injected into the LC-MS (LTQ-Orbitrap). The setting of **event 1** in Orbitrap HRMS is SIM at a mass range of 274 ± 3 Da with in-source CID under a negative ESI condition. So regardless of the structural classes of compounds, for example, benzylic, aromatic, aliphatic, and thioester [29] and regardless of adduct structures, whether they contain multiple GSH molecules or halogen molecules [32], an equal intensity of product ion pair at m/z 272.0888 and m/z 275.0926 at the same retention time can be detected, confirming the formation of GSH adducts (Figure 7.3). As a result of generating an easily recognizable doublet product ions, this GSH trapping assay exhibits better specificity while maintaining the great sensitivity of XoPI compared to the methods previously published [32]. The setting of Orbitrap HRMS in **event 2** is high-resolution full scan in negative mode, followed by isotope-pattern dependent MS/MS acquisition (triggered by mass difference of 3.00375 Da and ratio of 1.00) as described in **event 3** and **4.** These settings enable identification of GSH adducts which have no glutamyl dehydroalanylglycine fragmentation, including cysteinyl-glycine adducts due to the cleavage of the glutamyl residue. These adducts will, in fact, be captured and identified after examining spectra of NL of dehydroalanylglycine (144 Da) as shown in Figure 7.3. A template of streamlining data processing and analysis was also established in this study to assist in the identification and characterization of GSH adducts. An example using raloxifene is shown in Figure 7.4. After being bioactivated to form reactive intermediates, the raloxifene-GSH conjugates (**RaG**) were trapped and detected as **extracted ion chromatogram** (XIC) at 272.0888 and 275.0926 with 5 ppm mass accuracy (Figure 7.4a,b). **RaG1, RaG2, RaG3, RaG5, and RaG6** appeared in both isotopes' XIC with equal intensity at

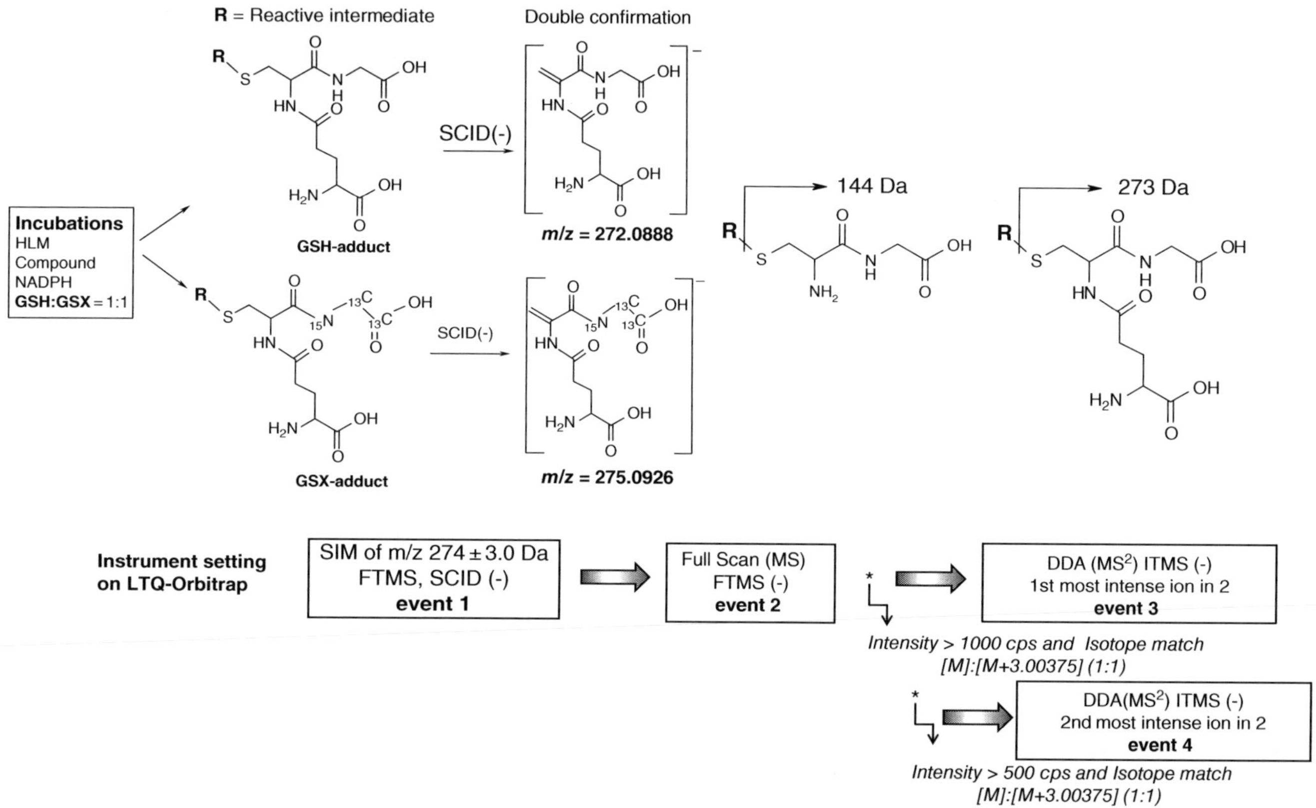

Figure 7.3 Assay strategy: Reactive intermediates (labeled as **R**) are captured from incubation of compounds with HLM, GSH and GSX at 1 : 1 ratio to form **R**-GSH adducts:**R**-GSX adducts at 1 : 1 ratio. **R**-GSH and **R**-GSX adducts are identified using XoPI-DDA-NL set up in Orbitrap HRMS.

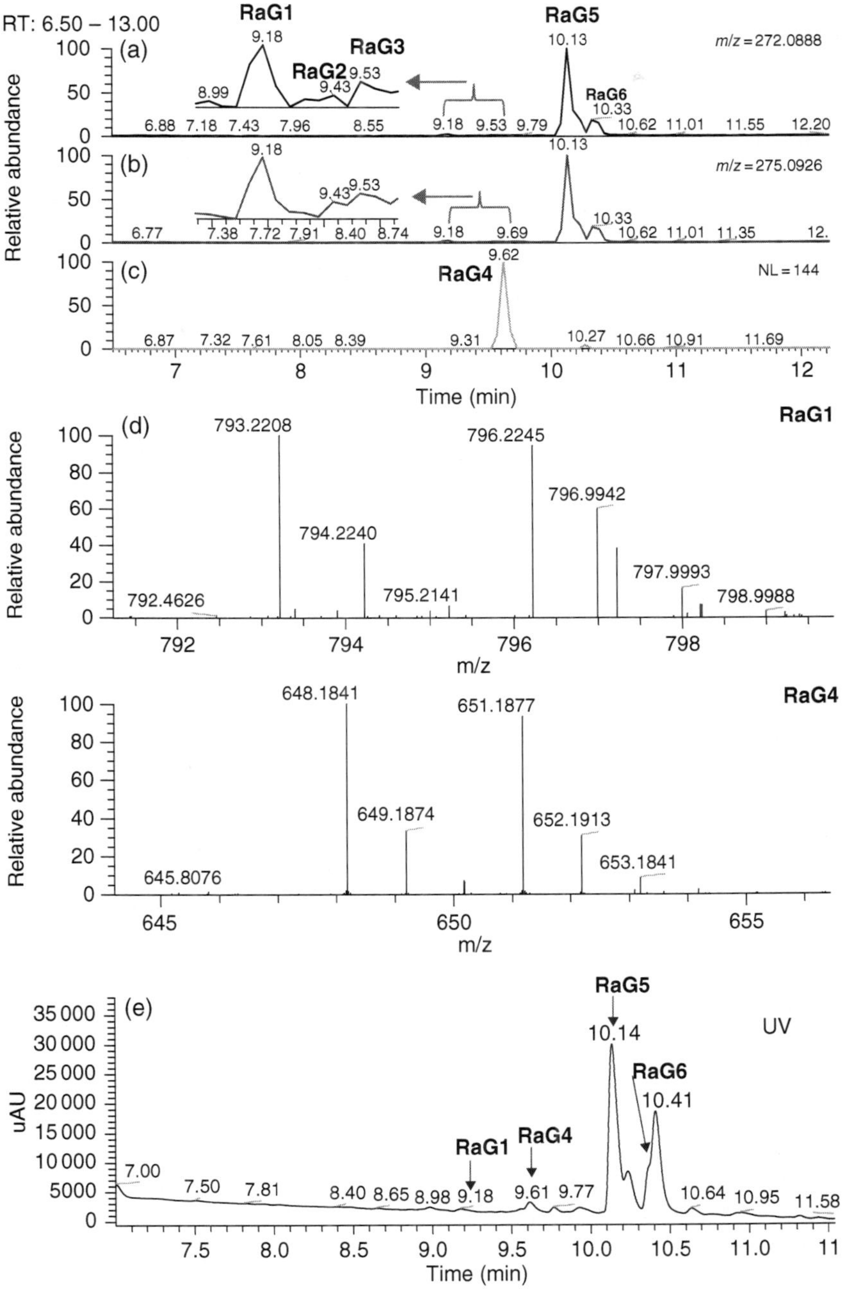

Figure 7.4 Data analysis of Raloxifene-GSH adducts. (a) XoPI *m/z* of 272.0888; (b) XoPI m/z of 275.0926; (c) DDA-NL of 144Da; (d) Full mass spectra of **RaG1** and **RaG4**; (e) UV spectrum.

the same retention time. By applying the NL of 144 Da, a novel GSH conjugate **RaG4** at the retention time of 9.62 minutes was identified (Figure 7.4c). Following the retention times of **RaG1, RaG2, RaG3, RaG4, RaG5, and RaG6**, the corresponding *m/z* of these GSH conjugates were obtained in the full scan and were confirmed by characteristic doublet peaks with 3.00375 Da (±5 ppm) difference (two of these are shown in Figure 7.4d, **RaG1**, 793.2208/796.2245, as proposed formula of P + GSH + O-2H/P + GSX + O-2H and **RaG4**, 648.1840/651.1877, as proposed formula of P + CysGly-2H/P + CysGly-$^{13}C_2{}^{15}N$-2H). **RaG4**, a unique GSH adduct that was not previously reported, was identified using the XoPI and DDA-NL approach. Thus, a total of six GSH adducts of raloxifene were identified in this study compared to three that were previously reported [28, 40, 41].

This GSH assay employed sample preparation with a 96-well plate format, high sensitivity and high specificity LC-MS method and a fast data processing algorithm to efficiently support compound design iteration cycles in drug discovery.

7.2.5 LC-MS Coupled with Software-Assisted Approach

As the modern HRMS technologies advance, great amounts of data with accurate mass can be generated in a shorter time for each experiment. Thus, data analysis and filtering a large set of raw data files could become the new bottleneck. Some software tools for rapid data collection and analysis have been developed for identification of reactive metabolites [38, 42–48]. A novel accurate mass post-data acquisition processing software, called **mass defect filter** (MDF) [42], was developed for the rapid detection of drug metabolites in complex biological matrices. The use of a mass defect filtering approach for the screening and identification of GSH adducts has been performed by acquiring high-resolution LC-MS data from LTQ Orbitrap and Fourier transform ion cyclotron resonance mass spectrometric instruments [43]. In this study, the method involved selecting GSH-adduct ions by processing acquired full scan LC-MS data using MDF templates (−0.040 to +0.040 Da around the masses of interest), and then characterizing the structure of the GSH adducts based on *m/z* values and using empirical formulas of calculating the mass of the adduct, and finally interpreting the product ion mass spectra [43]. By filtering and selecting GSH adducts for analysis, this approach provided higher sensitivity and detection of GSH-adducts that do not produce the 129 Da in CNL. Later, Lim et al. [38] applied MDF for isotope searching to identify GSH conjugates using 2 : 1 ratio of GSH and GSX. The acquired full-scan MS datasets were processed sequentially with the MDF, a control sample subtraction, and isotope pattern filtering software (MetWorks) to provide structural information of GSH trapped reactive metabolites. Recently, Barbara et al. [44] reported an LC-time-of-flight (TOF) MS^E post-acquisition data mining using MDF for reactive metabolite screening. Predicable ranges of the mass defect values, automated comparison of sample and control, and time-alignment of high- and low-energy MS^E data enabled the detection of reactive metabolites. A background-subtraction algorithm is very useful to simplify data analysis. Zhang et al. [45] demonstrated a comprehensive and reliable detection of GSH-trapped reactive metabolite through a background-subtraction algorithm, which provided better results when compared to several commercially available background-subtraction algorithms. The simplicity of the background-subtracted data allowed for facile identification of GSH adduct ions of interest which were then utilized to implement accurate mass triggered data-dependent MS/MS acquisition experiments to further confirm these metabolites.

Recently, Mass-MetaSite software-aided approaches [46, 47] were developed for post-acquisition analysis on data generated using TOF mass spectrometer in untargeted accurate mass MS^E. Mass-MetaSite software tool can perform sample control comparisons in an automated way in full scan mass spectra [48]. When set to focus on GSH adducts, Mass-MetaSite software can extract and identify GSH conjugates based on the presence of multiple collision-induced neutral losses and fragment ions specific for GSH conjugates in a high-energy untargeted full MS spectra (NL and fragment ions **d,** *m/z* 129.0426; **a**, *m/z* 308.0911, **c**, *m/z* 179.0485, **b,** *m/z* 275.1117 for singly charged adducts, Figure 7.2). Later, an automatic workflow was established, so the identified GSH adducts with NL in the MS/MS spectra were labeled by adding an asterisk sign (*) and FI in the MS/MS spectra by adding an exclamation mark sign (!) in the name of the compound. Hence, this automatic workflow could be a powerful tool for processing large dataset to identify drug-related GSH adducts.

7.2.6 Using GSH Derivatives as Trapping Reagents for Detection and Quantitation

Several GSH derivatives (chemical structures shown in Figure 7.5) such as glutathione-ethyl ester (GSH-EE), Dansyl-GSH and GSH-*N*-methylethylpiperidinium (QA-GSH), Glutamylcysteinlysine (GSK) have also been utilized to increase sensitivity, throughput, and quantitation [49–53]. Soglia et al. [49] reported a higher throughput sensitive method using GSH-EE (structure shown in Figure 7.5) to capture reactive intermediates, and microbore liquid chromatography–micro-electrospray ionization–tandem mass spectrometry (μLC–μ-ESI-MS–MS) to analyze the metabolites after a semi-automated 96-well solid-phase extraction (SPE) plate. Apart from a ~10-fold increased MS detection sensitivity for GSH-EE conjugates over GSH conjugates, the ethyl ester moiety in GSH-EE makes the molecule less polar than GSH, which resulted in a higher recovery on SPE and an increase in the retention time of the conjugates during reversed phase chromatography. In a recent report, GSH-EE was used together with hybrid triple quadrupole LTQ mass spectrometry, which a

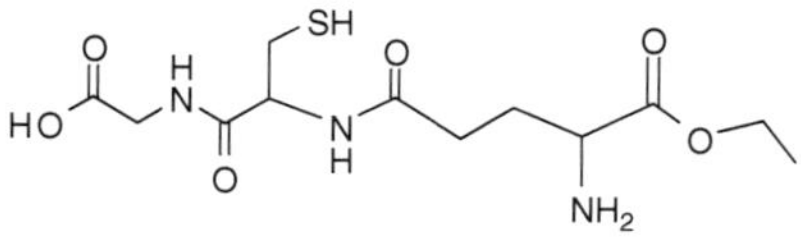

GSH-ethyl ester (GSH-EE)

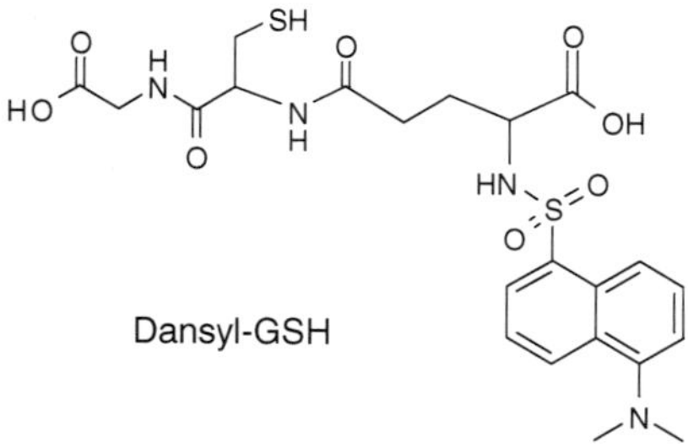

Dansyl-GSH

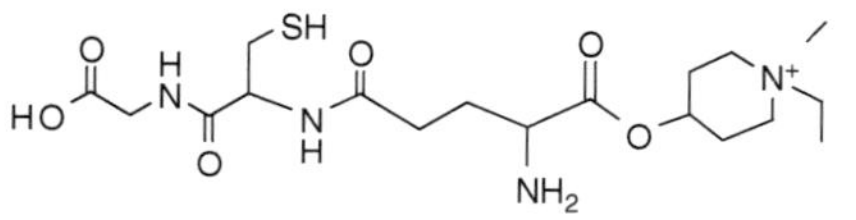

GSH-*N*-methylethylpiperidinium (QA-GSH)

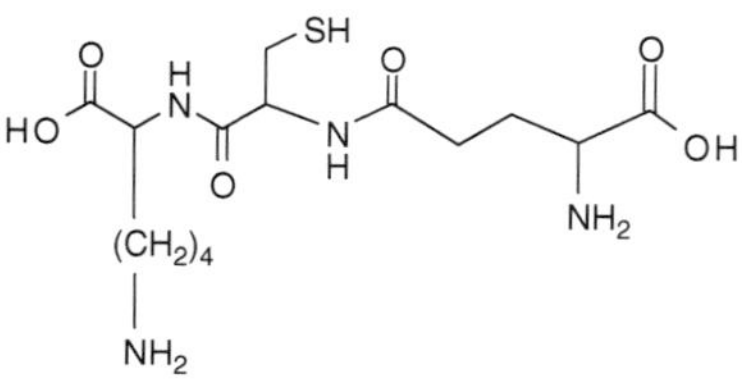

Glutamylcysteinlysine (GSK)

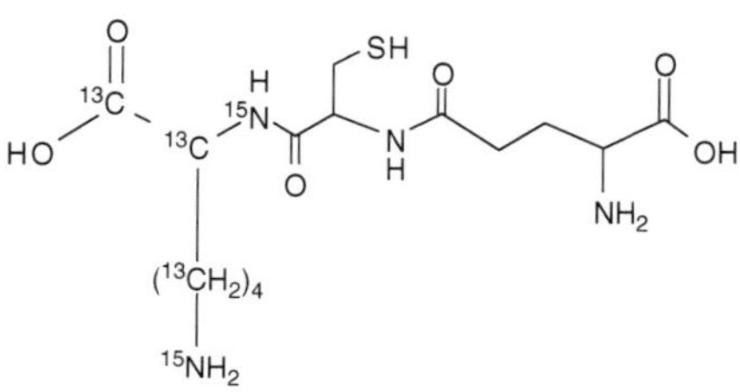

Glutamylcysteinlysine-($^{13}C_6{}^5N_2$) (Labeled GSK)

Figure 7.5 Chemical structure of glutathione derivatives.

negative PI scan at *m/z* 300 (deprotonated-glutamyl-dehydroalanyl-glycine ethyl ester) was employed as the survey scan to detect GSH-EE adducts, followed by a data-dependent EPI scan in the positive ion mode for structural characterization. Results showed highly sensitive and selective detection of GSH-EE adducts of acetaminophen, amodiaquine, carbamazepine, 4-ethylphenol, imipramine, and ticlopidine formed in HLM incubations [50]. Gan et al. [51] developed a quantitative method using Dansyl GSH in which the fluorescent Dansyl group was added to the free amino group of GSH (structure shown in Figure 7.5). Dansyl GSH was found to be equivalent to GSH in chemical reactivity, whereas Dansyl GSH adducts can be detected and quantitated in fluorescence detection at excitation 340 nm and emission 525 nm and confirmed by MS system after adequate chromatographic separation of the Dansyl GSH adducts from unreacted Dansyl GSH. However, this requires a long LC separation method (total analysis time 50 minutes) which limits its application as a high-throughput method.

Soglia et al. [53] proposed the use of a quaternary ammonium GSH analog (QA-GSH, structure shown in Figure 7.5) for semi-quantitative LC–MS/MS determination of reactive metabolites. A fixed positive charge on QA-GSH significantly improved ionization efficiency and enhanced detection sensitivity. Unlike the different mass spectrometric response from GSH adducts, QA-GSH adducts have similar MS response. The QA-GSH conjugate as an internal standard for semi-quantitation was tested by determining the level of QA-GS-acetaminophen conjugate formation at three different concentrations of acetaminophen and comparing the results to those from linear regression of authentic standards. The calculated levels of conjugate formed compared closely with those calculated from linear regression of authentic standard curves. As QA-GSH has a significant structural change compared to GSH (the carboxylic acid of GSH is blocked and substituted in QA-GSH to a charged amine moiety), QA-GSH would not mediate GST catalyzed adduct formation.

Recently, a bifunctional trapping agent, γ-GSK (Figure 7.5), was applied to simultaneous screening of both "soft" and "hard" reactive intermediates formed in microsomal incubations by conjugation to either the sulfhydryl or the aliphatic amine group, respectively. The resulting γGSK adducts are subsequently analyzed by LC-MS/MS. [54] The GSK and Labeled GSK (γ-GSK-$^{13}C_6{}^{15}N_2$, Figure 7.5) at a 1 : 1 mixture was used to capture reactive intermediates for several compounds, e.g. 2-(2-Thienyl) furan, both natural and labeled γGSK adducts (epoxidation of the thiophene ring to produce an epoxide known as a soft reactive intermediate and the ring scission of the furan moiety to form hard reactive intermediates) undergo the same NL of 129 Da under CID in a triple quadrupole mass spectrometer and thus display a distinct isotopic doublet with a mass difference of 8 Da. Therefore, CNL of 129 Da in combination with an isotopic doublet of the same intensity with *m/z* difference of 8 Da provided a rapid and selective way of detecting both "soft" and "hard" reactive metabolites.

7.3 Using Other Trapping Reagents

Hard reactive intermediates cannot be efficiently trapped by GSH or its derivatives, so alternatively, potassium cyanide (KCN), semicarbazide ($H_2NCONHNH_2$), methoxylamine (CH_3ONH_2), histidine, and lysine can be used to trap other hard electrophilic reactive intermediates such as iminium ion and aldehydes that result from the oxidative bioactivation of cyclic (or acylic) amines and primary alcohols [7, 55, 56]. These molecules have an advantage of being small in size and hence more penetrable to tissues. Screening methods utilizing a 1 : 1 mixture of cyanide and stable isotope cyanide ($^{13}C^{15}N$) to trap reactive metabolites with a "hard" iminium ion followed by the NL scans of 27 and 29 Da with a triple quadrupole mass spectrometer were reported [7, 56]. In the study of fourteen alicyclic amine compounds incubated with HLM, the cyanide trapping screen was demonstrated successfully for detection and identification of iminium ion intermediates [56]. Further Rousu et al. [57] used **time-of-flight mass spectrometry** (TOF MS) as an analytical tool, and used stable isotope KCN ($^{13}C^{15}N$) or stable isotope-semicarbazide ($H_2NCONHNH_2$-$^{13}C^{15}N_2$) to trap reactive metabolites generated from liver microsomal incubations with twelve structurally diversified compounds. Their study demonstrated that TOFMS provided unambiguous identification of a change in the molecular formula with formation of a reactive metabolite; and provided better detection sensitivity in comparison to traditional methods based on NL or PI scanning using a triple quadrupole mass spectrometer.

7.4 Identification and Characterization of Rearranged GSH Adducts

Although the above discussed screening strategies (NL, PI, and XoPI) will effectively identify most GSH adducts, in a few cases, structurally rearranged GSH-adducts could escape the identification and thus would be reported as false negatives [55, 58, 59]. As described in the recent paper by Kalgutkar [55], unusual Cys-Gly adducts (cyclized Cys-Gly thiazolidine) are adducts that escape identification. Cyclized Cys-Gly thiazolidine was generated in the incubation of Paroxetine, Serotonin, homomorpholine-containing or piperazine-containing compounds in GSH- and NADPH-supplemented HLM. A proposed mechanism for the formation of cyclized Cys-Gly thiazolidine in homomorpholine-containing compounds involves α-carbon oxidation of the homomorpholine ring by P450 to a carbinolamine intermediate, followed by ring-opening to an aminoaldehyde. This then reacts with the cysteinyl portion of dipeptide Cys-Gly, which is generated from a facile cleavage of the γ-glutamate residue in GSH by γ-glutamyltranspeptidase (GGT). These cyclized Cys-Gly thiazolidines were not detected by using mass spectrometer settings based on CID pathways [21], rather additional approaches or studies would be needed to identify and characterize them. For example, using an LC-MS/UV detector [21], LC–MS/radio-detection [60] and/or further isolation of the adducts, followed by LC-MS and ^{1}H-NMR studies would be required [58, 59]. Recently, Wang et al. [21] reported that combining UV information with HRMS mass spectrometry information enabled the identification of rearranged GSH adducts. In the study, Compound **1** (see Figure 7.6a) was incubated with 1 : 1 GSH : GSX. However, the screening in XoPI and DDA-NL settings failed to identify GSH adducts of Compound **1**, but an extra peak in UV spectra in the presence of GSH was observed, and their *m/z* of GSH adducts appearing as unique isotopic doublet peaks were obtained. Further MS^2 spectra were used for structure analysis and provided the support for a thiazolidine adduct as shown in Figure 7.6a. Therefore LC-UV-MS/MS approach was successful in facilitating the identification of this unique metabolite [21]. In the work of Lenz et al. [58] cyclized Cys-Gly adducts were derived from the oxidative bioactivation of a homomorpholine derivative AZX (Figure 7.6b) in GSH- and NADPH-supplemented HLM. Further isolation of adducts for structural characterization by ^{1}H NMR spectroscopy confirmed the ring contraction by formation of a thiazolidine-glycine. In addition, various structurally related trapping reagents were employed to further investigate the reaction mechanism along with a methoxylamine trapping experiment to confirm the structure of the postulated reactive intermediates of AZX.

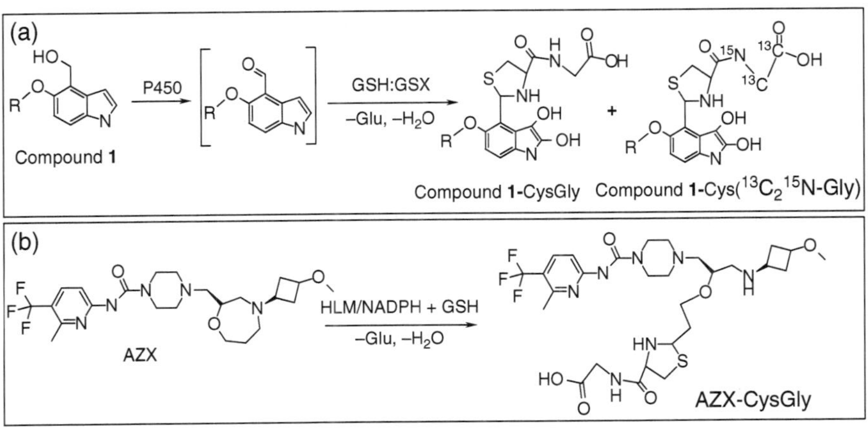

Figure 7.6 Proposed the formation of cyclized Cys-Gly adducts when incubated in HLM, NADPH, and GSH (a) Compound **1** (Ref. 21) and (b) AZX (Ref. 58).

7.5 Strategies for Optimization and Decision Tree

Strategies for minimizing reactive metabolite formation in drug discovery and risk assessment of reactive metabolites in drug development have been summarized in several recent reviews [61–67]. Generally, the screening for the formation of GSH conjugates will identify the structural liability of new chemotypes, however, if the structural moiety is required for pharmacological activity and formation of reactive metabolites cannot be completely eliminated, caution and judgment (e.g. a risk versus benefit assessment) are needed when assessing the impact of reactive metabolite formation in drug discovery and development. Thus, a decision tree is proposed in this chapter to help drug discovery projects to further assess the risk before selecting a drug development candidate (Figure 7.7). As illustrated in the GSH adduct reactive metabolite decision tree (Figure 7.7), when a new chemical scaffold shows *in vitro* pharmacology activity, a spot check of its risk of generating reactive metabolite(s) is needed. A GSH trapping assay can usually be applied for this purpose. The human liver microsomal incubation is usually carried out in phosphate buffer, pH 7.4, for 30 minutes, with GSH : GSX, NADPH, $MgCl_2$, and 1 mg ml^{-1} HLM [21]. A few compounds from the series can be re-evaluated in human hepatocytes in order to semi-quantitatively (UV detector) measure reactive metabolites in a system with natural abundance of GSH and cofactors. If a compound has <1% GSH-adducts formed, it would be considered to have little risk for further development. The 1% cut off value comes from two factors: (i) it can be reliably detected in a GSH adduct assay by LC-MS UV detection; (ii) assuming the maximum desired dose for a small molecular drug candidate is no more than 1000 mg day^{-1} and a complete absorption and metabolism *in vivo* (the worst-case scenario), <1% GSH adduct would suggest a < 10 mg day^{-1} of **covalent binding daily burden** (CBB), which is a cut off value cited in the literature from evaluating marketed drugs that carry covalent binding liability but are still in use. For those compounds that have greater than 1% of GSH adduct formation, a structural optimization is usually conducted to find compounds which carry the pharmacological activity but minimal reactive metabolite liability. For compounds that are first-in-class or a compound that is best-in-class in acute diseases such as rare or terminal illnesses including late stage cancers, the reactive metabolite risk benefit may be assessed with less stringent criteria. In the first-in-class situation, since there is no other therapy, that drug candidate can serve as a **proof-of-concept** (POC) with a backup compound with no reactive metabolite liability being brought forward. For other drug candidates in best-in-class and in a chronic disease category, a radio-labeled compound is required to be synthesized before candidate selection. A covalent binding assay using a radio-labeled compound is usually conducted under conditions similar to those described in the literature [68–71]. Once the fraction of covalent binding is determined in an

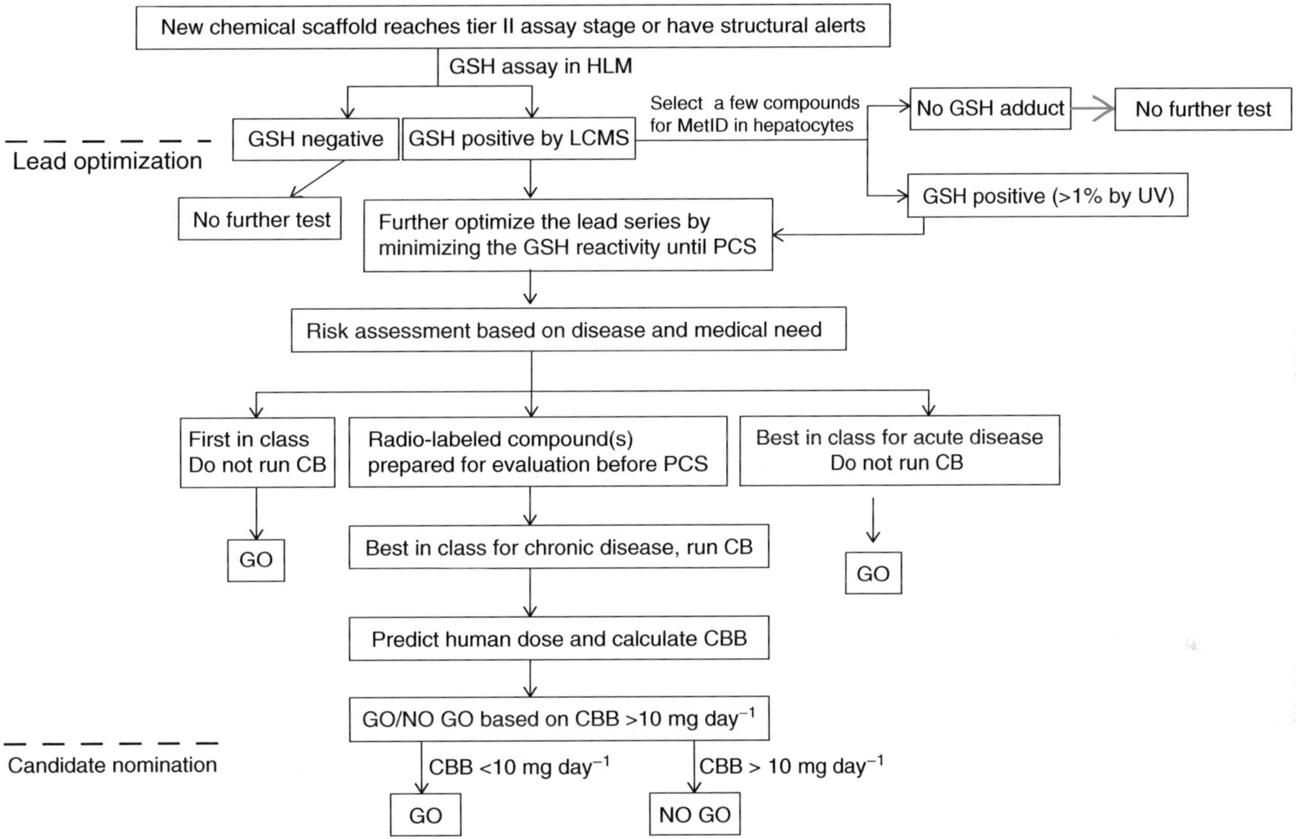

Figure 7.7 Reactive metabolite decision tree during the candidate identification phase of drug discovery.

in vitro experiment (f_{cb}), and a predicted human fraction of oral absorption (f_a), fraction of *in vivo* metabolism (f_m), and daily dose (D, mg) are obtained, CBB can be calculated as $CBB = D \times f_a \times f_m \times f_{cb}$ [68]. Studies on marketed drugs that generate reactive metabolites showed that the majority of drugs with CBB > 10 mg day^{-1} carry a black box warning or are already being withdrawn from the market. Thus, this decision tree also suggests stopping further development for drug candidates which carry CBB > 10 mg day^{-1}, especially when they are in development as best-in-class therapeutics in chronic diseases.

7.6 Summary

Identification and characterization of reactive metabolites are critical in designing new drug candidates with an improved toxicology profile in drug discovery and development. This chapter discusses mainly the GSH trapping assay methodologies for routine screening of reactive metabolites with evolving MS technology and data analysis tools. MS instruments discussed include the triple quadrupole, Q-trap, LTQ, as well as the more advanced HRMS such as Q-TOF and LTQ-Orbitrap. MS methods discussed include monitoring CNL of 129 Da in positive mode, XoPI of 272.0888 Da in negative mode and polar switching using either a data-dependent or data-independent method. In addition, MS methods discussed include applying a mixture of isotopic-labeled GSX and GSH as trapping reagents to enable rapid data analysis by focusing on the characteristic pattern of "twin ions" separated by 3.00375 Da in full scan mass spectra of GSH adducts. Software and data analysis tools discussed include MDF, background subtraction, and Mass-MetaSite, which have increased the speed and efficiency of data collection and interpretation. Thus, the combination of stable isotope-labeled GSH (GSX) and state-of-the-art LC-MS/MS techniques along with data acquisition and data mining technologies can be applied successfully to provide highly sensitive and selective identification of GSH adducts of drug candidates undergoing bioactivation. The most recent high-through-put GSH assay as shown in Figure 7.3 using LC-LTQ Orbitrap-UV system demonstrated high sensitivity, high specificity, and rapid data analysis for identification and characterization of reactive metabolite "all-in-one" [21].

The GSH screening assay has been utilized for ranking compounds with respect to the bioactivation liability, which is considered an important factor in the selection of small molecule candidates. In addition, information on the major bioactivation pathway of a given compound enables medicinal chemists to modify the chemical structure in order to minimize the degree of metabolic activation. However, in cases where the structural motif of a compound is required for its pharmacological activity, and complete elimination of reactive metabolites is not possible, further consideration of compounds as drug candidates should follow the decision trees proposed in Figure 7.7.

Overall, high-throughput sensitive and selective GSH trapping LC-MS assays have been established to evaluate the formation of reactive intermediates at the stage of lead optimization and candidate selection. Formation of reactive metabolites cannot always be eliminated; therefore, an integrated decision tree requiring caution and judgment (e.g. a risk versus benefit assessment) should be considered when assessing the impact of reactive metabolite formation in drug discovery and development.

Acknowledgment

The authors would like to acknowledge Dr. Jennifer Fretland, Dr. John Darbyshire, and Dr. Patrick Huang in their help with reviewing, reading, and editing this chapter.

Abbreviations

CBB	Covalent Binding daily Burden
DDA-NL	Data-Dependent Neutral Loss
FI	Fragment Ions
GSH	glutathione
GSX	stable-isotope-labeled glutathione
HLM	Human Liver Microsomes
HRMS	High Resolution Mass Spectrometry
HTS	High-Throughput Screening
LTQ	linear ion trap
LTQ-Orbitrap	linear ion trap Fourier transform
MRM	Multiple Reaction Monitoring
MS^2 or MS/MS	tandem MS
MS^E	full-scan MS acquisition using alternated low- and high-collision energy, which can be recorded by the Q-TOF instrument
MS^n	data-dependent multiple stage MS/MS acquisition, which can be recorded by the LTQ-Orbitrap instrument
m/z	mass-to-charge ratio
NADP+	β-nicotinamide adenine dinucleotide 2′-phosphate sodium salt
NADPH	β-nicotinamide adenine dinucleotide 2′-phosphate-reduced tetrasodium salt

P450 cytochrome P450
Q-Trap quadrupole linear ion trap
Q-TOF quadrupole time-of-flight
SCID in-Source Collision-Induced Dissociation
XIC eXtracted Ion Chromatogram
XoPI eXtraction of Product Ion

References

1 Stepan AF, Walker DP, Bauman J, Price DA, Baillie TA, Kalgutkar KS, and Aleo MD (2011) Structural alert/reactive metabolite concept as applied in medicinal chemistry to mitigate the risk of idiosyncratic drug toxicity: a perspective based on the critical examination of trends in the top 200 drugs marketed in the United States. *Chem Res Toxicol*, 24:1345–1410.
2 Thompson RA, Isin EM, Ogese MO, Mettetal JT, and Williams DP (2016) Reactive metabolites: current and emerging risk and hazard assessments. *Chem Res Toxicol*, 29:505–533.
3 Walgren JL, Mitchell MD, and Thompson DC (2015) Role of metabolism in drug-induced idiosyncratic hepatotoxicity. *Crit Rev Toxicol*, 35:325–361.
4 Grillo MP (2015) Detecting reactive drug metabolites for reducing the potential for drug toxicity. *Expert Opin Drug Metab Toxicol*, 11:1281–1302.
5 Park BK, Boobis A, Clarke S, Goldring CE, Jones D, Kenna JG, Lambert C, Laverty HG, Naisbitt DJ, Nelson S, Nicoll Griffith DA, Obach RS, Routledge P, Smith DA, Tweedie DJ, Vermeulen N, Williams DP, Wilson ID, Baillie TA, and Grillo MP (2011) Managing the challenge of chemically reactive metabolites in drug development. *Nat Rev Drug Discov*, 10:292–306.
6 Dalvie D, Kalgutkar AS, and Chen W (2015) Practical approaches to resolving reactive metabolite liabilities in early discovery. *Drug Metab Rev* 47:56–70.
7 Evans DC, Watt AP, Nicoll-Griffith DA, and Baillie TA (2004) Drug-protein adducts: an industry perspective on minimizing the potential for drug bioactivation in drug discovery and development. *Chem Res Toxicol*, 17:3–16.
8 Kumar S, Kassahun K, Tschirret-Guth RA, Mitra K, and Baillie TA (2008) Minimizing metabolic activation during pharmaceutical lead optimization: progress, knowledge gaps and future direction. *Curr Opin Drug Discove Devel*, 11:43–52.
9 Argikar UA, Mangold JB, and Harriman SP (2011) Strategies and chemical design approaches to reduce the potential for formation of reactive metabolic species. *Curr Top Med Chem*, 11:419–449.
10 Tang W and Lu AY (2010) Metabolic bioactivation and drug-related adverse effects: current status and future directions from a pharmaceutical research perspective. *Drug Metab Rev*, 42:225–249.

11 Kalgutkar AS, Gardner I. Obach RS, Shaffer CL, Callegari E, Henne KR, Mutlib AE, Dalvie DK, Lee JS, Nakai Y, O'Donnell JP, Boer J, and Harriman SP. (2005) A comprehensive listing of bioactivation pathways of organic functional groups. *Curr Drug Metab*, 6:161–225.

12 Park BK, Kitteringham NR, Maggs JL, Pirmohamed M, and Williams DP (2005) The role of metabolic activation in drug-induced hepatotoxicity. *Annu Rev Pharmacol Toxicol*, 45:177–202.

13 Itakura K, Furuhata A, Shibata N, Kobayashi M, and Uchida K (2003) Maillard reaction-like lysine modification by a lipid peroxidation product: immunochemical detection of protein-bound 2-hydroxyheptanal in vivo. *Biochem Biophys Res Commun*, 308:452–457;

14 Henderson AP, Bleasdale C, Clegg W, and Golding BT (2004) 2,6-Diarylaminotetrahydropyrans from reactions of glutaraldehyde with anilines: models for biomolecule cross-linking. *Chem Res Toxicol* 17:378–382.

15 Gorrod JW and Aislaitner G (1994) The metabolism of alicyclic amines to reactive iminium ion intermediates. *Eur J Drug Metab Pharmacokinet*, 19:209–217.

16 Gorrod JW, Whittlesea CM, and Lam SP (1991) Trapping of reactive intermediates by incorporation of ^{14}C-sodium cyanide during microsomal oxidation. *Adv Exp Med Biol*, 283:657–664.

17 Wen B and Zhu M (2015) Applications of mass spectrometry in drug metabolism: 50 years of progress. *Drug Metab Rev*, 47:71–87.

18 Ma S and Subramanian R (2006) Detecting and characterizing reactive metabolites by liquid chromatography/tandem mass spectrometry. *J Mass Spectrom*, 41:1121–1139.

19 Staack RF and Hopfgartner G (2007) New analytical strategies in studying drug metabolism. *Anal Bioanal Chem*, 388:1365–1380.

20 Ma S and Zhu M (2009) Recent advances in application of liquid chromatography-tandem mass spectrometry to the analysis of reactive drug metabolites. *Chem Biol Interact*, 179:25–37.

21 Wang Q, Liu H, Slavsky M, Fitzgerald M, Lu C, and O'Shea T (2019) A high-throughput glutathione trapping assay with combined high sensitivity and specificity in high resolution mass spectrometry by applying product ion extraction and data-dependent neutral loss. *J Mass Spectrom*, 54:158–166.

22 Baillie TA and Davis MR (1993) Mass spectrometry in the analysis of glutathione conjugates. *Biol Mass Spectrom*, 22:319–325.

23 Grillo MP, Hua F, Knutson CG, Ware JA, and Li C (2003) Mechanistic studies on the bioactivation of diclofenac: identification of diclofenac-S-acyl-glutathione in vitro in incubations with rat and human hepatocytes. *Chem Res Toxicol* 16:1410–1417.

24 Huang K, Huang L, and van Breemen RB (2015) Detection of reactive metabolites using isotope-labeled glutathione trapping and simultaneous neutral loss and precursor ion scanning with ultra-high-pressure liquid chromatography triple quadruple mass spectrometry. *Anal Chem* 87:3646–3654.

25 Liao S, Ewing NP, Boucher B, Materne O, and Brummel CL (2012) High-throughput screening for glutathione conjugates using stable-isotope labeling and negative electrospray ionization precursor-ion mass spectrometry. *Rapid Commun Mass Spectrom*, 26:659–669.

26 Yu LJ, Chen Y, Deninno MP, O'Connell TN, and Hop CE (2005) Identification of a novel glutathione adduct of diclofenac, 4′-hydroxy-2′-glutathion-deschloro-diclofenac, upon incubation with human liver microsomes. *Drug Metab Dispos*, 33:484–488.

27 Zheng J, Ma L, Xin B, Olah T, Humphreys WG, and Zhu M (2007) Screening and identification of GSH-trapped reactive metabolites using hybrid triple quadruple linear ion trap mass spectrometry. *Chem Res Toxicol*, 20:757–766.

28 Castro-Perez J, Plumb R, Liang L, and Yang E (2005) A high-throughput liquid chromatography/tandem mass spectrometry method for screening glutathione conjugates using exact mass neutral loss acquisition. *Rapid Commun Mass Spectrom*, 19:798–804.

29 Dieckhaus CM, Fernandez-Metzler CL, King R, Krolikowski PH, and Baillie TA (2005) Negative ion tandem mass spectrometry for the detection of glutathione conjugates. *Chem Res Toxicol*, 18:630–638.

30 Mahajan M.K and Evans CA (2008) Dual negative precursor ion scan approach for rapid detection of glutathione conjugates using liquid chromatography/tandem mass spectrometry. *Rapid Commun Mass Spectrom*, 22:1032–1040.

31 Wen B, Ma L, Nelson SD, and Zhu M (2008) High-throughput screening and characterization of reactive metabolites using polarity switching of hybrid triple quadrupole linear ion trap mass spectrometry. *Anal Chem*, 80:1788—1799.

32 Zhu X, Kalyanaraman N, and Subramanian R (2011) Enhanced screening of glutathione-trapped reactive metabolites by in-source collision induced dissociation and extraction of product ion using UHPLC-high resolution mass spectrometry. *Anal Chem*, 83:9516–9523.

33 Yan Z and Caldwell GW (2004) Stable-isotope trapping and high-throughput screenings of reactive metabolites using the isotope MS signature. *Anal Chem*, 76:6835–6847.

34 Yan Z, Maher N, Torres R, Caldwell GW, and Huebert N (2005) Rapid detection and characterization of minor reactive metabolites using stable-isotope trapping in combination with tandem mass spectrometry. *Rapid Commun Mass Spectrom*, 19:3322–3330.

35 Mutlib A, Lam W, Atherton J, Chen H, Galatsis P, and Stolle W (2005) Application of stable isotope labeled glutathione and rapid scanning mass spectrometers in detecting and characterizing reactive metabolites. *Rapid Commun Mass Spectrom*, 19:3482–3492.

36 Yan Z, Caldwell GW, and Maher N (2008) Unbiased high-throughput screening of reactive metabolites on the linear ion trap mass spectrometer using polarity switch and mass tag triggered data dependent acquisition. *Anal Chem*, 80:6410–6422.

37 Ma L, Wen B, Ruan Q, and Zhu M (2008) Rapid screening of GSH-trapped reactive metabolites by linear ion trap mass spectrometry with isotope pattern dependent scanning and post-acquisition data mining. *Chem Res Toxicol*, 21:1477–1483.

38 Lim HK, Chen J, Cook K, Sensenhauser C, Silva J, and Evans DC (2008) A generic method to detect electrophilic intermediates using isotopic pattern triggered data-dependent high-resolution accurate mass spectrometry. *Rapid Commun Mass Spectrom*, 22:1295–1311.

39 Mezine I, Bode C, Raughley B, Bhoopathy S, Roberts KJ, Owen AJ, and Hidalgo IJ (2013) Application of exogenous mixture of glutathione and stable isotope labeled glutathione for trapping reactive metabolites in cryopreserved human hepatocytes. Detection of the glutathione conjugates using high resolution accurate mass spectrometry. *Chem Biol Interact*, 204:173–184.

40 Dalvie D, Kang P, Zientek M, Xiang C, Zhou S, and Obach RS (2008) Effect of intestinal glucuronidation in limiting hepatic exposure and bioactivation of raloxifene in humans and rats. *Chem Res Toxicol*, 21:2260–2271.

41 Yu L, Liu H, and Li W (2004) Oxidation of raloxifene to quinoids: potential toxic pathways via a diquinone methide and o-quinones. *Chem Res Toxicol*, 17:879–888.

42 Zhang H, Zhang D, and Ray K (2003) A software filter to remove interference ions from drug metabolites in accurate mass liquid chromatography/mass spectrometric analyses. *J Mass Spectrom*, 38:1110–1112.

43 Zhu M, Ma L, Zhang H, and Humphreys WG (2007) Detection and structural characterization of glutathione-trapped reactive metabolites using liquid chromatography-high-resolution mass spectrometry and mass defect filtering. *Anal Chem*, 79:8333–8341.

44 Barbara, JE and Castro-Perez JM (2011) High-resolution chromatography/time-of-flight MSE with in silico data mining is an information-rich approach to reactive metabolite screening. *Rapid Commun Mass Spectrom*, 25:3029–3040.

45 Zhang H, and Yang H (2008) An algorithm for thorough background subtraction from high-resolution LC/MS data: application for detection of glutathione-trapped reactive metabolites. *J Mass Spectrom*, 43:1181–1190.

46 Brink A, Fontaine F, Marschmann M, Steinhuber B, Cece-Esencan EN, Zamora I, and Pähler A (2014) Post-acquisition analysis of untargeted accurate mass quadrupole time-of-flight MSE data for multiple collision-induced neutral losses and fragment ions of glutathione conjugates. *Rapid Commun Mass Spectrom*, 28: 2695–2703.

47 Cece-Esencan EN, Fontaine F, Plasencia G, Teppner M, Brink A, Pahler A, and Zamora I (2016) Enhancing throughput of glutathione adduct formation studies and structural identification using a software-assisted workflow based on high resolution mass spectrometery (HRMS) data. *SM Anal Bioanal Technique*, 1:1–12.

48 Bonn B, Leandersson C, Fontaine F, and Zamora I (2010) Enhanced metabolite identification with MSE and a semi-automated software for structural elucidation. *Rapid Commun Mass Spectrom*, 24:3127–3138.

49 Soglia JR, Harriman SP, Zhao S, Barberia J, Cole MJ, Boyd JG, and Contillo LG (2004) The development of a higher throughput reactive intermediate screening assay incorporating micro-bore liquid chromatography-micro-electrospray ionization-tandem mass spectrometry and glutathione ethyl ester as an in vitro conjugating agent. *J Pharm Biomed Anal*, 36:105–116.

50 Wen B and Fitch WL (2009) Screening and characterization of reactive metabolites using glutathione ethyl ester in combination with Q-trap mass spectrometry. *J Mass Spectrom*, 44:90–100.

51 Gan J, Harper TW, Hsueh MM, Qu Q, and Humphreys WG (2005) Dansyl glutathione as a trapping agent for the quantitative estimation and identification of reactive metabolites. *Chem Res Toxicol*, 18:896–903.

52 Leblanc A, Shiao TC, Roy R, and Sleno L (2010) Improved detection of reactive metabolites with a bromine-containing glutathione analog using mass defect and isotope pattern matching. *Rapid Commun Mass Spectrom*, 24:1241–1250.

53 Soglia JR, Contillo LG, Kalgutkar AS, Zhao S, Hop CE, Boyd JG, and Cole MJ (2006) A semiquantitative method for the determination of reactive metabolite conjugate levels in vitro utilizing liquid chromatography-tandem mass spectrometry and novel quaternary ammonium glutathione analogues. *Chem Res Toxicol*, 19:480–490.

54 Yan Z, Maher N, R. Torres, and Huebert N (2007) Use of a trapping agent for simultaneous capturing and high-throughput screening of both “soft” and “hard” reactive metabolites. *Anal Chem*, 79:4206–4214.

55 Kalgutkar AS (2017) Liabilities associated with the formation of “Hard” electrophiles in reactive metabolite trapping screens. *Chem Res Toxicol*, 30:220–238.

56 Argoti D, Liang L, Conteh A, Chen L, Bershas D, Yu CP, Vouros P, and Yang E (2005) Cyanide trapping of iminium ion reactive intermediates followed by detection and structure identification using liquid chromatography-tandem mass spectrometry (LC-MS/MS). *Chem Res Toxicol*, 18:1537–1544.

57 Rousu T, Pelkonen O, and Tolonen A (2009) Rapid detection and characterization of reactive drug metabolites in vitro using several isotope-labeled trapping agents and ultra-performance liquid chromatography/time-of-flight mass spectrometry. *Rapid Commun Mass Spectrom* 23:843–855.

58 Lenz EM, Martin S, Schmidt R, Morin PE, Smith R, Weston DJ, and Bayrakdarian M (2014) Reactive metabolite trapping screens and potential pitfalls: bioactivation of a homomorpholine and formation of an unstable thiazolidine adduct. *Chem Res Toxicol*, 27:968–980.

59 Doss GA, Miller RR, Zhang Z, Teffera Y, Nargund RP, Palucki B, Park MK, Tang YS, Evans DC, Baillie TA, and Stearns RA (2005) Metabolic activation of a 1,3-disubstituted piperazine derivative: evidence for a novel ring contraction to an imidazoline. *Chem Res Toxicol*, 18:271–276.

60 Zhao S, Dalvie DK, Kelly JM, Soglia JR, Frederick KS, Smith EB, Obach RS, and Kalgutkar AS (2007) NADPH dependent covalent binding of [^{3}H] paroxetine to human liver microsomes and S-9 fractions: identification of an electrophilic quinone metabolite of paroxetine. *Chem Res Toxicol*, 20:1649 –1657.

61 Uetrecht J (2008) Idiosyncratic drug reactions: past, present, and future. *Chem Res Toxicol*, 21:84–92.

62 Park KB, Dalton-Brown E, Hirst C, and Williams DP (2006) Selection of new chemical entities with decreased potential for adverse drug reactions. *Eur J Pharmacol*, 549:1–8.

63 Guengerich FP and MacDonald JS (2007) Applying mechanisms of chemical toxicity to predict drug safety. *Chem Res Toxicol*, 20:344–369.

64 Evans DC and Baillie TA (2005) Minimizing the potential for metabolic activation as an integral part of drug design. *Curr Opin Drug Discov Devel*, 8:44–50.

65 Caldwell GW and Yan Z (2006) Screening for reactive intermediates and toxicity assessment in drug discovery. *Curr Opin Drug Discov Devel*, 9:47–60.

66 Doss GA and Baillie TA (2006) Addressing metabolic activation as an integral component of drug design. *Drug Metab Rev*, 38:641–649.

67 Shu YZ, Johnson BM, and Yang TJ (2008) Role of biotransformation studies in minimizing metabolism-related liabilities in drug discovery. *AAPS J*, 10:178–192.

68 Gan J, Ruan Q, He B, Zhu M, Shyu WC, and Humphreys WG (2009) in vitro screening of 50 highly prescribed drugs for thiol adduct formations comparison of potential for drug-Induced toxicity and extent of adduct formation. *Chem Res Toxicol*, 22:690–698.

69 Bauman JN, Kelly JM, Tripathy S, Zhao SX, Lam WW, Kalgutkar AS, and Obach RS, (2009) Can in vitro metabolism-dependent covalent binding data distinguish hepatotoxic from nonhepatotoxic drugs? An analysis using human hepatocytes and liver S-9 fraction. *Chem Res Toxicol*, 22:332–340.

70 Dahal UP, Obach RS, and Gilbert AM, (2013) Benchmarking in vitro covalent binding burden as a tool to assess potential toxicity caused by nonspecific covalent binding of covalent drugs. *Chem Res Toxicol*, 26:1739–1745.

71 Thompson RA, Isin EM, Ogese MO, Mettetal JT, and Williams DP (2016) Reactive metabolites: current and emerging risk and hazard assessments. *Chem Res Toxicol*, 4:505–533.

8

Human-Based *In Vitro* Experimental Approaches for the Evaluation of Metabolism-Dependent Drug Toxicity

Albert P. Li

In Vitro ADMET Laboratories, Inc., Columbia, MD, USA

8.1 Introduction

A major challenge in drug development is the identification and removal of drug candidates with human-specific toxicity that cannot be detected in preclinical animal safety trials. Human-based *in vitro* systems with human-specific drug-metabolizing enzymes represent the only experimental systems for the evaluation of human-specific drug toxicity. As discussed in a Chapter 4, both P450 and non-P450 enzymes can participate in metabolism-dependent drug toxicity via the formation of toxic and reactive metabolites. An appropriate human-based *in vitro* system must be competent in all human drug-metabolizing enzyme pathways. As of this writing, primary human hepatocytes represent the most appropriate experimental model for this application. I will review below several human hepatocyte assays for the evaluation of drug toxicity, including metabolism-dependent drug toxicity.

8.2 Assays for Reactive Metabolites

One hallmark of a drug with metabolism-dependent toxicity is the formation of reactive metabolites. Identification of reactive metabolite formation thereby is one of the routine assays in drug development. Via structural elucidation of reactive metabolites, one can then modify the chemical structures of the drug candidates to block the site of metabolic activation. In general, the approach involves incubation of the chemical compounds with human liver microsomes (HLMs) in the presence of a trapping agent for the reactive metabolites, followed by identification of the trapped reactive metabolites by liquid chromatography-mass spectrometry (LCMS). An extensive review of LC-MS evaluation of reactive metabolites in provided in Chapter 7.

Transporters and Drug-Metabolizing Enzymes in Drug Toxicity, First Edition. Edited by Albert P. Li.

8.2.1 Glutathione Trapping Assay

A commonly used approach for the identification of reactive metabolites is the glutathione (gamma-GluCysGly; GSH) trapping assay. GSH conjugation, catalyzed by glutathione transferase (GST) is a cellular defense mechanism to alleviate the toxic effects of reactive metabolites. The GSH trapping assay in general is performed by incubating the drug in question with HLMs or post-mitochondrial supernatant (S9) in the presence of β-nicotinamide adenine dinucleotide 2′-phosphate reduced (NADPH) to allow oxidative metabolism, with GSH in the reaction mixture for conjugation with the reactive metabolites. The reactive metabolites are quantified and structure elucidated via liquid chromatography (LC)–mass spectrometry (MS) approaches monitoring characteristic product ions of gamma-glutamyl-dehydroalanyl-glycine, followed by full scan mode for acquisition of exact mass of GSH adducts and data-dependent MS(2) scan through isotopic matching for monitoring neutral loss fragments and for structural elucidation of the GSH adducts [1]. This approach has also been successfully applied with human hepatocytes using an exogenous mixture of unlabeled and stable isotope labeled (SIL: [1,2-(13)C, (15)N]-glycine) GSH (SIL-GSH) [2]. Drugs associated with drug induced liver injuries (DILI) known to form reactive metabolites using this assay include acetaminophen [3–5], nefazodone [6], tacrine [7], and troglitazone [8–10]. LCMS approaches for the detection of reactive metabolites have been thoroughly reviewed by others, including the use of trapping agents other than GSH [11] such as radiolabeled cysteine and cysteine analogues [12–14], cyanide [15], methoxylamine [16], glyoxal [17], semicarbazide [18], and peptides with cysteine and other nucleophilic amino acid residues [19, 20].

8.2.2 Covalent Binding Assay

That reactive metabolite formation and the subsequent covalent binding to cellular macromolecule is a mechanism of drug induced toxicity has been proposed several decades ago [21, 22], with covalent binding assay continuing to be applied toward the screening of new chemical entities (NCE) for toxic potential [23]. Treatment of experimental animals with a toxic drug followed by proteomics quantification of target proteins is one of the latest technologies in the identification of target proteins [24], allowing the identification of target organs and cell types. Quantification of covalent binding in general is performed with the incubation of radiolabeled drugs with HLMs, followed by precipitation of the protein and exhaustive washing to remove unbound radioactivity, and quantification of binding expressed as pmol bound/mg protein (>50 pmol bound mg^{-1} microsomal protein has been proposed as the criterion for unacceptable toxicity [23]). This assay is now considered to be an investigational but not a definitive decision tool due to the abundance of nontoxic and efficacious drug candidates with covalent binding potential. One interesting development is that intentional covalent binding to target proteins (targeted covalent inhibitor, TCI) is now being investigated for the development of new drugs [25].

8.3 Cell-Based Assays for Metabolism-Dependent Toxicity

In vitro cell culture systems have been applied toward the evaluation of drug toxicity. To evaluate metabolism-dependent toxicity, the experimental system should contain drug-metabolizing enzyme activities similar to that *in vivo*, including both organ-specific and species-specific properties.

In vitro screening for human hepatotoxicity using human hepatic cells as an approach to minimize hepatotoxic potential of drug candidates is now routinely used in drug development. The screening assays established for this application include primary cultured human hepatocytes [26–30], stem [31–35] derived human hepatocyte, and hepatocyte cell lines including HepG2 [36], HuH-7 [37], and HepaRG [38]. Of these systems, cryopreserved primary human hepatocytes represent the "gold standard" for the evaluation of human drug properties due to the expression of *in vivo* hepatic properties including complete, uninterrupted drug-metabolizing enzyme pathways, transporter-mediated drug uptake and efflux, and functional bile efflux [39–42]. Primary human hepatocytes are routinely applied in drug development for the evaluation of human drug properties, including drug metabolism, drug–drug interactions, toxicity, and pharmacology [30, 40, 43–49]. Successful cryopreservation [47] allows human hepatocytes to be used routinely. Virtually all pharmaceutical companies apply human hepatocytes in their drug discovery and development programs.

Deficiency in drug-metabolizing enzyme activities is the major drawback of stem cell-derived hepatocytes and hepatocyte cell lines [50] but can be improved via transfection of specific drug-metabolizing enzymes. For instance, HepG2 transfected with CYP isoforms [51, 52] have been applied toward the evaluation of the role of specific toxicity pathways. Of the hepatic cell lines, the most functional is HepaRG cells which have been shown to exhibit activities for several key P450 isoforms, but have been reported to be deficient in CYP2D6, CYP2C9, CYP2E1, and CYP3A5 gene expression and activities [53, 54].

Primary human hepatocytes thereby represent the most physiologically relevant experimental model for the evaluation of metabolism-dependent drug toxicity in humans.

8.4 Primary Human Hepatocyte Assays for Metabolism-Dependent Drug Toxicity

As described earlier, drugs with metabolism-dependent toxicity are likely to cause idiosyncratic drug toxicity due to individual differences in drug metabolism as well as fluctuations in drug-metabolism capacity in an individual due to environmental factors. Further, these drugs may exhibit species differences in toxicity due to species differences in drug metabolism. The following are assays developed with human hepatocytes in our laboratory for applications in early drug development to minimize hepatotoxic potential [55].

8.4.1 *In Vitro* Screening Assays for Hepatotoxicity

In this assay, palatable cryopreserved human hepatocytes are cultured for four hours, followed by treatment with the test articles for a designated time-period (e.g. 24 hours), followed by quantification of viability. This assay can be performed in 96-, 384-, and 1536-well plates. The endpoints used include cytoplasma enzyme leakage including lactate dehydrogenase (LDH, aspartate transaminase [AST], alanine transaminase [ALT]) for the evaluation of membrane damage; MTT/MTS/WST assay for mitochondrial metabolism, cellular adenosine triphosphate (ATP) contents for overall cell viability, GSH depletion for reactive metabolite generation, and caspase activation assay for apoptosis. For routine screening, cellular ATP assay represents the most useful as it covers virtually all cytotoxic mechanisms including mitochondrial damage, membrane disruption, and apoptosis. Further, luminescent quantification offers a low background signal, in contrast with LDH release assay which generally has a high basal level in untreated controls due to the high level of spontaneous release of cytoplasmic enzymes by cultured hepatocytes [56, 57]. Quantification of cellular ATP in general represents the most robust and sensitive endpoint for the screening of hepatotoxic potential of NCE in drug development. The *in vitro* hepatotoxicity assay can also be applied toward the evaluation of species differences and individual differences in drug toxicity. Examples of the application of human and nonhuman animal hepatocytes to demonstrate species differences is shown in Figure 8.1. Application of human hepatocytes from multiple individual to demonstrate that elevated CYP3A4 activity can be a risk factor for drug toxicity [59] is shown in Figure 8.2.

8.4.2 Cytotoxic Metabolic Pathway Identification Assay (CMPIA)

Cytotoxic metabolic pathway identification assay (CMPIA) is based on the concept that the cytotoxicity of a metabolism-dependent hepatotoxicant can be attenuated by an inhibitor of drug-metabolizing enzyme activity. In this assay, we evaluate the cytotoxicity of a toxicant in primary hepatocytes in the presence and absence

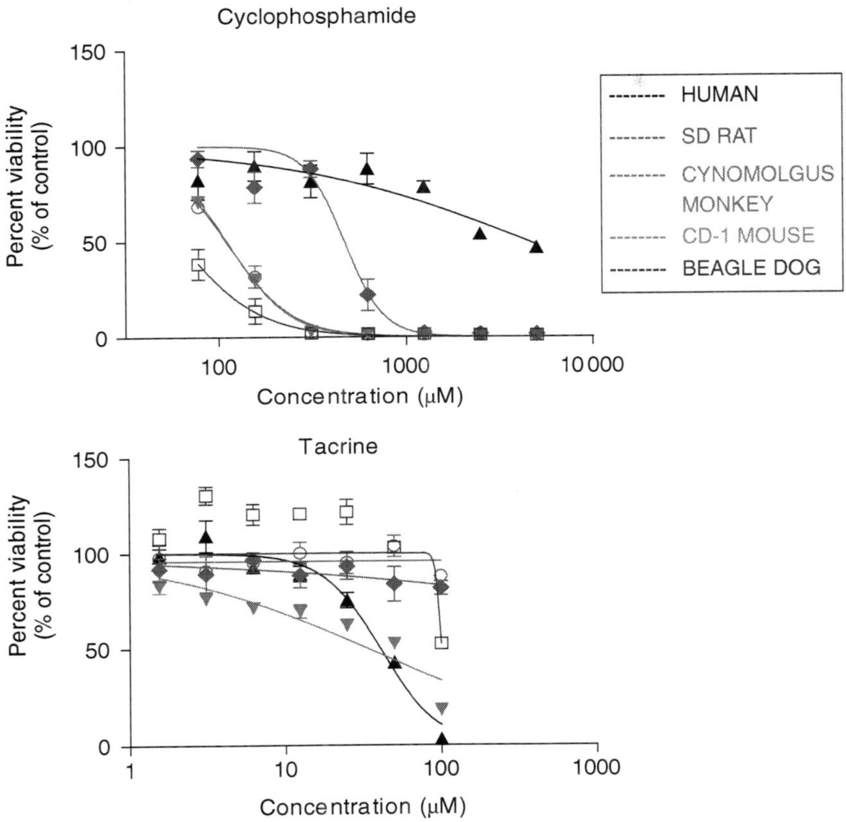

Figure 8.1 *In vitro* hepatotoxicity assay for species comparison. The assay is performed with human hepatocytes and hepatocytes from three animal species routinely used in preclinical toxicology studies. Data shown are results with hepatocytes from male CD1 mouse, male Sprague Dawley rat, male beagle dog, and male cynomolgus monkey for two hepatotoxic drugs, cyclophosphamide (all species except monkey) and tacrine (all species). The results showed clear species difference, with cyclophosphamide hepatotoxicity being lowest, and tacrine hepatotoxicity being higher in human and monkey hepatocytes. The result illustrates that preclinical animal species may overestimate or underestimate human drug toxicity. In the case of overestimation, a drug candidate which may have superior efficacy may not be selected for further development. In the case of underestimation, a drug candidate may be approved for clinical trial based on preclinical safety profiles but would fail in regulatory clinical trials due to unacceptable adverse effects. It is interesting to note that tacrine was the first drug approved for marketing for the treatment of Alzheimer's disease in 1993, but was withdrawn from the US market in 2012 due to its association with fatal liver toxicity [58]. *Source:* Based on Qizilbash et al. [58].

of a P450 inhibitor, 1-aminobenzotriazol, which is known to inhibit a majority of P450 isoforms [60]. A decrease in cytotoxicity in the absence or diminished drug metabolism would suggest that the drug evaluated exhibits metabolism-dependent toxicity (Figure 8.3) [55].

Donor	Lot	CL_{int}, hep (Intrinsic Hepatic Clearance (pmol/min/million cells))					
		CYP1A2 (Tacrine)	CYP2B6 (Bupropion)	CYP2C8 (Amodiaquine)	CYP2C9 (Diclofenac)	CYP2D6 (Dextromethorphan)	CYP3A4 (Midazolam)
1	1007	2.7	3.8	271.4	43.2	6.4	6.3
2	1023	3.0	10.2	159.6	36.2	4.6	26.2
3	1026	1.3	0.8	38.1	20.7	4.6	2.9
4	1031	1.0	2.4	154.6	13.1	2.9	3.3
5	1033	1.9	0.6	88.1	14.1	2.8	1.7
6	1036	1.5	1.1	107.2	18.3	2.6	4.7

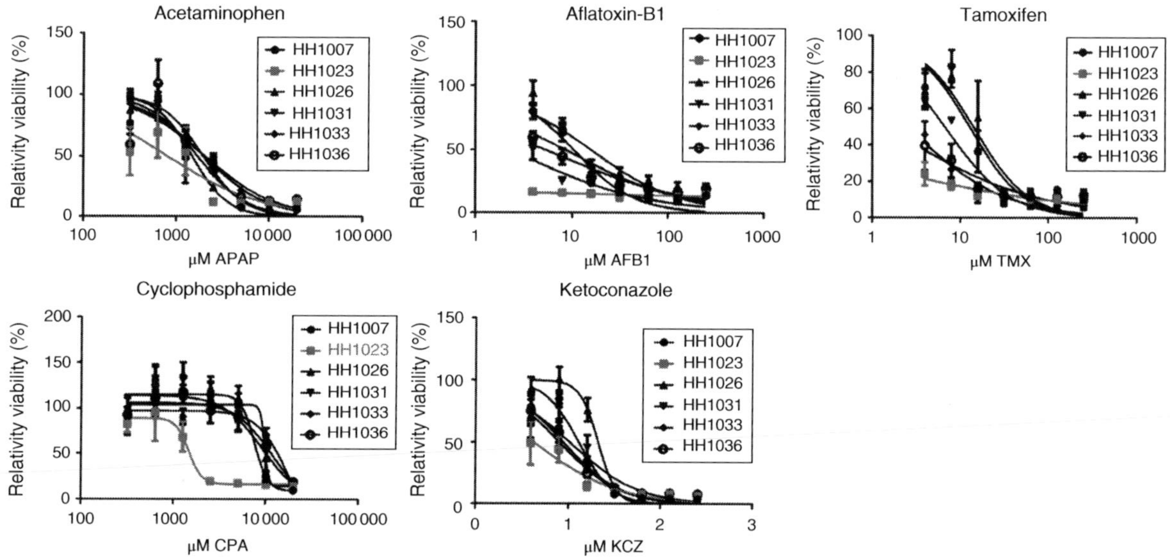

Figure 8.2 Evaluation of inter-individual difference in drug toxicity with human hepatocytes. In this study, elevated CYP3A4 activity was identified to be a potential risk factor for DILI. Evaluation of *in vitro* hepatotoxicity using hepatocytes from multiple donors show that the donor with the highest CYP3A4 activity (Lot 1023) was most sensitive to the cytotoxic effects of the metabolically activated hepatotoxicants: acetaminophen, aflatoxin B1, cyclophosphamide, ketoconazole, and tamoxifen. The results illustrate the possibility of exacerbation of drug toxicity when an individual has exaggerated activity of the activating drug-metabolizing enzyme pathways due to environmental and/or genetic factors at the time of administration of a drug that can be metabolized to toxic metabolites. The table illustrates the P450 isoform activities of the different human hepatocyte lots using isoform-selective substrates (shown in brackets). *Source:* Adapted from Utkarsh et al. [59].

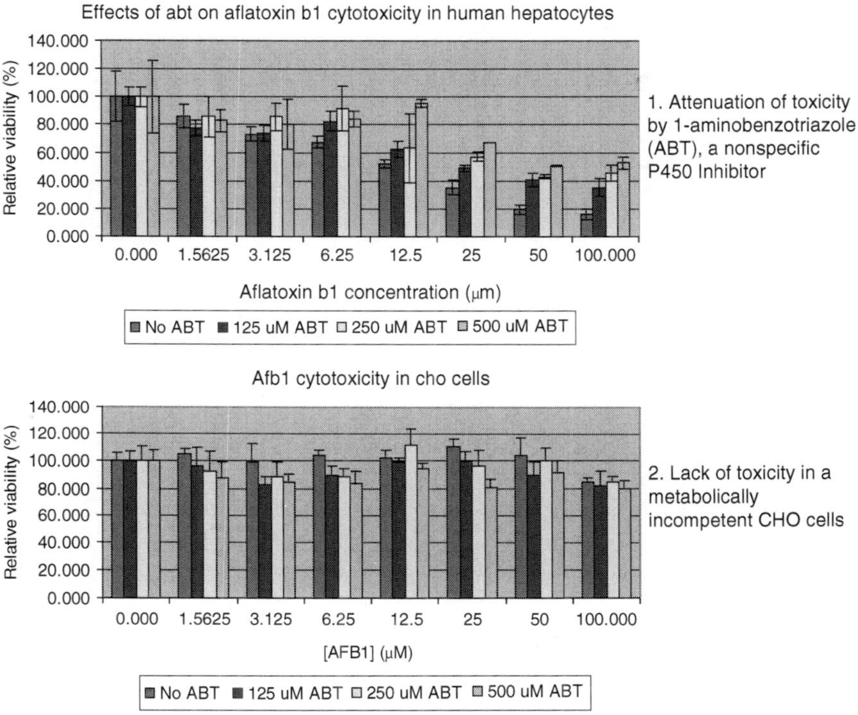

Figure 8.3 Cytotoxic Metabolic Pathway Identification Assay (CMPIA) and Metabolic Comparative Cytotoxicity Assay (MCCA). In this assay, a drug with metabolically activated toxicity is identified using two approaches as demonstrated here with aflatoxin B1 (AFB1). Firstly, in CMPIA (top chart), 1-aminobenzotriazole (IBT), a nonselective P450 inhibitor, was found to attenuate the cytotoxicity of AFB1 in human hepatocytes (top), demonstrating that P450 is involved in the metabolic activation of AFB1. Secondly, in MCCA, AFB1 was shown not to be cytotoxic toward a cell line with no drug metabolism enzymes, Chinese hamster ovary cells (bottom chart), while cytotoxicity was observed in the metabolically competent human hepatocytes (top chart), therefore confirming that drug metabolism is likely to be involved in AFB1 toxicity. The two assays can be readily performed in the same study for the identification of metabolism-dependent cytotoxicity. *Source:* Adapted from Li [55].

8.4.3 Metabolic Comparative Cytotoxicity Assay (MCCA)

The assay was developed in our laboratory based on the concept that a metabolism-dependent cytotoxic drug has higher cytotoxicity in a cell system with robust drug-metabolizing enzyme activities (metabolically competent), where the parent drug is metabolized to more toxic metabolites, than in a metabolically incompetent system. In this assay, we employ primary human hepatocytes as the metabolically competent system, and Chinese hamster ovary cells as the incompetent counterpart. A higher cytotoxicity observed in primary human hepatocytes versus CHO cells would indicate that metabolism may be involved in the manifestation of cytotoxicity, as demonstrated for the protoxicant aflatoxin B1 (Figure 8.3) [55].

8.4.4 MetMax™ Cryopreserved Human Hepatocytes (MMHH) Metabolic Activation Cytotoxicity Assay (MMACA)

Exogenous metabolic activation systems are routinely used in genotoxicity assays for the evaluation of promutagens – chemicals that are nongenotoxic but can be metabolized to genotoxic metabolites. A target cell population (e.g. *Salmonella typhimurium* [61], Chinese hamster ovary cells [62], mouse lymphoma cells [63]) which is not competent in drug metabolism is treated with a promutagen in the presence of the exogenous activating system such as rat liver supernatant (S9) [64] and hepatocytes [65]. The promutagen is metabolized by the exogenous metabolic system to genotoxic metabolites, causing mutations in the target organism. This approach is adopted in our laboratory for the evaluation of protoxicants – toxicants that require metabolic activation for the manifestation of toxicity. We have developed a novel exogenous metabolic activation system, namely MetMax™ cryopreserved human hepatocytes (MMHH) [66, 67]. MMHH are permeabilized human hepatocytes supplemented with phase 1 and phase 2 drug-metabolizing enzyme cofactors. In this MMHH metabolic activation cytotoxicity assay (MMACA), HEK293 cells are used as the metabolically incompetent target cells. Metabolism-dependent cytotoxicity is indicated by an increase in cytotoxicity in the presence of MMHH (Figure 8.4).

8.5 Emerging Hepatocyte Technologies for the Evaluation of Drug Toxicity

8.5.1 Human Hepatocytes ROS/ATP Assay for DILI Drugs

Drugs cause idiosyncratic drug toxicity which, due to the low incidence, cannot be readily detected in human clinical trials. I proposed previously that these drugs may be identified based on their common properties which can be used as biomarkers for their detection. Based on literature research, reactive metabolite formation, metabolism-dependent cytotoxicity, and drug interaction potential have been identified as likely biomarkers. My hypothesis has led to a collaboration with the U.S. FDA's National Center for Toxicology Research, resulting in the development of the reactive oxygen species (ROS)/ATP human hepatocyte assay with specificity and sensitivity values of near 90% in the detection of drugs known to cause severe liver injuries resulting in death or a need for liver transplantation [39].

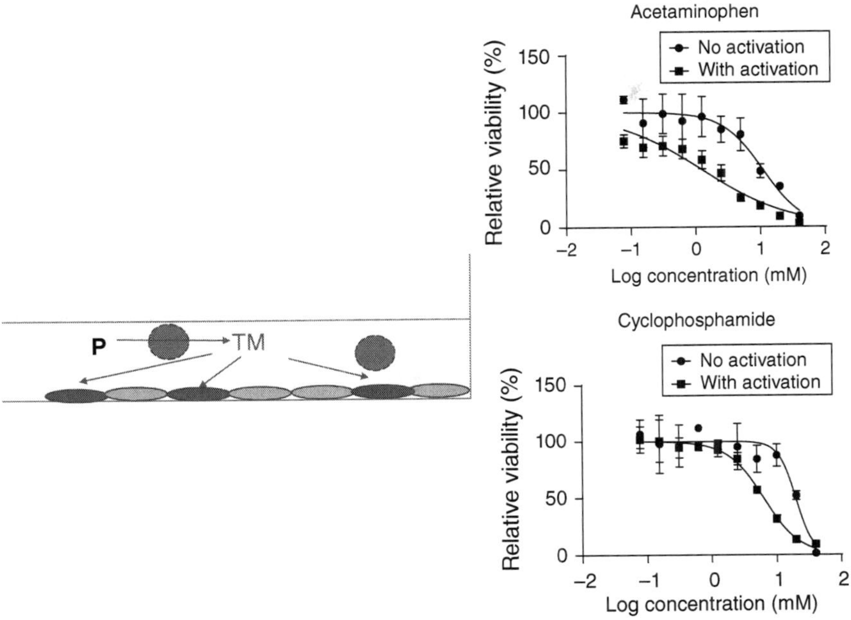

Figure 8.4 MetMax™ Cryopreserved Human Hepatocytes (MMHH) Metabolic Activation Cytotoxicity Assay (MMACA). The assay consists of a metabolically incompetent target cell line (HEK293) and an exogenous metabolic activating system, MMHH. MMHH (blue circles) are permeabilized human hepatocytes, which, in the presence of cofactors, can metabolize a protoxicant (P) to its toxic metabolites (TM) which in turn cause cytotoxicity in the target cells. We routinely use cellular ATP contents as an assay for the quantification of the viability of the target cells. The effectiveness of the MMMCA in metabolic activation is demonstrated by the increase of the cytotoxicity of acetaminophen and cyclophosphamide in the presence of MMHH.

The success of this assay is consistent with the theory that metabolism-dependent cytotoxicity involving reactive metabolites is a key property of drugs with severe hepatotoxicity. In this assay, human hepatocytes are treated with the drug in question up to 200 μM for 48 hours, followed by the quantification of cellular ATP contents and ROS formation. The area under the curve (AUC) is quantified plotting ROS/ATP ratio versus dose is compared to known toxic and nontoxic drugs. We routinely used ketoconazole and nefazodone as the DILI drugs, and cimetidine and ifosfamide as the non-DILI drugs. The DILI potential of the drugs evaluated is made based on the comparison of the ROS/ATP AUC value to that of the positive and negative controls (Figure 8.5). In our laboratory, a test article with ROS/ATP AUC equal or over 30% of that for ketoconazole is classified as having the potential to cause severe liver injuries.

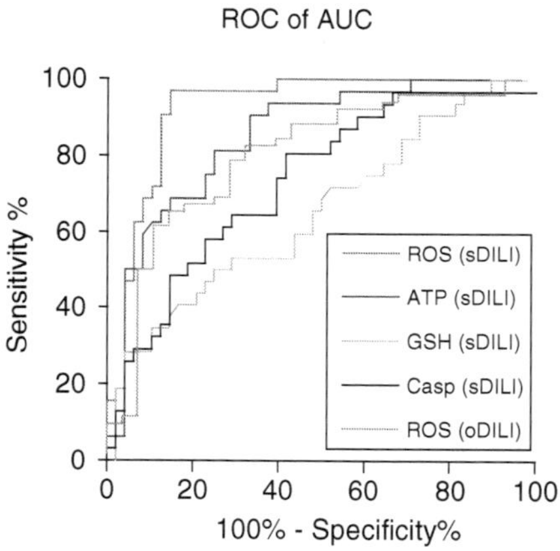

sDILI	ROS/ATP	GSH	ATP	Casp3/7
Area	0.9245	0.6501	0.8506	0.7325
P value	<0.0001	0.0236	<0.0001	0.00052
Cut-off	>111.3	<169.0	<163.3	>118.2
Sensitivity%	90	66	81	74
95% CI	74–98%	47–81%	64–93%	55–88%
Specificity%	86	52	75	60
95% CI	74–94.%	37–66%	60–86%	45–74%
LR	6.56	1.37	3.25	1.87

oDILI	ROS/ATP	GSH	ATP	Casp3/7
Area	0.8001	0.5151	0.7586	0.7005
P value	<0.0001	0.8244	0.00015	0.00324
Cut-off	>81.83	<169.0	<171.8	>98.99
Sensitivity%	79	66	75	75
95% CI	65–89%	47–81%	61–86%	61–86%
Specificity%	71	52	68	54
95% CI	51–87%	37–66%	48–84.%	34–73%
LR	2.76	1.37	2.33	1.62

Figure 8.5 ROS/ATP AUC Human Hepatocyte Assay for DILI Drugs. This assay, co-developed with the National Center for Toxicological Research of US FDA, is able to accurately identify DILI drugs with fatal consequences (sDILI). The assay was developed with 80 drugs in the testing set and 72 drugs in the validation set with similar contribution of DILI and non-DILI drugs. The sensitivity and specificity of the various endpoints in the identification of sDILI and oDILI (all DILI drugs, including drugs not causing liver failure) are shown. *Source:* Adapted from Zhang et al. [39].

8.5.2 Long-Term Hepatocyte Cultures

8.5.2.1 999Elite™ Long-Term Cultured Human Hepatocytes

In our laboratory, via optimization of cryopreservation and recovery procedures, we have developed the 999Elite™ cryopreserved human hepatocytes –>90% viability upon recovery from thawing, formation of >90% confluent cultures, and capability to be cultured without deterioration for >9 days (Figure 8.6). These long-term cultured human hepatocytes are now being applied toward the evaluation of various drug properties, including hepatotoxicity.

8.5.2.2 Hepatocyte/Non-Hepatocyte Cocultures

Stable long-term culture hepatocytes have been developed via coculturing of hepatocytes with non-hepatocytes, including mouse 3T3 fibroblasts (micropatterned cocultures) [68, 69] and liver nonparenchymal cells [70, 71]. These cocultured models are reported to maintain drug-metabolizing enzyme activities upon prolonged culturing [72, 73], thereby applicable for the evaluation of metabolism-dependent drug toxicity.

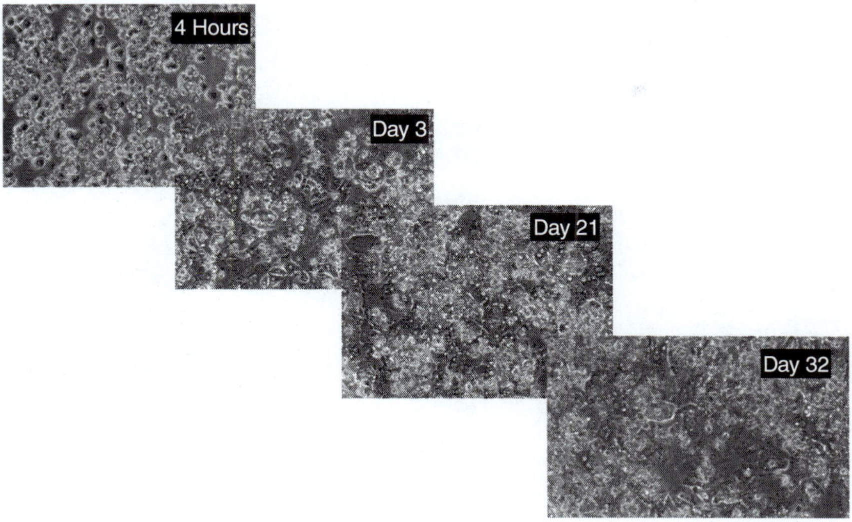

Figure 8.6 Long-term cultured 999Elite™ cryopreserved human hepatocytes. The hepatocytes are optimally cryopreserved human hepatocytes with >90% post-thaw viability, forming >90% confluent cultures, and can be cultured for a prolonged duration. The morphology of the hepatocytes upon culturing for various durations up to day 32 is shown. The long-term cultures are being applied toward the investigation of various aspects of drug development, including long-term drug toxicity and the efficacy and duration of gene therapy modalities.

8.5.2.3 Human Hepatocyte Spheroids

Culturing of cells as 3-dimensional (3-d) aggregates or spheroids has been developed decades ago with tumor cells to model solid tumors [74, 75] and have been applied toward hepatocytes [76–79]. A major advantage of hepatocyte spheroids is the longevity of the culture. In our laboratory, spheroids can be readily cultured from the 999Elite human hepatocytes and can be maintained for over three weeks in culture while retaining drug-metabolizing enzyme activities. Human hepatocyte spheroids have been applied toward the evaluation of hepatotoxicity [28, 80, 81]. In our laboratory, spheroids can be routinely cultured from the 999Elite™ human hepatocytes. The morphology of human hepatocyte spheroids and their application in the evaluation of drug toxicity upon prolonged treatment of 15 days is shown in Figure 8.7. This experimental system is now being applied toward the evaluation of various drug properties including drug metabolism, drug–drug interactions, and drug toxicity in our laboratory.

8.5.2.4 Microfluidic 3-Dimensional (3-d) Hepatocyte Cultures

Recently, microfluidic technologies have been applied toward 3-d cultures of hepatocytes, modeling *in vivo* perfusion of the liver by oxygenated blood from the

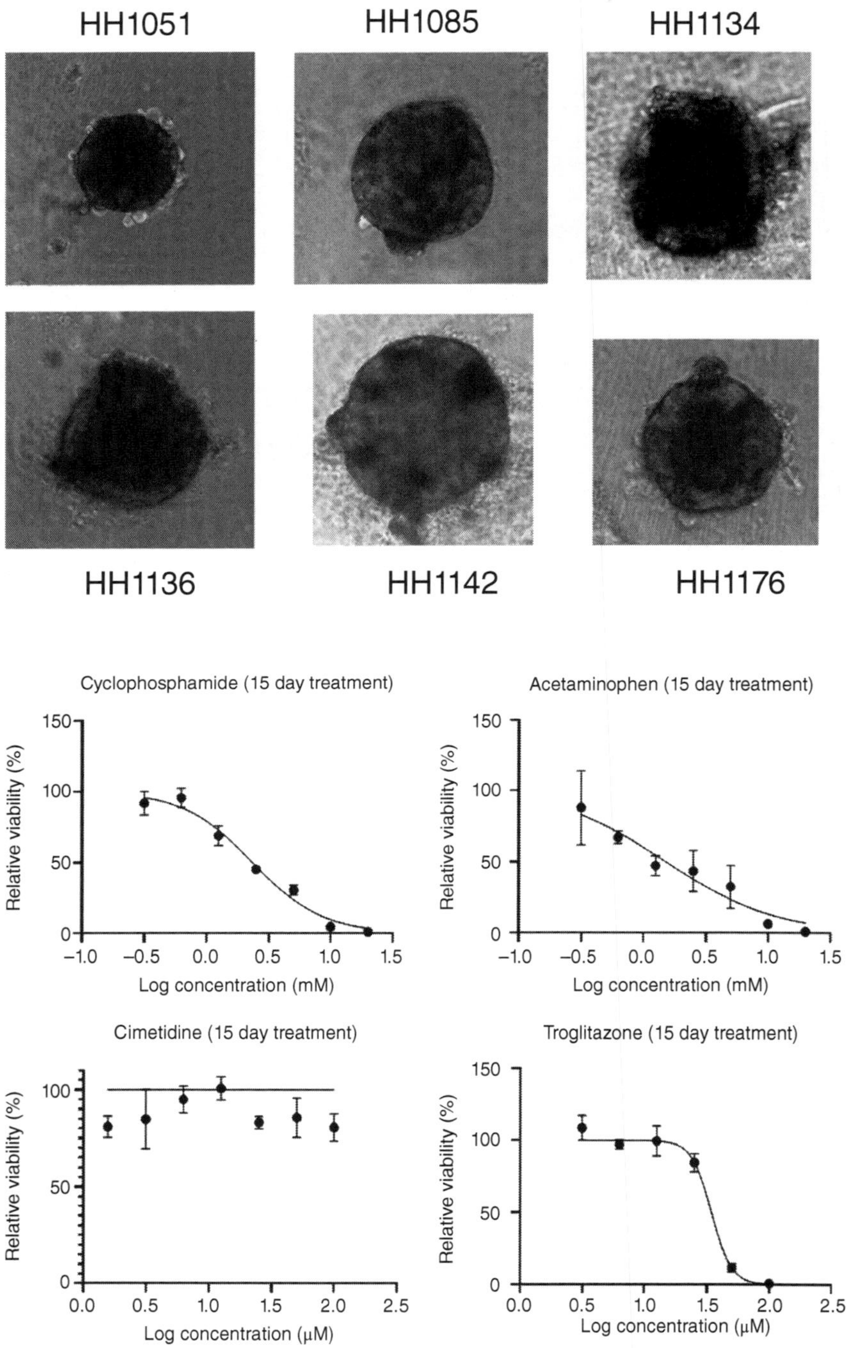

Figure 8.7 *In vitro* hepatotoxicity with human hepatocytes. Human hepatocyte spheroids developed from multiple lots of 999Elite™ human hepatocytes. The hepatocytes were cultured in ultra-low attachment 384-well plates, with approximately 2000 cells per well. The spheroids were formed in five to seven days. The human hepatocyte spheroids are now being characterized in our laboratory for hepatic functions upon prolonged culturing (>20 days) for the evaluation of their application in the evaluation of long-term drug metabolism, drug–drug interactions, and drug toxicity. The results of 15-day treatment of human hepatocytes with hepatotoxic (cyclophosphamide, acetaminophen, and troglitazone) and nonhepatotoxic (cimetidine) drugs, with viability quantified by cellular ATP contents, is shown.

portal circulation. The goal is to provide an *in vitro* model of the liver *in vivo* which may allow the evaluation of hepatotoxicity and which may not be reproduced with static 2-d or 3-d models. These models include 3-d cultures of hepatocytes [82–88] and cocultures of hepatocytes and non-hepatocytes in microfluidic perfusion chambers [89–92]. Microfluidic cocultures of hepatocytes and hepatic non-parenchymal cells such as stellate cells, endothelial cells, and Kupffer cells theoretically represent the best *in vitro* model of the liver. However, it is technically challenging to create this complex experimental model with the various parameters including the ratio of the multiple cell types, the physical positioning of the cells to mimic the liver *in vivo*, and to provision adequate oxygenation resembling the portal circulation.

8.6 Integrated Discrete Multiple Organ Coculture (IdMOC®)

Metabolism-dependent toxicity can be readily studied in hepatocytes due to their inherent robust drug-metabolizing enzyme activities. However, nonhepatic organs without significant drug-metabolizing enzyme activities are often targets of metabolism-dependent drug toxicity via exposure to toxic metabolites generated by the liver *in vivo*. Examples are the clinically manifested neurotoxicity of ifosfamide [93, 94] and 3,4-methylenedioxymethamphetamine (MDMA or "ecstasy") [95, 96], cardiotoxicity of cyclophosphamide [97] and doxorubicin [98, 99], and immunotoxicity of cocaine [100] and cyclophosphamide [101], where the target cells – neuronal cells, cardiomyocytes, and splenocytes/lymphocytes – are virtually devoid of drug-metabolism enzyme activities.

Integrated discrete multiple organ coculture (IdMOC) was developed in our laboratory as a practical *in vitro* experimental system for the evaluation of multiple organ toxicity in the presence of hepatic metabolism [102–104]. The experimental system is based on a wells-in-a well concept, where cells from individual organs are cultured in each of the inner wells, thereby physically separated ("discrete") from cells from other organs. After the cells are attached to the inner wells, medium is added to "flood" all the inner wells, thereby allowing the multiple cell types to communicate with each other via the overlying medium ("integrated") (Figure 8.8). This experimental system allows the simultaneous evaluation of toxicity of cells from different organs in the presence of hepatic metabolism. Applications include the evaluation of the efficacy and toxicity of anticancer agents as exemplified by the evaluation of the cytotoxicity of tamoxifen on cardiomyocytes (cardiotoxicity), proximal tubule epithelial cells (renal toxicity), aortic endothelial cells (vascular toxicity), hepatocytes (liver toxicity), and the intended target cells, the mammary adenocarcinoma MCF-7 cells cocultured in

Figure 8.8 Integrated Discrete Multiple Organ Coculture. A photograph of the 24-well IdMOC® plate is shown. Each plate consists of four containing wells, each containing six shallow inner wells. A typical application is to coculture two cell types, with three wells per cell type, such as the coculture of nonhepatic cells (e.g. splenocytes) with hepatocytes for the incorporation of hepatic metabolism in the evaluation of drug toxicity toward the nonhepatic cells.

IdMOC. The effectiveness of IdMOC to evaluate metabolic activation of protoxicants has been demonstrated with various protoxicants including cyclophosphamide, aflatoxin, and 4-aminophenol [105–107]. IdMOC has been applied to evaluate the cytotoxicity of cigarette smoke condensates toward various human pulmonary cell types including vascular endothelial cells, large airway epithelial cells, and small airway epithelial cells [108], and toward the hepatic metabolic activation of the cytotoxicity of botanical chemicals citrinin and ochratoxin [109]. The application of IdMOC to evaluate the immunotoxicity of cyclophosphamide in human splenocytes is shown in Figure 8.9.

We have employed IdMOC coculture of 3T3 cells and hepatocytes to classify toxicants based on metabolic activation [105] (Figure 8.10).

- *Type I: Direct-acting (e.g.* ***Tamoxifen****): These toxicants act directly on the target cells without a need for metabolic activation. For these toxicants, organ-specific toxicity is dependent of route of exposure.*
- *Type II: These are toxicants that are metabolically activated to highly reactive metabolites that mainly cause toxicity in the cells where they are formed (localized toxicity) (e.g.* ***Aflatoxin-B1****). The high reactivity of the metabolites precludes them from being transported by the systemic circulation to cause toxicity at distal sites. Organ-specific toxicity of these toxicants therefore is dependent on the site of metabolic activation.*

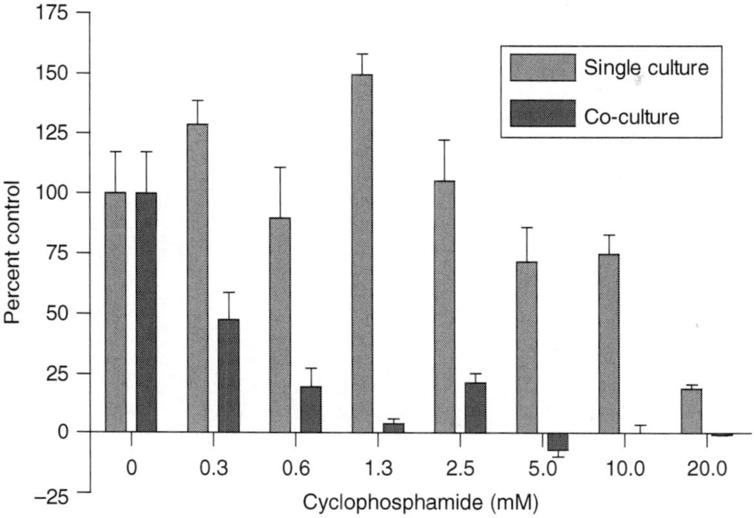

Figure 8.9 Protoxicant assay with IdMOC. The IdMOC system allows incorporation of a metabolically competent cell type such as hepatocytes in the evaluation of the cytotoxicity of a protoxicant of a metabolically incompetent cell type such as splenocytes. The hepatocytes metabolize the protoxicant to toxic metabolites which cause toxicity in the cocultured metabolically incompetent cells, thereby modeling the effects of liver metabolism on the toxicity of drugs toward extrahepatic tissues. The application of IdMOC as a model in the evaluation of the immunotoxicity of cyclophosphamide is shown here using freshly isolated human splenocytes as the target cells cocultured with human hepatocytes. Cyclophosphamide was shown to significantly inhibit splenocyte function quantified as phytohemagglutinin stimulation of cell proliferation in the presence of hepatocytes (coculture) but not in the absence of hepatocytes (single culture). The data were developed by a postdoctoral scientist in our laboratory, Hariharan Saminathan. *Source:* Adapted from Li et al. [105].

- *Type III: These are toxicants that form stable, diffusible toxic metabolites (e.g.* ***Cyclophosphamide****). For these toxicants, organ-specific toxicity dependent on metabolite distribution.*

8.7 Conclusion

Experimental evaluation of metabolism-dependent drug toxicity is an important activity to define drug toxicity. A drug that is metabolically activated to highly toxic metabolites may exhibit species differences in toxicity due to species difference in drug-metabolism enzymes, resulting in the inability of preclinical animal safety evaluation to accurately predict human toxicity, leading to failure in clinical safety trials due to unacceptable adverse effects. Furthermore, there may be individual differences in drug toxicity due to individual variations in drug-metabolic

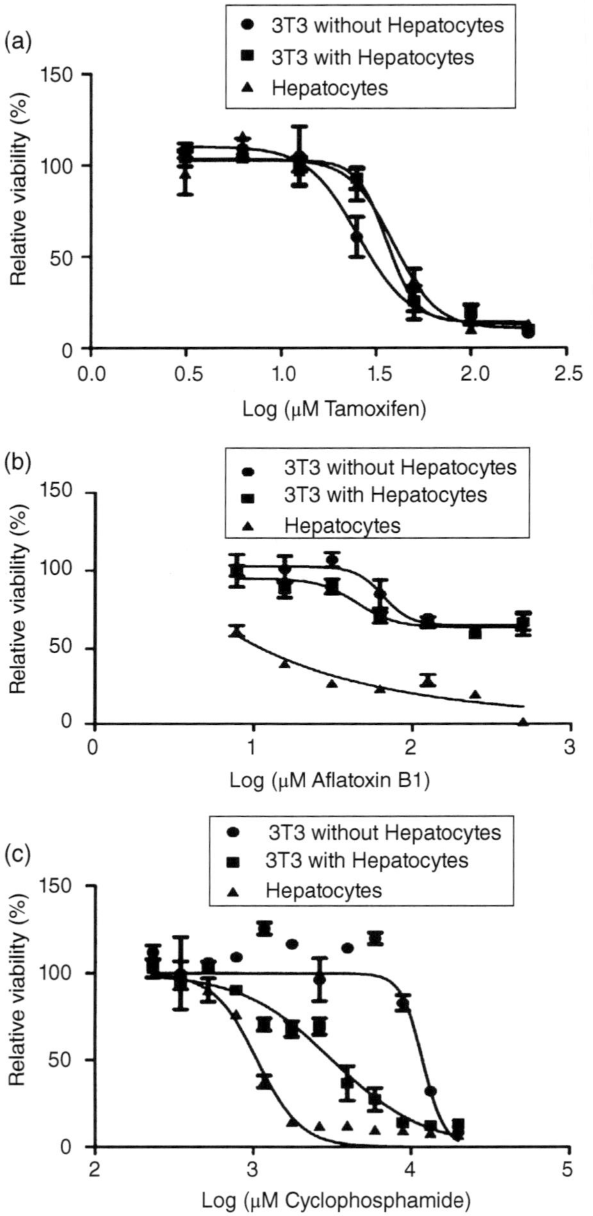

Figure 8.10 Classification of toxicants based on metabolic activation. A metabolically incompetent cell line, mouse 3T3 cells, is cultured with and without human hepatocytes in the IdMOC® system. Cytotoxicity was evaluated in the single cultured 3T3 cells, as well as in 3T3 cells cocultured with human hepatocytes, and in the human hepatocytes. (a) Tamoxifen represents a Class I toxicant – a direct acting toxicant where similar cytotoxicity was observed in 3T3 cells culture alone and cocultured with hepatocytes, and in hepatocytes, Aflatoxin B1 (b) represents a Class II toxicant, with higher cytotoxicity in hepatocytes and lower cytotoxicity in the 3T3 cells with and without hepatocyte coculture, demonstrating that the toxicant exerts its toxicity in the cells where the toxic metabolites are formed. Cyclophosphamide (c) represents a Class III toxicant which forms stable toxic metabolites that can cause toxicity in distant nonhepatic tissues, as demonstrated by a higher cytotoxicity in 3T3 cells cocultured with hepatocytes than without hepatocytes.

enzyme activities due to genetic and environmental factors. Reactive metabolite formation is a common property of a majority of drugs with idiosyncratic drug-induced liver injuries [110] so thereby should be seriously considered in drug development strategies. One plausible explanation for the low incidence of idiosyncratic drug toxicity, as stated by my Multiple Determinant Hypothesis [110], is that individuals succumb to serious drug toxicity due to the co-occurrence of various risk factors that enhanced formation of the toxic metabolites, compromised detoxification activities, and various physiological conditions potentiating the propagation of the toxic consequences.

References

1 Wang Q, Liu H, Slavsky M, Fitzgerald M, Lu C, O'Shea T. A high-throughput glutathione trapping assay with combined high sensitivity and specificity in high-resolution mass spectrometry by applying product ion extraction and data-dependent neutral loss. *J Mass Spectrom* 2019;54(2):158–166.

2 Mezine I, Bode C, Raughley B, Bhoopathy S, Roberts KJ, Owen AJ, et al. Application of exogenous mixture of glutathione and stable isotope labeled glutathione for trapping reactive metabolites in cryopreserved human hepatocytes. Detection of the glutathione conjugates using high resolution accurate mass spectrometry. *Chem Biol Interact* 2013;204(3):173–184.

3 Huang R, Okuno H, Takasu M, Takeda S, Kano H, Shiozaki Y, et al. Effects of rifampin on the glutathione depletion and cytochrome c reduction by acetaminophen reactive metabolites in an *in vitro* P450 enzyme system. *Jpn J Pharmacol* 2000;83(3):182–190.

4 Rashed MS, Nelson SD. Characterization of glutathione conjugates of reactive metabolites of 3'-hydroxyacetanilide, a nonhepatotoxic positional isomer of acetaminophen. *Chem Res Toxicol* 1989;2(1):41–45.

5 Lay JO, Jr., Potter DW, Hinson JA. Fast atom bombardment mass spectrometry and fast atom bombardment mass spectrometry/mass spectrometry of three glutathione conjugates of acetaminophen. *Biomed Environ Mass Spectrom* 1987;14(9):517–521.

6 Bauman JN, Frederick KS, Sawant A, Walsky RL, Cox LM, Obach RS, et al. Comparison of the bioactivation potential of the antidepressant and hepatotoxin nefazodone with aripiprazole, a structural analog and marketed drug. *Drug Metab Dispos* 2008;36(6):1016–1029.

7 Simon T, Becquemont L, Mary-Krause M, de Waziers I, Beaune P, Funck-Brentano C, et al. Combined glutathione-S-transferase M1 and T1 genetic polymorphism and tacrine hepatotoxicity. *Clin Pharmacol Ther* 2000;67(4):432–437.

8 Okada R, Maeda K, Nishiyama T, Aoyama S, Tozuka Z, Hiratsuka A, et al. Involvement of different human glutathione transferase isoforms in the glutathione conjugation of reactive metabolites of troglitazone. *Drug Metab Dispos* 2011;39(12):2290–2297.

9 Prabhu S, Fackett A, Lloyd S, McClellan HA, Terrell CM, Silber PM, et al. Identification of glutathione conjugates of troglitazone in human hepatocytes. *Chem Biol Interact* 2002;142(1–2):83–97.

10 Kassahun K, Pearson PG, Tang W, McIntosh I, Leung K, Elmore C, et al. Studies on the metabolism of troglitazone to reactive intermediates *in vitro* and *in vivo*. Evidence for novel biotransformation pathways involving quinone methide formation and thiazolidinedione ring scission. *Chem Res Toxicol* 2001;14(1):62–70.

11 Ma S, Subramanian R. Detecting and characterizing reactive metabolites by liquid chromatography/tandem mass spectrometry. *J Mass Spectrom* 2006;41(9):1121–1139.

12 Inoue K, Fukuda K, Yoshimura T, Kusano K. Comparison of the reactivity of trapping reagents toward electrophiles: cysteine derivatives can be bifunctional trapping reagents. *Chem Res Toxicol* 2015;28(8):1546–1555.

13 Harada H, Toyoda Y, Abe Y, Endo T, Takeda H. Quantitative evaluation of reactivity and toxicity of acyl glucuronides by [(35)S]cysteine trapping. *Chem Res Toxicol* 2019;32(10):1955–1964.

14 Penner N, Liu X, Prakash C. A new quaternary ammonium cysteine analogue as a trapping reagent for reactive metabolites screening. *Drug Metab Lett* 2014;8(1):36–42.

15 Inoue K, Shibata Y, Takahashi H, Ohe T, Chiba M, Ishii Y. A trapping method for semi-quantitative assessment of reactive metabolite formation using [35S]cysteine and [14C]cyanide. *Drug Metab Pharmacokinet* 2009;24(3):245–254.

16 Zhang C, Wong S, Delarosa EM, Kenny JR, Halladay JS, Hop CE, et al. Inhibitory properties of trapping agents: glutathione, potassium cyanide, and methoxylamine, against major human cytochrome p450 isoforms. *Drug Metab Lett* 2009;3(2):125–129.

17 Li X, Zheng T, Sang S, Lv L. Quercetin inhibits advanced glycation end product formation by trapping methylglyoxal and glyoxal. *J Agric Food Chem* 2014;62(50):12152–12158.

18 Rousu T, Pelkonen O, Tolonen A. Rapid detection and characterization of reactive drug metabolites *in vitro* using several isotope-labeled trapping agents and ultra-performance liquid chromatography/time-of-flight mass spectrometry. *Rapid Commun Mass Spectrom* 2009;23(6):843–855.

19 Mitchell MD, Elrick MM, Walgren JL, Mueller RA, Morris DL, Thompson DC. Peptide-based *in vitro* assay for the detection of reactive metabolites. *Chem Res Toxicol* 2008;21(4):859–868.

20 Laine JE, Auriola S, Pasanen M, Juvonen RO. D-Isomer of gly-tyr-pro-cys-pro-his-pro peptide: a novel and sensitive *in vitro* trapping agent to detect reactive metabolites by electrospray mass spectrometry. *Toxicol in Vitro* 2011;25(1):411–425.

21 Mitchell JR. Covalent binding: useful index of the formation of chemically reactive drug metabolites. *Biochem Soc Trans* 1975;3(5):622–626.

22 Gillette JR. Commentary. A perspective on the role of chemically reactive metabolites of foreign compounds in toxicity. I. Correlation of changes in covalent binding of reactivity metabolites with changes in the incidence and severity of toxicity. *Biochem Pharmacol* 1974;23(20):2785–2794.

23 Park BK, Boobis A, Clarke S, Goldring CE, Jones D, Kenna JG, et al. Managing the challenge of chemically reactive metabolites in drug development. *Nat Rev Drug Discov* 2011;10(4):292–306.

24 Whitby LR, Obach RS, Simon GM, Hayward MM, Cravatt BF. Quantitative chemical proteomic profiling of the *in vivo* targets of reactive drug metabolites. *ACS Chem Biol* 2017;12(8):2040–2050.

25 Baillie TA. Targeted covalent inhibitors for drug design. *Angew Chem Int Ed Eng* 2016;55(43):13408–13421.

26 Dambach DM, Andrews BA, Moulin F. New technologies and screening strategies for hepatotoxicity: use of *in vitro* models. *Toxicol Pathol* 2005;33(1):17–26.

27 Ullrich A, Berg C, Hengstler JG, Runge D. Use of a standardised and validated long-term human hepatocyte culture system for repetitive analyses of drugs: repeated administrations of acetaminophen reduces albumin and urea secretion. *ALTEX* 2007;24(1):35–40.

28 Gu X, Albrecht W, Edlund K, Kappenberg F, Rahnenfuhrer J, Leist M, et al. Relevance of the incubation period in cytotoxicity testing with primary human hepatocytes. *Arch Toxicol* 2018;92(12):3505–3515.

29 Li AP. Biomarkers and human hepatocytes. *Biomark Med* 2014;8(2):173–183.

30 Li AP. Evaluation of drug metabolism, drug-drug interactions, and *in vitro* hepatotoxicity with cryopreserved human hepatocytes. *Methods Mol Biol* 2010;640:281–294.

31 Maddah M, Mandegar MA, Dame K, Grafton F, Loewke K, Ribeiro AJS. Quantifying drug-induced structural toxicity in hepatocytes and cardiomyocytes derived from hiPSCs using a deep learning method. *J Pharmacol Toxicol Methods* 2020; 105:106895.

32 Lu J, Einhorn S, Venkatarangan L, Miller M, Mann DA, Watkins PB, et al. Morphological and functional characterization and assessment of iPSC-derived hepatocytes for *in vitro* toxicity testing. *Toxicol Sci* 2015;147(1):39–54.

33 Goldring C, Antoine DJ, Bonner F, Crozier J, Denning C, Fontana RJ, et al. Stem cell-derived models to improve mechanistic understanding and prediction of human drug-induced liver injury. *Hepatology* 2017;65(2):710–721.

34 Kim DE, Jang MJ, Kim YR, Lee JY, Cho EB, Kim E, et al. Prediction of drug-induced immune-mediated hepatotoxicity using hepatocyte-like cells derived from human embryonic stem cells. *Toxicology* 2017;387:1–9.

35 Natale A, Vanmol K, Arslan A, Van Vlierberghe S, Dubruel P, Van Erps J, et al. Technological advancements for the development of stem cell-based models for hepatotoxicity testing. *Arch Toxicol* 2019;93(7):1789–1805.

36 Rana P, Aleo MD, Gosink M, Will Y. Evaluation of *in vitro* mitochondrial toxicity assays and physicochemical properties for prediction of organ toxicity using 228 pharmaceutical drugs. *Chem Res Toxicol* 2019;32(1):156–167.

37 Liu Y. Study liver cytochrome P450 3A4 inhibition and hepatotoxicity using DMSO-differentiated HuH-7 cells. *Methods Mol Biol* 2016;1473:63–70.

38 Susukida T, Sekine S, Nozaki M, Tokizono M, Oizumi K, Horie T, et al. Establishment of a drug-induced, bile acid-dependent hepatotoxicity model using HepaRG cells. *J Pharm Sci* 2016;105(4):1550–1560.

39 Zhang J, Doshi U, Suzuki A, Chang CW, Borlak J, Li AP, et al. Evaluation of multiple mechanism-based toxicity endpoints in primary cultured human hepatocytes for the identification of drugs with clinical hepatotoxicity: results from 152 marketed drugs with known liver injury profiles. *Chem Biol Interact* 2016;255:3–11.

40 Hewitt NJ, Lechon MJ, Houston JB, Hallifax D, Brown HS, Maurel P, et al. Primary hepatocytes: current understanding of the regulation of metabolic enzymes and transporter proteins, and pharmaceutical practice for the use of hepatocytes in metabolism, enzyme induction, transporter, clearance, and hepatotoxicity studies. *Drug Metab Rev* 2007;39(1):159–234.

41 Li AP. *In vitro* approaches to evaluate ADMET drug properties. *Curr Top Med Chem* 2004;4(7):701–706.

42 Lloyd S, Hayden MJ, Sakai Y, Fackett A, Silber PM, Hewitt NJ, et al. Differential *in vitro* hepatotoxicity of troglitazone and rosiglitazone among cryopreserved human hepatocytes from 37 donors. *Chem Biol Interact* 2002;142(1–2):57–71.

43 Li AP. *In vitro* human hepatocyte-based experimental systems for the evaluation of human drug metabolism, drug-drug interactions, and drug toxicity in drug development. *Curr Top Med Chem* 2014;14(11):1325–1338.

44 Li AP. Human hepatocytes as an effective alternative experimental system for the evaluation of human drug properties: general concepts and assay procedures. *ALTEX* 2008;25(1):33–42.

45 Li AP. Overview: evaluation of metabolism-based drug toxicity in drug development. *Chem Biol Interact* 2009;179(1):1–3.

46 Li AP. Human-based *in vitro* experimental systems for the evaluation of human drug safety. *Curr Drug Saf* 2007;2(3):193–199.

47 Li AP. Human hepatocytes: isolation, cryopreservation and applications in drug development. *Chem Biol Interact* 2007;168(1):16–29.

48 Li AP. Accurate prediction of human drug toxicity: a major challenge in drug development. *Chem Biol Interact* 2004;150(1):3–7.

49 Li AP. Screening for human ADME/Tox drug properties in drug discovery. *Drug Discov Today* 2001;6(7):357–366.

50 Westerink WM, Schoonen WG. Cytochrome P450 enzyme levels in HepG2 cells and cryopreserved primary human hepatocytes and their induction in HepG2 cells. *Toxicol in Vitro* 2007;21(8):1581–1591.

51 Tolosa L, Donato MT, Perez-Cataldo G, Castell JV, Gomez-Lechon MJ. Upgrading cytochrome P450 activity in HepG2 cells co-transfected with adenoviral vectors for drug hepatotoxicity assessment. *Toxicol in Vitro* 2012;26(8):1272–1277.

52 Hosomi H, Fukami T, Iwamura A, Nakajima M, Yokoi T. Development of a highly sensitive cytotoxicity assay system for CYP3A4-mediated metabolic activation. *Drug Metab Dispos* 2011;39(8):1388–1395.

53 Kanebratt KP, Andersson TB. Evaluation of HepaRG cells as an *in vitro* model for human drug metabolism studies. *Drug Metab Dispos* 2008;36(7):1444–1452.

54 Jackson JP, Li L, Chamberlain ED, Wang H, Ferguson SS. Contextualizing hepatocyte functionality of cryopreserved HepaRG cell cultures. *Drug Metab Dispos* 2016;44(9):1463–1479.

55 Li AP. Metabolism comparative cytotoxicity assay (MCCA) and cytotoxic metabolic pathway identification assay (CMPIA) with cryopreserved human hepatocytes for the evaluation of metabolism-based cytotoxicity *in vitro*: proof-of-concept study with aflatoxin B1. *Chem Biol Interact* 2009;179(1):4–8.

56 Sewing S, Boess F, Moisan A, Bertinetti-Lapatki C, Minz T, Hedtjaern M, et al. Establishment of a predictive *in vitro* assay for assessment of the hepatotoxic potential of oligonucleotide drugs. *PLoS One* 2016;11(7):e0159431.

57 Feutren G, Lacour B, Bach JF. Immune lysis of hepatocytes in culture: accurate detection by aspartate aminotransferase release measurement. *J Immunol Methods* 1984;75(1):85–94.

58 Qizilbash N, Birks J, Lopez Arrieta J, Lewington S, Szeto S. WITHDRAWN: Tacrine for Alzheimer's disease. *Cochrane Database Syst Rev* 2007 (3):CD000202.

59 Utkarsh D, Loretz C, Li AP. *In vitro* evaluation of hepatotoxic drugs in human hepatocytes from multiple donors: identification of P450 activity as a potential risk factor for drug-induced liver injuries. *Chem Biol Interact* 2016;255:12–22.

60 de Montellano PRO. 1-Aminobenzotriazole: a mechanism-based cytochrome P450 inhibitor and probe of cytochrome P450 biology. *Med Chem (Los Angeles)* 2018;8(3): 038–065.

61 Ames BN, McCann J, Yamasaki E. Methods for detecting carcinogens and mutagens with the *Salmonella*/mammalian-microsome mutagenicity test. *Mutat Res* 1975;31(6):347–364.

62 Li AP, Carver JH, Choy WN, Hsie AW, Gupta RS, Loveday KS, et al. A guide for the performance of the Chinese hamster ovary cell/hypoxanthine-guanine phosphoribosyl transferase gene mutation assay. *Mutat Res* 1987;189(2):135–141.

63 Clive D, Flamm WG, Machesko MR, Bernheim NJ. A mutational assay system using the thymidine kinase locus in mouse lymphoma cells. *Mutat Res* 1972;16(7):77–87.

64 Li AP. Use of Aroclor 1254-induced rat liver homogenate in the assaying of promutagens in Chinese hamster ovary cells. *Environ Mutagen* 1984;6(4):539–544.

65 Teepe AG, Beck DJ, Li AP. Comparison of rat liver parenchymal and nonparenchymal cells in the activation of promutagens. *Environ Mol Mutagen* 1992;20(2):134–139.

66 Palacharla VRC, Chunduru P, Ajjala DR, Bhyrapuneni G, Nirogi R, Li AP. Development and validation of a higher-throughput cytochrome P450 inhibition assay with the novel cofactor-supplemented permeabilized cryopreserved human hepatocytes (MetMax human hepatocytes). *Drug Metab Dispos* 2019;47(10):1032–1039.

67 Li AP, Ho MD, Amaral K, Loretz C. A novel *in vitro* experimental system for the evaluation of drug metabolism: cofactor-supplemented permeabilized cryopreserved human hepatocytes (MetMax cryopreserved human hepatocytes). *Drug Metab Dispos* 2018;46(11):1608–1616.

68 Ware BR, McVay M, Sunada WY, Khetani SR. Exploring chronic drug effects on microengineered human liver cultures using global gene expression profiling. *Toxicol Sci* 2017;157(2):387–398.

69 Zarowna-Dabrowska A, McKenna EO, Schutte ME, Glidle A, Chen L, Cuestas-Ayllon C, et al. Generation of primary hepatocyte microarrays by piezoelectric printing. *Colloids Surf B: Biointerfaces* 2012;89:126–132.

70 Choi YY, Seok JI, Kim DS. Flow-based three-dimensional co-culture model for long-term hepatotoxicity prediction. *Micromachines (Basel)* 2019;11(1): 36.

71 Abu-Absi SF, Hansen LK, Hu WS. Three-dimensional co-culture of hepatocytes and stellate cells. *Cytotechnology* 2004;45(3):125–140.

72 Davidson MD, Kukla DA, Khetani SR. Microengineered cultures containing human hepatic stellate cells and hepatocytes for drug development. *Integr Biol (Camb)* 2017;9(8):662–677.

73 Lin C, Khetani SR. Micropatterned co-cultures of human hepatocytes and stromal cells for the assessment of drug clearance and drug-drug interactions. *Curr Protoc Toxicol* 2017;72:14.7.1–14.7.23.

74 Yuhas JM, Li AP, Martinez AO, Ladman AJ. A simplified method for production and growth of multicellular tumor spheroids. *Cancer Res* 1977;37(10):3639–3643.

75 Sutherland RM, MacDonald HR, Howell RL. Multicellular spheroids: a new model target for *in vitro* studies of immunity to solid tumor allografts. *J Natl Cancer Inst* 1977;58(6):1849–1853.

76 Li AP, Colburn SM, Beck DJ. A simplified method for the culturing of primary adult rat and human hepatocytes as multicellular spheroids. *In Vitro Cell Dev Biol* 1992;28A(9–10):673–677.

77 Brown LA, Arterburn LM, Miller AP, Cowger NL, Hartley SM, Andrews A, et al. Maintenance of liver functions in rat hepatocytes cultured as spheroids in a rotating wall vessel. *In Vitro Cell Dev Biol Anim* 2003;39(1–2):13–20.

78 Bell CC, Lauschke VM, Vorrink SU, Palmgren H, Duffin R, Andersson TB, et al. Transcriptional, functional, and mechanistic comparisons of stem cell-derived hepatocytes, HepaRG cells, and three-dimensional human hepatocyte spheroids as predictive *in vitro* systems for drug-induced liver injury. *Drug Metab Dispos* 2017;45(4):419–429.

79 Koide N, Shinji T, Tanabe T, Asano K, Kawaguchi M, Sakaguchi K, et al. Continued high albumin production by multicellular spheroids of adult rat hepatocytes formed in the presence of liver-derived proteoglycans. *Biochem Biophys Res Commun* 1989;161(1):385–391.

80 Ogihara T, Hosono M, Kojima H. [Further investigation of 3D culture spheroid models of human hepatocytes]. *Nihon Yakurigaku Zasshi* 2019;153(5):235–241.

81 Parmentier C, Hendriks DFG, Heyd B, Bachellier P, Ingelman-Sundberg M, Richert L. Inter-individual differences in the susceptibility of primary human hepatocytes towards drug-induced cholestasis are compound and time dependent. *Toxicol Lett* 2018;295:187–194.

82 Li L, Gokduman K, Gokaltun A, Yarmush ML, Usta OB. A microfluidic 3D hepatocyte chip for hepatotoxicity testing of nanoparticles. *Nanomedicine (London)* 2019;14(16):2209–2226.

83 Riahi R, Shaegh SA, Ghaderi M, Zhang YS, Shin SR, Aleman J, et al. Automated microfluidic platform of bead-based electrochemical immunosensor integrated with bioreactor for continual monitoring of cell secreted biomarkers. *Sci Rep* 2016;6:24598.

84 Legendre A, Jacques S, Dumont F, Cotton J, Paullier P, Fleury MJ, et al. Investigation of the hepatotoxicity of flutamide: pro-survival/apoptotic and necrotic switch in primary rat hepatocytes characterized by metabolic and transcriptomic profiles in microfluidic liver biochips. *Toxicol in Vitro* 2014;28(5):1075–1087.

85 Burkhardt B, Martinez-Sanchez JJ, Bachmann A, Ladurner R, Nussler AK. Long-term culture of primary hepatocytes: new matrices and microfluidic devices. *Hepatol Int* 2014;8(1):14–22.

86 Yeon JH, Na D, Park JK. Hepatotoxicity assay using human hepatocytes trapped in microholes of a microfluidic device. *Electrophoresis* 2010;31(18):3167–3174.

87 Toh YC, Lim TC, Tai D, Xiao G, van Noort D, Yu H. A microfluidic 3D hepatocyte chip for drug toxicity testing. *Lab Chip* 2009;9(14):2026–2035.

88 Zhang C, Chia SM, Ong SM, Zhang S, Toh YC, van Noort D, et al. The controlled presentation of TGF-beta1 to hepatocytes in a 3D-microfluidic cell culture system. *Biomaterials* 2009;30(23–24):3847–3853.

89 Cui J, Wang H, Zheng Z, Shi Q, Sun T, Huang Q, et al. Fabrication of perfusable 3D hepatic lobule-like constructs through assembly of multiple cell type laden hydrogel microstructures. *Biofabrication* 2018;11(1):015016.

90 Evenou F, Hamon M, Fujii T, Takeuchi S, Sakai Y. Gas-permeable membranes and co-culture with fibroblasts enable high-density hepatocyte culture as multilayered liver tissues. *Biotechnol Prog* 2011;27(4):1146–1153.

91 Lee PJ, Hung PJ, Lee LP. An artificial liver sinusoid with a microfluidic endothelial-like barrier for primary hepatocyte culture. *Biotechnol Bioeng* 2007;97(5):1340–1346.

92 Bale SS, Borenstein JT. Microfluidic cell culture platforms to capture hepatic physiology and complex cellular interactions. *Drug Metab Dispos* 2018;46(11):1638–1646.

93 Kupfer A, Aeschlimann C, Cerny T. Methylene blue and the neurotoxic mechanisms of ifosfamide encephalopathy. *Eur J Clin Pharmacol* 1996;50(4):249–252.

94 Brain EG, Yu LJ, Gustafsson K, Drewes P, Waxman DJ. Modulation of P450-dependent ifosfamide pharmacokinetics: a better understanding of drug activation *in vivo*. *Br J Cancer* 1998;77(11):1768–1776.

95 Capela JP, Macedo C, Branco PS, Ferreira LM, Lobo AM, Fernandes E, et al. Neurotoxicity mechanisms of thioether ecstasy metabolites. *Neuroscience* 2007;146(4):1743–1757.

96 Barbosa DJ, Capela JP, Oliveira JM, Silva R, Ferreira LM, Siopa F, et al. Pro-oxidant effects of Ecstasy and its metabolites in mouse brain synaptosomes. *Br J Pharmacol* 2012;165(4b):1017–1033.

97 Iqubal A, Iqubal MK, Sharma S, Ansari MA, Najmi AK, Ali SM, et al. Molecular mechanism involved in cyclophosphamide-induced cardiotoxicity: old drug with a new vision. *Life Sci* 2019;218:112–131.

98 Schaupp CM, White CC, Merrill GF, Kavanagh TJ. Metabolism of doxorubicin to the cardiotoxic metabolite doxorubicinol is increased in a mouse model of chronic glutathione deficiency: a potential role for carbonyl reductase 3. *Chem Biol Interact* 2015;234:154–161.

99 Edwardson DW, Narendrula R, Chewchuk S, Mispel-Beyer K, Mapletoft JP, Parissenti AM. Role of drug metabolism in the cytotoxicity and clinical efficacy of anthracyclines. *Curr Drug Metab* 2015;16(6):412–426.

100 Jeong TC, Matulka RA, Jordan SD, Yang KH, Holsapple MP. Role of metabolism in cocaine-induced immunosuppression in splenocyte cultures from B6C3F1 female mice. *Immunopharmacology* 1995;29(1):37–46.

101 Misra RR, Bloom SE. Roles of dosage, pharmacokinetics, and cellular sensitivity to damage in the selective toxicity of cyclophosphamide towards B and T cells in development. *Toxicology* 1991;66(3):239–256.

102 Li AP. The use of the integrated discrete multiple organ co-culture (IdMOC) system for the evaluation of multiple organ toxicity. *Altern Lab Anim* 2009;37(4):377–385.

103 Li AP. *In vitro* evaluation of human xenobiotic toxicity: scientific concepts and the novel integrated discrete multiple cell co-culture (IdMOC) technology. *ALTEX* 2008;25(1):43–49.

104 Li AP, Bode C, Sakai Y. A novel *in vitro* system, the integrated discrete multiple organ cell culture (IdMOC) system, for the evaluation of human drug toxicity: comparative cytotoxicity of tamoxifen towards normal human cells from five major organs and MCF-7 adenocarcinoma breast cancer cells. *Chem Biol Interact* 2004;150(1):129–136.

105 Li AP, Uzgare A, LaForge YS. Definition of metabolism-dependent xenobiotic toxicity with co-cultures of human hepatocytes and mouse 3T3 fibroblasts in the novel integrated discrete multiple organ co-culture (IdMOC) experimental system: results with model toxicants aflatoxin B1, cyclophosphamide and tamoxifen. *Chem Biol Interact* 2012;199(1):1–8.

106 Li AP. Evaluation of adverse drug properties with cryopreserved human hepatocytes and the integrated discrete multiple organ co-culture (IdMOC(TM)) system. *Toxicol Res* 2015;31(2):137–149.

107 Cole SD, Madren-Whalley JS, Li AP, Dorsey R, Salem H. High content analysis of an *in vitro* model for metabolic toxicity: results with the model toxicants 4-aminophenol and cyclophosphamide. *J Biomol Screen* 2014;19(10):1402–1408.

108 Richter PA, Li AP, Polzin G, Roy SK. Cytotoxicity of eight cigarette smoke condensates in three test systems: comparisons between assays and condensates. *Regul Toxicol Pharmacol* 2010;58(3):428–436.

109 Gayathri L, Karthikeyan BS, Rajalakshmi M, Dhanasekaran D, Li AP, Akbarsha MA. Metabolism-dependent cytotoxicity of citrinin and ochratoxin A alone and in combination as assessed adopting integrated discrete multiple organ co-culture (IdMOC). *Toxicol in Vitro* 2018;46:166–177.

110 Li AP. A review of the common properties of drugs with idiosyncratic hepatotoxicity and the "multiple determinant hypothesis" for the manifestation of idiosyncratic drug toxicity. *Chem Biol Interact* 2002;142(1–2):7–23.

Part III

Drug Transporters and Drug Toxicity

9

Mechanism-Based Experimental Models for the Evaluation of BSEP Inhibition and DILI

William A. Murphy[1], *Chitra Saran*[1,2], *Paavo Honkakoski*[1,3], *and Kim L.R. Brouwer*[1]

[1] *Division of Pharmacotherapy and Experimental Therapeutics, UNC Eshelman School of Pharmacy, University of North Carolina at Chapel Hill, Chapel Hill, NC, USA*
[2] *Department of Pharmacology, UNC School of Medicine, University of North Carolina at Chapel Hill, Chapel Hill, NC, USA*
[3] *School of Pharmacy, Faculty of Health Sciences, University of Eastern Finland, Kuopio, Finland*

9.1 Introduction

9.1.1 Drug-Induced Liver Injury

Drug-induced liver injury (DILI) is generally defined as unexpected damage or toxicological injury to the liver by drugs used in common clinical practice. In addition to being a significant impediment for drug developers, DILI can place patients at risk, sometimes resulting in acute liver failure (ALF) requiring transplant or even fatality [1, 2]. Population-based studies in France and Iceland project annual DILI occurrence rates of 13.9–19.1 cases per 100 000 people [3–5]. Other population studies in Europe and the US estimate annual DILI incidence rates of 2.4–2.7 cases per 100 000 people [6, 7]. The initial diagnosis of DILI is challenging and involves the exclusion of other underlying illnesses including acute or chronic liver disease [1, 8]. Before a drug enters clinical development, it is expected to have undergone a thorough DILI-risk assessment. When clinical testing begins, additional information regarding hepatotoxicity risk is rarely obtained because the incidence of DILI in patient populations of ~100 in Phase 1 and ~5000–10 000 subjects in Phases 2 and 3 is very low [2]. Therefore, accurate *in vitro*, *in vivo*, and *in silico* models are needed to predict clinical DILI potential in large-scale patient populations.

DILI can be classified into either intrinsic/dose-dependent or idiosyncratic injury. Idiosyncratic injury is typically unpredictable, and individual susceptibility is thought to be based on the interplay of genetic and environmental factors [2]. Initial classification of DILI depends on patient presentation, event chronology, the pharmacologic regimen, and whether or not the toxicological insult by the parent compound and/or metabolite(s) is reproducible in animal models [9].

Transporters and Drug-Metabolizing Enzymes in Drug Toxicity, First Edition. Edited by Albert P. Li.

Idiosyncratic DILI is particularly concerning because of its random onset, unpredictable clinical severity, and poor outcomes [10, 11]. Risk factors tend to be better defined and are mainly drug-specific for intrinsic DILI due to dose-dependency and mechanistic predictability (e.g., glutathione depletion in acetaminophen hepatotoxicity). Risk factors are generally undefined for idiosyncratic DILI, mainly owing to its low incidence, less thorough understanding of pathogenesis, and high patient variability in drug response. While the adaptive immune response has been implicated in the majority of cases [4, 9, 12], proposed toxicological mechanisms for idiosyncratic DILI, particularly for cases of delayed-onset, are complex and can vary immensely depending on drug- and patient-specific characteristics [10, 13]. Therefore, biomarkers as clinical predictors of idiosyncratic DILI are being developed [4, 11, 14]. For the remainder of this chapter, the term DILI will refer exclusively to idiosyncratic DILI.

Based on the interpretation of liver function tests upon clinical presentation, DILI can be categorized as hepatocellular, cholestatic, or mixed. Clinically speaking, hepatocellular DILI refers to direct injury of hepatocytes while cholestatic DILI refers to selective, direct injury of cholangiocytes that presents with elevated serum alkaline phosphatase levels (see Chapter 8 for further discussion on cholestatic DILI). Mixed DILI is defined as a combination of hepatocellular and cholestatic injury [2, 12]. However, in toxicological terms, cholestatic DILI is attributed to altered bile acid homeostasis [2]. Mechanistically, inhibition of the hepatic bile salt export pump (BSEP) may lead to accumulation of bile acids within hepatocytes and altered bile acid homeostasis that may result in cholestatic DILI [15, 16]. Therefore, BSEP inhibition is considered one mechanism of DILI. Other DILI mechanisms include mitochondrial toxicity, reactive oxygen species formation, and immune responses [2]. Recent work also has suggested that bile acid accumulation leads to hepatocyte injury by secondary mechanisms such as generation of reactive oxygen species, mitochondrial dysfunction, apoptosis or necrosis, and induction of proinflammatory cytokines [17–20].

9.1.2 Bile Acid Homeostasis and Role of Bile Salt Export Pump

Bile acids are amphipathic steroid-like molecules that are synthesized from cholesterol in the liver via cytochrome P450 (CYP)-mediated oxidation [21]. Cholic acid (CA) and chenodeoxycholic acid (CDCA) are the primary bile acid species produced in the liver. These bile acids are conjugated with glycine or taurine (taurocholate [TCA], glycocholate [GCA], taurochenodeoxycholate [TCDCA], glycocheodoxycholate [GCDCA]), and excreted via BSEP into bile, which flows into the intestinal tract to facilitate absorption of cholesterol and other nutrients [22]. The gut microbiome modifies conjugated bile acids by dehydroxylation and removal of glycine and taurine groups, thereby converting CA to deoxycholic acid (DCA) and CDCA to lithocholic acid (LCA) [22]. Bile acids are absorbed into enterocytes by the apical sodium bile salt transporter (ASBT) and transported into the portal circulation by the organic solute transporter α/β (OSTα/β), where they may be taken up into hepatocytes mainly via Na^+-dependent taurocholate cotransporting polypeptide (NTCP). This enterohepatic recirculation accounts for ~95% of the bile acid pool, while the remaining 5% lost in feces is replenished by *de novo* synthesis in the liver [21–23]. The primary bile acid-sensing nuclear receptor, farnesoid X receptor (FXR), transcriptionally regulates the expression of hepatic transporters NTCP, BSEP, and OSTα/β, and enzymes involved in bile acid synthesis and metabolism, thereby, maintaining bile acid homeostasis [22, 24].

BSEP was first described in 1991 as an adenosine triphosphate (ATP)-dependent bile acid transporter [25] in rat liver canalicular membrane vesicles. The complementary DNA (cDNA) of the protein was partially cloned from pig liver and was initially named sister of P-glycoprotein due to its sequence homology with P-glycoprotein [26, 27]. The full-length sequence was identified in 1998 [28] from rat liver and described as BSEP. BSEP is a member of the ATP-binding cassette (ABC) transporter family that consists of 12 transmembrane domains and two highly conserved cytoplasmic nucleotide-binding domains with Walker A and B motifs. The transmembrane domains confer substrate specificity and the Walker motifs contain ATP-binding sites [29]. The gene (*ABCB11*) located on chromosome 2 region q24–31 encodes an approximately 160 kDa molecular weight BSEP protein, which consists of 1321 amino acids and four N-linked glycosylation sites [30, 31]. BSEP is almost exclusively expressed in the liver and shows a narrow substrate specificity [32] as it primarily transports conjugated bile acids. Biliary excretion of bile acids via BSEP works against a 100- to 1000-fold concentration gradient and is the driving force for maintaining bile flow [33]. In 2002, Noé et al. [34] functionally characterized the transporter using membrane vesicles harvested from BSEP-overexpressing Sf9 insect cells and reported the rank order of intrinsic clearance (V_{max}/K_m) as TCDCA > TCA > TUDCA > GCA.

Reportedly, genetic defects or polymorphisms of *ABCB11* cause rare cholestatic diseases such as progressive familial intrahepatic cholestasis type 2 (PFIC2) [35, 36], benign recurrent intrahepatic cholestasis type 2 (BRIC2) [37, 38], and intrahepatic cholestasis of pregnancy (ICP) [39, 40]. While PFIC2 patients are most severely affected by the complete loss of BSEP function, BRIC2 and ICP patients retain some transport activity due to expression of partially active BSEP (*ABCB11*) variants. Several missense mutations (e.g., D482G, E297G, A590T, and R1050C) decrease cell surface expression of BSEP and may alter transport activity, thereby causing cholestatic diseases of varying severity [41–43]. Similarly, inhibition of BSEP function by drugs and/or metabolites could lead to cholestasis and liver injury. BSEP function can also be impaired by mechanisms beyond competitive inhibition, such as decreased expression and/or mis-localization of the transporter [44–46].

BSEP inhibition is associated with intrahepatic bile acid accumulation and liver damage. Although tolvaptan and troglitazone inhibit BSEP, toxicity from these drugs presents clinically as hepatocellular and not as cholestatic DILI due to secondary mechanisms that cause mitochondrial toxicity [47]. Significant BSEP inhibition has been documented for new chemical entities (NCEs) that either failed in clinical trials or were withdrawn from the market due to DILI (e.g., AMG009, TAK-875, CP-724,714, troglitazone) [48–52]. However, drugs or their metabolites may also inhibit one or more basolateral transporters (e.g., multidrug resistance-associated protein 3 [MRP3], MRP4, OSTα/β), thereby blocking an important compensatory route for elimination of hepatotoxic bile acids from hepatocytes into the systemic circulation. Drugs that inhibit BSEP may also inhibit MRP2 [51, 53], a canalicular transporter that mediates biliary excretion of glucuronidated and sulfated bile acids. Therefore, drug candidates need to be comprehensively tested for DILI risk.

As described above, BSEP function is intimately associated with hepatic bile acid homeostasis, intracellular bile acid accumulation, and hepatotoxicity. We recognize the importance of other DILI mechanisms and the involvement of additional bile acid transporters in these processes. However, due to substantial data that link BSEP inhibition directly to DILI occurrence, the focus of this chapter is primarily on methods to assess and predict BSEP inhibition.

9.2 Membrane Vesicles to Study BSEP Inhibition

Membrane vesicles are liposome-like structures composed of plasma membrane fragments isolated from whole cells. The plasma membrane of these vesicles can be experimentally manipulated to "flip" their natural orientation, which results in the cytosolic surface of the membrane facing outwards and the extracellular surface facing inwards (inside-out) [54]. Inside-out membrane vesicle assays have proven to be a cost-effective and high-throughput *in vitro* method to study substrate specificity and inhibition of unidirectional ATP-dependent efflux transporters [55, 56]. Inside-out vesicles are ideal for the study of BSEP transport because the substrate- and ATP-binding domains of the transporter are in direct contact with the assay buffer containing substrate and ATP. Additionally, membrane vesicles allow researchers to investigate BSEP substrate specificity/affinity, rate of transport, and inhibition mechanisms in isolation [15, 55, 57, 58]. Cytotoxic compounds can be tested readily without causing cell death that could confound experimental results [55]. A survey of pharmaceutical companies conducted by the International Transporter Consortium (ITC) indicated that initial investigation of BSEP inhibition for a NCE typically involves BSEP inhibition studies using membrane vesicles [47]. Since BSEP is expressed only on the canalicular membrane of hepatocytes, it is challenging to study BSEP interactions with a compound in intact BSEP-overexpressing cells without any efficient mechanisms of cellular uptake [55]. One way to overcome this issue is to use cells that co-express relevant uptake transporters such as NTCP. However, this method adds complexity in study design and data interpretation.

9.2.1 Membrane Vesicle Preparations

Membrane vesicle preparations from BSEP-expressing cells typically follow the same general procedure as described for intact erythrocytes using divalent cations, a hypotonic lysis buffer, and ultracentrifugation with a dextrose gradient to produce inside-out plasma membranes [56, 59]. The most commonly used cells for the isolation of membrane vesicles for BSEP transport studies are *Spodoptera frugiperda* (Sf9 or Sf21) insect cells. These cells are routinely used due to their high protein expression (~3% of total vesicular protein represents the transporter of interest) and relatively inexpensive preparation, making them ideal for high-throughput assays [54, 56]. These cells can be infected readily with a baculovirus containing human BSEP cDNA [60], followed by crude membrane extraction. Membrane vesicles also can be prepared from BSEP-transfected mammalian cells (e.g., HeLa, HEK293, MDCK, and CHO) [47, 54–56]. Although there is less cholesterol in membrane vesicles derived from insect cells, there is little difference in observed substrate K_m values or inhibitory potencies (IC_{50}) for BSEP; lower V_{max} values can be adjusted by loading cholesterol into Sf9 vesicles [61]. Differences in protein glycosylation patterns between mammalian and insect cells also have been documented; however, it is uncertain if this has any functional impact on BSEP [55, 56]. Canalicular membrane vesicles may be prepared from human liver tissue [54, 55]. This method is not suitable for the specific investigation of BSEP function due to the expression of numerous other transporters in whole tissue specimens.

The initial step to prepare membrane vesicles from intact cells involves homogenization of the transfected cells in a hypotonic lysis buffer under neutral pH conditions, followed by a low-speed centrifugation to remove any intracellular debris [57]. The supernatant containing crude membrane fractions then undergoes ultracentrifugation before resuspension in isotonic buffer. Chemicals can be added to the isotonic buffer to modify the vesicle content as needed for the assay (e.g., high sodium, high cholesterol) [54]. Then, the isotonic suspension is passed through a 27-gauge needle to obtain a final mixture containing right-side-out and inside-out membrane vesicles in roughly equal amounts (range: 30–70% inside-out) [54, 56, 57, 62, 63]. Separation of inside-out membrane vesicles from this mixture is typically unnecessary given that only intact inside-out vesicles will present BSEP in a conformation where it can bind the substrate and ATP for transport [47, 54, 55]. Purified inside-out vesicles also show little improvement in signal-noise ratios in BSEP assays [54, 56]. Of note, vesicle preparations can be stored at −80 °C in hypotonic or isotonic buffer solutions, allowing for generation of large batches; frequent freeze–thaw cycles are not recommended as this may impact protein function and assay performance [54–57]. Membrane vesicle systems containing BSEP isolated from insect or mammalian cells are commercially available, which allows for better reproducibility and comparison of data across different laboratories [47, 55].

9.2.2 Membrane Vesicle Assays and Data Interpretation

The vesicular transport assay for BSEP directly measures vesicular uptake of the test compound or probe substrate in the presence of ATP in the incubation buffer [54, 56]. TCA is commonly used as a model bile acid substrate for BSEP transport, but a number of structurally similar bile acids also have been utilized (e.g., GCA, TCDCA, and GCDCA) [47, 55]. It is important to recognize that TCA is not the most abundant or toxic bile acid species, and in some cases, transport inhibitors may be substrate-dependent. Typical inhibition assay conditions include an incubation of 0.5–2 μM TCA plus the test compound (e.g., putative inhibitor) and ATP at 37 °C for 2–10 minutes [47, 64]. The probe substrate concentration should be well below the K_m value (TCA K_m for BSEP ~11 μM) to assure that uptake is measured in the linear range for accurate calculation of the IC_{50} value of the test compound [47, 55]. Vesicular uptake is terminated by quickly diluting the incubation reaction 30- to 50-fold with ice-cold buffer. The diluted suspension is rapidly filtered through a glass fiber or membrane filter that entraps the vesicles containing the probe substrate for subsequent analysis by scintillation counting, fluorescence detection, or liquid chromatography coupled with tandem mass spectrometry (LC–MS/MS) [54, 55]. Transfer of the vesicle suspension must be performed quickly to avoid potential diffusion of probe substrate from the vesicles back into the buffer [55]. See Figure 9.1 for an overview of the membrane vesicle assay methodology for BSEP transport. Proper controls include the use of AMP in place of ATP in the incubation buffer (negative control), the use of a known BSEP inhibitor such as troglitazone or cyclosporine (positive control), and a control without the test compound as a reference value representing uninhibited probe substrate uptake [47, 54, 55]. Inclusion of a positive control is essential to account for differences between batches, allowing for accurate data comparison across multiple experiments or labs [47]. The rate of BSEP transport for the probe substrate is adjusted for total vesicular protein, which provides relative values for the substrate and control incubations assuming they were obtained from the same membrane preparation batch [55]. Data may be normalized for BSEP protein, but this requires correction for relative quantities of inside-out vs. right-side-out

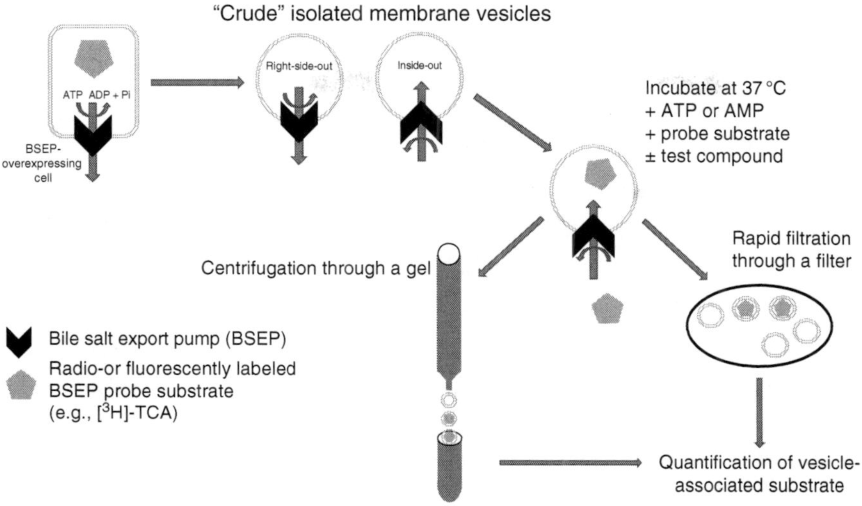

Figure 9.1 Schematic diagram depicting the membrane vesicle assay for BSEP transport. A "crude" mixture of plasma membranes isolated from a BSEP-overexpressing cell line contains both right-side-out and inside-out membrane vesicles. Only inside-out membrane vesicles actively transport BSEP substrates in the presence of ATP, while AMP serves as a negative control. After incubation with the BSEP probe substrate, with or without test compounds (e.g., BSEP inhibitors), vesicles containing the substrate can be isolated by gel centrifugation or membrane filtration. Vesicle-associated substrate can be quantified by radioactivity, fluorescence, or mass spectrometry-based detection methods. ADP, adenosine diphosphate; AMP, adenosine monophosphate; ATP, adenosine triphosphate; TCA, taurocholate. *Source:* Adapted from Brouwer et al. [55].

vesicles. If the test compound is identified as an "inhibitor", an IC_{50} value may be determined from assay data using a reference control and a range of inhibitor concentrations relevant to predicted or known clinical exposure.

IC_{50} values for BSEP inhibition have been documented for more than 600 drugs using membrane vesicles from BSEP-overexpressing cells [51]. BSEP inhibition potential, defined as C_{ss}/BSEP IC_{50} ratio ≥ 0.1, correlated with DILI at high specificity (70–80%) but modest sensitivity of ~50%, indicating a large number of false negative results [47, 51, 53]. Total and unbound concentrations at the BSEP inhibition site (canalicular membrane of hepatocytes) should be considered when determining relevant concentrations of the inhibitor [55]. Taking a more conservative approach, an IC_{50} of ≤25 μM is typically deemed significant and warrants further DILI-risk investigation [47, 53].

During initial screening, BSEP inhibition by test compounds is typically only evaluated at a single probe substrate concentration. Experimental determination of K_i to investigate inhibition mechanisms (e.g., competitive vs. noncompetitive) is labor-intensive and time-consuming, requiring several concentrations of both probe substrate and inhibitor, and is typically not investigated in early drug development. K_i determination has proven valuable in quantitative systems toxicology (QST) simulations and should be considered for drugs when a more comprehensive DILI risk assessment is required [47, 65].

One limitation of the vesicular transport assay is the inability to account for the diffusion of highly permeable substrates into vesicles [54, 55]. Most BSEP substrates such as hydrophilic bile acids exhibit a relatively low potential for free diffusion across plasma membranes [54]. Passive permeability, background transport, and nonspecific/specific binding can be accounted for by replacing ATP in the incubation buffer with AMP [54, 55]. The relatively high specificity but modest sensitivity of BSEP membrane vesicle assays to predict clinical DILI incidence may be attributed to inherent limitations with the model. Membrane vesicles lack metabolic enzymes, other compensatory transport proteins, subcellular organelles, and machinery for trafficking and regulation of transport proteins; these cellular components may influence the sequestration and excretion of drugs or derived metabolites that play key roles in DILI [55, 66]. Unbound intracellular concentrations of the drug at the transport site in humans may never reach BSEP IC_{50} values tested *in vitro*. An untested metabolite may be a more potent BSEP inhibitor than the parent compound. Thus, *in silico* modeling approaches and additional testing in whole-cell systems (e.g., sandwich-cultured hepatocytes [SCH]) may be useful to further assess *in vivo* BSEP inhibitory potential and bile acid-mediated DILI risk [47, 58, 67, 68].

9.3 Sandwich-Cultured Hepatocytes to Study BSEP Inhibition

9.3.1 B-CLEAR® Assay

SCH are another *in vitro* tool that can be used to measure BSEP-mediated inhibition of biliary excretion [69]. Unlike membrane vesicles, SCH are whole-cell systems expressing multiple transporters, metabolic enzymes for metabolite formation, and proteins involved in regulation of enzyme and transporter expression and transporter trafficking; these characteristics make SCH more physiologically relevant. Primary hepatocytes cultured in standard conditions show limited viability, rapid loss of cell polarity, and few bile canalicular structures [70]. Hepatocytes cultured between two layers of gelled collagen or extracellular matrix such as Matrigel™ [71, 72] in a sandwich configuration allow hepatocytes to regain cell polarity, express metabolic enzymes and transporters, and form bile canalicular structures with functional tight junctions [69, 73]. This culture technique can be used for rat, mouse, dog, and monkey hepatocytes as well as cryopreserved and freshly isolated human hepatocytes. Disruption of tight junctions by removal of extracellular Ca^{2+} ions (B-CLEAR® methodology) allows measurement of cell accumulation, biliary excretion and clearance of substrates [74, 75]. Additionally, *in vitro* intrinsic biliary clearance values calculated for several compounds using B-CLEAR® in rat and human SCH correlated well with *in vivo* biliary clearance data for the same compounds in rats and humans, respectively [76–78]. Therefore, SCH and B-CLEAR® are considered the gold standard to study hepatobiliary drug disposition. The application of these methods include examining species differences in hepatobiliary drug disposition, assessment of transporter inhibition and hepatotoxicity, and evaluation of transporter-mediated drug interactions. Data from SCH and B-CLEAR® assays can provide information regarding intracellular concentrations of xenobiotics [79], explain unexpected pharmacokinetic (PK) data, and provide useful information for mathematical modeling or physiologically based pharmacokinetic (PBPK) models [55]. Detailed methodology and considerations for SCH and B-CLEAR® are discussed in several articles [69, 80]. Briefly, SCH are preincubated in Ca^{2+}-containing Hanks' Balanced Salt Solution (HBSS) buffer (standard buffer) or Ca^{2+}-free HBSS buffer with ethylene glycol tetraacetic acid (EGTA). The depletion of Ca^{2+} ions by EGTA disrupts tight junctions that form the bile canalicular networks, thereby, allowing the comparison between standard buffer (cells + bile) and Ca^{2+}-free buffer (cells), as shown in Figure 9.2.

Hepatic accumulation is initiated by adding a probe substrate such as TCA in the presence or absence of a test compound. Based on the accumulation in the standard buffer (cells + bile) and the Ca^{2+}-free buffer (cells), the biliary excretion index (BEI) and *in vitro* biliary clearance ($CL_{biliary}$) can be calculated as follows:

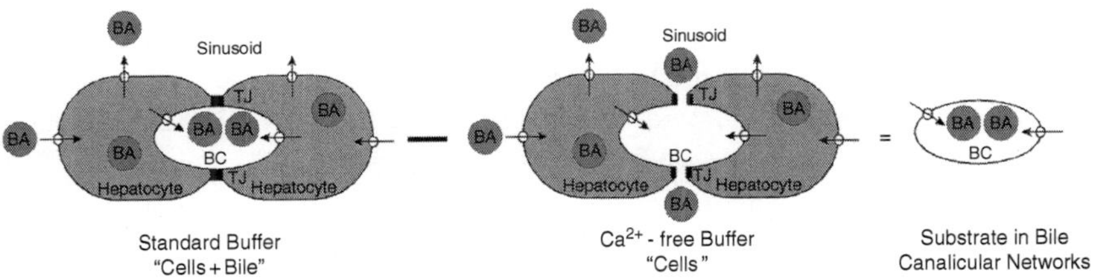

Figure 9.2 Schematic depicting the working principle of B-CLEAR® methodology to measure substrate accumulation in bile canalicular networks. Sandwich-cultured hepatocytes are pre-incubated in either Ca^{2+}-containing standard buffer or Ca^{2+}-free buffer that disrupts tight junctions sealing the canalicular networks. Subsequently, the cells are incubated with a probe substrate that is excreted into the bile canaliculi. In the standard buffer, the probe substrate is retained within the bile canaliculi, while it is released into the buffer in hepatocytes exposed to the Ca^{2+}-free buffer. The difference in substrate accumulation in both conditions (i.e., "cells+bile" and "cells") represents the substrate accumulation in the canalicular networks. BA, bile acid; BC, bile canaliculus; TJ, tight junctions.

$$\mathrm{BEI}(\%) = \frac{\mathrm{Accumulation}_{(\mathrm{Cells+Bile})} - \mathrm{Accumulation}_{(\mathrm{Cells})}}{\mathrm{Accumulation}_{(\mathrm{Cells+Bile})}} \times 100 \tag{1}$$

$$\mathrm{CL}_{\mathrm{biliary}} = \frac{\mathrm{Accumulation}_{(\mathrm{Cells+Bile})} - \mathrm{Accumulation}_{(\mathrm{Cells})}}{\mathrm{AUC}_{0-\mathrm{T}}} \tag{2}$$

Where, AUC_{0-T} represents the product of incubation time (T) and initial concentration of probe substrate in the medium. The BEI represents the fraction of probe substrate taken up into the hepatocytes that is excreted into the bile canaliculi. The *in vitro* $CL_{biliary}$ represents the clearance of the probe substrate from the medium to the bile canaliculi. The intracellular concentration can be estimated by measuring the mass of compound within the hepatocyte normalized to the hepatocyte volume (average hepatocellular volume determined using 3-*O*-methyl-D-glucose: rat hepatocytes = 6.2×10^{-6} μl/hepatocyte; human hepatocytes = 7.69 μL/mg protein) or by using hepatic $K_{p,uu,}$ i.e., the unbound partition coefficient [69, 81]. This estimated intracellular concentration can be used to estimate the intrinsic biliary clearance ($CL_{bile,int}$), which represents the clearance from cell to bile across the canalicular membrane [55].

$$\mathrm{CL}_{\mathrm{bile,int}} = \frac{\mathrm{Accumulation}_{(\mathrm{Cells+Bile})} - \mathrm{Accumulation}_{(\mathrm{Cells})}}{\mathrm{AUC}_{(\mathrm{cell,estimated})}} \tag{3}$$

When inhibition of *in vitro* $CL_{biliary}$ with no change in BEI or $CL_{bile,int}$ is observed, these findings suggest that the test compound inhibits hepatic uptake of the probe substrate. When inhibition of *in vitro* $CL_{biliary}$ and BEI or $CL_{bile,int}$ is observed, these findings suggest that the test compound inhibits both hepatic uptake and efflux of the probe substrate.

The most commonly used probe substrate to measure BSEP inhibition in SCH is TCA, similar to the membrane vesicle assay. This is because TCA transport kinetics is well-documented, and TCA is present in human and rodent hepatocytes. However, TCA is relatively more hydrophilic and is considered less toxic [82], and it is less abundant than the glycine-conjugated GCDCA and GCA in human hepatocytes [83]. These glycine-conjugated bile acids have been shown to play a more prominent role in hepatotoxicity in rat SCH [84]. The BEI of endogenous TCDCA in human SCH and endogenous GCDCA in rat SCH were twofold lower after treatment of SCH with troglitazone (10 μM, 24 hours) while the BEI for TCA remained unchanged in both human and rat SCH when compared to untreated control [85]. Therefore, although TCA is used as a model bile acid substrate to measure BSEP inhibition *in vitro*, other bile acids may better reflect hepatotoxicity [86]. Given this difference, other probe substrates such as GCDCA, TCDCA, and GCA have been employed for B-CLEAR® studies [85, 87], and can be used to evaluate BSEP inhibition. B-CLEAR® methodology has been used to study the uptake and efflux of a synthetic fluorescent bile acid derivative tauro-nor-THCA-24-DBD that is a substrate for both NTCP and BSEP [88]. This fluorescent probe substrate mimics TCA disposition and was used as a probe substrate to study BSEP inhibition by several drugs [89].

Yang et al. [90] used SCH, B-CLEAR®, and PK modeling to examine species differences in TCA disposition between human and rat hepatocytes. The authors showed that the basolateral efflux clearance of TCA was higher in rat compared to human hepatocytes; simulated data suggested that hepatic TCA exposure was 7 times higher due to a 10-fold decrease in both biliary and basolateral efflux clearance values in human SCH. Additionally, PK modeling showed that preincubation of human SCH with troglitazone decreased TCA uptake clearance ~10-fold along with a decreasing trend in biliary and basolateral efflux clearance [90]. Hepatocellular disposition and potential transporter-mediated drug–drug interactions (DDIs) were evaluated for tolvaptan and its metabolites using human SCH and B-CLEAR® [91]. In this study, higher tolvaptan concentrations (0.15–50 μM) decreased TCA accumulation in "cells + bile" and BEI when co-incubated for 10 min in the presence of 4% BSA [91]. Furthermore, tolvaptan was reported to be a noncompetitive BSEP inhibitor using the membrane vesicle assay and it decreased the biliary clearance of TCA as well as CDCA in human SCH [92].

B-CLEAR® studies provide more holistic data compared to membrane vesicle assays since this system accounts for the unbound drug concentration, drug metabolism, and compensatory mechanisms for hepatobiliary efflux via basolateral transporters [47]. This is important because some drugs (e.g., troglitazone, tolvaptan) produce metabolites that inhibit BSEP more potently than the parent drug [49, 92]. While SCH have many advantages over membrane vesicle assays, these studies are more time-consuming, costly, and have lower throughput. Another consideration is that interindividual differences in protein expression, genetic polymorphisms, age, gender, race, prior medication, and disease history may lead to variability in SCH data from different donors. In addition, SCH do not have the microvasculature, oxygenation, and zonal differences that arise from the mixing of portal and venous blood within the human liver.

9.3.2 Uptake and Efflux Studies with Mechanistic Modeling

SCH offer a robust method to study the interplay of multiple uptake and efflux transporters. Notably, SCH can be used to determine the net effect on hepatobiliary bile acid disposition due to drug-mediated inhibition of BSEP and other transport proteins [20, 93–95]. Mechanistic models can explicitly describe the uptake, hepatic efflux, and biliary clearance of a test compound, metabolite, or endogenous compound such as TCA using differential equations. Compounds that undergo extensive biliary clearance require mechanistic modeling to deconvolute overlapping pathways in SCH including basolateral efflux, biliary excretion, and flux from the bile compartment to the medium [20]. Mechanistic modeling of SCH data also allows for the estimation of *in vitro* transporter- and enzyme-related PK parameters as well as passive diffusion and intracellular protein binding of a compound [96]. Mechanistic PK modeling and simulation based on SCH data can be used to assess the relative impact of a test compound on various hepatobiliary clearance pathways and the overall influence on bile acid disposition [20, 93, 97, 98].

Published mechanistic PK models, as detailed by Yang et al. [20], are described mathematically by three separate compartments (medium, cells, and bile canalicular space). The simplest model assumes negligible basolateral efflux of the compound and/or metabolites. Model equations can be modified to account for both parent and metabolite with the assumption that only metabolites undergo efflux across the canalicular membrane into bile or across the basolateral membrane into the medium. A more complex model was used to describe TCA disposition in SCH [20] where TCA was assumed to exhibit biliary and basolateral efflux with minimal passive diffusion and no metabolism (see Figure 9.3). Parameter estimates obtained from mechanistic modeling of TCA disposition in the presence or absence of BSEP inhibitors include total uptake clearance (CL_{Uptake}), basolateral efflux clearance (CL_{BL}), total biliary clearance (CL_{Bile}), and a rate constant that accounts for flux from bile networks into the medium (K_{Flux}) [20, 90, 93, 98]. The CL_{Bile} parameter for TCA is based solely on BSEP function, given its major role in TCA biliary excretion [34, 99, 100]. Measurements at different time points during both the uptake and efflux phases are required to obtain accurate initial parameter estimates, which can then be used for sensitivity analyses and simulations. For example, mechanistic PK modeling of drug and TCA disposition data in SCH can be used to predict alterations in *in vitro* hepatocellular bile acid concentrations due to drug-mediated inhibition of multiple transporters. Mechanistic PK models can be adapted for the relevant pathways (e.g., intracellular sequestration, metabolism) depending on the specific substrates and inhibitors.

As discussed in Section 9.2, membrane vesicles provide useful BSEP substrate and inhibition parameters for a compound. However, these parameters are limited in clinical translatability without knowing the concentration of a compound *in vivo* at the canalicular membrane. PK parameters (e.g., $CL_{int, bile}$, $CL_{int, uptake}$) obtained from mechanistic modeling of human SCH data can be scaled-up to the whole-liver level and incorporated into PBPK models to predict the *in vivo* hepatobiliary disposition of drugs, their metabolites, and bile acids [98, 101–103]. *In vivo* PK profiles have been predicted successfully for several compounds by developing PBPK models with clearance estimates obtained from SCH [101, 104–108]. PBPK models also can be used to predict *in vivo* intracellular and cytosolic

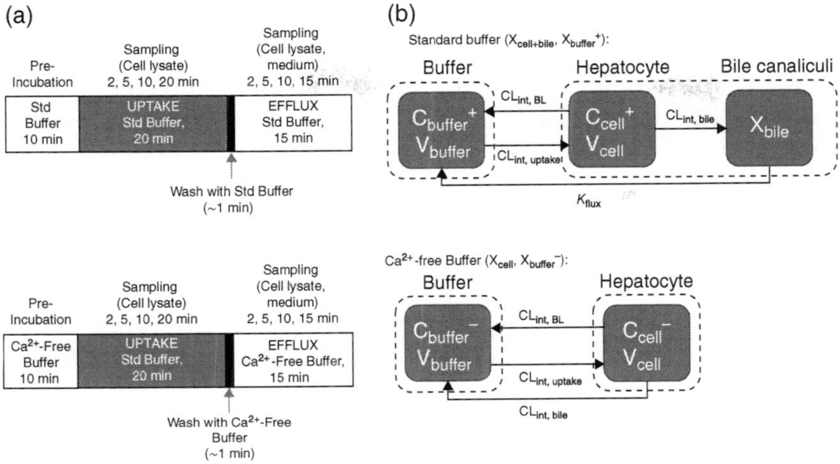

Figure 9.3 Schematic diagram coupling the experimental transport protocol (a) with the mechanistic pharmacokinetic (PK) model (b). Uptake and efflux studies were conducted in the presence of either standard (std) buffer or Ca^{2+}-free buffer. The sampling times and incubation duration shown in the experimental protocol may need to be adjusted depending on the PK characteristics of the substrate. In the PK model schemes, the sampling compartments are represented by dashed boxes, whereas V, C, and X denote compartmental volume, substrate concentration, and substrate mass, respectively. X in parentheses represents experimental measurements for model input. Subscripts denote the corresponding compartment as designated in the model scheme, and superscripts represent the presence (+) or absence (−) of Ca^{2+} in the preincubation and efflux buffer. Output parameters include intrinsic uptake clearance ($CL_{int, uptake}$), intrinsic basolateral efflux clearance ($CL_{int, BL}$), intrinsic biliary clearance ($CL_{int, bile}$), and a first-order rate constant for flux from bile networks into buffer (K_{flux}). *Source:* Adapted from Yang et al. [20].

hepatocyte concentrations of compounds, offering unique insight into the clinical translatability of transport parameters obtained during early drug development [20, 93, 101].

9.4 Other *In Vitro* Methods to Study BSEP Inhibition

ATP-dependent efflux of bile acids was first reported in canalicular membrane vesicles isolated from rat liver [25, 109, 110]. However, due to difficulties in preparation of high-quality vesicles and the presence of multiple ATP-dependent efflux transporters, this method has limited utility for the specific assessment of BSEP function. BSEP-expressing *Xenopus laevis* oocytes were used to elucidate rodent Bsep function for bile acid transport, and evaluation of Bsep substrates and inhibitors [28]. Applicability of this assay in drug discovery remained limited because the transporter kinetic parameters generated using oocytes are different from mammalian cells and lead to difficulty in extrapolating results to humans [54, 55]. Many uptake transporters have been studied using cell lines overexpressing the protein of interest, but the use of this approach to study efflux transporters is challenging, particularly for compounds that require a transporter for uptake into cells. To overcome these issues, double-transfected LLC-PK1 cells expressing both NTCP and BSEP were developed to allow transcellular transport of bile acids from the basolateral to canalicular membrane [111]. A similar double-expression HeLa cell line was developed to study polymorphic variants of BSEP (*ABCB11*) implicated in PFIC2 and BRIC2 [41]. BSEP-transfected HeLa, HEK293, and MDCK II cells have been used to explain the role of ubiquitination in the trafficking and degradation of cell surface BSEP [112–114]. While BSEP expression is inherently low in hepatic cell lines such as HepG2, HepaRG, and HuH-7 cells, expression can be induced by differentiation in HepaRG and HuH-7 cells [115, 116]. With induction of BSEP expression and metabolic activity, these hepatic cell models could be utilized to evaluate BSEP inhibition by a drug and/or metabolite.

Suspended or short-term cultured hepatocytes also have been purported to be a useful system to assess BSEP inhibition [80, 117, 118]. However, during hepatocyte isolation the polarity of hepatocytes is lost, except in remnant couplets, and canalicular proteins are internalized [119]; thus, hepatic basolateral transporters including MRP3, MRP4, and possibly OSTα/β would primarily drive substrate efflux into the medium. Accurate measurement of BSEP transport and inhibition requires the use of a polarized hepatocyte system. Pooled hepatocytes could be used to mitigate interindividual variability in SCH studies. However, not all hepatocytes have adequate transporter function. Therefore, care must be taken to use hepatocyte lots from commercial sources that are certified for transporter function, or validated in-house against established quality control data to assure that BSEP and other transport systems are functional. Primary hepatocytes cultured as 3D spheroids may express BSEP and have been used to accurately detect mechanisms of toxicity of drugs such as chlorpromazine, that include inhibition of *ABCB11* expression and intracellular bile acid accumulation [120]. Hepatocytes in 3D culture maintain expression of metabolic and transport proteins during long-term culture [121]; however, functional inhibition of BSEP remains to be investigated. A noteworthy platform that accounts for genetic variation and individual susceptibility factors to identify underlying mechanisms of hepatotoxicity is the use of cryopreserved primary hepatocytes from genetically diverse Collaborative Cross mouse strains either in sandwich-culture or as 3D spheroids [122]. The Collaborative Cross is a large panel of multi-parental recombinant inbred mouse strains that form highly sophisticated genetic reference populations [123]. Primary hepatocytes from such mouse strains offer the unique advantage of high genetic diversity and statistical power to detect underlying factors (e.g., BSEP (*ABCB11*) genetic polymorphisms) responsible for DILI.

More sophisticated *in vitro* models have been developed recently that show long-term metabolic viability, including micropatterned hepatocyte co-cultures (e.g., Hμrel hepatic co-cultures, HepatoPac®). Hμrel hepatocytes and HepatoPac® cultures utilize cryopreserved human hepatocytes co-cultured with non-parenchymal stromal cells and fibroblasts, respectively [124, 125]. These advanced *in vitro* models offer multiple advantages in predicting DDI and assessment of DILI liability, however, validation of transporter expression and function is required for future utility in measuring BSEP inhibition [47]. Organoid cultures, microfluidic liver models, and liver-on-a-chip models are being developed that may provide further insights into DILI and the role of BSEP inhibition [126–128].

9.5 Computational Methods Used to Predict BSEP Inhibition

Computational approaches to identify potential BSEP inhibitors would be valuable when designing and optimizing drug candidates. Based on numerous datasets that have classified compounds as BSEP inhibitors or non-inhibitors, primarily generated from *in vitro* membrane vesicle studies, several ligand-based modeling strategies have been utilized: pharmacophore, quantitative structure–activity relationship (QSAR), and Bayesian modeling [47]. Pharmacophore models can predict important structural features such as hydrophobic/aromatic groups and hydrogen bond acceptors among BSEP inhibitors, but they are often restricted by the chemical space covered by the training and external test sets of available compounds. Therefore, the ability of pharmacophore models to correctly predict BSEP inhibition is often limited to compounds that are closely related in structure. QSAR and Bayesian modeling use association of molecular descriptors with the extent of BSEP inhibition to reveal substructures or features that are important for BSEP inhibition [58, 129, 130]. These models often outperform simple pharmacophore models as evaluated by the Matthews correlation coefficient [47]. However, all ligand-based models suffer from limited structural information on the BSEP protein, and that the actual binding sites of inhibitors and modes of inhibition were unknown. Also, the definition of BSEP "inhibitors" and "non-inhibitors" vary, but typically are based on simple dichotomies of IC_{50} values, which are assay system- and laboratory-dependent [47].

Protein-based modeling introduces protein structural data to complement ligand-based information. The related mouse P-glycoprotein X-ray structure was used to develop a homology model for human BSEP [131] that allowed docking of chemicals into the putative substrate pocket and predicted BSEP inhibitors with reasonable accuracy. The very recent cryo-electron microscopy structure of human BSEP [132] is a promising start towards more precise docking-based identification of BSEP substrates and inhibitors, but currently the low resolution (3.5–4 Å) is insufficient for accurate binding studies.

Unfortunately, the current utility of these ligand- and protein-based models to identify compounds that inhibit BSEP remains limited. BSEP inhibitors are chemically diverse and they often inhibit other bile acid transporters, which complicates the identification of common structural features. The development of integrated *in silico* models to predict inhibition of other hepatic bile acid transporters in addition to BSEP might improve virtual detection of compounds with DILI liability.

9.6 *In Vitro* Models as a Predictor of Clinical DILI

9.6.1 The C-DILI™ Assay

The C-DILI™ assay, which uses human SCH, is a novel tool to predict cholestatic DILI. The C-DILI™ assay accounts for multiple mechanisms that lead to cholestatic DILI. It is increasingly recognized that intracellular accumulation of bile acids within hepatocytes can lead to hepatocellular or cholestatic toxicity. However, intracellular bile acid concentrations are modulated by a multitude of factors, and not solely by BSEP inhibition [133]. Bile acids are important signaling molecules that regulate their own synthesis, metabolism, and excretion via the fibroblast growth factor receptor (FGFR4)-FXR gut-liver axis [134]. FXR is activated directly by increased intracellular bile acid concentrations (e.g., CDCA). In turn, FXR regulates intracellular bile acid concentrations by induction of BSEP expression to increase biliary excretion and by induction of OSTα/β to facilitate basolateral efflux, an important hepatoprotective mechanism to prevent bile acid-mediated toxicity. FXR also downregulates bile acid synthesis via CYP7A1 in humans, and decreases NTCP expression to limit the uptake of bile acids into the hepatocyte [22]. Elevated intrahepatic bile acids stimulate proinflammatory cytokines and reactive oxygen species that cause oxidative stress and apoptosis [22, 135]. The C-DILI™ assay integrates BSEP inhibition, FXR antagonism, and inhibition of hepatoprotective or compensatory basolateral efflux transporters (e.g., OSTα/β, MRP3, MRP4) to predict DILI liability more accurately. Since the assay is based on human SCH, it also accounts for drug metabolism and derived metabolite-mediated toxic effects.

Detailed C-DILI™ methodology and assay considerations have been discussed previously [133, 136]. Briefly, cryopreserved human SCH are established in a 96-well format and exposed to test drugs and controls for 24 hours in standard medium and in sensitization medium containing free fatty acids and pooled bile acids [136]. Typically, test drugs are prepared at three concentrations scaled from the highest concentration of the drug reported in blood or plasma (C_{max}): C_{max},

10-fold C_{max}, and 50-fold C_{max} (or limit of solubility). After a 24-hour exposure, ATP content and LDH leakage are measured to reflect cytotoxicity and viability, and loss of membrane integrity, respectively. Test drugs that decrease ATP and increase LDH in sensitization medium relative to untreated controls (i.e., drugs showing differential toxicity between the standard and sensitization media) exhibit a risk of cholestatic hepatotoxicity. BSEP inhibitor and non-inhibitor controls are applied to serve as negative controls, positive controls, and direct toxicity controls to assess clinical risk of DILI. Cyclosporine A, a potent BSEP inhibitor, is used as a negative control because the time-dependent inhibition of BSEP is mitigated by activation of FXR by transiently elevated intracellular bile acids. Activated FXR is able to reduce bile acid synthesis and increase basolateral efflux via OSTα/β, thereby, preventing any cholestatic or hepatocellular DILI in both standard and sensitization media. On the other hand, troglitazone serves as a positive control for cholestatic hepatotoxicity. Troglitazone and its metabolite, troglitazone sulfate, are both potent BSEP inhibitors [49]. Additionally, troglitazone is a weak antagonist of FXR [137], and troglitazone sulfate inhibits OSTα/β as well as MRP4 [90, 138]. Due to the combination of these effects on bile acid homeostasis, cells are unable to recover from troglitazone-mediated toxicity in the sensitization medium. Therefore, troglitazone shows risk of cholestatic DILI and presents with lower ATP content and higher LDH release compared to untreated controls in the sensitization media only. Finally, imatinib is used as a control for direct or intrinsic hepatotoxicity with reduced ATP content and increased LDH release in both standard and sensitization media [133].

Overall, the C-DILI™ assay can be applied as an effective tool to assess risk of hepatotoxicity for novel drugs during development. Since the assay utilizes hepatocytes that are validated or certified for transporter activity, the intracellular concentration of the drug is the driving force for transporter inhibition and FXR antagonism that ultimately alters bile acid homeostasis and interferes with compensatory mechanisms. This assay provides a more accurate prediction of clinical DILI than membrane vesicles due to an integrated mechanistic approach to test DILI liability of compounds arising from BSEP inhibition, FXR antagonism, as well as inhibition of compensatory efflux transporters by the parent drug and/or metabolite(s) [133, 136].

9.7 Preclinical *In Vivo* Models for the Evaluation of BSEP/Bsep Inhibition and DILI

In vivo animal models offer the opportunity to closely assess the PK, pharmacodynamic (PD), and toxicological interaction of a compound with a biological system. Rodents are the most commonly used species for preclinical drug testing due to their physiological and genetic similarity to humans, and ease of maintenance [139]. In some cases, preclinical data obtained in rodents during drug development may suggest DILI liability. Rodent models are well established for numerous drugs associated with predictable, intrinsic DILI [140]. However, developing a rodent model to specifically measure BSEP/Bsep inhibition that would translate to clinical DILI has proven challenging, in part because bile acid composition is species-specific. Furthermore, measuring a drug interaction with a single transporter *in vivo* is extremely challenging and typically requires a specific probe substrate or other surrogate marker to reliably determine any alterations in transporter function. Unfortunately, there are currently no well-characterized synthetic probe substrates transported specifically by Bsep that can be utilized for preclinical animal models [47]. This issue compels researchers to rely on endogenous bile acids such as TCA or CDCA as *in vivo* biomarkers of Bsep function [141–143]. Bile acids were reliable indicators of Bsep inhibition and cholestasis in several rodent studies [142, 144–146]. However, altered serum bile acid concentrations in animal models should be interpreted cautiously with adequate consideration of other potential mechanisms. Several confounding factors (see Section 9.8.1) interfere with the reliability of serum bile acids as a marker of *in vivo* BSEP function in humans: inhibition of NTCP- and/or organic anion transporting polypeptide (OATP)-mediated basolateral bile acid uptake [147, 148] may lead to elevated serum bile acid concentrations, and inhibition of MRP3/4 and OSTα/β basolateral bile acid efflux [47, 149] may result in decreased serum bile acid concentrations.

While some drugs such as troglitazone have shown up to 10-fold more potent inhibition of human BSEP compared to rodent Bsep [118], the majority of drugs tested demonstrate similar IC_{50} values between species [53, 150]. These data suggest that species variability in BSEP/Bsep inhibition does not explain why drugs inhibiting BSEP and causing DILI in humans do not always show signs of liver toxicity in animals [47]. Nonetheless, investigators may still want to test a compound for differences in human and rodent BSEP/Bsep inhibition before interpreting data obtained from rodent models.

Another concern is the correlation between altered hepatobiliary bile acid disposition in rodents and human outcomes. Historically, preclinical rodent models have been poor predictors of clinical DILI incidence for drugs that alter BSEP/Bsep function, which can be attributed, in part, to species differences in bile acid composition and regulation [47]. Bile acid composition and regulation differ between humans and preclinical species in multiple ways: (i) the human bile acid pool is more hydrophobic than rodents [151], (ii) rodents synthesize bile acids from cholesterol via Cyp7a1 that is controlled by LXR, whereas, in humans, bile acid synthesis occurs mainly via CYP7A1 and CYP8B1, which are regulated by FXR [152], (iii) rats have no gall bladder to store bile, (iv) primary bile acids in humans are CDCA and CA, whereas, in mice, CDCA and CA are further metabolized by Cyp2c70 to muricholic acids (MCAs) and ursodeoxycholate (UDCA) [153, 154], (v) major differences in bile acid metabolism (e.g., sulfation and glycine amidation are higher in humans than in rodents, while taurine amidation and Cyp-mediated hydroxylation predominate in rodents) [155–157], (vi) bile acid synthesis rates are ~30% higher in healthy men than in women, while mice show the opposite sex difference [158], (vii) differences in gut microbiota between humans and rodents, and (viii) interspecies differences in hepatic BSEP/Bsep expression [159]. These remarkable differences in bile acid composition, homeostasis, and BSEP/Bsep expression make it challenging to predict human DILI using preclinical species.

Bile acid feeding is a method involving the oral administration of hydrophobic or toxic bile acids, found at higher levels in humans, to rodent models to mitigate species differences in bile acid composition. This method has been used to demonstrate the correlation between bile acid hydrophobicity and hepatotoxicity [19, 82, 160], but is seldom utilized to study DILI or Bsep inhibition. Yang et al. [161] recently demonstrated that rats concurrently fed the hydrophobic bile acid CDCA and ketoconazole, a known BSEP/Bsep inhibitor, showed significant increases in several hepatic bile acid species and plasma transaminases (aspartate transaminase [AST], alanine transaminase [ALT]) compared to rats receiving ketoconazole monotherapy. This approach may have the potential to assist in preclinical DILI assessment of BSEP/Bsep-inhibiting drugs [161].

Other methods have been used to assess Bsep inhibition *in vivo*. A novel rat model using siRNA knockdown of Bsep demonstrated enhanced hepatotoxicity when exposed to eight BSEP/Bsep inhibitors for only seven days [162]. Quantitative intravital microscopy of fluorescent bile acids in rat liver detected *in vivo* Bsep inhibition by bosentan and other known BSEP/Bsep inhibitors at doses significantly lower than required to increase serum bile acid concentrations in rats [163]. Precision-cut rat liver slices have been used to investigate biomarkers, including bile acid accumulation, as predictors of cholestasis [164]. Chimeric mouse models with humanized livers are being developed and have shown promising results [146, 165, 166]. Fialuridine, a nucleoside analogue that caused ALF without any signal of hepatotoxicity in preclinical studies, was tested in chimeric TK-NOG mice with humanized livers. These mice developed serological and clinical evidence of ALF in a preclinical model after only four days of treatment [167]. A major shortcoming of these chimeric models is their inability to express a fully integrated human innate and adaptive immunity [168]. Additionally, an immunocompromised state must be induced in these mice to prevent rejection of the transplanted human hepatocytes [12, 146]. Recent studies have shown that many clinical DILI cases are immune-mediated [1, 2, 12]. Thus, the inability of chimeric mouse models to express a functional, humanized immune system may lead to false negative results. While new preclinical methods and models to study BSEP/Bsep inhibition and DILI are emerging, there remains uncertainty surrounding their clinical relevance and a current lack of justification for their high cost in early drug development.

Typical preclinical animal models used for drug safety assessment may not be suitable for clinical translatability or prediction of DILI associated with BSEP interference in humans [53]. This is evidenced by BSEP-mediated DILI arising in clinical trials and post-marketing without any sign of preclinical liver toxicity [2, 47, 65]. However, animal models can provide valuable information regarding PK/PD and additional data for implementation into QST models (e.g., DILIsym).

9.8 *In Vivo* Clinical Biomarkers of BSEP Inhibition and DILI

Biomarkers for human subjects are analyzed from a variety of biological fluids and can offer unparalleled insight into disease detection, progression, and/or response to pharmacotherapy. For clinical development, biomarkers offer a more efficient and less expensive alternative to the direct measurement of clinical endpoints [169]. Biomarkers also can be incorporated into mechanistic modeling for more accurate and comprehensive predictions of clinical outcomes [11]. Clinical biomarkers of BSEP inhibition and DILI currently lack adequate sensitivity and specificity [14, 47]. Biomarkers capable of accurately predicting DILI and/or BSEP inhibition shortly after initiation of a medication would be a major advance in the field, and the search for such useful endogenous markers is ongoing [170]. This section briefly summarizes biomarkers that are used in current practice or are under investigation for clinical implementation, and their potential impact.

9.8.1 Serum Bile Acids as Clinical Biomarkers

Direct measurement of bile acid concentrations in the systemic circulation has been postulated to provide an indirect estimate of BSEP function in humans. Serum bile acids are measured routinely in patients with PFIC2, a genetic disease with markedly reduced BSEP function, to aid in diagnosis and evaluation of disease progression [171]. This method has been adapted by several researchers to assess clinical drug-BSEP interactions [144, 172]. However, the majority of *in vivo* bile acid biomarker studies have been performed in rodent models (see Section 9.7). Preclinical DILI studies detected elevated serum bile acids before changes in total bilirubin (TBIL) or transaminases were observed [142, 173], demonstrating the potential utility of bile acid biomarkers for early detection of DILI and/or BSEP/Bsep inhibition *in vivo*. In humans, the ratio of conjugated to unconjugated serum bile acids may be more useful than total serum bile acids when evaluating BSEP function and DILI. According to one study, elevated conjugated serum bile acids were detected before the onset of clinical transaminase elevations in a post hoc analysis of a Phase I compound failure due to hepatotoxicity [172].

Regrettably, serum bile acids may not always provide ideal specificity to detect *in vivo* BSEP inhibition due to a drug's potential to also modulate other bile acid uptake and/or efflux transporters and or metabolic pathways (see Section 9.7). Many clinical confounders may complicate the utility of serum bile acids as clinical biomarkers for BSEP inhibition including intraindividual variability in endogenous human bile acid concentrations [174], the effects of diet on the bile acid pool [175], and diurnal variations in bile acid concentrations [176]. Serum bile acids may be increased by extrahepatic biliary obstruction in the absence of BSEP inhibition [177], or due to decreased intestinal microbial metabolism in patients treated with antibiotics (e.g., clarithromycin) that do not inhibit human BSEP [47, 178]. Importantly, serum bile acid concentrations may not accurately reflect concentrations in the hepatocyte at the site of BSEP transport [149].

Despite great promise, serum bile acids are not used routinely as clinical biomarkers for BSEP inhibition and/or DILI. Urinary bile acids, which may exhibit smaller intra- and interindividual variability, have been suggested as an alternative [174]. Further investigations are required to validate bile acids as useful clinical biomarkers and establish the optimal biological sample, sampling conditions, and other key considerations. For example, it may be ideal to perform longitudinal sampling from baseline and/or during drug pretreatment to negate any intraindividual variability in bile acid concentrations [47]. The ability to directly measure the effect of a medication on BSEP function *in vivo* in humans would provide invaluable insight but remains elusive.

9.8.2 Clinical Biomarkers of DILI

Diagnosis and prediction of DILI is currently limited by the lack of sensitive and specific biomarkers [14, 179]. AST, ALT, alkaline phosphatase (ALP), and TBIL are the most commonly used serum biomarkers for DILI detection in clinical practice today [11]. Because these "traditional" biomarkers are not liver-specific and do not allow early detection of DILI, international collaborative efforts have been launched to identify and validate novel biomarkers [1, 14, 180]. Glutamate dehydrogenase (GLDH) and microRNA-122 (miR-122) are new biomarkers with earlier DILI detection than ALT that exhibit high hepatic specificity and sensitivity [11]. GLDH recently gained favor as miR-122 also can be released from healthy hepatocytes, resulting in higher intra- and interindividual variability [11]. Both novel biomarkers have been validated in separate cohorts in which GLDH showed better DILI correlation than miR-122 [180]. These new biomarkers warrant further validation through clinical testing before their routine use can be universally accepted [170].

A clinical biomarker to better predict and reliably measure DILI would allow drug developers to evaluate a compound's DILI risk without having to rely on observations of extremely rare clinical endpoints. Methodologies currently being evaluated as more specific indicators of DILI at an earlier stage in drug development include pharmacogenomic analysis [181], proteomics [182], and metabolomics [183].

9.9 Quantitative Systems Toxicology to Predict DILI

The application of QST methods in pharmacological research has grown immensely over the past decade. QST takes a translational approach to quantitatively assess the toxic effects of a xenobiotic on the molecular and physical levels of a living organism through integration of computational methods and *in vitro*/*in vivo* experimental data [184]. This is accomplished by using differential equations to describe drug or chemical interactions with complex biological pathways that result in stress and/or death to cells, tissues, and organs [65, 185].

Organ-specific QST models are being developed to address high rates of drug attrition in preclinical and clinical development due to unanticipated toxicity [186, 187]. The multifactorial nature of DILI and the challenges associated with accurate predictions warrant a QST approach [184] to account for the interplay of complex mechanisms and extensive interindividual variability. A liver-specific model known as DILIsym® that focuses on hepatotoxicity outcome (see Figure 9.4) is in use by pharmaceutical companies and regulatory agencies to guide drug safety assessment. DILIsym incorporates submodels based on three common mechanisms of DILI: interference with mitochondrial respiration, bile acid accumulation, and oxidative stress [65, 185]. Additional submodels describing pathways of hepatocellular toxicity have been built into DILIsym and connected to outcomes such as hepatocyte death and biomarker release into serum.

Clinically relevant drug concentrations at the site of toxicity predicted by PBPK models are an essential input component for QST models [47, 184]. For DILIsym software, PBPK models provide time course estimates of hepatocellular drug concentrations for model input [65, 185]. This information is crucial for appropriate model-based *in vivo* translation of drug-enzyme and drug-transporter parameters such as IC_{50} values initially obtained from *in vitro* data. *In vitro* assays are necessary to assess the concentration-dependent ability of a drug, and any major metabolite(s), to interact with the three key mechanistic submodels of DILI [65, 185]. Hence, *in vitro* data are needed to obtain essential input parameters for drug-specific DILIsym hepatotoxicity simulations. Once all data and parameter estimates are input into the model, DILIsym provides predictions of time-dependent hepatocyte death and subsequent biomarker release into serum, among numerous other outputs [65, 185]. ALT is the most sensitive and specific of the

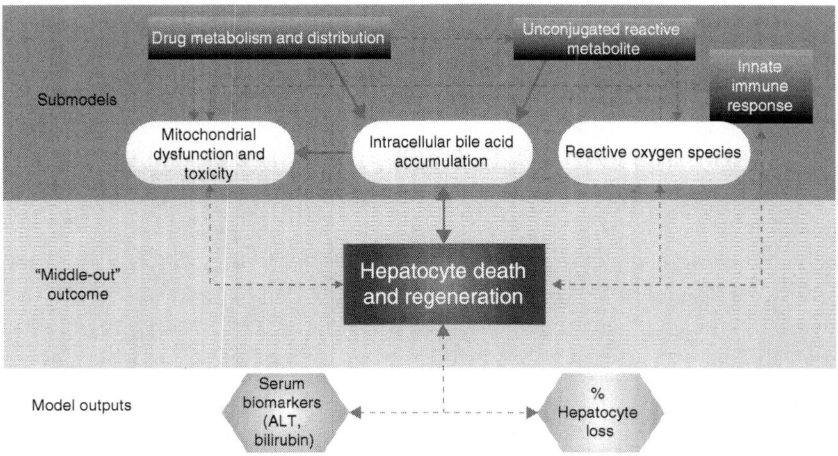

Figure 9.4 Diagram depicting bile acid accumulation as a key submodel in the DILIsym software. This model was developed using a "middle-out" approach, starting with the ultimate result of hepatocyte death/regeneration and building outwards with the development of submodels and clinical outcomes. DILIsym contains three "key" submodels: intracellular bile acid accumulation, generation of reactive oxygen species, and mitochondrial dysfunction. Production of reactive metabolites and innate immune response activation also are included in the DILIsym model. Each process is integrated with the outcome of hepatocyte death, which results in the release of serum biomarkers. ALT, alanine transaminase. *Source:* Adapted from Watkins [65].

"traditional" biomarkers (see Section 9.8.2) [65]. A unique feature of DILIsym is the ability of the model to predict the percentage of hepatocyte death based on time-dependent serum ALT levels and vice versa [11]. Quantitative data on hepatocyte death obtained from DILIsym allows assessment of the predicted impact of ALT signals on animal or human liver health. Other DILIsym outputs include serum TBIL and novel serum biomarkers (see Section 9.8.2): GLDH, miR-122, and full length and caspase-cleaved cytokeratin 18 [65, 185]. Simulated patient populations encompassing a high degree of inter-individual variability, based on genetic and nongenetic factors observed in the general population, can be included in the model to simulate clinical trials. Although these simulated populations typically contain only ~300 patients, the input variation for DILI susceptibility is proposed to exceed that encountered in a far larger patient population [65]. Therefore, this feature can provide liver safety data, typically obtained from Phase III trials and post-marketing patient populations, with parameter inputs acquired from *in vitro* and PK studies. In addition to human models, mouse, rat, and dog versions of DILIsym are available to confirm or predict preclinical findings [188, 189]. To date, many drug-related bile acid-mediated hepatotoxicity events observed in preclinical, clinical trial, and post-marketing populations have been successfully validated through incorporation of *in vitro*, *in vivo* animal, and *in vivo* human data into DILIsym [190–192].

In addition to BSEP, multiple transporters contribute to hepatic bile acid disposition, as detailed above [93]. Drug-interaction parameters must be obtained from various *in vitro* assays (see Sections 9.2–9.4) on each bile acid transport mechanism for input into the DILIsym model. SCH data with mechanistic modeling (see Section 9.3.2) arguably provides the most valuable *in vitro* biliary disposition DILIsym parameters, which can be used to predict a drug's cumulative impact on hepatocellular bile acid accumulation [20, 93]. A bile acid homeostasis model was previously established in DILIsym [193] and can be used for drugs and metabolites shown to inhibit BSEP *in vitro* to predict or validate their influence on bile acid disposition and hepatotoxicity [64, 190, 192]. DILIsym also can provide mechanistic insights of DILI, which can be particularly valuable. Applications of this model have provided strong confirmatory evidence of the role of bile acid accumulation in human hepatotoxicity, and have identified key knowledge gaps in the field of bile acid-mediated DILI [193]. DILIsym modeling also supports the hypothesis that rats have a less toxic bile acid pool than humans, consistent with their lower susceptibility to hepatotoxicity due to bile acid accumulation [192, 194]. The importance of BSEP inhibitory mechanisms in hepatotoxic events has been demonstrated consistently by DILIsym modeling [64, 190]. When bile acids begin to accumulate in the hepatocyte, they can displace competitive inhibitors from the BSEP substrate binding site, but they do not affect binding of noncompetitive inhibitors, which bind at allosteric sites. Hence, there is a higher likelihood of toxic bile acid accumulation in hepatocytes with a noncompetitive than a competitive BSEP inhibitor of similar or higher inhibitory potency [65, 185]. Originally, an IC_{50} "cut-off" of $\leq 25\,\mu M$ for BSEP inhibition by a compound was proposed to predict a high risk for bile acid-mediated DILI [47, 53]. However, DILIsym modeling has shown that bile acid accumulation can still play a significant role in hepatotoxicity for less potent BSEP inhibitors if at least one other mechanism of DILI is present [65, 185]. For instance, even though the BSEP IC_{50} value for tolvaptan is $31.6\,\mu M$ [92], DILIsym modeling implicated bile acid accumulation, supplemented by mitochondrial impairment, as a core mechanism underlying tolvaptan-mediated DILI [191].

A shortcoming of DILIsym, and QST modeling in general, is the need for multiple data inputs that can only be obtained through time-consuming *in vivo* and *in vitro* experimental assays [47]. This requirement may limit the practicality of this tool in early drug development. Nevertheless, QST modeling has accelerated mechanistic understanding and prediction of DILI at an unprecedented rate over the last decade, and provides a promising outlook for the advancement of precision medicine and toxicology.

9.10 Conclusions

Evaluation of BSEP inhibition is recommended to aid in thorough DILI-risk assessment during drug discovery and development [47]. The clinical translation of BSEP/Bsep inhibition and predictive DILI data derived from mechanism-based *in vitro* and *in vivo* experimental models requires a holistic approach. Initial efforts to utilize *in vitro* and *in vivo* data to predict DILI incidence in large-scale patient populations have provided great insight but lack adequate predictive power. For example, BSEP IC_{50} values obtained from membrane vesicle studies are poor predictors of DILI when evaluated in isolation [65]. SCH and other *in vitro* methods provide additional information to aid in DILI prediction, but have their own limitations. Differences between rodent and human bile acid pools limit the translatability of preclinical models, which may be improved by further development of humanized rodent models [146]. Current *in vivo* biomarkers lack specificity and sensitivity [47], so novel clinical biomarkers are being sought to aid in the prediction and diagnosis of DILI [170, 180].

The vast array of complex biological interactions coupled with large interpatient variability in the general population calls for computational approaches to better interpret and predict DILI incidence. Hence, present-day efforts are heavily geared toward the improvement of QSAR and QST models to predict BSEP inhibition potential and clinical DILI outcomes, respectively, for compounds in development. Current QSAR models cannot adequately guide compound design or selection to avoid inhibition of BSEP and other bile acid transporters by the parent drug and/or metabolites [47]. Current QST models, such as DILIsym, have demonstrated utility in drug development, but require prior collection of *in vitro* and *in vivo* data. DILIsym models are being refined towards a more efficient and less resource-intensive drug development process.

Over the past decade, computational modeling has gradually bridged the gap between *in vitro*/*in vivo* experimental data and clinical outcomes. Continued development of novel and accurate *in vitro*, *in vivo*, and *in silico* methods to predict BSEP inhibition and DILI will have profound effects on drug development and the medical field.

Funding Information

This work was supported, in part, by the National Institutes of Health under award number R35 GM122576 from the National Institute of General Medical Sciences (Kim L.R. Brouwer). Dr. William Murphy was supported by an Eshelman Fellowship from the UNC Eshelman School of Pharmacy. Dr. Paavo Honkakoski was supported by the Nannerl O. Keohane Distinguished Visiting Professorship from UNC-Chapel Hill and Duke University. Any opinions, findings, conclusions, or recommendations expressed in this publication are those of the authors and do not necessarily reflect the views of the National Institutes of Health.

Conflict of Interest

Kim L.R. Brouwer is a coinventor of the sandwich-cultured hepatocyte technology for quantification of biliary excretion (B-CLEAR®) and related technologies, which have been licensed exclusively to Qualyst Transporter Solutions, recently acquired by BioIVT.

Acknowledgments

The authors thank Dr. James Beaudoin for his insightful comments and Kylie Lewis for assistance in preparation of this chapter.

References

1 Andrade, R. J., Chalasani, N., Björnsson, E. S., Suzuki, A., Kullak-Ublick, G. A., Watkins, P. B., Devarbhavi, H., Merz, M., Lucena, M. I., Kaplowitz, N. & Aithal, G. P. 2019. Drug-Induced Liver Injury. *Nat Rev Dis Primers*, 5, 58.

2 Mosedale, M. & Watkins, P. B. 2017. Drug-Induced Liver Injury: Advances in Mechanistic Understanding That Will Inform Risk Management. *Clin Pharmacol Ther*, 101, 469–480.

3 Björnsson, E. S., Bergmann, O. M., Björnsson, H. K., Kvaran, R. B. & Olafsson, S. 2013. Incidence, Presentation, and Outcomes in Patients with Drug-Induced Liver Injury in the General Population of Iceland. *Gastroenterology*, 144, 1419-25, 1425.e1.

4 Nicoletti, P., Aithal, G. P., Bjornsson, E. S., Andrade, R. J., Sawle, A., Arrese, M., Barnhart, H. X., Bondon-Guitton, E., Hayashi, P. H., Bessone, F., Carvajal, A., Cascorbi, I., Cirulli, E. T., Chalasani, N., Conforti, A., Coulthard, S. A., Daly, M. J., Day, C. P., Dillon, J. F., Fontana, R. J., Grove, J. I., Hallberg, P., Hernández, N., Ibáñez, L., Kullak-Ublick, G. A., Laitinen, T., Larrey, D., Lucena, M. I., Maitland-Van Der Zee, A. H., Martin, J. H., Molokhia, M., Pirmohamed, M., Powell, E. E., Qin, S., Serrano, J., Stephens, C., Stolz, A., Wadelius, M., Watkins, P. B., Floratos, A., Shen, Y., Nelson, M. R., Urban, T. J., Daly, A. K. & International Drug-Induced Liver Injury Consortium, Drug-Induced Liver Injury Network Investigators, and International Serious Adverse Events Consortium 2017. Association of Liver

Injury from Specific Drugs, or Groups of Drugs, with Polymorphisms in HLA and Other Genes in A Genome-Wide Association Study. *Gastroenterology*, 152, 1078–1089.

5 Sgro, C., Clinard, F., Ouazir, K., Chanay, H., Allard, C., Guilleminet, C., Lenoir, C., Lemoine, A. & Hillon, P. 2002. Incidence of Drug-Induced Hepatic Injuries: A French Population-Based Study. *Hepatology*, 36, 451–455.

6 De Abajo, F. J., Montero, D., Madurga, M. & García Rodríguez, L. A. 2004. Acute and Clinically Relevant Drug-Induced Liver Injury: A Population Based Case-Control Study. *Br J Clin Pharmacol*, 58, 71–80.

7 Vega, M., Verma, M., Beswick, D., Bey, S., Hossack, J., Merriman, N., Shah, A., Navarro, V. & Drug-Induced Liver Injury Network Investigators. 2017. The Incidence of Drug- and Herbal and Dietary Supplement-Induced Liver Injury: Preliminary Findings from Gastroenterologist-Based Surveillance in the Population of the State of Delaware. *Drug Saf*, 40, 783–787.

8 Garzel, B., Zhang, L., Huang, S.-M. & Wang, H. 2019. A Change in Bile Flow: Looking Beyond Transporter Inhibition in the Development of Drug-Induced Cholestasis. *Curr Drug Metab*, 20, 621–632.

9 Katarey, D. & Verma, S. 2016. Drug-Induced Liver Injury. *Clin Med*, 16, s104–s109.

10 Chalasani, N. & Björnsson, E. 2010. Risk Factors for Idiosyncratic Drug-Induced Liver Injury. *Gastroenterology*, 138, 2246–2259.

11 Church, R. J. & Watkins, P. B. 2019. Serum Biomarkers of Drug-Induced Liver Injury: Current Status and Future Directions. *J Dig Dis*, 20, 2–10.

12 Mosedale, M. & Watkins, P. B. 2020. Understanding Idiosyncratic Toxicity: Lessons Learned from Drug-Induced Liver Injury. *J Med Chem*, 63 (12): 6436–6461.

13 Gonzalez-Jimenez, A., Mceuen, K., Chen, M., Suzuki, A., Robles-Diaz, M., Medina-Caliz, I., Bessone, F., Hernandez, N., Arrese, M., Parana, R., Lucena, M. I., Stephens, C. & Andrade, R. J. 2019. The Influence of Drug Properties and Host Factors on Delayed Onset of Symptoms in Drug-Induced Liver Injury. *Liver Int*, 39, 401–410.

14 Fu, S., Wu, D., Jiang, W., Li, J., Long, J., Jia, C. & Zhou, T. 2019. Molecular Biomarkers in Drug-Induced Liver Injury: Challenges and Future Perspectives. *Front Pharmacol*, 10, 1667.

15 Stieger, B., Fattinger, K., Madon, J., Kullak-Ublick, G. A. & Meier, P. J. 2000. Drug- and Estrogen-Induced Cholestasis Through Inhibition of the Hepatocellular Bile Salt Export Pump (BSEP) of Rat Liver. *Gastroenterology*, 118, 422–430.

16 Wagner, M., Zollner, G. & Trauner, M. 2009. New Molecular Insights into the Mechanisms of Cholestasis. *J Hepatol*, 51, 565–580.

17 Cai, S. Y., Ouyang, X., Chen, Y., Soroka, C. J., Wang, J., Mennone, A., Wang, Y., Mehal, W. Z., Jain, D. & Boyer, J. L. 2017. Bile Acids Initiate Cholestatic Liver

Injury by Triggering a Hepatocyte-Specific Inflammatory Response. *JCI Insight*, 2, e90780.

18 Hofmann, A. F. 1999. The Continuing Importance of Bile Acids in Liver and Intestinal Disease. *Arch Intern Med*, 159, 2647–2658.

19 Perez, M. J. & Briz, O. 2009. Bile-Acid-Induced Cell Injury and Protection. *World J Gastroenterol*, 15, 1677–1689.

20 Yang, K., Guo, C., Woodhead, J. L., St Claire, R. L., Watkins, P. B., Siler, S. Q., Howell, B. A. & Brouwer, K. L. R. 2016. Sandwich-Cultured Hepatocytes as a Tool to Study Drug Disposition and Drug-Induced Liver Injury. *J Pharm Sci*, 105, 443–459.

21 Russell, D. W. 2003. The Enzymes, Regulation, and Genetics of Bile Acid Synthesis. *Annu Rev Biochem*, 72, 137–174.

22 Chiang, J. Y. 2009. Bile Acids: Regulation of Synthesis. *J Lipid Res*, 50, 1955–1966.

23 Chiang, J. Y. 2013. Bile Acid Metabolism and Signaling. *Compr Physiol*, 3, 1191–1212.

24 Ananthanarayanan, M., Balasubramanian, N., Makishima, M., Mangelsdorf, D. J. & Suchy, F. J. 2001. Human Bile Salt Export Pump Promoter Is Transactivated by the Farnesoid X Receptor/Bile Acid Receptor. *J Biol Chem*, 276, 28857–28865.

25 Nishida, T., Gatmaitan, Z., Che, M. & Arias, I. M. 1991. Rat Liver Canalicular Membrane Vesicles Contain an ATP-Dependent Bile Acid Transport System. *Proc Natl Acad Sci USA*, 88, 6590–6594.

26 Childs, S., Yeh, R. L., Georges, E. & Ling, V. 1995. Identification of a sister gene to P-glycoprotein. *Cancer Research*, 55(10), 2029–2034.

27 Kosters, A. & Karpen, S. J., 2008. Bile acid transporters in health and disease. *Xenobiotica*, 38(7–8), 1043–1071.

28 Gerloff, T., Stieger, B., Hagenbuch, B., Madon, J., Landmann, L., Roth, J., Hofmann, A. F. & Meier, P. J. 1998. The Sister of P-Glycoprotein Represents the Canalicular Bile Salt Export Pump of Mammalian Liver. *J Biol Chem*, 273, 10046–10050.

29 Meier, P. J. & Stieger, B. 2002. Bile Salt Transporters. *Annu Rev Physiol*, 64, 635–661.

30 Soroka, C. J. & Boyer, J. L. 2014. Biosynthesis and Trafficking of the Bile Salt Export Pump, Bsep: Therapeutic Implications of Bsep Mutations. *Mol Aspects Med*, 37, 3–14.

31 Strautnieks, S. S., Kagalwalla, A. F., Tanner, M. S., Knisely, A. S., Bull, L., Freimer, N., Kocoshis, S. A., Gardiner, R. M. & Thompson, R. J. 1997. Identification of a Locus for Progressive Familial Intrahepatic Cholestasis PFIC2 on Chromosome 2q24. *Am J Hum Genet*, 61, 630–633.

32 Stieger, B., Meier, Y. & Meier, P. J. 2007. The Bile Salt Export Pump. *Pflugers Arch*, 453, 611–620.

33 Trauner, M. & Boyer, J. L. 2003. Bile Salt Transporters: Molecular Characterization, Function, and Regulation. *Physiol Rev*, 83, 633–671.

34 Noé, J., Stieger, B. & Meier, P. J. 2002. Functional Expression of the Canalicular Bile Salt Export Pump of Human Liver. *Gastroenterology*, 123, 1659–1666.

35 Jansen, P. L., Strautnieks, S. S., Jacquemin, E., Hadchouel, M., Sokal, E. M., Hooiveld, G. J., Koning, J. H., De Jager-Krikken, A., Kuipers, F., Stellaard, F., Bijleveld, C. M., Gouw, A., Van Goor, H., Thompson, R. J. & Müller, M. 1999. Hepatocanalicular Bile Salt Export Pump Deficiency in Patients with Progressive Familial Intrahepatic Cholestasis. *Gastroenterology*, 117, 1370–1379.

36 Strautnieks, S. S., Bull, L. N., Knisely, A. S., Kocoshis, S. A., Dahl, N., Arnell, H., Sokal, E., Dahan, K., Childs, S., Ling, V., Tanner, M. S., Kagalwalla, A. F., Németh, A., Pawlowska, J., Baker, A., Mieli-Vergani, G., Freimer, N. B., Gardiner, R. M. & Thompson, R. J. 1998. A Gene Encoding a Liver-Specific ABC Transporter Is Mutated in Progressive Familial Intrahepatic Cholestasis. *Nat Genet*, 20, 233–238.

37 Kubitz, R., Keitel, V., Scheuring, S., Köhrer, K. & Häussinger, D. 2006. Benign Recurrent Intrahepatic Cholestasis Associated with Mutations of the Bile Salt Export Pump. *J Clin Gastroenterol*, 40, 171–175.

38 Van Mil, S. W., Van Der Woerd, W. L., Van Der Brugge, G., Sturm, E., Jansen, P. L., Bull, L. N., Van Den Berg, I. E., Berger, R., Houwen, R. H. & Klomp, L. W. 2004. Benign Recurrent Intrahepatic Cholestasis Type 2 Is Caused by Mutations in ABCB11. *Gastroenterology*, 127, 379–384.

39 Eloranta, M. L., Häkli, T., Hiltunen, M., Helisalmi, S., Punnonen, K. & Heinonen, S. 2003. Association of Single Nucleotide Polymorphisms of the Bile Salt Export Pump Gene with Intrahepatic Cholestasis of Pregnancy. *Scand J Gastroenterol*, 38, 648–652.

40 Keitel, V., Vogt, C., Häussinger, D. & Kubitz, R. 2006. Combined Mutations of Canalicular Transporter Proteins Cause Severe Intrahepatic Cholestasis of Pregnancy. *Gastroenterology*, 131, 624–629.

41 Ho, R. H., Leake, B. F., Kilkenny, D. M., Meyer Zu Schwabedissen, H. E., Glaeser, H., Kroetz, D. L. & Kim, R. B. 2010. Polymorphic Variants in the Human Bile Salt Export Pump (BSEP; ABCB11): Functional Characterization and Interindividual Variability. *Pharmacogenet Genomics*, 20, 45–57.

42 Lam, P., Pearson, C. L., Soroka, C. J., Xu, S., Mennone, A. & Boyer, J. L. 2007. Levels of Plasma Membrane Expression in Progressive and Benign Mutations of the Bile Salt Export Pump (BSEP/ABCB11) Correlate with Severity of Cholestatic Diseases. *Am J Physiol Cell Physiol*, 293, C1709–C1716.

43 Lam, P., Soroka, C. J. & Boyer, J. L. 2010. The Bile Salt Export Pump: Clinical and Experimental Aspects of Genetic and Acquired Cholestatic Liver Disease. *Semin Liver Dis*, 30, 125–133.

44 Crawford, R. R., Potukuchi, P. K., Schuetz, E. G. & Schuetz, J. D. 2018. Beyond Competitive Inhibition: Regulation of ABC Transporters by Kinases and Protein-Protein Interactions as Potential Mechanisms of Drug–Drug Interactions. *Drug Metab Dispos*, 46, 567–580.

45 Garzel, B., Yang, H., Zhang, L., Huang, S. M., Polli, J. E. & Wang, H. 2014. The Role of Bile Salt Export Pump Gene Repression in Drug-Induced Cholestatic Liver Toxicity. *Drug Metab Dispos*, 42, 318–322.

46 Sissung, T. M., Huang, P. A., Hauke, R. J., Mccrea, E. M., Peer, C. J., Barbier, R. H., Strope, J. D., Ley, A. M., Zhang, M., Hong, J. A., Venzon, D., Jackson, J. P., Brouwer, K. R., Grohar, P., Glod, J., Widemann, B. C., Heller, T., Schrump, D. S. & Figg, W. D. 2019. Severe Hepatotoxicity of Mithramycin Therapy Caused by Altered Expression of Hepatocellular Bile Transporters. *Mol Pharmacol*, 96, 158–167.

47 Kenna, J. G., Taskar, K. S., Battista, C., Bourdet, D. L., Brouwer, K. L. R., Brouwer, K. R., Dai, D., Funk, C., Hafey, M. J., Lai, Y., Maher, J., Pak, Y. A., Pedersen, J. M., Polli, J. W., Rodrigues, A. D., Watkins, P. B., Yang, K., Yucha, R. W. & On behalf of the International Transporter Consortium. 2018. Can Bile Salt Export Pump Inhibition Testing in Drug Discovery and Development Reduce Liver Injury Risk? An International Transporter Consortium Perspective. *Clin Pharmacol Ther*, 104, 916–932.

48 Feng, B., Xu, J. J., Bi, Y. A., Mireles, R., Davidson, R., Duignan, D. B., Campbell, S., Kostrubsky, V. E., Dunn, M. C., Smith, A. R. & Wang, H. F. 2009. Role of Hepatic Transporters in the Disposition and Hepatotoxicity of a HER2 Tyrosine Kinase Inhibitor Cp-724,714. *Toxicol Sci*, 108, 492–500.

49 Funk, C., Pantze, M., Jehle, L., Ponelle, C., Scheuermann, G., Lazendic, M. & Gasser, R. 2001a. Troglitazone-Induced Intrahepatic Cholestasis by an Interference with the Hepatobiliary Export of Bile Acids in Male and Female Rats. Correlation with the Gender Difference in Troglitazone Sulfate Formation and the Inhibition of the Canalicular Bile Salt Export Pump (BSEP) by Troglitazone and Troglitazone Sulfate. *Toxicology*, 167, 83–98.

50 Funk, C., Ponelle, C., Scheuermann, G. & Pantze, M. 2001b. Cholestatic Potential of Troglitazone as a Possible Factor Contributing to Troglitazone-Induced Hepatotoxicity: In Vivo and in Vitro Interaction at the Canalicular Bile Salt Export Pump (BSEP) in the Rat. *Mol Pharmacol*, 59, 627–635.

51 Morgan, R. E., Van Staden, C. J., Chen, Y., Kalyanaraman, N., Kalanzi, J., Dunn, R. T., Afshari, C. A. & Hamadeh, H. K. 2013. A Multifactorial Approach to Hepatobiliary Transporter Assessment Enables Improved Therapeutic Compound Development. *Toxicol Sci*, 136, 216–241.

52 Otieno, M. A., Snoeys, J., Lam, W., Ghosh, A., Player, M. R., Pocai, A., Salter, R., Simic, D., Skaggs, H., Singh, B. & Lim, H. K. 2018. Fasiglifam (TAK-875): Mechanistic Investigation and Retrospective Identification of Hazards for Drug Induced Liver Injury. *Toxicol Sci*, 163, 374–384.

53 Morgan, R. E., Trauner, M., Van Staden, C. J., Lee, P. H., Ramachandran, B., Eschenberg, M., Afshari, C. A., Qualls, C. W., Lightfoot-Dunn, R. & Hamadeh, H. K. 2010. Interference with Bile Salt Export Pump Function Is a Susceptibility Factor for Human Liver Injury in Drug Development. *Toxicol Sci*, 118, 485–500.

54 Cheng, Y., Woolf, T. F., Gan, J. & He, K. 2016. in vitro Model Systems to Investigate Bile Salt Export Pump (BSEP) Activity and Drug Interactions: A Review. *Chem Biol Interact*, 255, 23–30.

55 Brouwer, K. L., Keppler, D., Hoffmaster, K. A., Bow, D. A., Cheng, Y., Lai, Y., Palm, J. E., Stieger, B., Evers, R. & On behalf of the International Transporter Consortium. 2013. in Vitro Methods to Support Transporter Evaluation in Drug Discovery and Development. *Clin Pharmacol Ther*, 94, 95–112.

56 Glavinas, H., Méhn, D., Jani, M., Oosterhuis, B., Herédi-Szabó, K. & Krajcsi, P. 2008. Utilization of Membrane Vesicle Preparations to Study Drug-ABC Transporter Interactions. *Expert Opin Drug Metab Toxicol*, 4, 721–732.

57 Ali, I., Khalid, S., Stieger, B. & Brouwer, K. L. R. 2019. Effect of a Common Genetic Variant (p.V444A) in the Bile Salt Export Pump on the Inhibition of Bile Acid Transport by Cholestatic Medications. *Mol Pharm*, 16, 1406–1411.

58 Warner, D. J., Chen, H., Cantin, L.-D., Kenna, J. G., Stahl, S., Walker, C. L. & Noeske, T. 2012. Mitigating the Inhibition of Human Bile Salt Export Pump by Drugs: Opportunities Provided by Physicochemical Property Modulation, in Silico Modeling, and Structural Modification. *Drug Metab Dispos*, 40, 2332–2341.

59 Steck, T. L., Weinstein, R. S., Straus, J. H. & Wallach, D. F. 1970. Inside-Out Red Cell Membrane Vesicles: Preparation and Purification. *Science*, 168, 255–257.

60 Akita, H., Suzuki, H., Ito, K., Kinoshita, S., Sato, N., Takikawa, H. & Sugiyama, Y. 2001. Characterization of Bile Acid Transport Mediated by Multidrug Resistance Associated Protein 2 and Bile Salt Export Pump. *Biochim Biophys Acta*, 1511, 7–16.

61 Kis, E., Ioja, E., Nagy, T., Szente, L., Herédi-Szabó, K. & Krajcsi, P. 2009. Effect of Membrane Cholesterol on BSEP/BSEP Activity: Species Specificity Studies for Substrates and Inhibitors. *Drug Metab Dispos*, 37, 1878–1886.

62 Loe, D. W., Almquist, K. C., Deeley, R. G. & Cole, S. P. 1996. Multidrug Resistance Protein (MRP)-Mediated Transport of Leukotriene C4 and Chemotherapeutic Agents in Membrane Vesicles. Demonstration of Glutathione-Dependent Vincristine Transport. *J Biol Chem*, 271, 9675–9682.

63 Volk, E. L. & Schneider, E. 2003. Wild-Type Breast Cancer Resistance Protein (BCRP/ABCG2) Is a Methotrexate Polyglutamate Transporter. *Cancer Res*, 63, 5538–5543.

64 Woodhead, J. L., Yang, K., Siler, S. Q., Watkins, P. B., Brouwer, K. L., Barton, H. A. & Howell, B. A. 2014b. Exploring Bsep Inhibition-Mediated Toxicity with a Mechanistic Model of Drug-Induced Liver Injury. *Front Pharmacol*, 5, 240.

65 Watkins, P. B. 2019. The Dili-Sim Initiative: Insights into Hepatotoxicity Mechanisms and Biomarker Interpretation. *Clin Transl Sci*, 12, 122–129.

66 Köck, K., Ferslew, B. C., Netterberg, I., Yang, K., Urban, T. J., Swaan, P. W., Stewart, P. W. & Brouwer, K. L. 2014. Risk Factors for Development of Cholestatic Drug-Induced Liver Injury: Inhibition of Hepatic Basolateral Bile Acid Transporters Multidrug Resistance-Associated Proteins 3 and 4. *Drug Metab Dispos*, 42, 665–674.

67 Hirano, H., Kurata, A., Onishi, Y., Sakurai, A., Saito, H., Nakagawa, H., Nagakura, M., Tarui, S., Kanamori, Y., Kitajima, M. & Ishikawa, T. 2006. High-Speed Screening and QSAR Analysis of Human ATP-Binding Cassette Transporter ABCB11 (Bile Salt Export Pump) to Predict Drug-Induced Intrahepatic Cholestasis. *Mol Pharm*, 3, 252–265.

68 Ritschel, T., Hermans, S. M. A., Schreurs, M., Van Den Heuvel, J. J. M. W., Koenderink, J. B., Greupink, R. & Russel, F. G. M. 2014. In Silico Identification and in Vitro Validation of Potential Cholestatic Compounds through 3D Ligand-Based Pharmacophore Modeling of BSEP Inhibitors. *Chem Res Toxicol*, 27, 873–881.

69 Swift, B., Pfeifer, N. D. & Brouwer, K. L. 2010. Sandwich-Cultured Hepatocytes: An in Vitro Model to Evaluate Hepatobiliary Transporter-Based Drug Interactions and Hepatotoxicity. *Drug Metab Rev*, 42, 446–471.

70 Treyer, A. & Müsch, A. 2013. Hepatocyte Polarity. *Compr Physiol*, 3, 243–287.

71 Liu, X., Brouwer, K. L., Gan, L. S., Brouwer, K. R., Stieger, B., Meier, P. J., Audus, K. L. & Lecluyse, E. L. 1998. Partial Maintenance of Taurocholate Uptake by Adult Rat Hepatocytes Cultured in a Collagen Sandwich Configuration. *Pharm Res*, 15, 1533–1539.

72 Liu, X., Lecluyse, E. L., Brouwer, K. R., Gan, L. S., Lemasters, J. J., Stieger, B., Meier, P. J. & Brouwer, K. L. 1999b. Biliary Excretion in Primary Rat Hepatocytes Cultured in a Collagen-Sandwich Configuration. *Am J Physiol*, 277, G12–G21.

73 Lecluyse, E. L., Audus, K. L. & Hochman, J. H. 1994. Formation of Extensive Canalicular Networks by Rat Hepatocytes Cultured in Collagen-Sandwich Configuration. *Am J Physiol*, 266, C1764–C1774.

74 Hoffmaster, K. A., Turncliff, R. Z., Lecluyse, E. L., Kim, R. B., Meier, P. J. & Brouwer, K. L. 2004. P-Glycoprotein Expression, Localization, and Function in Sandwich-Cultured Primary Rat and Human Hepatocytes: Relevance to the Hepatobiliary Disposition of a Model Opioid Peptide. *Pharm Res*, 21, 1294–1302.

75 Liu, X., Lecluyse, E. L., Brouwer, K. R., Lightfoot, R. M., Lee, J. I. & Brouwer, K. L. 1999c. Use of Ca^{2+} Modulation to Evaluate Biliary Excretion in Sandwich-Cultured Rat Hepatocytes. *J Pharmacol Exp Ther*, 289, 1592–1599.

76 Abe, K., Bridges, A. S., Yue, W. & Brouwer, K. L. 2008. In Vitro Biliary Clearance of Angiotensin II Receptor Blockers and 3-Hydroxy-3-Methylglutaryl-Coenzyme a Reductase Inhibitors in Sandwich-Cultured Rat Hepatocytes: Comparison with in Vivo Biliary Clearance. *J Pharmacol Exp Ther*, 326, 983–990.

77 Ghibellini, G., Vasist, L. S., Leslie, E. M., Heizer, W. D., Kowalsky, R. J., Calvo, B. F. & Brouwer, K. L. 2007. In Vitro-in Vivo Correlation of Hepatobiliary Drug Clearance in Humans. *Clin Pharmacol Ther*, 81, 406–413.

78 Liu, X., Chism, J. P., Lecluyse, E. L., Brouwer, K. R. & Brouwer, K. L. 1999. Correlation of Biliary Excretion in Sandwich-Cultured Rat Hepatocytes and in Vivo in Rats. *Drug Metab Dispos*, 27, 637–644.

79 Pfeifer, N. D., Harris, K. B., Yan, G. Z. & Brouwer, K. L. 2013b. Determination of Intracellular Unbound Concentrations and Subcellular Localization of Drugs in Rat Sandwich-Cultured Hepatocytes Compared with Liver Tissue. *Drug Metab Dispos*, 41, 1949–1956.

80 Lecluyse, E. L., Bullock, P. L., Parkinson, A. & Hochman, J. H. 1996. Cultured Rat Hepatocytes. *Pharm Biotechnol*, 8, 121–159.

81 Chu, X., Korzekwa, K., Elsby, R., Fenner, K., Galetin, A., Lai, Y., Matsson, P., Moss, A., Nagar, S., Rosania, G. R., Bai, J. P., Polli, J. W., Sugiyama, Y., Brouwer, K. L. & On behalf of the International Transporter Consortium. 2013. Intracellular Drug Concentrations and Transporters: Measurement, Modeling, and Implications for the Liver. *Clin Pharmacol Ther*, 94, 126–141.

82 Attili, A. F., Angelico, M., Cantafora, A., Alvaro, D. & Capocaccia, L. 1986. Bile Acid-Induced Liver Toxicity: Relation to the Hydrophobic-Hydrophilic Balance of Bile Acids. *Med Hypotheses*, 19, 57–69.

83 Alvaro, D., Cantafora, A., Attili, A. F., Ginanni Corradini, S., De Luca, C., Minervini, G., Di Biase, A. & Angelico, M. 1986. Relationships between Bile Salts Hydrophilicity and Phospholipid Composition in Bile of Various Animal Species. *Comp Biochem Physiol B*, 83, 551–554.

84 Chatterjee, S., Bijsmans, I. T., Van Mil, S. W., Augustijns, P. & Annaert, P. 2014a. Toxicity and Intracellular Accumulation of Bile Acids in Sandwich-Cultured Rat Hepatocytes: Role of Glycine Conjugates. *Toxicol in Vitro*, 28, 218–230.

85 Marion, T. L., Perry, C. H., St Claire, R. L. & Brouwer, K. L. 2012. Endogenous Bile Acid Disposition in Rat and Human Sandwich-Cultured Hepatocytes. *Toxicol Appl Pharmacol*, 261, 1–9.

86 Marion, T. L., Perry, C. H., St Claire, R. L., Yue, W. & Brouwer, K. L. 2011. Differential Disposition of Chenodeoxycholic Acid Versus Taurocholic Acid in Response to Acute Troglitazone Exposure in Rat Hepatocytes. *Toxicol Sci*, 120, 371–380.

87 Beaudoin, J. J., Bezençon, J., Sjöstedt, N., Fallon, J. K. & Brouwer, K. L. R. 2020. Role of Organic Solute Transporter Alpha/Beta in Hepatotoxic Bile Acid Transport and Drug Interactions. *Toxicol Sci* 176 (1): 34–45

88 De Bruyn, T., Sempels, W., Snoeys, J., Holmstock, N., Chatterjee, S., Stieger, B., Augustijns, P., Hofkens, J., Mizuno, H. & Annaert, P. 2014. Confocal Imaging with a Fluorescent Bile Acid Analogue Closely Mimicking Hepatic Taurocholate Disposition. *J Pharm Sci*, 103, 1872–1881.

89 Chatterjee, S., Richert, L., Augustijns, P. & Annaert, P. 2014b. Hepatocyte-Based in Vitro Model for Assessment of Drug-Induced Cholestasis. *Toxicol Appl Pharmacol*, 274, 124–136.

90 Yang, K., Pfeifer, N. D., Köck, K. & Brouwer, K. L. 2015. Species Differences in Hepatobiliary Disposition of Taurocholic Acid in Human and Rat Sandwich-Cultured Hepatocytes: Implications for Drug-Induced Liver Injury. *J Pharmacol Exp Ther*, 353, 415–423.

91 Lu, Y., Slizgi, J. R., Brouwer, K. R., Claire, R. L., Freeman, K. M., Pan, M., Brock, W. J. & Brouwer, K. L. 2016. Hepatocellular Disposition and Transporter Interactions with Tolvaptan and Metabolites in Sandwich-Cultured Human Hepatocytes. *Drug Metab Dispos*, 44(6), 867–870.

92 Slizgi, J. R., Lu, Y., Brouwer, K. R., St Claire, R. L., Freeman, K. M., Pan, M., Brock, W. J. & Brouwer, K. L. 2016. Inhibition of Human Hepatic Bile Acid Transporters by Tolvaptan and Metabolites: Contributing Factors to Drug-Induced Liver Injury? *Toxicol Sci*, 149, 237–250.

93 Guo, C., Yang, K., Brouwer, K. R., St Claire, R. L. & Brouwer, K. L. R. 2016. Prediction of Altered Bile Acid Disposition Due to Inhibition of Multiple Transporters: An Integrated Approach Using Sandwich-Cultured Hepatocytes, Mechanistic Modeling, and Simulation. *J Pharmacol Exp Ther*, 358, 324–333.

94 Lepist, E. I., Gillies, H., Smith, W., Hao, J., Hubert, C., St Claire, R. L., Brouwer, K. R. & Ray, A. S. 2014. Evaluation of the Endothelin Receptor Antagonists Ambrisentan, Bosentan, Macitentan, and Sitaxsentan as Hepatobiliary Transporter Inhibitors and Substrates in Sandwich-Cultured Human Hepatocytes. *PLoS One*, 9, e87548.

95 Pedersen, J. M., Matsson, P., Bergström, C. A. S., Hoogstraate, J., Norén, A., Lecluyse, E. L. & Artursson, P. 2013. Early Identification of Clinically Relevant Drug Interactions with the Human Bile Salt Export Pump (BSEP/ABCB11). *Toxicol Sci*, 136, 328–343.

96 Matsunaga, N., Fukuchi, Y., Imawaka, H. & Tamai, I. 2018. Sandwich-Cultured Hepatocytes for Mechanistic Understanding of Hepatic Disposition of Parent Drugs and Metabolites by Transporter-Enzyme Interplay. *Drug Metab Dispos*, 46, 680–691.

97 Matsunaga, N., Wada, S., Nakanishi, T., Ikenaga, M., Ogawa, M. & Tamai, I. 2014. Mathematical Modeling of the in Vitro Hepatic Disposition of Mycophenolic Acid and Its Glucuronide in Sandwich-Cultured Human Hepatocytes. *Mol Pharm*, 11, 568–579.

98 Pfeifer, N. D., Bridges, A. S., Ferslew, B. C., Hardwick, R. N. & Brouwer, K. L. 2013. Hepatic Basolateral Efflux Contributes Significantly to Rosuvastatin Disposition II: Characterization of Hepatic Elimination by Basolateral, Biliary, and Metabolic Clearance Pathways in Rat Isolated Perfused Liver. *J Pharmacol Exp Ther*, 347, 737–745.

99 Chandra, P. & Brouwer, K. L. 2004. The Complexities of Hepatic Drug Transport: Current Knowledge and Emerging Concepts. *Pharm Res*, 21, 719–735.

100 Hayashi, H., Takada, T., Suzuki, H., Onuki, R., Hofmann, A. F. & Sugiyama, Y. 2005. Transport by Vesicles of Glycine- and Taurine-Conjugated Bile Salts and Taurolithocholate 3-Sulfate: A Comparison of Human BSEP with Rat BSEP. *Biochim Biophys Acta*, 1738, 54–62.

101 Jones, H. M., Barton, H. A., Lai, Y., Bi, Y.-A., Kimoto, E., Kempshall, S., Tate, S. C., El-Kattan, A., Houston, J. B., Galetin, A. & Fenner, K. S. 2012. Mechanistic Pharmacokinetic Modeling for the Prediction of Transporter-Mediated Disposition in Humans from Sandwich Culture Human Hepatocyte Data. *Drug Metab Dispos*, 40, 1007–1017.

102 Yan, G. Z., Brouwer, K. L., Pollack, G. M., Wang, M. Z., Tidwell, R. R., Hall, J. E. & Paine, M. F. 2011. Mechanisms Underlying Differences in Systemic Exposure of Structurally Similar Active Metabolites: Comparison of Two Preclinical Hepatic Models. *J Pharmacol Exp Ther*, 337, 503–512.

103 Yan, G. Z., Generaux, C. N., Yoon, M., Goldsmith, R. B., Tidwell, R. R., Hall, J. E., Olson, C. A., Clewell, H. J., Brouwer, K. L. & Paine, M. F. 2012. A Semiphysiologically Based Pharmacokinetic Modeling Approach to Predict the Dose-Exposure Relationship of an Antiparasitic Prodrug/Active Metabolite Pair. *Drug Metab Dispos*, 40, 6–17.

104 Bi, Y. A., Kimoto, E., Sevidal, S., Jones, H. M., Barton, H. A., Kempshall, S., Whalen, K. M., Zhang, H., Ji, C., Fenner, K. S., El-Kattan, A. F. & Lai, Y. 2012. in vitro Evaluation of Hepatic Transporter-Mediated Clinical Drug-Drug Interactions: Hepatocyte Model Optimization and Retrospective Investigation. *Drug Metab Dispos*, 40, 1085–1092.

105 Jamei, M., Bajot, F., Neuhoff, S., Barter, Z., Yang, J., Rostami-Hodjegan, A. & Rowland-Yeo, K. 2014. A Mechanistic Framework for in Vitro-in Vivo Extrapolation of Liver Membrane Transporters: Prediction of Drug-Drug Interaction between Rosuvastatin and Cyclosporine. *Clin Pharmacokinet*, 53, 73–87.

106 Jones, H. M., Mayawala, K. & Poulin, P. 2013. Dose Selection Based on Physiologically Based Pharmacokinetic (PBPK) Approaches. *AAPS J*, 15, 377–387.

107 Li, R., Ghosh, A., Maurer, T. S., Kimoto, E. & Barton, H. A. 2014.Physiologically Based Pharmacokinetic Prediction of Telmisartan in Human. *Drug Metab Dispos*, 42, 1646–1655.

108 Varma, M. V., Scialis, R. J., Lin, J., Bi, Y. A., Rotter, C. J., Goosen, T. C. & Yang, X. 2014. Mechanism-Based Pharmacokinetic Modeling to Evaluate Transporter-Enzyme Interplay in Drug Interactions and Pharmacogenetics of Glyburide. *AAPS J*, 16, 736–748.

109 Adachi, Y., Kobayashi, H., Kurumi, Y., Shouji, M., Kitano, M. & Yamamoto, T. 1991. ATP-Dependent Taurocholate Transport by Rat Liver Canalicular Membrane Vesicles. *Hepatology*, 14, 655–659.

110 Stieger, B., O'neill, B. & Meier, P. J. 1992. ATP-Dependent Bile-Salt Transport in Canalicular Rat Liver Plasma-Membrane Vesicles. *Biochem J*, 284 (Pt 1), 67–74.

111 Mita, S., Suzuki, H., Akita, H., Hayashi, H., Onuki, R., Hofmann, A. F. & Sugiyama, Y. 2006. Vectorial Transport of Unconjugated and Conjugated Bile Salts by Monolayers of LLC-PK1 Cells Doubly Transfected with Human NTCP and BSEP or with Rat NTCP and BSEP. *Am J Physiol Gastrointest Liver Physiol*, 290, G550–G556.

112 Aida, K., Hayashi, H., Inamura, K., Mizuno, T. & Sugiyama, Y. 2014. Differential Roles of Ubiquitination in the Degradation Mechanism of Cell Surface-Resident Bile Salt Export Pump and Multidrug Resistance-Associated Protein 2. *Mol Pharmacol*, 85, 482–491.

113 Hayashi, H., Inamura, K., Aida, K., Naoi, S., Horikawa, R., Nagasaka, H., Takatani, T., Fukushima, T., Hattori, A., Yabuki, T., Horii, I. & Sugiyama, Y. 2012. AP2 Adaptor Complex Mediates Bile Salt Export Pump Internalization and Modulates Its Hepatocanalicular Expression and Transport Function. *Hepatology*, 55, 1889–1900.

114 Hayashi, H. & Sugiyama, Y. 2009. Short-Chain Ubiquitination Is Associated with the Degradation Rate of a Cell-Surface-Resident Bile Salt Export Pump (BSEP/ABCB11). *Mol Pharmacol*, 75, 143–150.

115 Kang, H. E., Malinen, M. M., Saran, C., Honkakoski, P. & Brouwer, K. L. R. 2019. Optimization of Canalicular ABC Transporter Function in Huh-7 Cells by Modification of Culture Conditions. *Drug Metab Dispos*, 47, 1222–1230.

116 Le Vee, M., Jigorel, E., Glaise, D., Gripon, P., Guguen-Guillouzo, C. & Fardel, O. 2006. Functional Expression of Sinusoidal and Canalicular Hepatic Drug Transporters in the Differentiated Human Hepatoma Heparg Cell Line. *Eur J Pharm Sci*, 28, 109–117.

117 Lundquist, P., Englund, G., Skogastierna, C., Lööf, J., Johansson, J., Hoogstraate, J., Afzelius, L. & Andersson, T. B. 2014. Functional ATP-Binding Cassette Drug Efflux Transporters in Isolated Human and Rat Hepatocytes Significantly Affect Assessment of Drug Disposition. *Drug Metab Dispos*, 42, 448–458.

118 Zhang, J., He, K., Cai, L., Chen, Y.-C., Yang, Y., Shi, Q., Woolf, T. F., Ge, W., Guo, L., Borlak, J. & Tong, W. 2016. Inhibition of Bile Salt Transport by Drugs Associated with Liver Injury in Primary Hepatocytes from Human, Monkey, Dog, Rat, and Mouse. *Chem Biol Interact*, 255, 45–54.

119 Bow, D. A., Perry, J. L., Miller, D. S., Pritchard, J. B. & Brouwer, K. L. 2008. Localization of P-Gp (Abcb1) and MRP2 (ABCC2) in Freshly Isolated Rat Hepatocytes. *Drug Metab Dispos*, 36, 198–202.

120 Hendriks, D. F., Fredriksson Puigvert, L., Messner, S., Mortiz, W. & Ingelman-Sundberg, M. 2016. Hepatic 3D Spheroid Models for the Detection and Study of Compounds with Cholestatic Liability. *Sci Rep*, 6, 35434.

121 Bell, C. C., Lauschke, V. M., Vorrink, S. U., Palmgren, H., Duffin, R., Andersson, T. B. & Ingelman-Sundberg, M. 2017. Transcriptional, Functional, and Mechanistic Comparisons of Stem Cell-Derived Hepatocytes, Heparg Cells, and Three-Dimensional Human Hepatocyte Spheroids as Predictive in Vitro Systems for Drug-Induced Liver Injury. *Drug Metab Dispos*, 45, 419–429.

122 Mosedale, M. 2018. Mouse Population-Based Approaches to Investigate Adverse Drug Reactions. *Drug Metab Dispos*, 46, 1787–1795.

123 Collaborative Cross Consortium 2012. The Genome Architecture of the Collaborative Cross Mouse Genetic Reference Population. *Genetics*, 190, 389–401.

124 Bonn, B., Svanberg, P., Janefeldt, A., Hultman, I. & Grime, K. 2016. Determination of Human Hepatocyte Intrinsic Clearance for Slowly Metabolized Compounds: Comparison of a Primary Hepatocyte/Stromal Cell Co-Culture with Plated Primary Hepatocytes and Heparg. *Drug Metab Dispos*, 44, 527–533.

125 Chan, T. S., Yu, H., Moore, A., Khetani, S. R. & Tweedie, D. 2019. Meeting the Challenge of Predicting Hepatic Clearance of Compounds Slowly Metabolized by Cytochrome P450 Using a Novel Hepatocyte Model, Hepatopac. *Drug Metab Dispos*, 47, 58–66.

126 Deng, J., Wei, W., Chen, Z., Lin, B., Zhao, W., Luo, Y. & Zhang, X. 2019. Engineered Liver-on-a-Chip Platform to Mimic Liver Functions and Its Biomedical Applications: A Review. *Micromachines (Basel)*, 10 (10): 676.

127 Godoy, P., Hewitt, N. J., Albrecht, U., Andersen, M. E., Ansari, N., Bhattacharya, S., Bode, J. G., Bolleyn, J., Borner, C., Bottger, J., Braeuning, A., Budinsky, R. A., Burkhardt, B., Cameron, N. R., Camussi, G., Cho, C. S., Choi, Y. J., Craig Rowlands, J., Dahmen, U., Damm, G., Dirsch, O., Donato, M. T., Dong, J., Dooley, S., Drasdo, D., Eakins, R., Ferreira, K. S., Fonsato, V., Fraczek, J., Gebhardt, R., Gibson, A., Glanemann, M., Goldring, C. E., Gomez-Lechon, M. J., Groothuis, G. M., Gustavsson, L., Guyot, C., Hallifax, D., Hammad, S., Hayward, A., Haussinger, D., Hellerbrand, C., Hewitt, P., Hoehme, S., Holzhutter, H. G., Houston, J. B., Hrach, J., Ito, K., Jaeschke, H., Keitel, V., Kelm, J. M., Kevin Park, B., Kordes, C., Kullak-Ublick, G. A., Lecluyse, E. L., Lu, P., Luebke-Wheeler, J., Lutz, A., Maltman, D. J., Matz-Soja, M., Mcmullen, P., Merfort, I., Messner, S., Meyer, C., Mwinyi, J., Naisbitt, D. J., Nussler, A. K., Olinga, P., Pampaloni, F., Pi, J., Pluta, L., Przyborski, S. A., Ramachandran, A., Rogiers, V., Rowe, C., Schelcher, C., Schmich, K., Schwarz, M., Singh, B., Stelzer, E. H., Stieger, B., Stober, R., Sugiyama, Y., Tetta, C., Thasler, W. E., Vanhaecke, T., Vinken, M., Weiss, T. S., Widera, A., Woods, C. G., Xu, J. J., Yarborough, K. M. & Hengstler, J. G. 2013. Recent Advances in

2D and 3D in Vitro Systems Using Primary Hepatocytes, Alternative Hepatocyte Sources and Non-Parenchymal Liver Cells and Their Use in Investigating Mechanisms of Hepatotoxicity, Cell Signaling and Adme. *Arch Toxicol*, 87, 1315–1530.

128 Prior, N., Inacio, P. & Huch, M. 2019. Liver Organoids: From Basic Research to Therapeutic Applications. *Gut*, 68, 2228–2237.

129 Welch, M. A., Köck, K., Urban, T. J., Brouwer, K. L. & Swaan, P. W. 2015. Toward Predicting Drug-Induced Liver Injury: Parallel Computational Approaches to Identify Multidrug Resistance Protein 4 and Bile Salt Export Pump Inhibitors. *Drug Metab Dispos*, 43, 725–734.

130 Xi, L., Yao, J., Wei, Y., Wu, X., Yao, X., Liu, H. & Li, S. 2017. The in Silico Identification of Human Bile Salt Export Pump (ABCB11) Inhibitors Associated with Cholestatic Drug-Induced Liver Injury. *Mol Biosyst*, 13, 417–424.

131 Jain, S., Grandits, M., Richter, L. & Ecker, G. F. 2017. Structure Based Classification for Bile Salt Export Pump (BSEP) Inhibitors Using Comparative Structural Modeling of Human BSEP. *J Comput Aided Mol Des*, 31, 507–521.

132 Wang, L., Hou, W. T., Chen, L., Jiang, Y. L., Xu, D., Sun, L., Zhou, C. Z. & Chen, Y. 2020. Cryo-EM Structure of Human Bile Salts Exporter ABCB11. *Cell Res* 30, 623–625.

133 Jackson, J. P. & Brouwer, K. R. 2019. The C-Dili™ Assay: An Integrated in vitro Approach to Predict Cholestatic Hepatotoxicity. *Methods Mol Biol*, 1981, 75–85.

134 Byun, S., Kim, D. H., Ryerson, D., Kim, Y. C., Sun, H., Kong, B., Yau, P., Guo, G., Xu, H. E., Kemper, B. & Kemper, J. K. 2018. Postprandial FGF19-Induced Phosphorylation by SRC Is Critical for FXR Function in Bile Acid Homeostasis. *Nat Commun*, 9, 2590.

135 Guicciardi, M. E. & Gores, G. J. 2002. Bile Acid-Mediated Hepatocyte Apoptosis and Cholestatic Liver Disease. *Dig Liver Dis*, 34, 387–392.

136 Jackson, J. P., Freeman, K. M., St Claire, R. L., Black, C. B. & Brouwer, K. R. 2018. Cholestatic Drug Induced Liver Injury: A Function of Bile Salt Export Pump Inhibition and Farnesoid X Receptor Antagonism. *Appl in Vitro Toxicol*, 4, 265–279.

137 Kaimal, R., Song, X., Yan, B., King, R. & Deng, R. 2009. Differential Modulation of Farnesoid X Receptor Signaling Pathway by the Thiazolidinediones. *J Pharmacol Exp Ther*, 330, 125–134.

138 Malinen, M. M., Ali, I., Bezençon, J., Beaudoin, J. J. & Brouwer, K. L. R. 2018. Organic Solute Transporter OSTA/B Is Overexpressed in Nonalcoholic Steatohepatitis and Modulated by Drugs Associated with Liver Injury. *Am J Physiol Gastrointest Liver Physiol*, 314, G597–G609.

139 Bryda, E. C. 2013. The Mighty Mouse: The Impact of Rodents on Advances in Biomedical Research. *Mo Med*, 110, 207–211.

140 Pan, Y., Cao, M., You, D., Qin, G. & Liu, Z. 2019. Research Progress on the Animal Models of Drug-Induced Liver Injury: Current Status and Further Perspectives. *Biomed Res Int*, 2019, 1283824.

141 Alnouti, Y., Csanaky, I. L. & Klaassen, C. D. 2008. Quantitative-Profiling of Bile Acids and Their Conjugates in Mouse Liver, Bile, Plasma, and Urine Using LC-MS/MS. *J Chromatogr B Analyt Technol Biomed Life Sci*, 873, 209–217.

142 Cepa, S., Potter, D., Wong, L., Schutt, L., Tarrant, J., Pang, J., Zhang, X., Andaya, R., Salphati, L., Ran, Y., An, L., Morgan, R. & Maher, J. 2018. Individual Serum Bile Acid Profiling in Rats Aids in Human Risk Assessment of Drug-Induced Liver Injury Due to BSEP Inhibition. *Toxicol Appl Pharmacol*, 338, 204–213.

143 Kenna, J. G. 2014. Current Concepts in Drug-Induced Bile Salt Export Pump (BSEP) Interference. *Curr Protoc Toxicol*, 61, 23.7.1–23.7.15.

144 Fattinger, K., Funk, C., Pantze, M., Weber, C., Reichen, J., Stieger, B. & Meier, P. J. 2001. The Endothelin Antagonist Bosentan Inhibits the Canalicular Bile Salt Export Pump: A Potential Mechanism for Hepatic Adverse Reactions. *Clin Pharmacol Ther*, 69, 223–231.

145 Qu, X., Zhang, Y., Zhang, S., Zhai, J., Gao, H., Tao, L. & Song, Y. 2018. Dysregulation of BSEP and MRP2 May Play an Important Role in Isoniazid-Induced Liver Injury Via the SIRT1/FXR Pathway in Rats and HEPG2 Cells. *Biol Pharm Bull*, 41, 1211–1218.

146 Sanoh, S., Tamura, Y., Fujino, C., Sugahara, G., Yoshizane, Y., Yanagi, A., Kisoh, K., Ishida, Y., Tateno, C., Ohta, S. & Kotake, Y. 2019. Changes in Bile Acid Concentrations after Administration of Ketoconazole or Rifampicin to Chimeric Mice with Humanized Liver. *Biol Pharm Bull*, 42, 1366–1375.

147 Stieger, B. 2011. The Role of the Sodium-Taurocholate Cotransporting Polypeptide (NTCP) and of the Bile Salt Export Pump (BSEP) in Physiology and Pathophysiology of Bile Formation. *Handb Exp Pharmacol*, 205–259.

148 Vaz, F. M., Paulusma, C. C., Huidekoper, H., De Ru, M., Lim, C., Koster, J., Ho-Mok, K., Bootsma, A. H., Groen, A. K., Schaap, F. G., Oude Elferink, R. P., Waterham, H. R. & Wanders, R. J. 2015. Sodium Taurocholate Cotransporting Polypeptide (SLC10A1) Deficiency: Conjugated Hypercholanemia without a Clear Clinical Phenotype. *Hepatology*, 61, 260–267.

149 Rodrigues, A. D., Lai, Y., Cvijic, M. E., Elkin, L. L., Zvyaga, T. & Soars, M. G. 2014. Drug-Induced Perturbations of the Bile Acid Pool, Cholestasis, and Hepatotoxicity: Mechanistic Considerations Beyond the Direct Inhibition of the Bile Salt Export Pump. *Drug Metab Dispos*, 42, 566–574.

150 Dawson, S., Stahl, S., Paul, N., Barber, J. & Kenna, J. G. 2012. In Vitro Inhibition of the Bile Salt Export Pump Correlates with Risk of Cholestatic Drug-Induced Liver Injury in Humans. *Drug Metab Dispos*, 40, 130–138.

151 Heuman, D. M. 1989. Quantitative Estimation of the Hydrophilic-Hydrophobic Balance of Mixed Bile Salt Solutions. *J Lipid Res*, 30, 719–730.

152 Sanyal, S., Båvner, A., Haroniti, A., Nilsson, L. M., Lundåsen, T., Rehnmark, S., Witt, M. R., Einarsson, C., Talianidis, I., Gustafsson, J. A. & Treuter, E. 2007. Involvement of Corepressor Complex Subunit GPS2 in Transcriptional Pathways Governing Human Bile Acid Biosynthesis. *Proc Natl Acad Sci USA*, 104, 15665–15670.

153 Sayin, S. I., Wahlström, A., Felin, J., Jäntti, S., Marschall, H. U., Bamberg, K., Angelin, B., Hyötyläinen, T., Orešič, M. & Bäckhed, F. 2013. Gut Microbiota Regulates Bile Acid Metabolism by Reducing the Levels of Tauro-Beta-Muricholic Acid, a Naturally Occurring FXR Antagonist. *Cell Metab*, 17, 225–235.

154 Takahashi, S., Fukami, T., Masuo, Y., Brocker, C. N., Xie, C., Krausz, K. W., Wolf, C. R., Henderson, C. J. & Gonzalez, F. J. 2016. Cyp2c70 Is Responsible for the Species Difference in Bile Acid Metabolism between Mice and Humans. *J Lipid Res*, 57, 2130–2137.

155 Schadt, S., Simon, S., Kustermann, S., Boess, F., Mcginnis, C., Brink, A., Lieven, R., Fowler, S., Youdim, K., Ullah, M., Marschmann, M., Zihlmann, C., Siegrist, Y. M., Cascais, A. C., Di Lenarda, E., Durr, E., Schaub, N., Ang, X., Starke, V., Singer, T., Alvarez-Sanchez, R., Roth, A. B., Schuler, F. & Funk, C. 2015. Minimizing Dili Risk in Drug Discovery—A Screening Tool for Drug Candidates. *Toxicol In Vitro*, 30, 429–437.

156 Thakare, R., Alamoudi, J. A., Gautam, N., Rodrigues, A. D. & Alnouti, Y. 2018a. Species Differences in Bile Acids I. Plasma and Urine Bile Acid Composition. *J Appl Toxicol*, 38, 1323–1335.

157 Thakare, R., Alamoudi, J. A., Gautam, N., Rodrigues, A. D. & Alnouti, Y. 2018. Species Differences in Bile Acids II. Bile Acid Metabolism. *J Appl Toxicol*, 38, 1336–1352.

158 Gälman, C., Angelin, B. & Rudling, M. 2011. Pronounced Variation in Bile Acid Synthesis in Humans Is Related to Gender, Hypertriglyceridaemia and Circulating Levels of Fibroblast Growth Factor 19. *J Intern Med*, 270, 580–588.

159 Wang, L., Prasad, B., Salphati, L., Chu, X., Gupta, A., Hop, C. E., Evers, R. & Unadkat, J. D. 2015. Interspecies Variability in Expression of Hepatobiliary Transporters across Human, Dog, Monkey, and Rat as Determined by Quantitative Proteomics. *Drug Metab Dispos*, 43, 367–374.

160 Delzenne, N. M., Calderon, P. B., Taper, H. S. & Roberfroid, M. B. 1992. Comparative Hepatotoxicity of Cholic Acid, Deoxycholic Acid and Lithocholic Acid in the Rat: In Vivo and in Vitro Studies. *Toxicol Lett*, 61, 291–304.

161 Yang, F., Takeuchi, T., Tsuneyama, K., Yokoi, T. & Oda, S. 2019. Experimental Evidence of Liver Injury by BSEP-Inhibiting Drugs with a Bile Salt Supplementation in Rats. *Toxicol Sci*, 170, 95–108.

162 Li, Y., Evers, R., Hafey, M. J., Cheon, K., Duong, H., Lynch, D., Lafranco-Scheuch, L., Pacchione, S., Tamburino, A. M., Tanis, K. Q., Geddes, K., Holder, D., Zhang, N. R., Kang, W., Gonzalez, R. J., Galijatovic-Idrizbegovic, A., Pearson, K. M., Lebron, J. A., Glaab, W. E. & Sistare, F. D. 2019. Use of a Bile Salt Export Pump Knockdown Rat Susceptibility Model to Interrogate Mechanism of Drug-Induced Liver Toxicity. *Toxicol Sci*, 170, 180–198.

163 Ryan, J., Morgan, R. E., Chen, Y., Volak, L. P., Dunn, R. T. & Dunn, K. W. 2018. Intravital Multiphoton Microscopy with Fluorescent Bile Salts in Rats as an in Vivo Biomarker for Hepatobiliary Transport Inhibition. *Drug Metab Dispos*, 46, 704–718.

164 Starokozhko, V., Greupink, R., Van De Broek, P., Soliman, N., Ghimire, S., Inge A M de Graaf & Groothuis, G. M. M. 2017. Rat Precision-Cut Liver Slices Predict Drug-Induced Cholestatic Injury. *Arch Toxicol*, 91, 3403–3413.

165 Ballet, F. 2015. Preventing Drug-Induced Liver Injury: How Useful Are Animal Models? *Dig Dis*, 33, 477–485.

166 Kohara, H., Bajaj, P., Yamanaka, K., Miyawaki, A., Harada, K., Miyamoto, K., Matsui, T., Okai, Y., Wagoner, M. & Shinozawa, T. 2020. High-Throughput Screening to Evaluate Inhibition of Bile Acid Transporters Using Human Hepatocytes Isolated from Chimeric Mice. *Toxicol Sci*, 173, 347–361.

167 Xu, D., Nishimura, T., Nishimura, S., Zhang, H., Zheng, M., Guo, Y. Y., Masek, M., Michie, S. A., Glenn, J. & Peltz, G. 2014. Fialuridine Induces Acute Liver Failure in Chimeric TK-NOG Mice: A Model for Detecting Hepatic Drug Toxicity Prior to Human Testing. *PLoS Med*, 11, e1001628.

168 Weaver, R. J., Blomme, E. A., Chadwick, A. E., Copple, I. M., Gerets, H. H. J., Goldring, C. E., Guillouzo, A., Hewitt, P. G., Ingelman-Sundberg, M., Jensen, K. G., Juhila, S., Klingmüller, U., Labbe, G., Liguori, M. J., Lovatt, C. A., Morgan, P., Naisbitt, D. J., Pieters, R. H. H., Snoeys, J., Van De Water, B., Williams, D. P. & Park, B. K. 2020. Managing the Challenge of Drug-Induced Liver Injury: A Roadmap for the Development and Deployment of Preclinical Predictive Models. *Nat Rev Drug Discov*, 19, 131–148.

169 Aronson, J. K. & Ferner, R. E. 2017. Biomarkers—A General Review. *Curr Protoc Pharmacol*, 76, 9.23.1–9.23.17.

170 Roth, S. E., Avigan, M. I., Bourdet, D., Brott, D., Church, R., Dash, A., Keller, D., Sherratt, P., Watkins, P. B., Westcott-Baker, L., Lentini, S., Merz, M., Ramaiah, L., Ramaiah, S. K., Stanley, A. M. & Marcinak, J. 2020. Next-Generation Dili Biomarkers: Prioritization of Biomarkers for Qualification and Best Practices for Biospecimen Collection in Drug Development. *Clin Pharmacol Ther*, 107, 333–346.

171 Gunaydin, M. & Bozkurter Cil, A. T. 2018. Progressive Familial Intrahepatic Cholestasis: Diagnosis, Management, and Treatment. *Hepat Med*, 10, 95–104.

172 Aleo, M. D., Aubrecht, J., Bonin, PD., Burt, D. A., Colangelo, J., Luo, L., Schomaker, S., Swiss, R., Kirby, S., Rigdon, GC & Dua, P. 2019. Phase I Study of

PF-04895162, a Kv7 Channel Opener, Reveals Unexpected Hepatotoxicity in Healthy Subjects, but Not Rats or Monkeys: Clinical Evidence of Disrupted Bile Acid Homeostasis. *Pharmacol Res Perspect*, 7, e00467.

173 Yamazaki, M., Miyake, M., Sato, H., Masutomi, N., Tsutsui, N., Adam, K. P., Alexander, D. C., Lawton, K. A., Milburn, M. V., Ryals, J. A., Wulff, J. E. & Guo, L. 2013. Perturbation of Bile Acid Homeostasis Is an Early Pathogenesis Event of Drug Induced Liver Injury in Rats. *Toxicol Appl Pharmacol*, 268, 79–89.

174 Bathena, S. P., Thakare, R., Gautam, N., Mukherjee, S., Olivera, M., Meza, J. & Alnouti, Y. 2015. Urinary Bile Acids as Biomarkers for Liver Diseases I. Stability of the Baseline Profile in Healthy Subjects. *Toxicol Sci*, 143, 296–307.

175 Steiner, C., Othman, A., Saely, C. H., Rein, P., Drexel, H., Von Eckardstein, A. & Rentsch, K. M. 2011. Bile Acid Metabolites in Serum: Intraindividual Variation and Associations with Coronary Heart Disease, Metabolic Syndrome and Diabetes Mellitus. *PLoS One*, 6, e25006.

176 Gälman, C., Angelin, B. & Rudling, M. 2005. Bile Acid Synthesis in Humans Has a Rapid Diurnal Variation That Is Asynchronous with Cholesterol Synthesis. *Gastroenterology*, 129, 1445–1453.

177 Trottier, J., Białek, A., Caron, P., Straka, R. J., Milkiewicz, P. & Barbier, O. 2011. Profiling Circulating and Urinary Bile Acids in Patients with Biliary Obstruction before and after Biliary Stenting. *PLoS One*, 6, e22094.

178 Li, Y., Hafey, M. J., Duong, H., Evers, R., Cheon, K., Holder, D. J., Galijatovic-Idrizbegovic, A., Sistare, F. D. & Glaab, W. E. 2017. Antibiotic-Induced Elevations of Plasma Bile Acids in Rats Independent of BSEP Inhibition. *Toxicol Sci*, 157, 30–40.

179 Robles-Díaz, M., Medina-Caliz, I., Stephens, C., Andrade, R. J. & Lucena, M. I. 2016. Biomarkers in Dili: One More Step Forward. *Front Pharmacol*, 7, 267.

180 Church, R. J., Kullak-Ublick, G. A., Aubrecht, J., Bonkovsky, H. L., Chalasani, N., Fontana, R. J., Goepfert, J. C., Hackman, F., King, N. M. P., Kirby, S., Kirby, P., Marcinak, J., Ormarsdottir, S., Schomaker, S. J., Schuppe-Koistinen, I., Wolenski, F., Arber, N., Merz, M., Sauer, J. M., Andrade, R. J., Van Bömmel, F., Poynard, T. & Watkins, P. B. 2019. Candidate Biomarkers for the Diagnosis and Prognosis of Drug-Induced Liver Injury: An International Collaborative Effort. *Hepatology*, 69, 760–773.

181 Urban, T. J., Daly, A. K. & Aithal, G. P. 2014. Genetic Basis of Drug-Induced Liver Injury: Present and Future. *Semin Liver Dis*, 34, 123–133.

182 Dragoi, D., Benesic, A., Pichler, G., Kulak, N. A., Bartsch, H. S. & Gerbes, A. L. 2018. Proteomics Analysis of Monocyte-Derived Hepatocyte-Like Cells Identifies Integrin Beta 3 as a Specific Biomarker for Drug-Induced Liver Injury by Diclofenac. *Front Pharmacol*, 9, 699.

183 Xie, Z., Chen, E., Ouyang, X., Xu, X., Ma, S., Ji, F., Wu, D., Zhang, S., Zhao, Y. & Li, L. 2019. Metabolomics and Cytokine Analysis for Identification of Severe Drug-Induced Liver Injury. *J Proteome Res*, 18, 2514–2524.

184 Bloomingdale, P., Housand, C., Apgar, J. F., Millard, B. L., Mager, D. E., Burke, J. M. & Shah, D. K. 2017. Quantitative Systems Toxicology. *Curr Opin Toxicol*, 4, 79–87.

185 Watkins, P. B. 2020. Quantitative Systems Toxicology Approaches to Understand and Predict Drug-Induced Liver Injury. *Clin Liver Dis*, 24, 49–60.

186 Ferreira, S., Fisher, C., Furlong, L. I., Laplanche, L., Park, B. K., Pin, C., Saez-Rodriguez, J. & Trairatphisan, P. 2020. Quantitative Systems Toxicology Modeling to Address Key Safety Questions in Drug Development: A Focus of the TransQST Consortium. *Chem Res Toxicol*, 33, 7–9.

187 Waring, M. J., Arrowsmith, J., Leach, A. R., Leeson, P. D., Mandrell, S., Owen, R. M., Pairaudeau, G., Pennie, W. D., Pickett, S. D., Wang, J., Wallace, O. & Weir, A. 2015. An Analysis of the Attrition of Drug Candidates from Four Major Pharmaceutical Companies. *Nat Rev Drug Discov*, 14, 475–486.

188 Howell, B. A., Yang Y, Kumar, R., Woodhead, J. L., Harrill, A. H., Clewell, H. J., 3rd, Andersen, M. E., Siler, S. Q., Watkins, P. B. 2012. In Vitro to in Vivo Extrapolation and Species Response Comparisons for Drug-Induced Liver Injury (Dili) Using Dilisym™: A Mechanistic, Mathematical Model of Dili. *J Pharmacokinet Pharmacodyn*, 39, 527–541.

189 Shoda, L. K., Woodhead, J. L., Siler, S. Q., Watkins, P. B. & Howell, B. A. 2014. Linking Physiology to Toxicity Using Dilisym®, a Mechanistic Mathematical Model of Drug-Induced Liver Injury. *Biopharm Drug Dispos*, 35, 33–49.

190 Longo, D. M., Woodhead, J. L., Walker, P., Herédi-Szabó, K., Mogyorósi, K., Wolenski, F. S., Dragan, Y. P., Mosedale, M., Siler, S. Q., Watkins, P. B. & Howell, B. A. 2019. Quantitative Systems Toxicology Analysis of in Vitro Mechanistic Assays Reveals Importance of Bile Acid Accumulation and Mitochondrial Dysfunction in Tak-875-Induced Liver Injury. *Toxicol Sci*, 167, 458–467.

191 Woodhead, J. L., Brock, W. J., Roth, S. E., Shoaf, S. E., Brouwer, K. L., Church, R., Grammatopoulos, T. N., Stiles, L., Siler, S. Q., Howell, B. A., Mosedale, M., Watkins, P. B. & Shoda, L. K. 2017. Application of a Mechanistic Model to Evaluate Putative Mechanisms of Tolvaptan Drug-Induced Liver Injury and Identify Patient Susceptibility Factors. *Toxicol Sci*, 155, 61–74.

192 Yang, K., Woodhead, J. L., Watkins, P. B., Howell, B. A. & Brouwer, K. L. 2014. Systems Pharmacology Modeling Predicts Delayed Presentation and Species Differences in Bile Acid-Mediated Troglitazone Hepatotoxicity. *Clin Pharmacol Ther*, 96, 589–598.

193 Woodhead, J. L., Yang, K., Brouwer, K. L. R., Siler, S. Q., Stahl, S. H., Ambroso, J. L., Baker, D., Watkins, P. B. & Howell, B. A. 2014a. Mechanistic Modeling Reveals the Critical Knowledge Gaps in Bile Acid-Mediated Dili. *CPT Pharmacometrics Syst Pharmacol*, 3, e123.

194 Generaux, G., Lakhani, V. V., Yang, Y., Nadanaciva, S., Qiu, L., Riccardi, K., Di, L., Howell, B. A., Siler, S. Q., Watkins, P. B., Barton, H. A., Aleo, M. D. & Shoda, L. K. M. 2019. Quantitative Systems Toxicology (QST) Reproduces Species Differences in Pf-04895162 Liver Safety Due to Combined Mitochondrial and Bile Acid Toxicity. *Pharmacol Res Perspect*, 7, e00523.

10

Hepatic Bile Acid Transporters in Drug-Induced Cholestasis

Tao Hu and Hongbing Wang

Department of Pharmaceutical Sciences, University of Maryland School of Pharmacy, Baltimore, Maryland, USA

Abbreviations

ABC	ATP-binding cassette
ALP	alkaline phosphatase
ALT	alanine aminotransferase
ASBT	apical sodium-dependent bile salt transporter
AMP	adenosine monophosphate
BEI	biliary excretion index
BSEP	bile salt export pump
CA	cholic acid
CDCA	chenodeoxycholic acid
DCA	deoxycholic acid
DIC	drug-induced cholestasis
DILI	drug-induced liver injury
FDA	food and drug administration
FXR	farnesoid X receptor
HEK	human embryonic kidney
HNF	hepatocyte nuclear factor
IL	interleukin
LCA	lithocholic acid
LRH	liver receptor homolog
MAPK	mitogen-activated protein kinase
MRP	multidrug resistance-associated protein
Nrf2	nuclear factor erythroid 2-related factor 2
NTCP	sodium-taurocholate cotransporting polypeptide
OATP	organic anion transporting polypeptide
OSTα/β	organic solute transporter α/β

Transporters and Drug-Metabolizing Enzymes in Drug Toxicity, First Edition. Edited by Albert P. Li.

PI3K	phosphoinositide-3-kinase
PKC	protein kinase C
PXR	pregnane X receptor
RAR	retinoic acid receptor
ROS	reactive oxygen species
RXR	retinoid X receptor
SHP	small heterodimer partner
TCDC	taurochenodeoxycholic acid
TNFα	tumor necrosis factor alpha
VDR	vitamin D receptor

10.1 Introduction

The liver is the central detoxification organ in the body responsible for the uptake, metabolism and excretion of numerous drugs and other xenobiotics. This essential function predisposes the liver to various chemical hazards, making drug-induced liver injury (DILI) a critical liability during the development of pharmaceuticals. To date, more than a thousand drugs and herbal remedies have been reported to be associated with different types of DILI [1]. In fact, DILI is the most common cause of acute liver failure and is a primary reason for failures of otherwise successful drug candidates and the withdrawals of marketed drugs [2]. For instance, drugs including troglitazone, tolcapone, trovafloxacin, bromfenac, nefazodone, and ximelagatran were all withdrawn from the market due to severe hepatotoxicity, while bosentan and diclofenac received black box warnings because of the same reasons [3, 4].

Among others, two most severe manifestations of DILI are cholestatic and mixed cholestatic hepatic injuries that account for ~50% of all hepatic drug toxicity [5]; both are associated with impaired bile flow and bile acid transport in the liver. To provide a comprehensive understanding of the drug-induced cholestasis (DIC) and its underlying mechanisms, the present chapter introduces bile acid synthesis and transport in the body, discusses hepatic bile acid uptake and efflux transporters, and reviews their association with the development of DIC.

10.2 Bile Acid and DIC

DIC is characterized by a drug-induced impairment of canalicular bile flow leading to a detrimental accumulation of bile constituents in the liver. As a hallmark of DIC, the disruption of bile flow in the liver can be caused by impaired formation of bile due to drugs that interact with bile acid synthesis or hepatic bile acid transport, leading to intrahepatic cholestasis [6]. After being secreted from the hepatocytes, the bile flow can also be interrupted by obstruction resulting from bile duct damage by drugs causing extrahepatic cholestasis [7].

10.2.1 Bile Acid

Bile acids are a group of hydroxylated steroids synthesized from cholesterol in the liver. It is well known that bile acids exert physiologically important functions in the body by aiding in the digestion and absorption of lipids and fat-soluble nutrients from the small intestine, as well as stimulating bile flow and cholesterol secretion from the liver [2]. Additionally, bile acids have been reported to be the ligands of several nuclear receptors, including the farnesoid X receptor (FXR), vitamin D receptor (VDR), and pregnane X receptor (PXR), which regulate the expression of a plethora of genes associated with glucose and lipid metabolism, xenobiotic detoxification, and the homeostasis of bile acids itself [8, 9]. Indeed, mounting evidence has demonstrated that bile acids function as hormones in the regulation of their own biosynthesis, transport, and metabolism [10]. Thus, alteration of bile acids homeostasis in the body may result in significant clinical consequences.

10.2.1.1 Bile Acid Synthesis

Bile acids are synthesized from cholesterol primarily in the hepatocytes, conjugated to glycine or taurine, and actively secreted along with cholesterol and phospholipids into the bile. The classic bile acid synthetic pathway initiates with 7α-hydroxylation of cholesterol, the rate-limiting step catalyzed by cholesterol-7α-hydroxylase (CYP7A1) [11]. Bile acid synthesis also occurs through an alternative pathway, which is governed by the mitochondrial sterol 27-hydroxylase (CYP27A1) and CYP7B1 [11]. Both pathways lead to the formation of cholic acid (CA) and chenodeoxycholic acid (CDCA), the two most common primary bile acids in humans, with the classic pathway accounting for, at least, 75% production of the total bile acids [3, 12]. The primary bile acids are then conjugated with glycine or taurine in the liver to form bile salts with increased hydrophilicity, thereby facilitating their secretion into the gallbladder through bile canaliculi [13]. Upon ingestion of foods, bile salts are released from the gallbladder into the lumen of small intestine, where dehydroxylation of the primary bile acids by gut bacteria produces secondary bile acids, including deoxycholic acid (DCA) and lithocholic acid (LCA), which are derived from 7-dehydroxylation of CA and CDCA, respectively [3]. Importantly, most bile acids are reabsorbed in the intestine and are actively transported back to the liver via the portal vein. This pivotal process known as enterohepatic circulation is highly efficient, resulting in the reabsorption of 95% bile acids released from the liver, while the remaining 5% is excreted into feces [13]. Under normal physiological conditions, a healthy adult recycles approximately 2–3 g of bile acids from the intestine to the liver, limiting hepatic de novo synthesis of bile acids to about 500 mg daily [14]. Thus, reabsorbed bile acids constitute a key negative feedback loop and play a crucial role in maintaining the bile acid homeostasis in the body.

10.2.1.2 Bile Acid Transport

Bile acids are normally synthesized in the hepatocytes, stored in the gallbladder, and circulated between the intestine and liver via enterohepatic circulation. Since conjugation of bile acids to glycine or taurine makes them less likely to undergo passive transport, enterohepatic recycling of bile acids across the hepatocytes and enterocytes involves active uptake and excretion of bile acids by a cluster of membrane transporter proteins.

Hepatic uptake of bile salts starts at the basolateral membrane of hepatocytes, uptake transporters including sodium-taurocholate cotransporting polypeptide (NTCP) and organic anion transporting polypeptides (OATPs) mediate the transport of bile salts from portal blood into the hepatocytes. Within hepatocytes, bile salts reach the canalicular membrane by either rapid diffusion in the cytoplasm or binding to specific intracellular binding proteins and moving to the canalicular membrane in a protein-bound form [15]. Canalicular excretion of biliary constituents is mediated by efflux transporters, including bile salt export pump (BSEP) and multidrug resistance-associated protein 2 (MRP2), against a high concentration gradient into bile, which represents the rate-limiting step of bile secretion from the liver.

Once released from the gallbladder into the lumen of intestine, bile acids are de-conjugated by gut bacteria and converted into secondary bile acids. Many constituents of bile, including bile acids, cholesterol, and phospholipids, are reabsorbed by active transport in the terminal ileum and passive absorption in the proximal small intestine and the colon. The active transport of bile acids from intestine is primarily mediated by the apical sodium-dependent bile salt transporter (ASBT). From enterocytes, bile acids enter the mesenteric blood through basolateral efflux mediated by multidrug resistance-associated protein 3 (MRP3) and organic solute transporter α/β (OSTα/β) and eventually return to the liver from the portal circulation by hepatic basolateral uptake transporters, completing the enterohepatic circulation (Figure 10.1) [3].

10.2.2 Cytotoxicity of Bile Acids and DIC

Despite the pivotal roles of bile acids in assisting the elimination of cholesterol, promoting digestion, and absorption of lipids and fat-soluble nutrients, as well as stimulating bile flow and biliary phospholipid excretion, excessive accumulation

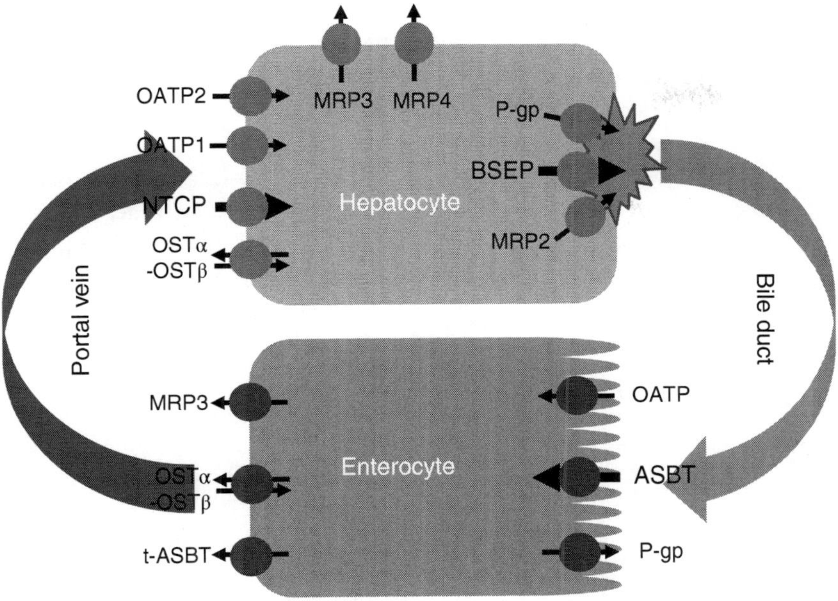

Figure 10.1 Schematic illustration of transporters control bile flow in the liver and intestine. *Source:* Adapted from Garzel et al. [2].

of bile acids in the hepatocytes is cytotoxic. Impaired bile acid homeostasis can be caused by bile acid biosynthesis disorders and/or disruption of bile acid flow. A decrease in bile flow, defined as cholestasis, is caused by impaired secretion from hepatocytes or obstruction of bile flow through intrahepatic or extrahepatic bile ducts, leading to subsequent accumulation of toxic bile components in the liver [16]. The cytotoxicity of a bile acid is strongly associated with its hydrophobicity, where higher hydrophobicity indicates greater cytotoxicity [17]. The long retention time and high accumulation of the natural dihydroxy bile acids CDCA and DCA in the hepatocytes have been implicated as a major cause of liver damage in cholestasis [18]. The mechanisms of the cytotoxic action of bile acids in the hepatocytes include the generation of reactive oxygen species (ROS), DNA injury, mitochondrial damage, and induction of apoptosis and necrosis [19–21].

Clinically, DIC accounts for most of the cases of DILI reported thus far. Characterized by impaired formation of bile or disruption of bile flow in the liver after drug exposure [3], the clinical presentation of cholestatic DILI is variable, ranging from asymptomatic elevation in alkaline phosphatase (ALP) to symptoms including jaundice, pruritus, and fever [22]. In clinical practice, while DIC displays altered serological parameters including an increase in serum ALP greater than twofold of its upper limit of normal and/or an alanine aminotransferase

(ALT)/ALP ratio of less than 2 [22], without a gold standard for diagnosis, DIC is often a diagnosis of exclusion [23]. Based on the pathological changes and clinical syndromes, DIC can be categorized into the following groups [1]: (i) Acute DIC without hepatitis: This type of DIC is rare and is typically induced by estrogen or anabolic steroids, leading to minimal or no hepatic parenchymal involvement. This disease histologically manifests pure canalicular cholestasis and is also called bland cholestasis. (ii) Acute DIC with hepatitis: Acute DIC with hepatitis is a mixed cholestatic hepatitis and is the most common form of DILI leading to cholestasis [24]. It is often associated with portal inflammation and varying degrees of hepatocyte injury. (iii) Acute DIC with bile duct injury: Also known as cholangiolytic cholestasis, this type of DIC exhibits bile duct injury with minimal involvement of parenchymal liver cell injury. And (iv) chronic drug-induced cholangiopathies: The chronic drug-induced cholangiopathies vary from asymptomatic patients with mild bile duct disarray and elevated ALP or gamma glutamyl transferase (GGT) to progressive forms of the vanishing bile duct syndrome (VBDS) [25].

Hepatic canalicular and basolateral transport proteins have been shown to be involved in the pathophysiology of DIC. Drug-mediated functional disturbance of hepatic bile acid transporters can lead to intracellular accumulation of harmful bile acids in the hepatocytes and subsequent development of cholestatic damage [3]. Indeed, a number of drug substrates and modulators of hepatic bile acid transporters are known to cause DIC, including antihypertensives, anticonvulsants, antidiabetics, nonsteroidal anti-inflammatory drugs, lipid-lowering agents, and psychotropic drugs [1]. In this regard, DIC represents an important reason for drug attrition during development as well as black box warnings and withdrawal of drugs from the market.

10.3 Hepatic Bile Acid Uptake Transporters in DIC

In the liver, the polarized hepatocytes are arranged in linear chords separated by spaces or sinusoids. This structural feature considerably facilitates bile flow in and out of the hepatocytes through several hepatic canalicular and basolateral transporter proteins as shown in Table 10.1. Mounting evidence has demonstrated that dysfunction of hepatic bile acid transporters by drugs can lead to increased intracellular accumulation of toxic bile acids in the liver and the development of DIC.

10.3.1 Sodium-Taurocholate Cotransporting Polypeptide (NTCP)

NTCP was initially identified from the rat liver in 1991 and is encoded by the *SLC10A1* (solute carrier family 10 member 1) gene in humans [41]. The rat and

Table 10.1 Major hepatic bile acid transporters and their bile salt substrates.

Transporter	Gene	Hepatic localization	Transport mode	Bile salt substrates	Reference
NTCP	*SLC10A1*	Basolateral	Sodium-dependent uptake	Cholate, glycocholate, taurocholate, ursodeoxycholate, chenodeoxycholate, glycoursodeoxycholate, glycochenodeoxycholate, tauroursodeoxycholate, taurochenodeoxycholate	[26–29]
OATP1B1	*SLCO1B1*	Basolateral	Sodium-independent uptake	Cholate, glycocholate, glycolithocholate, glycodeoxycholate, glycochenodeoxycholate, taurolithocholate, taurodeoxycholate, taurochenodeoxycholate	[30]
OATP1B3	*SLCO1B3*	Basolateral	Sodium-independent uptake	Cholate, glycocholate, glycolithocholate, glycodeoxycholate, glycochenodeoxycholate, taurolithocholate, taurodeoxycholate, taurochenodeoxycholate	[30]
BSEP	*ABCB11*	Canalicular	ATP-dependent efflux	Taurocholate, glycocholate, glycochenodeoxycholate, taurolithocholate, taurodeoxycholate, tauroursodeoxycholate, taurochenodeoxycholate	[31–34]
MRP2	*ABCC2*	Canalicular	ATP-dependent efflux	Sulfate conjugated bile acids: taurolithocholate sulfate, etc., glucuronidated bile salt conjugates	[35, 36]
MRP3	*ABCC3*	Basolateral	ATP-dependent efflux	Taurocholate, glycocholate, taurolithocholate-3-sulfate, taurochenodeoxycholate-3-sulfate	[37]
MRP4	*ABCC4*	Basolateral	ATP-dependent efflux	Cholate, taurocholate, glycocholate, tauroursodeoxycholate, taurochenodeoxycholate, glycodeoxycholate, glycoursodeoxycholate, glycochenodeoxycholate, sulfated conjugates of bile acids	[38, 39]
OSTα/β	*SLC51A/B*	Basolateral	Bidirectional transport	Taurine conjugates of bile acids: taurocholate, etc.	[40]

human NTCPs consist of 362 and 349 amino acids, respectively, and share 77% amino acid sequence identity [42]. In the liver, expression of NTCP is restricted to the basolateral membrane of hepatocytes, and its primary function involves the recycling of bile acids from portal blood to the hepatocytes in a sodium-dependent manner, accounting for more than 80% of conjugated bile salts uptake [43, 44]. NTCP is an electrogenic transporter that transfers one bile salt molecule together with two sodium ions unidirectionally [45]. The activity of this transporter is driven by both the transmembrane sodium gradient and the intracellular negative membrane potential, allowing its uptake of bile salts from portal blood into the hepatocytes against a 5- to 10-fold concentration gradient [46].

10.3.1.1 Substrates of NTCP

As the driving force of hepatic uptake of bile acids, NTCP transports glycine and taurine conjugated bile acids with high affinity, while taking up unconjugated bile acids at a relatively moderate capacity [47]. To date, many bile salts, including cholate, ursodeoxycholate, glycocholate, chenodeoxycholate, glycoursodeoxycholate, chenodeoxycholate-3-sulfateglycocholate, glycochenodeoxycholate, taurocholate, tauroursodeoxycholate, and taurochenodeoxycholate, have been reported to be substrates of NTCP [26–29]. Comparatively, the affinity of NTCP to dihydroxy bile salts (conjugates of chenodeoxycholate and deoxycholate) is higher than its affinity to trihydroxy bile salts (conjugates of cholate) [48]. The functional properties of human NTCP and its rodent orthologues are similar in general, although human NTCP exhibits a relatively higher affinity for taurocholate. The Km value of taurocholate toward rat Ntcp ranges from 12 to 61 μM, as determined using different cellular systems, whereas the published Km values for NCTP-mediated taurocholate transport in human hepatocytes vary considerably (5–84 μM) [49–52], presumably due to the pathophysiological status and genetic backgrounds of the liver donors.

In addition to bile salts, NTCP interacts with estrone-3-sulfate, bromosulfophthalein, thyroid hormones and their metabolites, as well as a narrow panel of drugs. For example, approximately 35% of the hepatic uptake of rosuvastatin, a HMG-CoA reductase inhibitor, is mediated by NTCP, and alteration of NTCP activity may lead to unexpected drug interactions with rosuvastatin [53]. Additionally, atorvastatin, fluvastatin, and pitavastatin are substrates of human NTCP based on *in vitro* findings [54, 55]. In 2012, a novel function of NTCP was revealed by Yan et al. in which NTCP mediates the hepatic entry of hepatitis B virus (HBV) and hepatitis D virus (HDV) and was designated as a new functional receptor of HBV and HDV responsible for virus entry into hepatocytes [56]. Given the diverse physiological roles of NTCP, identification of novel NTCP substrates is of potential clinical significance. Recently, Dong et al. developed a quantitative pharmacophore to elucidate the structural requirements of NTCP substrates, this model consisting of three hydrophobes, one hydrogen bond donor, one negative ionizable feature, and three excluded volumes [57], provide a useful tool for the identification of NTCP substrates.

10.3.1.2 Regulation of NTCP

Given that most bile acids excreted into the intestine are reabsorbed back to the hepatocytes via NTCP, controlling the expression of NTCP is a key step in maintaining homeostasis of hepatic bile acids. The transcriptional mechanisms of NTCP regulation involve a complex interacting network of numerous nuclear receptors and other transcription factors, including but not limited to FXR, retinoic acid receptor alpha:retinoid X receptor alpha (RARα:RXRα) heterodimer, small heterodimer partner (SHP), liver receptor homolog (LRH)-1, hepatocyte nuclear factor-1α (HNF1α), HNF3β, HNF4α, glucocorticoid receptor (GR), PGC-1α, forkhead box A2 (FOXA2), and the signal transducer and transactivator 5 (Stat5) [58, 59]. DNA response elements specific for GR, HNF4α, HNF1α, HNF3β, and RARα:RXRα have been identified in the promoter region of the *NTCP* gene, while others, such as PGC-1α, FXR, and SHP, indirectly alter the expression of NTCP. Negative feedback regulation of NTCP in response to bile acids serves as an adaptive response upon impaired bile secretion, to protect the hepatocytes from accumulating cytotoxic bile salts. As ligands of FXR, bile acids provoke FXR-mediated upregulation of SHP, a transcriptional repressor, which in turn downregulates RARα:RXRα- and HNF4α-mediated NTCP transcription by competing with their respective co-activator binding [60]. In addition to bile salts, inflammatory cytokines, including tumor necrosis factor alpha (TNFα), interleukin (IL)-1β, and IL-6, downregulate NTCP expression, as a response to inflammation-associated cholestasis. For instance, IL-1β markedly reduced NTCP mRNA expression in cultured rat and human hepatocytes and inhibited NTCP promoter activity in transfected HepG2 cells [61, 62]. This observation was further linked mechanistically to JNK-dependent RXR phosphorylation with reduced RXRα:RARα transcriptional activity. Additionally, cholestatic liver injury associated with physiological obstruction, drugs, and toxins exposure, as well as hormonal perturbation during pregnancy rapidly downregulates NTCP expression as a compensatory measure [63].

The NTCP activity is also subject to posttranslational regulations. In mammals, after translation, most proteins must be translocated to specific cellular sections, such as the plasma membrane, in order to exhibit their biological functions. In response to physiological/pathophysiological stimuli, many hepatic transporters' traffic between the cytoplasm and the cell membrane, is regulated at the posttranslational level involving various signaling pathways. Known cellular factors influencing posttranslational regulation of NTCP include cyclic adenosine monophosphate (cAMP), intracellular Ca^{2+}, nitric oxide, phosphoinositide-3-kinase (PI3K), protein kinase Cs (PKCs), and protein phosphatases [64–66]. In rat hepatocytes, cAMP rapidly stimulates sodium/taurocholate cotransport with increased transport maximal rate [67], partially due to the cAMP-induced movement of NTCP from endosomes to the plasma membrane [68]. Mechanistically, the sinusoidal membrane expression of NTCP in rat liver is regulated by serine/threonine phosphorylation [69], by which NTCP is dephosphorylated at Serine 226 in response to cAMP, leading to an increase in its plasma membrane retention and transport activity [70]. In contrast, hepatic bile acids, such as taurochenodeoxycholic acid (TCDC), can inhibit TC uptake by internalizing membrane NTCP, a mechanism in addition to the negative feedback regulation of NTCP expression [71].

10.3.1.3 NTCP and Cholestasis

Although exogenous overexpression of NTCP results in increased bile acid uptake in cultured cells, the clinical significance of NTCP upregulation is rather obscure. In contrast, downregulation of NTCP has been observed in nearly all experimental cholestasis models and in patients with cholestatic liver disease. In liver samples from infants with chronic obstructive cholestasis due to biliary atresia, the mRNA expression of NTCP was downregulated [72]. Decreased expression of NTCP was also associated with elevated serum bile salt levels in patients with extrahepatic biliary atresia, and restoration of bile flow in these patients increased the mRNA level of NTCP [73]. NTCP mRNA expression was also decreased by 41% in patients with inflammation-induced cholestasis [74]. Analysis of the liver samples from patients with progressive familial intrahepatic cholestasis showed downregulation of NTCP at the protein level with no mRNA changes, suggesting the existence of a posttranslational regulation of NTCP in this disease [75]. While it is clear that NTCP expression is often decreased in human cholestatic liver diseases as an adaptation to protect the hepatocytes from overloaded cytotoxic bile salts, NTCP inhibition alone may not be a major causative factor for DILI and DIC potentials. To date, a variety of drugs have been identified as NTCP inhibitors, including cyclosporine A, ketoconazole, propanolol, furosemide, rifamycin, saquinavir, and ritonavir [76]. Given its role in uptake at the basolateral membrane of hepatocytes, inhibition of NTCP is supposed to lead to an increase of serum bile salt levels and a decreased accumulation of bile salts in the liver. However, the biological consequence may be variable, as sodium-independent basolateral uptake of bile salts by OATPs as well as alteration of biliary clearance, may partially compensate for the inhibition of NTCP. Cyclosporine A, for example, inhibits both rat Ntcp and human NTCP in heterologous expression systems with an IC_{50} value around 1 µM [77], but it leads to cholestatic liver injury [78]. This is likely due to the multiple actions of cyclosporine A on the compensatory mechanisms of bile salt uptake and efflux in hepatocytes. Pharmacophore analysis of 94 Food and Drug Administration (FDA) approved drugs, including NTCP inhibitors and non-inhibitors, revealed that they are approximately equally distributed across the drugs of most DILI concern, less DILI concern, and no DILI concern, indicating that there is no relationship between NTCP inhibition and DILI risk [79].

10.3.2 Other Hepatic Bile Acid Uptake Transporters

In addition to sodium-dependent uptake of bile acids from blood into hepatocytes by NTCP, the basolateral uptake of bile acids can be mediated in a sodium-independent manner, via several OATPs. In human liver, OATP1A2 (*SLCO1A2*, formerly OATP-A), OATP1B1 (*SLCO1B1*, formerly OATP-C) and OATP1B3 (*SLCO1B3*, formerly OATP-8) are able to transport bile acids with OATP1B1 being the main player in this class [15].

OATP1A2 is the first human OATP cloned when a human liver complementary DNA (cDNA) library was screened with a rat Oatp-derived cDNA probe [80]. *In vitro* overexpression of OATP1A2 in frog oocytes results in sodium-independent uptake of bile salts such as taurocholate, cholate, and tauroursodeoxycholate. However, with its relatively low expression in the liver, OATP1A2 only moderately contributes to the overall hepatic uptake of bile salts [15]. OATP1B1, on the other hand, represents the most important sodium-independent bile acid uptake system in human liver. It uptakes taurocholate with an affinity only slightly lower than that of NTCP [81]. OATP1B3 shares 80% amino acid identity with OATP1B1 and is also expressed at the basolateral membrane of human hepatocytes [82]. Using stable transfected human embryonic kidney (HEK) 293 cells, OATP1B1, and OATP1B3 were found to transport a number of bile acids (CA, glycocholic acid, glycochenodeoxycholic acid, TCDC, glycodeoxycholic acid, taurodeoxycholic acid, glycolithocholic acid, and taurolithocholic acid) with similar Km values, ranging from 0.74 to 14.7 μM for OATP1B1 and 0.47 to 15.3 μM for OATP1B3, respectively [30]. Moreover, both transporters exhibit higher affinity toward conjugated bile acids in comparison to unconjugated bile acids [30].

Interestingly, along with NTCP repression, hepatic expression of OATP1B1 and OATP1B3 was decreased in patients with inflammation-induced icteric cholestasis, type 2 and 3 progressive familial intrahepatic cholestasis, primary sclerosing cholangitis, and extrahepatic biliary atresia [75, 83]. Like the regulation of NTCP under cholestasis, downregulation of OATP1B1 and OATP1B3 also plays a role in decreasing the accumulation of bile acids in hepatocytes and protecting the hepatocytes against cholestatic cytotoxicity.

10.4 Hepatic Bile Acid Efflux Transporters in DIC

Bile acid secretion from the hepatocytes is required not only for efficient lipid absorption, but also for the maintenance of bile salt homeostasis in the liver. Efflux of bile acids into the bile or back to the blood is a tightly controlled biological process, involving multiple efflux transporters, such as BSEP, MRP2, MRP3, and MRP4, located on the apical or basolateral surface of the hepatocytes. Led by BSEP, canalicular membrane transporters mediate a highly efficient uphill bile acid secretion. Impaired function of these transporters is associated with the development of cholestasis.

10.4.1 Bile Salt Export Pump (BSEP)

Bile canalicular efflux is the rate-limiting step in hepatic clearance of bile salts into the bile duct, driven by a number of canalicular membrane transporter proteins. Among others, BSEP is the primary efflux transporter responsible for canalicular bile salt secretion [84]. Originally named "sister of P-glycoprotein," BSEP is a canalicular ATP-binding cassette (ABC) transporter that shares 88% sequence homology with P-glycoprotein [85]. In human, BSEP is encoded by the ABC transporter subfamily B member 11 (*ABCB11*) gene and exhibits high mRNA expression in both the liver and testis, with low expression levels in lung, trachea, thymus, colon, kidney, and prostate [86, 87]. In line with its biological role, functional BSEP is exclusively localized at the canalicular membrane of hepatocytes. As a member of the ABC transporter superfamily, BSEP consists of 1321 amino acids and is predicted to have 12 transmembrane helices, 2 large nucleotide binding domains [88]. It requires ATP as an energy source to transport bile salts from the liver to the bile against steep concentration gradients across the canalicular membrane.

10.4.1.1 Substrates of BSEP

BSEP mediates the canalicular excretion of conjugated bile acids with high affinity, including glycine and taurine conjugates of the primary bile acids CA and CDCA as well as the secondary bile acid DCA, while showing weak affinity for unconjugated bile acids [89]. A number of bile acids and their metabolites, such as taurocholate, glycocholate, tauroursodeoxycholate, taurochenodeoxycholate, glycochenodeoxycholate, taurodeoxycholate, taurolithocholate, and taurolithocholate-3-sulfate, were transported by BSEP [31–34]. As determined in membrane vesicles from transfected HEK293 cells, the Km values of human BSEP toward taurocholate, taurochenodeoxycholate, glycochenodeoxycholate, and glycocholate are 6.2, 6.6, 7.5, and 21.7 μM, respectively [31]. Comparison of initial uptake rates indicated that taurine conjugated bile salts were transported more rapidly than glycine conjugated bile salts by BSEP [31]. Of note, the intrinsic clearance of rodent Bsep and human BSEP toward bile salts showed the similar rank order and comparable Km values [32], suggesting the well-preserved transport properties of BSEP in different species. One exception, however, is taurolithocholate 3-sulfate, which is significantly transported by human BSEP with a Km value of 9.5 μM, but not by rat Bsep [31]. Besides bile salts, BSEP transports a limited number of drugs with low affinity. For example, vinblastine and calcein AM, a fluorescent substrate of Mdr1, were transported by mouse Bsep in transfected LLC-PK1 cells [90]. Pravastatin was found to be removed by BSEP across the apical membrane with a Km value of 124 μM [91]. Nevertheless, the limited capacity of BSEP in drug transport suggests this transporter may not represent a major mechanism of xenobiotic detoxification.

10.4.1.2 Regulation of BSEP

As the major determinant of biliary bile secretion, hepatic expression of BSEP is coordinately regulated by a number of transcription factors such as FXR, LRH1, the nuclear factor erythroid 2-related factor 2 (Nrf2), and the steroid receptor coactivator-2 (SRC2) [92, 93]. FXR, a bile acid sensor, is the predominant transcription factor governing the expression of BSEP and facilitating a feed forward regulation of canalicular bile secretion [94]. Upon activation by bile salts, FXR binds as a heterodimer with RXRα to a highly conserved inverted repeat 1 (IR-1) response element (farnesoid X receptor response element [FXRE]) located in the proximal promoter region of BSEP, leading to an increase in BSEP transcription and canalicular bile salt secretion [95]. Both free and conjugated bile acids are endogenous FXR ligands with high affinity. CDCA represents the most potent natural agonist for FXR (EC_{50} value around 10 μM), followed by DCA and CA, in terms of their activation of the BSEP promoter via FXR [96]. In contrast, LCA, a hydrophobic secondary bile acid and an intestinal metabolite of CDCA, was identified as an antagonist of FXR [97]. Using FXR-knockout mouse model, studies further demonstrated that losing FXR leads to diminished basal expression of mouse Bsep and abolished induction of this transporter by CA feeding [98]. Collectively, FXR-mediated regulation of BSEP by bile salts is a highly conserved mechanism across multiple species protecting hepatocytes against cholestasis-associated liver injury.

LRH1 is another transcriptional regulator involving hepatic homeostasis of cholesterol and bile acids. Song et al., reported that LRH1 transcriptionally regulates BSEP expression through the functional liver receptor homolog response element 1 (LRHRE1) in the BSEP promoter and functions as a modulator in the bile acid-FXR axis most likely by enhancing the basal expression of BSEP [99]. This finding is consistent with animal studies, where expression of Bsep was decreased in LRH1-knockout mice in comparison to their wild-type littermates [100].

Nrf2 is a key stress sensor that protects cells against oxidative damage by coordinating the expression of many genes, including those encoding hepatic metabolism enzymes and efflux transporters. A functional Nrf2 binding site has been identified in the promoter of human *BSEP* gene. Activation of Nrf2 by oltipraz positively regulated the expression of BSEP in both HepG2 and human hepatocytes, while this induction was efficiently diminished by small interfering RNA targeting Nrf2 [101]. Notably, bile acids are not direct activators of Nrf2. Accumulation of toxic bile acids in cholestasis triggers hepatic generation of ROS, which subsequently leads to Nrf2 protein stabilization, forming an indirect protective response [102]. A schematic illustration depicting the regulation of BSEP through FXR, LRH1, and Nrf2 is shown in Figure 10.2.

In addition to upregulation, BSEP expression can be downregulated under certain physiological/pathophysiological conditions or after drug challenges. Intrahepatic cholestasis of pregnancy is a common disorder accompanied by elevated serum levels of bile acids during pregnancy. Song et al., reported that BSEP transcription was markedly repressed in the later stages of pregnancy [103]. Studies in human primary hepatocytes, Huh7 cells, and mice *in vivo,* revealed that 17β-estradiol (E2) repressed BSEP expression in an estrogen receptor α (ERα)-dependent manner. Mechanistic evidence further indicated that ERα directly interacted with FXR and repressed FXR recruitment to the promoter of BSEP [104]. Our previous analysis of the mRNA expression of BSEP in sandwich-cultured human primary hepatocytes treated with 30 BSEP inhibitors ($IC_{50} \leq 25\,\mu M$) [105], demonstrated that BSEP inhibitors that also reduced the expression of BSEP were often associated with severe DILI [106]. Recently, we also reported that relatively long-term exposure to metformin or tamoxifen significantly decreased both mRNA and protein expression of BSEP in sandwich-cultured human primary hepatocytes and led to a decrease in BSEP-mediated biliary secretion [107]. However, neither compounds were direct inhibitors of BSEP as shown in this and other reports [107, 108]. Hence, downregulation of BSEP expression is expected to

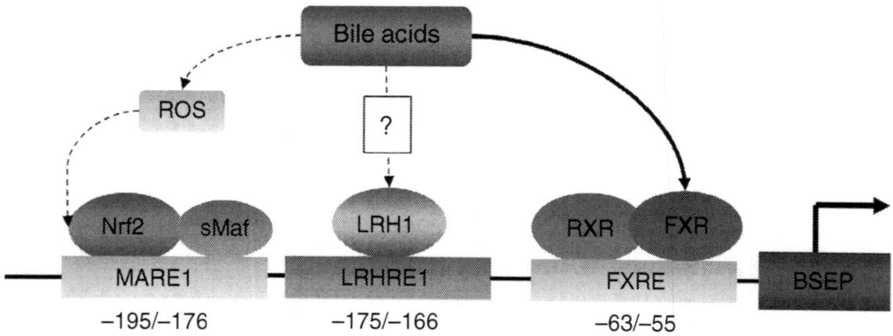

Figure 10.2 Transcriptional regulation of bile salt export pump (BSEP) by FXR, LRH1, and Nrf2. Upon bile acid stimulation, FXR and RXRα form a heterodimer and bind to the IR-1 FXR response element (FXRE) located −63/−55 of BSEP promoter. LRH1 binds to a functional liver receptor homolog-1 response element (LRHRE1) at −175/−166. Bile acid retention can also lead to the generation of reactive oxygen species (ROS) in hepatocytes, which in turn stabilize Nrf2. Nrf2 and a small Maf (sMaf) protein partner together and bind to a Maf recognition element (MARE1) at −195/−176 adjacent to the LRHRE1.

contribute to the DIC caused by metformin and tamoxifen, as well as potentially other BSEP non-inhibitors with cholestatic features.

10.4.1.3 Internalization of BSEP

Intracellular shuffling of BSEP between its cytoplasmic pool and canalicular membrane is a rapid adaptation of biliary excretion in response to changes in cellular hydration, bile salt balance, hormone perturbation, oxidative stress, and drug challenges [93]. Accumulating evidence demonstrated that hepatic trafficking of BSEP protein is regulated at the posttranslational level through processes, including glycosylation, phosphorylation, and ubiquitination [109–111]. It is clear now that while increasing canalicular BSEP represents a protective mechanism for bile secretion, reduced expression of this transporter at the apical surface is a common feature of cholestasis [112].

In addition to direct inhibition, cyclosporin A, a known BSEP inhibitor, was reported to induce internalization of this transporter in isolated rat hepatocyte couplets [113]. In several animal models of liver injury, Bsep was only mildly changed at the expression level, but showed marked alterations in its subcellular localization. For instance, in a rat model of 17α-ethinylestradiol-induced cholestasis, functional analysis of canalicular vesicles isolated from the liver indicated a significant decrease in ATP-dependent taurocholate transport, which is associated with endocytic internalization of Bsep, however, without changing its overall expression level [114, 115]. Additionally, induction of cholestasis in rodents by lithocholate, lipopolysaccharide, and hypoxia leads to internalization of Bsep into subapical membrane vesicles and loss of Bsep staining at the canalicular membrane [116]. The mechanisms of BSEP internalization during cholestasis remain to be fully elucidated, although posttranslational protein modifications regulated by multiple signaling molecules, such as PI3K, PKC, mitogen-activated protein kinases (MAPKs), and cytokines, may contribute to the trafficking and localization of BSEP [117, 118]. Regardless of the mechanisms involved, internalization of BSEP plays an important role in cholestasis.

10.4.1.4 BSEP and Cholestasis

Direct inhibition of the efflux activity of BSEP by drugs is known to play a role in DIC, inhibition of BSEP reduces canalicular bile salt secretion and consequently leads to liver injury as a result of acquired cholestasis. Many drugs that cause DILI were reported to inhibit BSEP activity *in vitro* at pharmacologically relevant concentrations (Table 10.2); a correlation between BSEP inhibition and the DIC liability of pharmaceuticals in humans has been demonstrated [119]. The list of chemicals that are BSEP inhibitors is continuously growing, various drugs including cyclosporin A, rifampicin, bosentan, troglitazone, and glibenclamide are competitive inhibitors of BSEP [48], and many were shown to cause clinical cholestatic manifestations.

Table 10.2 A non-exhaustive list of non-cholestatic and cholestatic drugs with or without BSEP inhibition activity.

Effects on cholestasis	Effects on BSEP	Drugs
Non-cholestatic	Non-BSEP inhibitors	5-Fluorouracil, alprenolol, antipyrine, aspirin, caffeine, chlorpheniramine, clopamide, dexamethasone, diphenhydramine, doxorubicin, etoposide, fluorescein, metoclopramide, nadolol, naloxone, phenformin, probenecid, quinidine, tacrine, terbutaline, theophylline, timolol, triamterene, vinblastine
	BSEP inhibitors	Alpidem, benzbromarone, buspirone, clobetasol propionate, finasteride, flupirtine, glafenine, lopinavir, mibefradil, oxybutynin, praziquantel, primaquine, sorafenib, taxol, tolcapone, valinomycin
Cholestatic	Non-BSEP inhibitors	Bezafibrate, carbamazepine, chloramphenicol, chlorpromazine, chlorpropamide, cimetidine, desipramine, D-penicillamine, famotidine, fluoxetine, furosemide, haloperidol, ibuprofen, maprotiline, metformin, nitrofurantoin, nortriptyline, promethazine, quinine, ranitidine, sulfasalazine, sulindac, tamoxifen, tolbutamide, trimethoprim, verapamil
	BSEP inhibitors	Acitretin, clozapine, cyclosporin A, dicloxacillin, erythromycin estolate, fenofibrate, fluvastatin, glyburide, indinavir, indomethacin, nifedipine, nitrendipine, 19-norethindrone, omeprazole, pioglitazone, rifampicin, rifamycin SV, ritonavir, rosiglitazone, simvastatin, ticlopidine, troglitazone

Source: Modified from Kock et al. [108].

Bosentan, for example, was shown to be a competitive BSEP inhibitor with a Ki value of 12 μM in BSEP-expressing Sf9-cell vesicles; this drug also caused dose-dependent and reversible liver injury in 2–18% of patients, with a significant increase of serum bile salt levels [120]. Such liver injury by bosentan is mediated, at least in part, by its inhibition of BSEP and subsequent intracellular accumulation of cytotoxic bile salts in hepatocytes. Recently, Morgan et al. evaluated the potential of BSEP inhibition by more than 200 benchmark compounds using membrane vesicles harvested from BSEP-transfected Sf9 insect cells [105]. Results from this study demonstrated a relatively strong association between the potency of BSEP inhibition and the liability of human liver injury. Notably, BSEP can also be inhibited through the less well-understood indirect mechanisms. For example, although the cholestatic estrogen metabolite, estradiol-17β-glucuronide, did not directly inhibit

ATP-dependent taurocholate transport in Bsep-expressing Sf9 cell vesicles [121], it demonstrated potent cholestatic effects in animal studies [48]. A further study revealed that estradiol-17β-glucuronide-mediated inhibition of Bsep requires the co-expression of Mrp2 by which the compound was secreted into the canalicular lumen before exerting its inhibition of Bsep and cholestatic effects [121]. Both direct and indirect inhibition of BSEP by drugs can contribute significantly to DIC in the clinic. However, it is worth noting that inhibition of BSEP is not always associated with cholestasis. As shown in Table 10.2, several known BSEP inhibitors exhibit no clinical/experimental cholestatic features, presumably due to the compensatory mechanisms of bile acid transport by other transporters.

Investigation on the expression of BSEP in specimens from patients with liver disease after percutaneous transhepatic biliary drainage showed that BSEP mRNA expression was significantly downregulated in patients with poor drainage in comparison to those who were well-drained [122]. The serum bile salt level in poorly drained patients was sixfold higher than that in well-drained patients. In patients with inflammation-induced icteric cholestasis, BSEP mRNA level was decreased by 34% when compared to controls [74]. Decreased BSEP expression was also found in human liver slices after incubation with lipopolysaccharide or in human hepatocytes after treatment with IL-1β [62, 123]. Additionally, BSEP internalization was found in most human cholestatic liver diseases, including obstructive extrahepatic cholestasis, inflammatory cholestasis associated with autoimmune hepatitis, mixed (obstructive plus inflammatory) cholestatic disease, and acute cholestasis induced by drugs [124]. Hence, suppression of transporter expression and internalization may contribute to the cholestatic features of drugs that are not BSEP inhibitors. Collectively, reduced BSEP efflux activity by drugs through direct/indirect inhibition, repression of transporter expression, and internalization from cellular membrane contributes significantly to clinical DIC.

10.4.2 Other Hepatic Bile Acid Efflux Transporters

In addition to BSEP, the predominant driving force for biliary excretion, bile acids can be removed from the hepatocytes by several multidrug resistance-associated proteins (MRPs) as alternative mechanisms. MRPs belong to the C subfamily of ABC membrane transporters. Three members of MRPs including MRP2, MRP3, and MRP4 have been found to transport bile acids in the liver with distinct expression localizations and cellular functions.

10.4.2.1 MRP2

Expressed on the canalicular membrane of hepatocytes, MRP2 (*ABCC2*) has a broad substrate spectrum involving the transport of various conjugated organic anions, including drugs, toxicants, and other xenobiotics. It also mediates ATP-dependent biliary excretion of glucuronide and sulfate conjugated bile acids, bilirubin glucuronides, and glutathione conjugates [35, 36, 125]. As an alternative mechanism, MRP2 extracts bile salts from the hepatocytes, mitigating retention of bile acids, particularly under cholestatic challenges. Inherited complete loss of human MRP2 function results in the Dubin-Johnson syndrome characterized by hyperbilirubinemia and a selective abnormality in the excretion of conjugated anions into the bile canaliculus; however, bile acid excretion into bile is usually normal in these patients [126]. Downregulation of Mrp2 has been reported in a rat model of intrahepatic and obstructive cholestasis, suggesting its involvement in biliary excretion of amphiphilic anionic conjugates [127]. In patients with obstructive cholestasis caused by gallstone blockage of bile ducts, liver MRP2 protein expression decreased to 25% of the non-cholestatic controls, with internalization in the cholestatic livers [128].

10.4.2.2 MRP3 and MRP4

In contrast to MRP2, MRP3 (*ABCC3*) and MRP4 (*ABCC4*) are localized on the basolateral membrane in liver cells and mediate the efflux of bile acids into the systemic circulation. Human MRP3 transports taurocholate and glycocholate with low affinity [37, 129]. Under normal physiological conditions, MRP3 protein expression is similar or slightly lower than BSEP expression in the liver [130]. However, hepatic MRP3 protein expression is highly variable and highly inducible [131]. For example, MRP3 expression is upregulated in patients with obstructive cholestasis when the biliary excretion of organic anions across the canalicular membrane of hepatocytes is impaired [132]. A strong MRP3 protein expression was also observed in the basolateral membrane of hepatocytes from patients with MRP2 deficiency (Dubin-Johnson syndrome) [133]. In comparison to MRP3, hepatic expression of MRP4 is maintained at a relatively low level [131]. MRP4 transports sulfated conjugates of bile acids and steroids with high affinity [38, 39, 134]. In human and rat livers under cholestatic conditions, expression of MRP4 was also increased [135, 136]. Under normal physiological conditions, basolateral efflux of bile salts from the hepatocytes into portal blood is negligible, however, both MRP3 and MRP4 serve as a compensatory system of bile acids efflux under conditions when canalicular excretion is impaired.

10.5 Bidirectional Bile Acid Transporter OSTα/β

Unlike most bile acid uptake and efflux transporters which are unidirectional, organic solute transporter (OST) α/β is a bi-directional heterodimeric transporter expressed at the basolateral membrane of hepatocytes, enterocytes, and renal proximal tubule. The transporter is comprised of two subunits encoded by *SLC51A* and *SLC51B* genes in humans and transports bile acids, sulfate conjugates of steroid hormones, and drugs across the basolateral plasma membranes [40, 137].

OSTα/β-mediated transport is bi-directional and the direction of substrate movement depends on the alteration of electrochemical gradient, which results in either efflux or uptake of substrates [138]. Like NTCP and BSEP, hepatic OSTα/β expression is regulated by FXR in response to bile acid perturbation. Two functional FXR binding motifs in *SLC51A* and one in *SLC51B* have been identified and targeted mutation of these elements reduced FXR-mediated transactivation of OSTα/β [139]. CA-induced Ostα/β expression in the liver, kidney, and ileum in wild-type but not FXR-knockout mice [140]. In human primary hepatocyte cultures, FXR agonist obeticholic acid significantly induced the mRNA expression of OSTα and OSTβ by 6.4- and 42.9-fold, respectively [141]. The expression of OSTβ was also increased by ~10-fold in human liver with obstructive cholestasis caused by gallstone biliary obstruction [142]. In addition to bile acids, OSTα/β expression can be stimulated by hypoxic conditions. Hepatic hypoxic stress is associated with several forms of liver injury and is often accompanied by increased expression of OSTα/β. Mechanistically, putative hypoxia responsive elements have been identified in the promoters of both *SLC51A* and *SLC51B* genes. Activation of and binding to these response elements by hypoxia-inducible factor 1 alpha (HIF-1α) has been observed in luciferase reporter and electrophoretic mobility shift assays, respectively [143]. Thus, transcriptional upregulation of OSTα/β is speculated to serve as an adaptive response to protect the hepatocytes against accumulated bile acids and other stresses.

10.6 Summary

This chapter has systematically reviewed bile acid synthesis and transport in the liver, examined major hepatic bile acid uptake and efflux transporters, and discussed their association with the development of DIC. Initially synthesized in the hepatocytes, the majority of secreted bile acids are eventually recycled back to the liver through a long journey, the enterohepatic circulation. Numerous cellular factors contribute to the maintenance of hepatic bile acid homeostasis, and disruption of this balance often leads to cholestatic liver damages. To date, despite the improved understanding of the mechanisms underlying DIC and the evolved approaches to predicting such liver injury, DIC continues to be a critical concern during drug development and clinical usage. Dysfunction of hepatic bile acid transporters contributes significantly to the development of DIC, involving drug-mediated inhibition and repression of transporter proteins. It is worth noting that although a solid correlation has been established between bile acid biliary secretion and the liability of drug-induced DIC, some cholestatic drugs do not inhibit bile acid transporters, reciprocally not all inhibitors cause cholestasis. Due to the compensatory mechanisms of hepatic bile acid uptake/efflux transporters and the known species-specific bile acid pool, the extrapolation of *in vitro* data to *in vivo* settings and the translation of preclinical findings to humans are complicated and challenging. In this regard, more intensive human studies are warranted in the future to develop physiologically relevant models to predict DIC and improve drug safety.

References

1 M.S. Padda, M. Sanchez, A.J. Akhtar, J.L. Boyer, Drug-induced cholestasis, *Hepatology*, 53 (2011) 1377–1387.

2 B. Garzel, L. Zhang, S.M. Huang, H. Wang, A change in bile flow: looking beyond transporter inhibition in the development of drug-induced cholestasis, *Curr Drug Metab*, 20 (2019) 621–632.

3 K. Yang, K. Kock, A. Sedykh, A. Tropsha, K.L. Brouwer, An updated review on drug-induced cholestasis: mechanisms and investigation of physicochemical properties and pharmacokinetic parameters, *J Pharm Sci*, 102 (2013) 3037–3057.

4 S. Babai, L. Auclert, H. Le-Louet, Safety data and withdrawal of hepatotoxic drugs, *Therapie*, (2018) https://doi.org/10.1016/j.therap.2018.02.004.

5 E. Bjornsson, R. Olsson, Outcome and prognostic markers in severe drug-induced liver disease, *Hepatology*, 42 (2005) 481–489.

6 M.L. Fernandez-Murga, P.D. Petrov, I. Conde, J.V. Castell, M.J. Gomez-Lechon, R. Jover, Advances in drug-induced cholestasis: Clinical perspectives, potential mechanisms and in vitro systems, *Food Chem Toxicol*, 120 (2018) 196–212.

7 M.J. Canet, N.J. Cherrington, Drug disposition alterations in liver disease: extrahepatic effects in cholestasis and nonalcoholic steatohepatitis, *Expert Opin Drug Metab Toxicol*, 10 (2014) 1209–1219.

8 A. Wilson, A. Almousa, W.A. Teft, R.B. Kim, Attenuation of bile acid-mediated FXR and PXR activation in patients with Crohn's disease, *Sci Rep*, 10 (2020) 1866.

9 S. Han, T. Li, E. Ellis, S. Strom, J.Y. Chiang, A novel bile acid-activated vitamin D receptor signaling in human hepatocytes, *Mol Endocrinol*, 24 (2010) 1151–1164.

10 B.L. Zwicker, L.B. Agellon, Transport and biological activities of bile acids, *Int J Biochem Cell Biol*, 45 (2013) 1389–1398.

11 B. Staels, V.A. Fonseca, Bile acids and metabolic regulation: mechanisms and clinical responses to bile acid sequestration, *Diabetes Care*, 32 Suppl 2 (2009) S237–S245.

12 H.S. Schadt, A. Wolf, F. Pognan, S.D. Chibout, M. Merz, G.A. Kullak-Ublick, Bile acids in drug induced liver injury: key players and surrogate markers, *Clin Res Hepatol Gastroenterol*, 40 (2016) 257–266.

13 J.Y. Chiang, Bile acids: regulation of synthesis, *J Lipid Res*, 50 (2009) 1955–1966.

14 A.F. Hofmann, L.R. Hagey, Bile acids: chemistry, pathochemistry, biology, pathobiology, and therapeutics, *Cell Mol Life Sci*, 65 (2008) 2461–2483.

15 M. Trauner, J.L. Boyer, Bile salt transporters: molecular characterization, function, and regulation, *Physiol Rev*, 83 (2003) 633–671.

16 E. Sticova, M. Jirsa, J. Pawlowska, New insights in genetic cholestasis: from molecular mechanisms to clinical implications, *Can J Gastroenterol Hepatol*, 2018 (2018) 2313675.

17 A.F. Hofmann, The continuing importance of bile acids in liver and intestinal disease, *Arch Intern Med*, 159 (1999) 2647–2658.

18 A.F. Attili, M. Angelico, A. Cantafora, D. Alvaro, L. Capocaccia, Bile acid-induced liver toxicity: relation to the hydrophobic-hydrophilic balance of bile acids, *Med Hypotheses*, 19 (1986) 57–69.

19 B. Yerushalmi, R. Dahl, M.W. Devereaux, E. Gumpricht, R.J. Sokol, Bile acid-induced rat hepatocyte apoptosis is inhibited by antioxidants and blockers of the mitochondrial permeability transition, *Hepatology*, 33 (2001) 616–626.

20 H. Yu, X. Zhang, R. Liu, H. Li, X. Xiao, Y. Zhou, C. Wei, M. Yang, M. Liao, J. Zhao, Z. Xia, Q. Liao, Mcl-1 suppresses abasic site repair following bile acid-induced hepatic cellular DNA damage, *Tumour Biol*, 39 (2017) 1010428317712102.

21 B.L. Woolbright, K. Dorko, D.J. Antoine, J.I. Clarke, P. Gholami, F. Li, S.C. Kumer, T.M. Schmitt, J. Forster, F. Fan, R.E. Jenkins, B.K. Park, B. Hagenbuch, M. Olyaee, H. Jaeschke, Bile acid-induced necrosis in primary human hepatocytes and in patients with obstructive cholestasis, *Toxicol Appl Pharmacol*, 283 (2015) 168–177.

22 V. Sundaram, E.S. Bjornsson, Drug-induced cholestasis, *Hepatol Commun*, 1 (2017) 726–735.

23 K. Tajiri, Y. Shimizu, Practical guidelines for diagnosis and early management of drug-induced liver injury, *World J Gastroenterol*, 14 (2008) 6774–6785.

24 D.E. Kleiner, The pathology of drug-induced liver injury, *Semin Liver Dis*, 29 (2009) 364–372.

25 C. Degott, G. Feldmann, D. Larrey, A.M. Durand-Schneider, D. Grange, J.P. Machayekhi, A. Moreau, F. Potet, J.P. Benhamou, Drug-induced prolonged cholestasis in adults: a histological semiquantitative study demonstrating progressive ductopenia, *Hepatology*, 15 (1992) 244–251.

26 S. Mita, H. Suzuki, H. Akita, H. Hayashi, R. Onuki, A.F. Hofmann, Y. Sugiyama, Vectorial transport of unconjugated and conjugated bile salts by monolayers of LLC-PK1 cells doubly transfected with human NTCP and BSEP or with rat Ntcp and Bsep, *Am J Physiol Gastrointest Liver Physiol*, 290 (2006) G550–G556.

27 S. Mita, H. Suzuki, H. Akita, H. Hayashi, R. Onuki, A.F. Hofmann, Y. Sugiyama, Inhibition of bile acid transport across Na+/taurocholate cotransporting polypeptide (SLC10A1) and bile salt export pump (ABCB 11)-coexpressing LLC-PK1 cells by cholestasis-inducing drugs, *Drug Metab Dispos*, 34 (2006) 1575–1581.

28 A.L. Craddock, M.W. Love, R.W. Daniel, L.C. Kirby, H.C. Walters, M.H. Wong, P.A. Dawson, Expression and transport properties of the human ileal and renal sodium-dependent bile acid transporter, *Am J Phys*, 274 (1998) G157–G169.

29 K. Maeda, M. Kambara, Y. Tian, A.F. Hofmann, Y. Sugiyama, Uptake of ursodeoxycholate and its conjugates by human hepatocytes: role of Na(+)-taurocholate cotransporting polypeptide (NTCP), organic anion transporting polypeptide (OATP) 1B1 (OATP-C), and oatp1B3 (OATP8), *Mol Pharm*, 3 (2006) 70–77.

30 T. Suga, H. Yamaguchi, T. Sato, M. Maekawa, J. Goto, N. Mano, Preference of conjugated bile acids over unconjugated bile acids as substrates for OATP1B1 and OATP1B3, *PLoS One*, 12 (2017) e0169719.

31 H. Hayashi, T. Takada, H. Suzuki, R. Onuki, A.F. Hofmann, Y. Sugiyama, Transport by vesicles of glycine- and taurine-conjugated bile salts and taurolithocholate 3-sulfate: a comparison of human BSEP with rat Bsep, *Biochim Biophys Acta*, 1738 (2005) 54–62.

32 J. Noe, B. Stieger, P.J. Meier, Functional expression of the canalicular bile salt export pump of human liver, *Gastroenterology*, 123 (2002) 1659–1666.

33 E. Kis, E. Ioja, T. Nagy, L. Szente, K. Heredi-Szabo, P. Krajcsi, Effect of membrane cholesterol on BSEP/Bsep activity: species specificity studies for substrates and inhibitors, *Drug Metab Dispos*, 37 (2009) 1878–1886.

34 K. Yamaguchi, T. Murai, H. Yabuuchi, T. Kurosawa, Measurement of the transport activities of bile salt export pump using LC-MS, *Anal Sci*, 25 (2009) 1155–1158.

35 H. Akita, H. Suzuki, K. Ito, S. Kinoshita, N. Sato, H. Takikawa, Y. Sugiyama, Characterization of bile acid transport mediated by multidrug resistance associated protein 2 and bile salt export pump, *Biochim Biophys Acta*, 1511 (2001) 7–16.

36 G.A. Kullak-Ublick, B. Stieger, B. Hagenbuch, P.J. Meier, Hepatic transport of bile salts, *Semin Liver Dis*, 20 (2000) 273–292.

37 T. Hirohashi, H. Suzuki, H. Takikawa, Y. Sugiyama, ATP-dependent transport of bile salts by rat multidrug resistance-associated protein 3 (Mrp3), *J Biol Chem*, 275 (2000) 2905–2910.

38 M. Rius, J. Hummel-Eisenbeiss, A.F. Hofmann, D. Keppler, Substrate specificity of human ABCC4 (MRP4)-mediated cotransport of bile acids and reduced glutathione, *Am J Physiol Gastrointest Liver Physiol*, 290 (2006) G640–649.

39 N. Zelcer, G. Reid, P. Wielinga, A. Kuil, I. van der Heijden, J.D. Schuetz, P. Borst, Steroid and bile acid conjugates are substrates of human multidrug-resistance protein (MRP) 4 (ATP-binding cassette C4), *Biochem J*, 371 (2003) 361–367.

40 N. Ballatori, W.V. Christian, J.Y. Lee, P.A. Dawson, C.J. Soroka, J.L. Boyer, M.S. Madejczyk, N. Li, OSTalpha-OSTbeta: a major basolateral bile acid and steroid transporter in human intestinal, renal, and biliary epithelia, *Hepatology*, 42 (2005) 1270–1279.

41 B. Hagenbuch, B. Stieger, M. Foguet, H. Lubbert, P.J. Meier, Functional expression cloning and characterization of the hepatocyte Na+/bile acid cotransport system, *Proc Natl Acad Sci U S A*, 88 (1991) 10629–10633.

42 P.J. Meier, U. Eckhardt, A. Schroeder, B. Hagenbuch, B. Stieger, Substrate specificity of sinusoidal bile acid and organic anion uptake systems in rat and human liver, *Hepatology*, 26 (1997) 1667–1677.

43 B. Stieger, B. Hagenbuch, L. Landmann, M. Hochli, A. Schroeder, P.J. Meier, In situ localization of the hepatocytic Na+/Taurocholate cotransporting polypeptide in rat liver, *Gastroenterology*, 107 (1994) 1781–1787.

44 G.A. Kullak-Ublick, J. Glasa, C. Boker, M. Oswald, U. Grutzner, B. Hagenbuch, B. Stieger, P.J. Meier, U. Beuers, W. Kramer, G. Wess, G. Paumgartner, Chlorambucil-taurocholate is transported by bile acid carriers expressed in human hepatocellular carcinomas, *Gastroenterology*, 113 (1997) 1295–1305.

45 B. Hagenbuch, Molecular properties of hepatic uptake systems for bile acids and organic anions, *J Membr Biol*, 160 (1997) 1–8.

46 J.L. Boyer, J. Graf, P.J. Meier, Hepatic transport systems regulating pHi, cell volume, and bile secretion, *Annu Rev Physiol*, 54 (1992) 415–438.

47 S. Hata, P. Wang, N. Eftychiou, M. Ananthanarayanan, A. Batta, G. Salen, K.S. Pang, A.W. Wolkoff, Substrate specificities of rat oatp1 and ntcp: implications for hepatic organic anion uptake, *Am J Physiol Gastrointest Liver Physiol*, 285 (2003) G829–G839.

48 B. Stieger, The role of the sodium-taurocholate cotransporting polypeptide (NTCP) and of the bile salt export pump (BSEP) in physiology and pathophysiology of bile formation, *Handb Exp Pharmacol*, (2011) 205–259.

49 A. Dietmaier, R. Gasser, J. Graf, M. Peterlik, Investigations on the sodium dependence of bile acid fluxes in the isolated perfused rat liver, *Biochim Biophys Acta*, 443 (1976) 81–91.

50 W.F. Kuhn, D.A. Gewirtz, Stimulation of taurocholate and glycocholate efflux from the rat hepatocyte by arginine vasopressin, *Am J Phys*, 254 (1988) G732–G740.

51 S. Mita, H. Suzuki, H. Akita, B. Stieger, P.J. Meier, A.F. Hofmann, Y. Sugiyama, Vectorial transport of bile salts across MDCK cells expressing both rat Na+-taurocholate cotransporting polypeptide and rat bile salt export pump, *Am J Physiol Gastrointest Liver Physiol*, 288 (2005) G159–G167.

52 Y. Shitara, A.P. Li, Y. Kato, C. Lu, K. Ito, T. Itoh, Y. Sugiyama, Function of uptake transporters for taurocholate and estradiol 17beta-D-glucuronide in cryopreserved human hepatocytes, *Drug Metab Pharmacokinet*, 18 (2003) 33–41.

53 R.H. Ho, R.G. Tirona, B.F. Leake, H. Glaeser, W. Lee, C.J. Lemke, Y. Wang, R.B. Kim, Drug and bile acid transporters in rosuvastatin hepatic uptake: function, expression, and pharmacogenetics, *Gastroenterology*, 130 (2006) 1793–1806.

54 H. Fujino, T. Saito, S. Ogawa, J. Kojima, Transporter-mediated influx and efflux mechanisms of pitavastatin, a new inhibitor of HMG-CoA reductase, *J Pharm Pharmacol*, 57 (2005) 1305–1311.

55 R. Greupink, L. Dillen, M. Monshouwer, M.T. Huisman, F.G. Russel, Interaction of fluvastatin with the liver-specific Na+ -dependent taurocholate cotransporting polypeptide (NTCP), *Eur J Pharm Sci*, 44 (2011) 487–496.

56 H. Yan, G. Zhong, G. Xu, W. He, Z. Jing, Z. Gao, Y. Huang, Y. Qi, B. Peng, H. Wang, L. Fu, M. Song, P. Chen, W. Gao, B. Ren, Y. Sun, T. Cai, X. Feng, J. Sui, W. Li, Sodium taurocholate cotransporting polypeptide is a functional receptor for human hepatitis B and D virus, *eLife*, (2012) 1:e00049.

57 Z. Dong, S. Ekins, J.E. Polli, A substrate pharmacophore for the human sodium taurocholate co-transporting polypeptide, *Int J Pharm*, 478 (2015) 88–95.

58 S.J. Karpen, A.Q. Sun, B. Kudish, B. Hagenbuch, P.J. Meier, M. Ananthanarayanan, F.J. Suchy, Multiple factors regulate the rat liver basolateral sodium-dependent bile acid cotransporter gene promoter, *J Biol Chem*, 271 (1996) 15211–15221.

59 D. Jung, B. Hagenbuch, M. Fried, P.J. Meier, G.A. Kullak-Ublick, Role of liver-enriched transcription factors and nuclear receptors in regulating the human, mouse, and rat NTCP gene, *Am J Physiol Gastrointest Liver Physiol*, 286 (2004) G752–G761.

60 G. Zollner, M. Wagner, P. Fickert, A. Geier, A. Fuchsbichler, D. Silbert, J. Gumhold, K. Zatloukal, A. Kaser, H. Tilg, H. Denk, M. Trauner, Role of nuclear receptors and hepatocyte-enriched transcription factors for Ntcp repression in biliary obstruction in mouse liver, *Am J Physiol Gastrointest Liver Physiol*, 289 (2005) G798–G805.

61 D. Li, T.L. Zimmerman, S. Thevananther, H.Y. Lee, J.M. Kurie, S.J. Karpen, Interleukin-1 beta-mediated suppression of RXR:RAR transactivation of the Ntcp promoter is JNK-dependent, *J Biol Chem*, 277 (2002) 31416–31422.

62 M. Le Vee, P. Gripon, B. Stieger, O. Fardel, Down-regulation of organic anion transporter expression in human hepatocytes exposed to the proinflammatory cytokine interleukin 1beta, *Drug Metab Dispos*, 36 (2008) 217–222.

63 A. Geier, M. Wagner, C.G. Dietrich, M. Trauner, Principles of hepatic organic anion transporter regulation during cholestasis, inflammation and liver regeneration, *Biochim Biophys Acta*, 1773 (2007) 283–308.

64 B. Bouscarel, S.D. Kroll, H. Fromm, Signal transduction and hepatocellular bile acid transport: cross talk between bile acids and second messengers, *Gastroenterology*, 117 (1999) 433–452.

65 A. Kosters, S.J. Karpen, Bile acid transporters in health and disease, *Xenobiotica*, 38 (2008) 1043–1071.

66 M.S. Anwer, B. Stieger, Sodium-dependent bile salt transporters of the SLC10A transporter family: more than solute transporters, *Pflugers Arch*, 466 (2014) 77–89.

67 S. Grune, L.R. Engelking, M.S. Anwer, Role of intracellular calcium and protein kinases in the activation of hepatic Na+/taurocholate cotransport by cyclic AMP, *J Biol Chem*, 268 (1993) 17734–17741.

68 S. Mukhopadhayay, M. Ananthanarayanan, B. Stieger, P.J. Meier, F.J. Suchy, M.S. Anwer, cAMP increases liver Na+-taurocholate cotransport by translocating transporter to plasma membranes, *Am J Phys*, 273 (1997) G842–G848.

69 S. Mukhopadhyay, M. Ananthanarayanan, B. Stieger, P.J. Meier, F.J. Suchy, M.S. Anwer, Sodium taurocholate cotransporting polypeptide is a serine, threonine phosphoprotein and is dephosphorylated by cyclic adenosine monophosphate, *Hepatology*, 28 (1998) 1629–1636.

70 M.S. Anwer, H. Gillin, S. Mukhopadhyay, N. Balasubramaniyan, F.J. Suchy, M. Ananthanarayanan, Dephosphorylation of Ser-226 facilitates plasma membrane retention of Ntcp, *J Biol Chem*, 280 (2005) 33687–33692.

71 S. Muhlfeld, O. Domanova, T. Berlage, C. Stross, A. Helmer, V. Keitel, D. Haussinger, R. Kubitz, Short-term feedback regulation of bile salt uptake by bile salts in rodent liver, *Hepatology*, 56 (2012) 2387–2397.

72 H.L. Chen, Y.J. Liu, H.L. Chen, S.H. Wu, Y.H. Ni, M.C. Ho, H.S. Lai, W.M. Hsu, H.Y. Hsu, H.C. Tseng, Y.M. Jeng, M.H. Chang, Expression of hepatocyte transporters and nuclear receptors in children with early and late-stage biliary atresia, *Pediatr Res*, 63 (2008) 667–673.

73 B.L. Shneider, V.L. Fox, K.B. Schwarz, C.L. Watson, M. Ananthanarayanan, S. Thevananther, D.M. Christie, W. Hardikar, K.D. Setchell, G. Mieli-Vergani, F.J. Suchy, A.P. Mowat, Hepatic basolateral sodium-dependent-bile acid transporter expression in two unusual cases of hypercholanemia and in extrahepatic biliary atresia, *Hepatology*, 25 (1997) 1176–1183.

74 G. Zollner, P. Fickert, R. Zenz, A. Fuchsbichler, C. Stumptner, L. Kenner, P. Ferenci, R.E. Stauber, G.J. Krejs, H. Denk, K. Zatloukal, M. Trauner, Hepatobiliary transporter expression in percutaneous liver biopsies of patients with cholestatic liver diseases, *Hepatology*, 33 (2001) 633–646.

75 V. Keitel, M. Burdelski, U. Warskulat, T. Kuhlkamp, D. Keppler, D. Haussinger, R. Kubitz, Expression and localization of hepatobiliary transport proteins in progressive familial intrahepatic cholestasis, *Hepatology*, 41 (2005) 1160–1172.

76 Z. Dong, S. Ekins, J.E. Polli, Structure-activity relationship for FDA approved drugs as inhibitors of the human sodium taurocholate cotransporting polypeptide (NTCP), *Mol Pharm*, 10 (2013) 1008–1019.

77 R.B. Kim, B. Leake, M. Cvetkovic, M.M. Roden, J. Nadeau, A. Walubo, G.R. Wilkinson, Modulation by drugs of human hepatic sodium-dependent bile acid transporter (sodium taurocholate cotransporting polypeptide) activity, *J Pharmacol Exp Ther*, 291 (1999) 1204–1209.

78 I.M. Arias, Cyclosporin, the biology of the bile canaliculus, and cholestasis, *Gastroenterology*, 104 (1993) 1558–1560.

79 Z. Dong, S. Ekins, J.E. Polli, Quantitative NTCP pharmacophore and lack of association between DILI and NTCP Inhibition, *Eur J Pharm Sci*, 66 (2015) 1–9.

80 G.A. Kullak-Ublick, B. Hagenbuch, B. Stieger, C.D. Schteingart, A.F. Hofmann, A.W. Wolkoff, P.J. Meier, Molecular and functional characterization of an organic anion transporting polypeptide cloned from human liver, *Gastroenterology*, 109 (1995) 1274–1282.

81 G.A. Kullak-Ublick, M.G. Ismair, B. Stieger, L. Landmann, R. Huber, F. Pizzagalli, K. Fattinger, P.J. Meier, B. Hagenbuch, Organic anion-transporting polypeptide B (OATP-B) and its functional comparison with three other OATPs of human liver, *Gastroenterology*, 120 (2001) 525–533.

82 J. Konig, Y. Cui, A.T. Nies, D. Keppler, Localization and genomic organization of a new hepatocellular organic anion transporting polypeptide, *J Biol Chem*, 275 (2000) 23161–23168.

83 E. Sticova, A. Lodererova, E. van de Steeg, S. Frankova, M. Kollar, V. Lanska, R. Kotalova, T. Dedic, A.H. Schinkel, M. Jirsa, Down-regulation of OATP1B proteins correlates with hyperbilirubinemia in advanced cholestasis, *Int J Clin Exp Pathol*, 8 (2015) 5252–5262.

84 B. Stieger, B. O'Neill, P.J. Meier, ATP-dependent bile-salt transport in canalicular rat liver plasma-membrane vesicles, *Biochem J*, 284 (Pt 1) (1992) 67–74.

85 S. Childs, R.L. Yeh, E. Georges, V. Ling, Identification of a sister gene to P-glycoprotein, *Cancer Res*, 55 (1995) 2029–2034.

86 T. Langmann, R. Mauerer, A. Zahn, C. Moehle, M. Probst, W. Stremmel, G. Schmitz, Real-time reverse transcription-PCR expression profiling of the complete human ATP-binding cassette transporter superfamily in various tissues, *Clin Chem*, 49 (2003) 230–238.

87 C. Hilgendorf, G. Ahlin, A. Seithel, P. Artursson, A.L. Ungell, J. Karlsson, Expression of thirty-six drug transporter genes in human intestine, liver, kidney, and organotypic cell lines, *Drug Metab Dispos*, 35 (2007) 1333–1340.

88 R.M. Green, F. Hoda, K.L. Ward, Molecular cloning and characterization of the murine bile salt export pump, *Gene*, 241 (2000) 117–123.

89 B. Stieger, Recent insights into the function and regulation of the bile salt export pump (ABCB11), *Curr Opin Lipidol*, 20 (2009) 176–181.

90 V. Lecureur, D. Sun, P. Hargrove, E.G. Schuetz, R.B. Kim, L.B. Lan, J.D. Schuetz, Cloning and expression of murine sister of P-glycoprotein reveals a more discriminating transporter than MDR1/P-glycoprotein, *Mol Pharmacol*, 57 (2000) 24–35.

91 M. Hirano, K. Maeda, H. Hayashi, H. Kusuhara, Y. Sugiyama, Bile salt export pump (BSEP/ABCB11) can transport a nonbile acid substrate, pravastatin, *J Pharmacol Exp Ther*, 314 (2005) 876–882.

92 M. Ananthanarayanan, N. Balasubramanian, M. Makishima, D.J. Mangelsdorf, F.J. Suchy, Human bile salt export pump promoter is transactivated by the farnesoid X receptor/bile acid receptor, *J Biol Chem*, 276 (2001) 28857–28865.

93 R. Kubitz, C. Droge, J. Stindt, K. Weissenberger, D. Haussinger, The bile salt export pump (BSEP) in health and disease, *Clin Res Hepatol Gastroenterol*, 36 (2012) 536–553.

94 J.J. Eloranta, G.A. Kullak-Ublick, Coordinate transcriptional regulation of bile acid homeostasis and drug metabolism, *Arch Biochem Biophys*, 433 (2005) 397–412.

95 T. Gerloff, A. Geier, I. Roots, P.J. Meier, C. Gartung, Functional analysis of the rat bile salt export pump gene promoter, *Eur J Biochem*, 269 (2002) 3495–3503.

96 J.L. Lew, A. Zhao, J. Yu, L. Huang, N. De Pedro, F. Pelaez, S.D. Wright, J. Cui, The farnesoid X receptor controls gene expression in a ligand- and promoter-selective fashion, *J Biol Chem*, 279 (2004) 8856–8861.

97 J. Yu, J.L. Lo, L. Huang, A. Zhao, E. Metzger, A. Adams, P.T. Meinke, S.D. Wright, J. Cui, Lithocholic acid decreases expression of bile salt export pump through farnesoid X receptor antagonist activity, *J Biol Chem*, 277 (2002) 31441–31447.

98 C.J. Sinal, M. Tohkin, M. Miyata, J.M. Ward, G. Lambert, F.J. Gonzalez, Targeted disruption of the nuclear receptor FXR/BAR impairs bile acid and lipid homeostasis, *Cell*, 102 (2000) 731–744.

99 X. Song, R. Kaimal, B. Yan, R. Deng, Liver receptor homolog 1 transcriptionally regulates human bile salt export pump expression, *J Lipid Res*, 49 (2008) 973–984.

100 C. Mataki, B.C. Magnier, S.M. Houten, J.S. Annicotte, C. Argmann, C. Thomas, H. Overmars, W. Kulik, D. Metzger, J. Auwerx, K. Schoonjans, Compromised intestinal lipid absorption in mice with a liver-specific deficiency of liver receptor homolog 1, *Mol Cell Biol*, 27 (2007) 8330–8339.

101 J. Weerachayaphorn, S.Y. Cai, C.J. Soroka, J.L. Boyer, Nuclear factor erythroid 2-related factor 2 is a positive regulator of human bile salt export pump expression, *Hepatology*, 50 (2009) 1588–1596.

102 K.P. Tan, M. Yang, S. Ito, Activation of nuclear factor (erythroid-2 like) factor 2 by toxic bile acids provokes adaptive defense responses to enhance cell survival at the emergence of oxidative stress, *Mol Pharmacol*, 72 (2007) 1380–1390.

103 X. Song, A. Vasilenko, Y. Chen, L. Valanejad, R. Verma, B. Yan, R. Deng, Transcriptional dynamics of bile salt export pump during pregnancy: mechanisms and implications in intrahepatic cholestasis of pregnancy, *Hepatology*, 60 (2014) 1993–2007.

104 Y. Chen, A. Vasilenko, X. Song, L. Valanejad, R. Verma, S. You, B. Yan, S. Shiffka, L. Hargreaves, C. Nadolny, R. Deng, Estrogen and estrogen receptor-alpha-mediated transrepression of bile salt export pump, *Mol Endocrinol*, 29 (2015) 613–626.

105 R.E. Morgan, M. Trauner, C.J. van Staden, P.H. Lee, B. Ramachandran, M. Eschenberg, C.A. Afshari, C.W. Qualls, Jr., R. Lightfoot-Dunn, H.K. Hamadeh, Interference with bile salt export pump function is a susceptibility factor for human liver injury in drug development, *Toxicol Sci*, 118 (2010) 485–500.

106 B. Garzel, H. Yang, L. Zhang, S.M. Huang, J.E. Polli, H. Wang, The role of bile salt export pump gene repression in drug-induced cholestatic liver toxicity, *Drug Metab Dispos*, 42 (2014) 318–322.

107 B. Garzel, T. Hu, L. Li, Y. Lu, S. Heyward, J. Polli, L. Zhang, S.M. Huang, J.P. Raufman, H. Wang, Metformin disrupts bile acid efflux by repressing bile salt export pump expression, *Pharm Res*, 37 (2020) 26.

108 K. Kock, B.C. Ferslew, I. Netterberg, K. Yang, T.J. Urban, P.W. Swaan, P.W. Stewart, K.L. Brouwer, Risk factors for development of cholestatic drug-induced liver injury: inhibition of hepatic basolateral bile acid transporters multidrug resistance-associated proteins 3 and 4, *Drug Metab Dispos*, 42 (2014) 665–674.

109 K. Mochizuki, T. Kagawa, A. Numari, M.J. Harris, J. Itoh, N. Watanabe, T. Mine, I.M. Arias, Two N-linked glycans are required to maintain the transport activity of the bile salt export pump (ABCB11) in MDCK II cells, *Am J Physiol Gastrointest Liver Physiol*, 292 (2007) G818–828.

110 R. Kubitz, G. Sutfels, T. Kuhlkamp, R. Kolling, D. Haussinger, Trafficking of the bile salt export pump from the Golgi to the canalicular membrane is regulated by the p38 MAP kinase, *Gastroenterology*, 126 (2004) 541–553.

111 L. Wang, H. Dong, C.J. Soroka, N. Wei, J.L. Boyer, M. Hochstrasser, Degradation of the bile salt export pump at endoplasmic reticulum in progressive familial intrahepatic cholestasis type II, *Hepatology*, 48 (2008) 1558–1569.

112 M.S. Anwer, Cellular regulation of hepatic bile acid transport in health and cholestasis, *Hepatology*, 39 (2004) 581–590.

113 I.D. Roman, M.D. Fernandez-Moreno, J.A. Fueyo, M.G. Roma, R. Coleman, Cyclosporin A induced internalization of the bile salt export pump in isolated rat hepatocyte couplets, *Toxicol Sci*, 71 (2003) 276–281.

114 R. Bossard, B. Stieger, B. O'Neill, G. Fricker, P.J. Meier, Ethinylestradiol treatment induces multiple canalicular membrane transport alterations in rat liver, *J Clin Invest*, 91 (1993) 2714–2720.

115 J.M. Lee, M. Trauner, C.J. Soroka, B. Stieger, P.J. Meier, J.L. Boyer, Expression of the bile salt export pump is maintained after chronic cholestasis in the rat, *Gastroenterology*, 118 (2000) 163–172.

116 C.J. Soroka, J.L. Boyer, Biosynthesis and trafficking of the bile salt export pump, BSEP: therapeutic implications of BSEP mutations, *Mol Aspects Med*, 37 (2014) 3–14.

117 A.C. Boaglio, A.E. Zucchetti, F.D. Toledo, I.R. Barosso, E.J. Sanchez Pozzi, F.A. Crocenzi, M.G. Roma, ERK1/2 and p38 MAPKs are complementarily involved in estradiol 17ss-D-glucuronide-induced cholestasis: crosstalk with cPKC and PI3K, *PLoS One*, 7 (2012) e49255.

118 A.C. Boaglio, A.E. Zucchetti, E.J. Sanchez Pozzi, J.M. Pellegrino, J.E. Ochoa, A.D. Mottino, M. Vore, F.A. Crocenzi, M.G. Roma, Phosphoinositide 3-kinase/ protein kinase B signaling pathway is involved in estradiol 17beta-D-glucuronide-induced cholestasis: complementarity with classical protein kinase C, *Hepatology*, 52 (2010) 1465–1476.

119 S. Dawson, S. Stahl, N. Paul, J. Barber, J.G. Kenna, In vitro inhibition of the bile salt export pump correlates with risk of cholestatic drug-induced liver injury in humans, *Drug Metab Dispos*, 40 (2012) 130–138.

120 K. Fattinger, C. Funk, M. Pantze, C. Weber, J. Reichen, B. Stieger, P.J. Meier, The endothelin antagonist bosentan inhibits the canalicular bile salt export pump: a potential mechanism for hepatic adverse reactions, *Clin Pharmacol Ther*, 69 (2001) 223–231.

121 B. Stieger, K. Fattinger, J. Madon, G.A. Kullak-Ublick, P.J. Meier, Drug- and estrogen-induced cholestasis through inhibition of the hepatocellular bile salt export pump (Bsep) of rat liver, *Gastroenterology*, 118 (2000) 422–430.

122 J. Shoda, M. Kano, K. Oda, J. Kamiya, Y. Nimura, H. Suzuki, Y. Sugiyama, H. Miyazaki, T. Todoroki, S. Stengelin, W. Kramer, Y. Matsuzaki, N. Tanaka, The expression levels of plasma membrane transporters in the cholestatic liver of patients undergoing biliary drainage and their association with the impairment of biliary secretory function, *Am J Gastroenterol*, 96 (2001) 3368–3378.

123 M.G. Elferink, P. Olinga, A.L. Draaisma, M.T. Merema, K.N. Faber, M.J. Slooff, D.K. Meijer, G.M. Groothuis, LPS-induced downregulation of MRP2 and BSEP in human liver is due to a posttranscriptional process, *Am J Physiol Gastrointest Liver Physiol*, 287 (2004) G1008–1016.

124 M.G. Roma, F.A. Crocenzi, A.D. Mottino, Dynamic localization of hepatocellular transporters in health and disease, *World J Gastroenterol*, 14 (2008) 6786–6801.

125 K. Jemnitz, K. Heredi-Szabo, J. Janossy, E. Ioja, L. Vereczkey, P. Krajcsi, ABCC2/ Abcc2: a multispecific transporter with dominant excretory functions, *Drug Metab Rev*, 42 (2010) 402–436.

126 T. Iyanagi, Y. Emi, S. Ikushiro, Biochemical and molecular aspects of genetic disorders of bilirubin metabolism, *Biochim Biophys Acta*, 1407 (1998) 173–184.

127 M. Trauner, M. Arrese, C.J. Soroka, M. Ananthanarayanan, T.A. Koeppel, S.F. Schlosser, F.J. Suchy, D. Keppler, J.L. Boyer, The rat canalicular conjugate export pump (Mrp2) is down-regulated in intrahepatic and obstructive cholestasis, *Gastroenterology*, 113 (1997) 255–264.

128 J. Chai, S.Y. Cai, X. Liu, W. Lian, S. Chen, L. Zhang, X. Feng, Y. Cheng, X. He, Y. He, L. Chen, R. Wang, H. Wang, J.L. Boyer, W. Chen, Canalicular membrane MRP2/ABCC2 internalization is determined by Ezrin Thr567 phosphorylation in human obstructive cholestasis, *J Hepatol*, 63 (2015) 1440–1448.

129 N. Zelcer, T. Saeki, I. Bot, A. Kuil, P. Borst, Transport of bile acids in multidrug-resistance-protein 3-overexpressing cells co-transfected with the ileal Na+-dependent bile-acid transporter, *Biochem J*, 369 (2003) 23–30.

130 H.J. Burt, A.E. Riedmaier, M.D. Harwood, H.K. Crewe, K.L. Gill, S. Neuhoff, Abundance of hepatic transporters in caucasians: a meta-analysis, *Drug Metab Dispos*, 44 (2016) 1550–1561.

131 T. Lang, M. Hitzl, O. Burk, E. Mornhinweg, A. Keil, R. Kerb, K. Klein, U.M. Zanger, M. Eichelbaum, M.F. Fromm, Genetic polymorphisms in the multidrug resistance-associated protein 3 (ABCC3, MRP3) gene and relationship to its mRNA and protein expression in human liver, *Pharmacogenetics*, 14 (2004) 155–164.

132 J. Chai, Y. He, S.Y. Cai, Z. Jiang, H. Wang, Q. Li, L. Chen, Z. Peng, X. He, X. Wu, T. Xiao, R. Wang, J.L. Boyer, W. Chen, Elevated hepatic multidrug resistance-associated protein 3/ATP-binding cassette subfamily C 3 expression in human obstructive cholestasis is mediated through tumor necrosis factor alpha and c-Jun NH2-terminal kinase/stress-activated protein kinase-signaling pathway, *Hepatology*, 55 (2012) 1485–1494.

133 J. Konig, D. Rost, Y. Cui, D. Keppler, Characterization of the human multidrug resistance protein isoform MRP3 localized to the basolateral hepatocyte membrane, *Hepatology*, 29 (1999) 1156–1163.

134 D. Keppler, Multidrug resistance proteins (MRPs, ABCCs): importance for pathophysiology and drug therapy, *Handb Exp Pharmacol*, (2011) 299–323.

135 G.U. Denk, C.J. Soroka, Y. Takeyama, W.S. Chen, J.D. Schuetz, J.L. Boyer, Multidrug resistance-associated protein 4 is up-regulated in liver but down-regulated in kidney in obstructive cholestasis in the rat, *J Hepatol*, 40 (2004) 585–591.

136 U. Gradhand, T. Lang, E. Schaeffeler, H. Glaeser, H. Tegude, K. Klein, P. Fritz, G. Jedlitschky, H.K. Kroemer, I. Bachmakov, B. Anwald, R. Kerb, U.M. Zanger, M. Eichelbaum, M. Schwab, M.F. Fromm, Variability in human hepatic MRP4 expression: influence of cholestasis and genotype, *Pharm J*, 8 (2008) 42–52.

137 T. Suga, H. Yamaguchi, J. Ogura, N. Mano, Characterization of conjugated and unconjugated bile acid transport via human organic solute transporter alpha/beta, *Biochim Biophys Acta Biomembr*, 1861 (2019) 1023–1029.

138 K. Kock, K.L. Brouwer, A perspective on efflux transport proteins in the liver, *Clin Pharmacol Ther*, 92 (2012) 599–612.

139 J.F. Landrier, J.J. Eloranta, S.R. Vavricka, G.A. Kullak-Ublick, The nuclear receptor for bile acids, FXR, transactivates human organic solute transporter-alpha and -beta genes, *Am J Physiol Gastrointest Liver Physiol*, 290 (2006) G476–G485.

140 G. Zollner, M. Wagner, T. Moustafa, P. Fickert, D. Silbert, J. Gumhold, A. Fuchsbichler, E. Halilbasic, H. Denk, H.U. Marschall, M. Trauner,

Coordinated induction of bile acid detoxification and alternative elimination in mice: role of FXR-regulated organic solute transporter-alpha/beta in the adaptive response to bile acids, *Am J Physiol Gastrointest Liver Physiol*, 290 (2006) G923–G932.

141 Y. Zhang, J.P. Jackson, R.L. St Claire, 3rd, K. Freeman, K.R. Brouwer, J.E. Edwards, Obeticholic acid, a selective farnesoid X receptor agonist, regulates bile acid homeostasis in sandwich-cultured human hepatocytes, *Pharmacol Res Perspect*, 5(4): e00329 (2017).

142 J. Chai, X. Feng, L. Zhang, S. Chen, Y. Cheng, X. He, Y. Yang, Y. He, H. Wang, R. Wang, W. Chen, Hepatic expression of detoxification enzymes is decreased in human obstructive cholestasis due to gallstone biliary obstruction, *PLoS One*, 10 (2015) e0120055.

143 C.A. Schaffner, J. Mwinyi, Z. Gai, W.E. Thasler, J.J. Eloranta, G.A. Kullak-Ublick, The organic solute transporters alpha and beta are induced by hypoxia in human hepatocytes, *Liver Int*, 35 (2015) 1152–1161.

11

Role of Renal Transporters in Drug–Drug Interactions and Nephrotoxicity

Yan Zhang[1] *and Donald Miller*[2]

[1] *DMPK, Nuvation Bio, Inc., 1500 Broadway, New York, 10036, USA*
[2] *Department of Pharmacology and Theurapetics, Max Rady College of Medicine, University of Manitoba, Winnipeg, R3E 0T6, Canada*

11.1 Overview of Renal Transporters

The kidney is one of the most important organs in our body for secretion and elimination of drugs, drug metabolites, environmental toxins, and waste, and therefore, has a critical role in maintaining homeostasis and limiting xenobiotic exposure. The three processes involved in renal excretion include (i) passive glomerular filtration, (ii) active tubular secretion, and (iii) tubular reabsorption. While many substances can be excreted through glomerular filtration, others rely on active secretion processes mediated by various renal transporters. Transporters expressed in the renal proximal tubular cells mediate the active transport of many endogenous and exogenous substances. Renal transporters play important roles in drug pharmacokinetics and contribute to drug–drug interactions (DDIs) and renal toxicity to therapeutic agents [1–4]. Renal secretion of drugs involves multiple transporters expressed on the apical and basolateral membranes of the renal proximal tubular cells working in concert. Drug and drug metabolites are first taken up from the blood side by uptake transporters localized on the basolateral membrane of the renal tubular epithelial cells, and consequently, removed by the efflux transporters localized on the apical membrane into the tubular lumen (Figure 11.1).

The extent to which drugs are secreted or reabsorbed in the kidney depend on distinct transporters on each side of the tubular cell membrane of the nephron that belong to two super transporter families, the ATP binding cassette (ABC) and the solute carrier (SLC) family. Key members of the SLC family that are expressed in the renal proximal tubular cells include organic cation transporters OCT2 (*SLC22A2*), OCT3 (*SLC22A3*), organic anion transporters (OAT) 1 (*SLC22A6*), OAT2 (*SLC22A7*), and OAT3 (*SLC22A8*) on the basolateral membrane and multidrug and toxin extrusion (MATE) transporter 1 (*SLC47A1*) and MATE2K (*SLC47A2*) on the apical membrane [3–5] (Figure 11.1). The ABC transporters expressed on the proximal tubular cells are all localized to the apical membrane

Transporters and Drug-Metabolizing Enzymes in Drug Toxicity, First Edition. Edited by Albert P. Li.

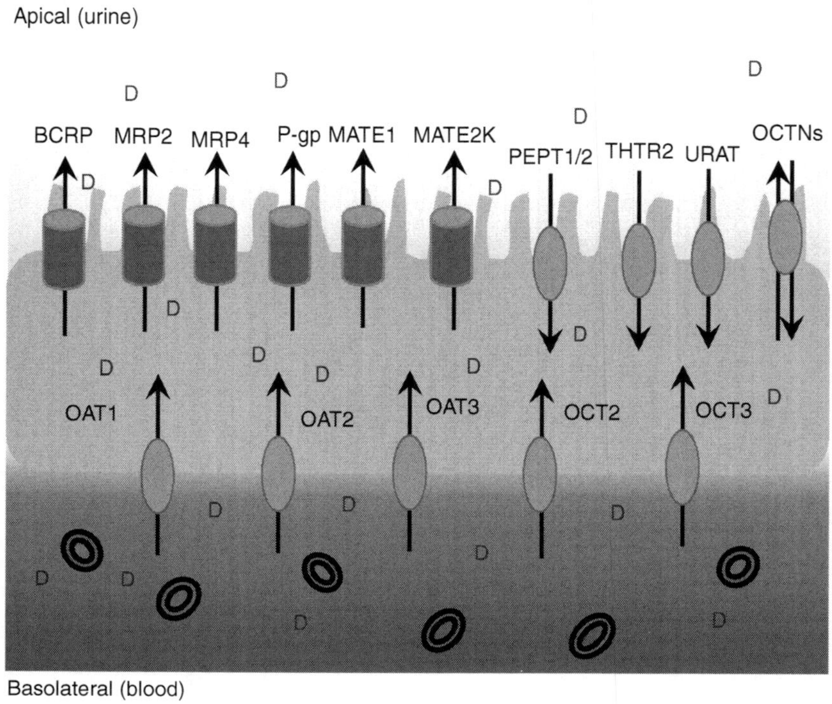

Figure 11.1 Expression of major drug transporters in kidney proximal tubular cells and their roles in transporting drug substances. Drug and drug metabolites are first taken up from the blood side by uptake transporters localized on the basolateral membrane of the renal tubular epithelial cells and consequently removed by the efflux transporters localized on the apical membrane into the tubular lumen. Drugs and nutrients can also be reabsorbed by transporters localized on the apical side of the renal tubular cells such as PEPTs, OCTNs, URAT, and THTR2.

where they facilitate the exit of solutes from inside the tubular epithelial cell to the tubular lumen. These transporters include P-glycoprotein (P-gp) (*ABCB1*), multidrug resistance associated protein 2 (MRP2, *ABCC2*), MRP 4 (*ABCC4*), and breast cancer resistance protein (BCRP, *ABCG2*) [3, 5] (Figure 11.1). A few examples of other transporters expressed on the apical membrane include peptide transporters (PEPT1 (*SLC15A1*) and PEPT2 (*SLC15A2*)) and organic cation and carnitine transporter (OCTN1 (*SLC22A4*) and OCTN2 (*SLC22A5*)), which will not be included in the scope of the current discussion.

11.1.1 Basolateral Transporters

Transporters expressed on the basolateral membrane of the renal tubular cells are responsible for the uptake of various organic anions and cations from the blood side into the tubular epithelial cells and are the first step in the secretion process. These functionally distinct basolateral transporters include but are not limited to OAT1, OAT2, OAT3, OCT2, and OCT3 (Figure 11.1). Both OCT2 and OCT3 transport organic cations across the basolateral membrane down their electrochemical gradient, while OAT1, OAT2, and OAT3 transport organic anions across the basolateral membrane against an electrochemical gradient by counter transport of intracellular α-ketoglutarte [6]. In addition to the transport of anions, organic anion transporters may also interact with cations and neutral compounds [6]. OAT1 and OAT3 are highly expressed and transport substrate drugs including antibiotics (penicillin), nonsteroidal anti-inflammatory drugs (NSAIDs), antivirals (cidofovir), antacids and antihistamines (famotidine), loop diuretics (furosemide), and *para*-aminohippurate (PAH) (Table 11.1). OCT2 is one of the most predominant basolateral organic cation transporters. It is responsible for the uptake of a variety of drugs including metformin, dofetilide, pindolol, ranitidine, and the platin-based drugs such as oxaliplatin [3, 7, 8] (Table 11.1).

11.1.2 Apical Transporters

The transporters localized on the apical membrane of the renal proximal tubular cells are in direct contact with the tubular lumen. Some of these transporters have important roles in the reabsorption of nutrients, amino acid derivatives, and metabolites. Examples include the peptide transporters (PEPT1 (*SLC15A1*) and PEPT2 (*SLC15A2*)), OCTNs (OCTN1 (*SLC22A4*) and OCTN2 (*SLC22A5*)), URAT1 (*SLC22A12*), and potentially thiamine transporter 2 (THTR2, *SLC19A3*) [5]. Others including the efflux transporters MATE1, MATE2K, P-gp, MRP2, and MRP4 are involved in the active secretion of drug and drug metabolites in the kidney. These apically localized efflux transporters play a crucial role in removing drugs and drug metabolites to the tubular lumen, therefore, are important in drug elimination and detoxification.

The ABC transporters P-gp, MRP2, and MRP4 transport a wide range of drugs and their metabolites. MRP2 and MRP4 also share many common substrates. They work in concert with the basolateral localized OAT1 and OAT3 to efficiently remove weakly acidic drugs to the renal tubular lumen. MATE1 and MATE2K transporters belong to the SLC transporter family. Although MATE2K is expressed most abundantly and specifically in the kidney, MATE1 is also expressed in the liver and localized on the canalicular membrane of hepatocytes [9]. In conjunction with OCT2 on the basolateral membrane of the tubular cells, MATE1 and MATE2K involve in the secretion of many cationic drugs to the tubular lumens. The physiological functions of MATE1 and MATE2K can also be affected by similar inhibitors, such as quinidine, cimetidine, trimethoprim, and pyrimethamine, therefore, potentially alter the pharmacokinetics of their substrates and cause their tubular accumulation, thus, result in renal toxicity.

Table 11.1 Selected renal transporters on apical and basolateral membrane of renal tubule cells.

Protein (*Ggne*)	Tissue localization	Substrates	Inhibitors	Role in PK
MDR1/P-gp (*ABCB1*)	Intestine, liver, kidney, BBB	Digoxin, fexofenadine, dabigatran, vincristine, loperimide, doxorubicin, paclitaxel	Cyclosporine A, quinidine, verapamil, amiodarone, clarithromycin, itraconazole, lapatinib, quinidine, ritonavir	Absorption, distribution, elimination, DDI
BCRP (*ABCG2*)	Intestine, liver, kidney, BBB	Mitoxantrone, methotrexate, topotecan, imatinib, rosuvastatin, sulfasalazine, doxorubicin	Oestrone, 17β-oestradiol, curcumin, cyclosporine A, eltrombopag	Absorption, distribution, elimination, DDI
MRP2 (*ABCC2*)	Intestine, liver, kidney	Glutathione and glucuronide conjugates, methotrexate, etoposide, mitoxantrone, valsartan, olmesartan, indinavir, cisplatin	Cyclosporine A, delaviridine, efavirenz, emtricitabine	Absorption, distribution, elimination
MRP4 (*ABCC4*)	Kidney, BBB	Glutathione and glucuronide conjugates, methotrexate, cyclic nucleotide analogs, antiviral nucleoside analogs	Cyclosporine A, delaviridine, efavirenz, emtricitabine	Absorption, distribution, elimination
OAT1 (*SLC22A6*)	Kidney	Para-aminohippurate (PAH), adefovir, cidofovir, zidovudine, acyclovir, famotidine, furosemide, ganciclovir, penicillin G	Probenecid, novobiocin, PAH, teriflunomide	Distribution, elimination, DDI
OAT3 (*SLC22A8*)	Kidney, BBB	Non-steroidal anti-inflammatory drugs, ceftizoxime, furosemide, adefovir, cefaclor, famotidine, ganciclovir, penicillin G	Probenecid, novobiocin, PAH, teriflunomide	Distribution, elimination, DDI
OCT2 (*SLC22A2*)	Kidney	Metformin, pindolol, procainamide, ranitidine, oxaliplatin, varenicline, dofetilide	Cimetidine, cetirizine, quinidine	Distribution, elimination, DDI
MATE1 (*SLC47A1*)	Liver, kidney	Metformin, dofetilide	Quinidine, cimetidine, procainamide, dolutegravir, trimethoprim, pyrimethamine	Distribution, elimination, DDI
MATE2K (*SLC47A2*)	Kidney	Metformin, dofetilide	Quinidine, cimetidine, procainamide, dolutegravir, trimethoprim, pyrimethamine	Distribution, elimination, DDI

Source: Based on FDA https://www.fda.gov/Drugs/DevelopmentApprovalProcess/DevelopmentResources/DrugInteractionsLabeling/ucm093664.htm#table5-1; International Transporter Consortium et al. [7]; Morrissey et al. [3].

11.2 Renal Transporters and Drug–Drug Interactions

Accumulating evidence suggests that renal drug transporters can significantly influence pharmacokinetic, pharmacodynamic, and DDIs. Although transporters localized on both apical and basolateral membranes have their distinct functions in the elimination of drugs and drug metabolites, they often interact in a cooperative manner to remove toxins and waste products from the body into the tubular lumen. As an example, tubular secretion of drugs that are weak bases is a two-step process involving organic cation transporter (OCT)-mediated uptake into the tubular epithelial cell and MATE-dependent efflux into the tubular lumen. In this case, inhibiting transporter activity on the apical membrane may result in reduced excretion of drug and drug metabolites into the tubular lumen and an intracellular accumulation of drug or drug metabolite in the proximal tubular epithelium (Figure 11.2). Depending on the drug and the extent of inhibition of the apical transporters, there is the potential for increased intracellular accumulation of drug within the tubular epithelial cell and resulting renal toxicity. Thus, alterations in apical membrane transporters in the kidney have a major impact on

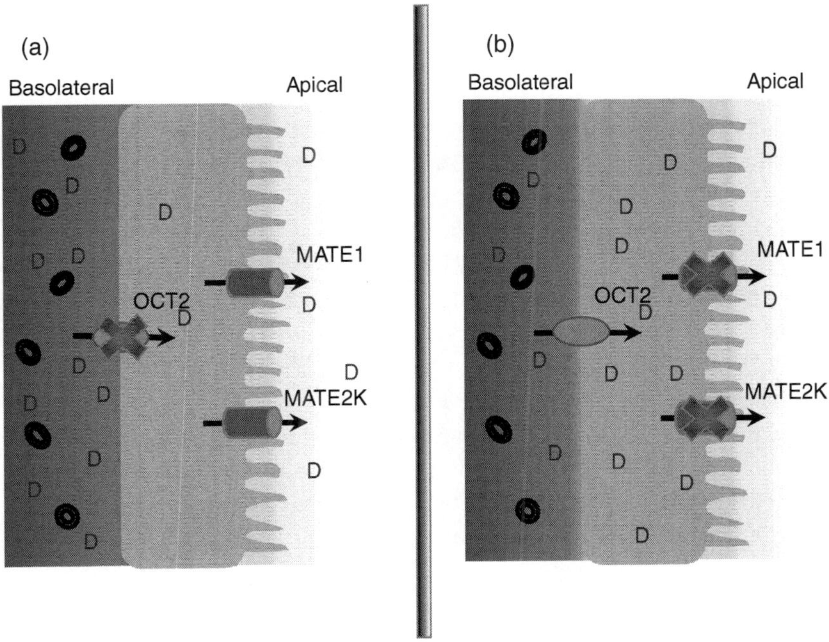

Figure 11.2 Inhibition of organic cation transporters in the renal tubular cells and the effect on tubular drug secretion and intracellular accumulation. (a) Inhibiting basolateral uptake mediated by OCT2 results in reduction of intracellular drug accumulation and net secretion. (b) Inhibiting apical efflux transporters such as MATE1 and MATE2K results in a reduction of tubular secretion but increasing intracellular drug concentration.

the renal secretion of drugs and resulting drug concentrations in the tubular epithelium and drug-induced nephrotoxicity. In contrast, inhibiting the transporters on the basolateral side that are involved in renal secretion may have a protective effect reducing intracellular drug levels in the kidney (Figure 11.2). However, while the kidney may be spared, inhibition of the basolateral transporters involved in renal secretion could potentially lead to increased systemic toxicity due to elevations in plasma concentrations of the drug and its metabolite as a result of the reductions in renal clearance (Figure 11.2).

11.2.1 Impact on the Pharmacokinetics of Drugs

In terms of transporter mediated DDIs, an interacting drug may inhibit the function of certain transporters (perpetrator drug) or may be a substrate of a given transporter (victim drug). The DDI potential of a perpetrator drug via transporter inhibition can be determined by comparison of the maximum unbound plasma concentration of the perpetrator drug (I) and its inhibitory activity on the transporter expressed as the half maximal inhibitory concentration (IC_{50}). The IC_{50} values can be determined by *in vitro* studies. Recent FDA guidance indicates an I/IC_{50} value ≥ 0.1 as a criterion for consideration of conducting a clinical transporter based DDI for the renal transporters including OAT1, OAT3, OCT2, MATE1, and MATE2K (Table 11.2) [10].

DDIs involving the inhibition of transporters localized on the basolateral membrane of the proximal tubular cells may enhance the plasma concentrations of a substrate of the transporter by reducing the renal clearance and therefore, prolonging its elimination half-lives. An example is metformin, a biguanide antidiabetic drug and prototypical substrate of OCTs and MATEs. Metformin is a hydrophilic drug and is not metabolized but rather excreted unchanged via the kidneys [11]. As metformin is a strong base ($pK_a = 11.5$ at pH 8) [12] and hence a cation at physiological pH, it is dependent on drug transporters to be absorbed, distributed and excreted [13]. In humans, OCT1 on the basolateral membrane of hepatocytes is involved in the uptake of metformin from the blood into the liver, which is its major site of pharmacological action [14]. Similarly, OCT2 on the basolateral membrane of renal tubular cells mediates the uptake of metformin from blood into the kidney, which is the major eliminating organ for the drug [15]. The final efflux step for metformin from the tubular cells to the urine is mediated by MATE1 and MATE2K on the apical membrane of renal tubular cells [13, 16, 17]. DDIs involving inhibition of OCT2, MATE1, and/or MATE2K, alone or in combination, can have significant impact on metformin exposure and clearance. Clinical studies between metformin and potent inhibitors of OCT2 have demonstrated significant impact on metformin plasma exposure. One such OCT2 inhibitor is dolutegravir, an antiretroviral drug that blocks the activity of the HIV viral

Table 11.2 Relative *in vitro* inhibition on human OCT2, MATE1, and MATE2K by selective compounds at clinically relevant dose.

		Transporter inhibition Ki or IC_{50} (μM)			$C_{max,ss,u}/IC_{50}$ (or Ki)			
Drugs	$C_{max,ss,u}$ (μM)	**OCT2**	**MATE1**	**MATE2K**	**OCT2**	**MATE1**	**MATE2K**	**References**
Trimethoprim (200 mg BID)	5.0	137–1327	2.6–6.2	0.35	< 0.04	0.8–1.9	14	[31, 32, 35]
Pyrimethamine (50 mg QD)	0.3	10	0.093	0.059	0.03	3.2	5.1	[19]
Cimetidine (400 mg QHS)	7.7	124	3.8	6.9	0.06	2.0	1.1	[29, 30]
Dolutegravir (50 mg)	0.109	0.21	3.6	12.5	0.06 or 0.52	0.03	0.01	[13, 18]
Bictegravir (50 mg QD)	0.13	0.42	8	NA	0.3	0.02	NA	[28, 117]
Famotidine (200 mg QD)	1	66	0.25	2.5	0.015	4	0.4	[36]
Vandetanib (300 mg)	0.038	8.8	0.16	0.30	< 0.01	0.24	0.13	[69, 118, 119]

integrase enzyme [13]. In patients receiving both metformin and dlutegravir, the pharmacokinetics of metformin were significantly altered. Coadministration of dolutegravir 50 mg twice daily increased metformin area under the curve (AUC) and C_{max} values by 145 and 111%, respectively, and decreased the clearance of metformin by 59% (Table 11.3) [13]. Mechanistically, dolutegravir is a potent inhibitor of OCT2-mediated transport of metformin with reported inhibition IC_{50} values ranging from 0.21 to 1.9 μM [13, 18]. Although it also inhibits MATE1 and MATE2K, the inhibition was not clinically relevant using the I/IC_{50} value ≥ 0.1 criterion proposed by FDA [13, 18] (Table 11.2). The magnitude of increase in metformin exposure in the presence of dolutegravir exceeded those observed for cimetidine and pyrimethamine (to be discussed in later sections), two prototypical inhibitors of renal cationic transporters including OCT2, MATE1, and MATE2K [19–21]. It should be noted that both cimetidine and pyrimethamine are more potent inhibitors of MATE1 and MATE2K transporters compared to OCT2 (Table 11.2). Therefore, the more robust effects of dolutegravir on metformin pharmacokinetics may be attributable to a more potent inhibition of the OCT2 on the basolateral side of the renal tubular epithelial cells compared to either cimetidine or pyrimethamine.

The effect of dolutegravir on serum creatinine concentration and creatinine clearance was determined in a phase 1 clinical study [22]. Treatment with dolutegravir, at a 50 mg once daily or twice daily dose, decreased creatinine clearance by 10 and 14%, respectively, after 14 days of dosing (Table 11.4). Comparatively, the impact of dolutegravir on creatinine clearance was much less significant than its effects on metformin clearance [22]. This is likely due to the fact that in addition to OCT2 [22, 23], renal secretion of creatinine also involves both OAT2 [24, 25] and OCT3 [26] expressed on the basolateral membrane of renal tubule. Therefore, inhibition of OCT2 by dolutegravir alone would not have as much effect on creatinine clearance compared to metformin which largely relies on OCT2 for the renal uptake.

Another potent OCT2 inhibitor is Bictegravir. Bictegravir is a novel, potent, HIV integrase strand transfer inhibitor (INSTI) being co-formulated with the nucleoside analog reverse transcriptase inhibitors emtricitabine (FTC) and tenofovir alafenamide (TAF) as a complete single-tablet regimen for the treatment of HIV-1 infection [27]. Similar to dolutegravir, bictegravir is also a potent inhibitor of OCT2 (IC_{50} = 0.42 μM) with less inhibition on MATE1 (IC_{50} = 8 μM) based on *in vitro* assays [28]. Unlike dolutegravir, the potent inhibition of basolateral transporter OCT2 by bictegravir at clinically relevant concentrations (Table 11.2), does not alter metformin pharmacokinetics as much as that of dolutegravir. Bictegravir increased metformin plasma exposure ~ 39 vs. 145% increase seen with dolutegravir in a blinded, placebo-controlled, crossover Phase 1 study in 32 healthy subjects (Table 11.3) [28]. Since the published transporter data regarding bictegravir DDI

Table 11.3 Effects of selective renal transporter inhibition on metformin pharmacokinetics and pharmacodynamics.

Drugs	Dose (mg)	Transporter selectivity	Effect on metformin PK	Effect on metformin PD	References
Bictegravir	50 QD	OCT2 selective	AUC $\uparrow$ 39%, CL_R $\downarrow$ 31%	No change on metformin PD	[28]
Dolutegravir	50 BID	OCT2 selective	AUC $\uparrow$ 145%, CL_R $\downarrow$ 59%	No hypoglycemia or lactic acidosis	[13]
Cimetidine	400 BID	MATEs > OCT2	AUC $\uparrow$ 46%, CL_R $\downarrow$ 28%	Lactate/pyruvate ratio $\uparrow$ at 8 h	[20]
Pyrimethamine	50 QD	MATEs >> OCT2	AUC $\uparrow$ 39%, CL_R $\downarrow$ 35%	Lactate/pyruvate ratio $\leftrightarrow$	[19]
Trimethoprim	200 BID	MATEs >> OCT2	AUC $\uparrow$ 37%, CL_R $\downarrow$ 32%	Transient lactate plasma conc $\uparrow$	[34]
Famotidine	200 mg QD	MATE1 > MATE2K >> OCT2	AUC $\leftrightarrow$, CL_R $\uparrow$ 28%	$\downarrow$9% of glucose AUC	[36]
Vandetanib	300 mg	MATEs >> OCT2	AUC $\uparrow$ 74%, CL_R $\downarrow$ 52%	NA	[118]

NA, not available.

Table 11.4 Effects of selective renal transporter inhibition on creatinine changes.

Drug	Dosing regimen	SCr ↑ (%)	CLcr ↓ (%)	References
Trimethoprim	200 mg BID	23	25	[34]
Pyrimethamine	50 mg QD	21	45	[19]
Dolutegravir	50 mg QD or BID	10–15	10 or 14	[13, 18]
Famotidine	200 mg QD	6	12	[36]
Bictegravir	50 mg QD	ND	39	[28]

is limited, the mechanism behind the different effect on metformin plasma exposures between bictegravir and dolutegravir is unclear.

In addition to the potent inhibitors of basolateral transporter OCT2, DDIs involving the inhibition of the apical organic cation transporters have also been studied extensively. One of the early studies was carried out between cimetidine, a H_2 receptor antagonist, and metformin. Cimetidine, administered as a 400 mg twice daily dose, reduced the renal clearance of metformin by an average of 28% [20]. As a result, blood metformin AUC was increased by 46% (Table 11.3) [20]. Cimetidine is an inhibitor of multiple organic cation transporters including OCT2, MATE1, and MATE2K with more potent inhibition toward the apical transporters MATE1 and MATE2K. This was demonstrated in HEK293 cells overexpressing the various transporters [29]. The inhibition constant (K_i) of cimetidine for metformin uptake mediated by OCT2 was 124 μM. In contrast, cimetidine was a more potent inhibitor of MATE1 and MATE2K with the K_i values of 3.8 and 6.9 μM, respectively [29]. The mean unbound steady-state plasma concentrations of cimetidine following oral administration of 400 mg dose at night (QHS) were reported to be 7.7 μM with average of ~20% plasma protein binding [30]. These steady-state plasma levels are higher than the K_i values for MATE1 and MATE2-K inhibition but below the K_i values for inhibition of OCT2 (Table 11.2). As the resulting I/IC_{50} values are ≥0.1 for MATEs but less than 0.1 for OCT2, these findings suggest that at therapeutic doses of cimetidine, transporter mediated DDIs by cimetidine mainly result from MATE inhibition. These studies suggested that inhibition of the apical efflux mediated by MATEs, rather than the basolateral uptake by OCT2, is the likely mechanism underlying the pharmacokinetic DDI caused by cimetidine in the kidney [29].

Similar to cimetidine, the commonly used antibiotic trimethoprim inhibits both OCT2 and MATEs *in vitro* with a more potent inhibition of MATE transporters. The inhibitory IC_{50} values were 137–1327 μM for OCT2 and 2.6–6.2 μM for MATE1 and 0.35 μM for MATE2K (Table 11.2) [31–33]. In a recently conducted clinical study in 24 healthy volunteers, 500 mg metformin was given three times daily along with trimethoprim at 200 mg twice daily dose. Trimethoprim significantly reduced the renal clearance of metformin by 32% and increased plasma AUC for metformin by 37% (Table 11.3) [34]. In the same study, co-treatment with trimethoprim also significantly increased the plasma concentration and decreased renal clearance of creatinine by 23 and 25%, respectively (Tables 11.3 and 11.4). At 200 mg twice daily dose, the observed Css_{ave} plasma concentration of trimethoprim is 9.14 μM correlating to an unbound concentration value of 5.0 μM (fraction unbound for trimethoprim is 0.55) [35]. Therefore, the changes in metformin pharmacokinetics following co-treatment of trimethoprim, are likely due to the inhibition of MATEs rather than OCT2, similar as in the case for cimetidine (Table 11.2).

Another well-known example of kidney-based DDI is pyrimethamine, an anti-parasite drug used to fight protozoal infections. Pyrimethamine is a potent inhibitor of MATE1 and MATE2K with little effect on OCT2. In transporter transfected human embryonic kidney (HEK) cells, the K_i values of pyrimethamine for MATE1 and MATE2K were 93 and 59 nM, respectively. In contrast, the K_i for OCT2 was 10 μM [19]. Following standard 50 mg oral dose, pyrimethamin has a long elimination half-life (~ 96 hours) and an estimated unbound plasma concentration of ~300 nM, which is significantly greater than its K_i values for MATE1 and MATE2K but much lower than that for OCT2 (Table 11.2) [19]. In healthy subjects treated with 250 mg metformin, 50 mg pyrimethamine was able to significantly increase the C_{max} and AUC values of metformin by 42 and 39%, respectively. In addition, metformin renal clearance was reduced by 35% (Table 11.3) [19]. In this same study, pyrimethamine treatment also significantly decreased the creatinine renal clearance by 45% in those subjects treated with 250 mg metformin (Table 11.4).

Unlike the other MATEs transporters discussed above, famotidine, an H_2-receptor antagonist used to reduce gastric acid secretion, had shown a distinct effect on metformin pharmacokinetics. Famotidine is a potent inhibitor of MATE1 but also inhibits MATE2 and OCT2. The mean inhibitory IC_{50} values for famotidine against metformin transport were 0.25, 2.5, and 66 μM in MATE1, MATE2, and OCT2 cells, respectively [36]. When given at 160–200 mg dose, famotidine is considered a much more selective inhibitor to MATE1 than that of OCT2 (Table 11.2). In a crossover study, healthy subjects received metformin alone or with famotidine. Consistent with MATE1 inhibition, famotidine administration (200 mg) significantly decreased urinary creatinine clearance and increased creatinine plasma concentrations. In the same study, famotidine transiently improved the glucose lowering effects of metformin which agreed with the fact that famotidine increased the estimated bioavailability of metformin. Interestingly, despite a potent inhibitory effect on MATE1, metformin plasma concentration was not altered by famotidine treatment (Table 11.3). This could be explained by the opposing effect that an increase in both metformin absorption and metformin renal clearance, which resulted in no change in metformin plasma exposure [36]. The increase in total metformin renal clearance may due in part to famotidine inhibition of transporter-mediated reabsorption of metformin through plasma membrane monoamine transporter (PMAT, *SLC29A4*) and/or thiamine transporter 2 (THTR2, *SLC19A3*) [36, 37].

11.2.2 Impact on the Drug PD

DDIs involving a renal transporter in the inhibition of excretion and/or elimination not only affect the pharmacokinetics but also the pharmacodynamics of drugs or endogenous substances. One of the examples is the change in plasma concentrations and renal clearance of creatinine as shown in some of the cases discussed in the previous section. Creatinine is a catabolic end-product of creatine breakdown in the skeletal muscle. It is predominantly eliminated by glomerular filtration and serum creatinine is commonly used to assess kidney function. However, creatinine also undergoes active tubular secretion, which can account for 10–40% of creatinine clearance [38]. Renal tubular secretion of creatinine was reported to be driven by uptake via OCT2 [22, 23], OCT3 [26], and OAT2 [24, 25]. In addition, MATE1 and MATE2-K have been shown to play a role in the tubular efflux of creatinine [17]. Therefore, drugs that interact with one or more of these transporters can cause nonpathological elevations in serum creatinine. Such is the case with elevations in serum creatinine reported in patients and healthy subjects with several drugs including, trimethoprim [39], dolutegravir [22], itacitinib [40], and abemaciclib [41]. It is important to note as serum creatinine is widely used to monitor kidney function, these changes can lead to false positives for kidney injury.

In addition to the impact on creatinine, DDIs mediated by renal transporter can also affect the pharmacodynamics of the interacted drug. One of the well-studied examples is metformin. Coadministration of pyrimethamine was able to significantly increase the C_{max} and AUC values of metformin by 42 and 39%, respectively, and decreased the renal clearance by 35% [19]. In addition, pyrimethamine cotreatment also significantly decreased the creatinine renal clearance by 45% in those subjects treated with metformin (Table 11.4). Metformin inhibits mitochondrial cellular respiration, which increases anaerobic metabolism and lactate levels [42]. Elevation of plasma lactate levels may cause lactic acidosis, which is the biguanide-related adverse effect of most concern [12]. Therefore, not surprisingly, the significant increase in metformin plasma concentrations following pyrimethamine treatment also increased the plasma lactate concentrations. However, the ratio of lactate to pyruvate did not differ significantly between the pyrimethamine-treated group and the nontreatment group [19].

Somogyi et al. discovered that 400 mg cimetidine twice daily dose was able to increase the metformin AUC by 46% and decrease metformin renal clearance by an average of 28% [20]. Despite the significant changes observed on metformin pharmacokinetics in this study, cimetidine had no effect on the renal clearance of creatinine in the healthy subjects. Blood lactate to pyruvate ratios were also calculated among the three treatment groups, i.e. cimetidine, metformin, and metformin plus cimetidine. Among the four time points accessed, the lactate to pyruvate ratios were much higher when metformin was administered either alone (except for at eight hours) or with cimetidine, compared with cimetidine alone. At eight hours, the ratio was significantly higher when metformin was combined with cimetidine, compared with metformin alone [20]. Despite the significantly increased lactate to pyruvate ratio with the co-treatment of cimetidine, no subject experienced any adverse effects [20]. Therefore, in this specific case, although the renal clearance of creatinine was not altered following cimetidine co-treatment, pharmacodynamics of metformin was slightly affected (Table 11.3).

Among the metformin DDIs, the largest impact on metformin plasma exposures came from the co-treatment with dolutegravir, a potent OCT2 inhibitor, which resulted in an increase of metformin AUC and C_{max} values by 145 and 111%, respectively, and decrease of the metformin clearance by 59% [17]. Since dolutegravir did not have an effect on other transporters, such as PMAT, OCT3, MATE1, and MATE2K, involving in metformin absorption, distribution, and excretion, the authors concluded that the observed dolutegravir-metformin DDI was mainly due to OCT2 inhibition [13]. Despite the significant enhancement of the metformin plasma exposures, no pattern of AE was seen. No cases of drug-related hypoglycemia or lactic acidosis were reported (Table 11.3). Furthermore, metformin was well tolerated when used in combination with dolutegravir in healthy subjects [13]. Despite a significant increase in exposure, metformin was well tolerated when used in combination with dolutegravir.

Although the extent by various renal transporter inhibitors was moderate on metformin pharmacokinetics and pharmacodynamics, it may be clinically relevant in patients with borderline renal function or in patients given high-dose metformin that can result in lactic acidosis. In this regard, in a study by Phillips et al. [43], the authors noted that high doses of metformin in patients with declining renal function were risk factors in metformin associated lactic acidosis. Therefore, organic cation transporter inhibitors may have the propensity of elevating metformin plasma concentrations by reducing its tubular secretion. Patients being prescribed other drugs with transporter inhibitor properties along with metformin should consider having the dose of metformin reduced. The interaction would be more significant in patients with impaired renal function especially in older patient population.

For the inhibitors of OCT2, when a co-dosed substrate drug has a wide therapeutic window, such as metformin, it can be titrated on the dose of the substrate drugs. However, if the co-dosed substrate drug has a narrow therapeutic index, it is contraindicated and should not be given at the same time. Dofetilide, an antiarrhythmic agent, is one such example. Dofetilide undergoes both renal and hepatic clearance, with approximately 80% of a single dose of dofetilide being excreted in urine, of which approximately 80% is excreted as unchanged dofetilide. Renal elimination involves both glomerular filtration and active tubular secretion via the cation transport system [44–46]. In healthy male subjects, when dofetilide was administered with 100 and 400 mg of cimetidine, the AUC of dofetilide increased by 11 and 48% and the C_{max} increased by 11 and 29%, respectively. In addition, the renal clearance of dofetilide was reduced by 13 and 33%, respectively [44]. Most importantly, dofetilide-induced prolongation of the QTc interval was enhanced by cimetidine and the mean maximum change in QTc interval from baseline was increased by 22 and 33% with 100 and 400 mg of cimetidine, respectively. Although in the same study dofetilide was well tolerated and the linear relationship between the prolongation of the QTc interval and plasma dofetilide concentrations was unaffected by cimetidine treatment, it is recommended that cimetidine at all doses be avoided due to the narrow therapeutic index of dofetilide. In the drug label for dofetilide, in addition to cimetidine, the concomitant use of cation transporter inhibitors such as verapamil, trimethoprim, and dolutegravir with dofetilide is contraindicated [46].

11.3 Renal Transporters and Nephrotoxicity

The kidney is a common site for drug related toxicity. This is due in part to the important role the kidney plays in elimination of drugs and the ability of the kidney through a combination of glomerular filtration and tubular secretion to concentrate solutes within the nephron. Depending on the drug, concentrations within the nephron can be several-fold greater than the systemic circulation. As drug-induced kidney toxicity has been implicated in as much as 25% of the cases of acute kidney injury (AKI) in the critically ill [47], there is an increased focus on identification of renal toxicity in drug development. Drug-induced nephrotoxicity arises from a variety of mechanisms. These include direct effects on the kidney through altered intraglomerular hemodynamics, tubular cell toxicity, immune, and inflammatory responses, as well as indirectly through drug-induced rhabdomyolysis, and thrombotic microangiopathy [48]. The following section provides specific examples of drug-induced AKI and the mechanism involved. The intent is to provide mechanistic representation of glomerular, proximal, and distal tubule drug toxicity and the cellular factors mediating AKI. Nephrotoxicity either related or unrelated to renal transporters will be discussed.

11.3.1 Nephrotoxicity Unrelated to Drug Transporters

The extent to which a drug can cause nephrotoxicity is based on several factors including the pharmacologic action, metabolism, and excretion of the administered drug. Preclinically, those drugs causing direct nephrotoxicity can be histologically assessed and grouped according to the anatomical area most effected (see Table 11.5 for examples). Drugs causing glomerular injury typically alter renal microvascular blood flow or have direct cytotoxic effects within the glomerular endothelial and epithelial cells. Drugs causing proximal tubule cytotoxicity usually result from high intracellular drug concentrations. As such, drugs that undergo substantial renal clearance through active secretion mechanisms are particularly susceptible and can be influenced greatly by DDIs and potentially transporter polymorphism. In contrast, distal tubule cytotoxicity often occurs due to high luminal concentrations of drugs that exceed the saturation solubility resulting in drug precipitation and intraluminal injury. In such cases factors such as volume depleted state and altered pH environment in the lumen of the tubule influence the extent of toxicity observed. A final category of nephrotoxicity is drug-induced actue tubulointerstitial nephritis (DTIN) that involves eliciting an immune response within the kidney (Table 11.5). Drugs that affect the renal blood flow can potentially induce AKI in severely ill patients. The kidney receives 25% of the cardiac output, and the majority flows into the kidney cortex [49]. Therefore, even a slight decline in renal blood flow results in hypoxic injury of the medullary region [50]. Combination therapy with drugs having propensity of causing renal damage may result in synergistic nephrotoxicity, thus increasing the risk of renal injury.

NSAIDs are one of the most commonly used over-the-counter (OTC) medications in the United States and are known to have adverse effects on kidney function, especially when used chronically [51, 52]. NSAIDs inhibit the function of cyclooxygenase (COX) enzyme, interfering with arachidonic acid metabolism into prostaglandins E2, prostacyclins, and thromboxanes [53]. In the kidney, prostaglandins act as vasodilators and suppress the effects of vasoconstrictor hormones, therefore increasing renal perfusion and ensuring adequate blood flow to the kidney. NSAIDs inhibit this mechanism by decreasing the synthesis of prostaglandins, thereby, reducing renal blood flow and increasing the chances for AKI [52]. Similar to NSAIDs, drugs with anti-angiotensin II activity (e.g. angiotensin-converting enzyme (ACE) inhibitors, angiotensin receptor blockers (ARBs)) can interfere with the kidneys' ability to regulate glomerular pressure and decrease glomerular filtration rate (GFR). Other drugs, such as calcineurin inhibitors (e.g. cyclosporine, tacrolimus) promote afferent arteriolar constriction by stimulating production of vasoconstrictors and inhibiting synthesis and secretion of endothelial-derived nitric oxide, cause dose-dependent vasoconstriction of the afferent arterioles, leading to renal impairment in at-risk patients [48].

Table 11.5 Examples of drug-induced nephrotoxicity and biomarkers used to assess injury.

Type of injury	Drugs involved	Cofactors/confounders	Useful biomarkers[a]
Glomerular injury	• Doxorubicin • Puromycin • NSAIDs		Urine- CysC, Albumin, B2M Blood – CysC, B2M, creatinine
Proximal tubule injury • Apical exposure	• Aminoglycosides: – Gentamicin – Vancomycin	– DDIs – primarily those involving inhibition of apical efflux transporters – Transporter polymorphism – Reduced GFR	Urine – CysC, NGAL, ***NAG***, KIM-1, αGST
• Basolateral exposure	• Cisplatin • Tenofovir		Blood – NGAL
Distal tubule injury • Crystalline-induced tubular injury	• Methotrexate • Amphotericin Indinavir • AcyclovirAtazanavir • Ciprofloxacin • Triamterene	– Urine pH – Dehydration – High dose/rapid infusion	Urine – Clusterin, NGAL, αGST Blood- NGAL
Immune-related injury • Drug-induced acute tubulointerstitial nephritis (ATIN)	Immune checkpoint inhibitors including: • Nivolumab, Pembrolizumab, Atezolizumab • Antimicrobials • NSAIDs • Proton pump inhibitors	– Drug hypersensitivity – Polypharmacy	Urine – NGAL, ***MCP-1,*** Il-18

[a] Biomarkers in bold italics show selectivity toward specific type of injury indicated.

Renal toxicity may also result from the use of drugs that form crystals when they are insoluble in human urine. The crystals precipitate, usually within the distal tubular lumen, interrupting urine flow. The likelihood of crystal precipitation depends on the concentration of the drug in the urine and the urinary pH. Commonly prescribed drugs associated with production of crystals include antibiotics (e.g. ampicillin, ciprofloxacin, and sulfonamides); antivirals (e.g. acyclovir, ganciclovir); methotrexate; and triamterene [48].

Renal tubule cells, in particular proximal tubule cells, are vulnerable to the toxic effects of drugs because their role in concentrating and reabsorbing glomerular filtrate constantly exposes them to high levels of circulating toxins. Drugs cause tubular cell toxicity by impairing mitochondrial function, interfering with tubular transport, increasing oxidative stress, or forming free radicals [54]. Drugs associated with this mechanism of injury include, but are not limited to, aminoglycosides, antiretrovirals (adefovir, cidofovir, and tenofovir), chemotherapeutic agents (methotrexate and cisplatin), antifungal (amphotericin B), and other reagents [49, 54]. Cisplatin is thought to stimulate oxygen free radical formation, reduce renal perfusion and disrupt DNA and RNA synthesis [55, 56]. Free radical formation also plays a role in aminoglycoside nephrotoxicity. Early studies at Tulane University School of Medicine suggested a role for hydroxyl radical in gentamicin- and puromycin-induced acute renal failure and proteinuria, respectively, in preclinical animals [57, 58].

11.3.2 Nephrotoxicity Related to Drug Transporters

Various physiological factors determine the extent of drug-induced nephrotoxicity. Transporters expressed on the apical and basolateral membrane of the renal tubular membranes are one of them. Some drugs are nephrotoxic only after they are transported and accumulated in the proximal tubular cells. Those agents that may cause direct cellular toxicity include antibiotics, antivirals, antifungal, and chemotherapeutic agents. Examples of interactions with renal transporters involving some of these drugs are discussed in the section below.

Drug-induced renal toxicity may be caused by the enhanced kidney tissue accumulation mediated by transporters. The organic anion transporters, OAT1 and OAT3 expressed on the basolateral proximal tubular cells, are responsible for the renal tubular secretion of many anionic compounds including cephalosporin antibiotics and some antiviral agents. For example, the acyclic nucleotide phosphonates adefovir, cidofovir, and tenofovir enter the cell via basolateral OAT1 and promote cellular injury primarily through disturbing mitochondrial function [49]. Cidofovir is an injectable antiviral medication primarily used as a treatment for cytomegalovirus (CMV) retinitis (an infection of the retina of the eye) in AIDS patients [59]. It is known that cidofovir can cause renal failure by accumulating in proximal tubule cells. Clinical interactions have been observed between probenecid and cidofovir. Following probenecid coadministration, it was observed that the renal clearance of cidofovir was reduced by 38% and plasma concentration was increased by 78% when cidofovir was given at higher dose [60], consequently minimizing the concentration of cidofovir in proximal tubule cells [61]. Probenecid mitigates nephrotoxicity of cidofovir by inhibiting organic anion transports such as OAT1 and OAT3 on the basolateral side of kidney proximal tubule cells. Therefore, cidofovir is to be administered with probenecid which decreases its side effects to the kidney. In this case, targeted inhibition of cellular uptake may reduce nephrotoxicity risks. While inhibition of the basolateral localized transporters, such as OAT1 and OAT3, may result in elevated systemic exposures, which could be nephroprotective, inhibition of the MATEs that are localized on the apical proximal tubular cells, may potentially increase the liability of causing nephrotoxicity.

Among those potential nephrotoxic drugs, such as aminoglycoside antibiotics, NSAIDs, immunosuppressive drugs, and chemotherapy agents, cisplatin is known to cause severe AKI during cancer treatment. Cisplatin has been widely used in combination chemotherapy for the treatment of various cancers. It is cleared by the kidney through both glomerular filtration and tubular secretion [62]. Renal elimination accounts for more than 90% of the dose in human subjects [63] and the most severe adverse effect caused by cisplatin is dose-dependent nephrotoxicity. Cisplatin is transported by OCTs and MATEs, which have contributed to its active renal clearance [64, 65]. It has been reported that the nephrotoxic effect of cisplatin can be potentiated by coadministration with several drugs through the interaction of transporters localized in the renal tubular cells. One of the examples is ondansetron. Both preclinical and clinical studies have demonstrated that coadministration with ondansetron could potentiate the nephrotoxicity of cisplatin [66, 67].

Ondansetron is one of the 5-hudroxytryptamine-3 receptor (5-HT_3) antagonists, which has been widely used as an antiemetic agent in the prevention and treatment of chemotherapy-induced nausea and vomiting, especially in patients receiving highly emetogenic cisplatin-based regimen [68]. In HEK-293 cell lines overexpressing human OCT2, MATE1, and MATE2-K, ondansetron inhibited metformin accumulation in a concentration-dependent manner. Ondansetron is a much more potent inhibitor of human MATE1 and MATE2K transporters than that of human OCT2 *in vitro*, with the observed inhibitory constant (K_i) being 0.035, 0.015, and 3.85 μM, respectively [67]. It was demonstrated that Mate1 was the major cationic drug transporter being inhibited by ondansetron in comparing studies carried out between *Mate1* wild type and *Mate1-/-* mice. Ondansetron and cisplatin co-treatment in wild type mice and cisplatin treatment *Mate1-/-* mice had similar effects on cisplatin-induced elevation on blood urea nitrogen (BUN) and serum creatinine levels. In addition, similar patterns were observed for the elevation of *Kim-1* and *Lcn2* gene transcripts, two molecular biomarkers of kidney injury, in the ondansetron treatment and *Mate1-/-* mice [67]. These results suggested ondansetron may cause undesirable DDIs and enhance cisplatin-induced nephrotoxicity by inhibiting Mate1 in mice. Similarly, in the *Mate1-/-* mice, Nakamura et al. also demonstrated that Mate1 is involved in cisplatin-induced nephrotoxicity [65]. These data have important clinical implications, as these 5-HT_3 receptor antagonists may serve as perpetrators to cause undesirable DDIs. In a retrospective analysis of clinical studies from 600 cancer patients treated with cisplatin, Kou and colleagues [66] concluded that ondansetron can enhance cisplatin-induced nephrotoxicity, which is likely to be attributed to inhibition of MATEs. In addition to ondansetron, vandetanib, an oral tyrosine kinase inhibitor has also been characterized in transporter transfected HEK-293 cell lines. Vandetanib is a potent inhibitor of MATE1 and MATE2K versus OCT2 with the IC_{50} values of 0.16, 0.30, and 8.8 μM, respectively (Table 11.2) [69]. Inhibition of the two MATE transporters on the apical membrane of tubule cells may explain some of the clinical observations of decreased creatinine clearance and increased cisplatin nephrotoxicity in some patients receiving vandetanib [70].

Therefore, examples from the two independent research laboratories discussed herein [65, 67] demonstrated that loss-of-function mutations in apical efflux transporters such as MATE1 that reduce drug excretion from the cell into the urine may impact drug elimination and promote nephrotoxicity by elevating intracellular drug concentrations. In addition to MATE1, functional alterations caused by genetic polymorphism of other renal efflux transporters, such as MRP2, can also lead to a build-up of drugs or metabolic wastes in renal proximal tubular cells. Accumulation of these substances may potentially result in mitochondrial DNA synthesis inhibition and causing Fanconi syndrome. Tenofovir-induced Fanconi syndrome represents one such example. Izzedine and colleagues [71] discovered that MRP2 haplotypes were associated with tenofovir-induced renal proximal tubulopathy likely due to accumulation of tenofovir in the proximal tubular cells. Patients with HIV receiving tenofovir who developed Fanconi syndrome were noted to have a single nucleotide polymorphism in the *ABCC2* gene encoding MRP2, which transports tenofovir out of the cell into the urine. In contrast, HIV patients treated with tenofovir who did not develop Fanconi syndrome did not have the gene polymorphism [71].

One other organic cation transporter expressed in the proximal tubular cells that deserves comment in terms of drug nephrotoxicity is OCT2. Despite sharing a wide range of substrates as MATEs, inhibition on OCT2 may have protective effect on renal tissue, similar effects as described with the OAT1/OAT3 transporters, by reducing the renal tissue accumulation. This is illustrated by studies examining erlotinib and nilotinib and the platinum drugs. In contrast to ondansetron and vandetanib, other tyrosine kinase inhibitors such as erlotinib and nilotinib may provide renal protection from cisplatin. In an *in vitro* study, cellular platinum uptake after incubation of HEK-OCT2 with oxaliplatin was decreased in the presence of erlotinib and nilotinib in a concentration-dependent manner [72]. In addition, erlotinib potently inhibited OCT2 function and significantly reduced cellular uptake of two known OCT2 substrates, 4-[4(dimethylamino) styryl]-N-methylpyridinium-iodide (ASP^+) and tetraethylammonium (TEA) [73]. In a randomized phase II clinical trial to determine whether there is a therapeutic benefit of adding erlotinib to cisplatin and radiotherapy in squamous cell carcinoma of the head and neck (SCCHN), there was an apparent absence of severe renal impairment and auditory toxicities, two of the major devastating toxicities associated with the use of cisplatin, in the patients who received erlotinib [74]. It is possible that by inhibiting the function of OCT2, erlotinib restricted the cellular accumulation of cisplatin in renal tubular cells and thus prevent tissue injury in those patients who received the cotreatment of erlotinib [73].

In addition to the tyrosine kinase inhibitors such as erlotinib, another example of an OCT2 inhibitor which can reduce the nephrotoxicity of cisplatin is cimetidine. The functional effects of cisplatin treatment on kidney were studied in both wild type and *Oct1/2* knockout mice [75]. Comparing to the wild type mice, only mild nephrotoxicity was observed after cisplatin treatment in the knockout mice based on the observation of a marked reduction of glucosuria and an absence of proteinuria or apoptosis of renal cells in the knockout mice. In addition, coadministration of cimetidine also demonstrated nephroprotective effect in wild-type mice given cisplatin [75]. These findings are especially important to establish chemotherapeutical protocols aiming to maximize the antineoplastic effect of cisplatin while reducing its nephrotoxic risks. The roles of OCT2 in cisplatin-induced nephrotoxicity have been widely investigated over the last decades and direct evidence came from those studies involving transporter polymorphisms. In cancer patients receiving cisplatin treatment, it was reported that a nonsynonymous single-nucleotide polymorphism (rs3316019) in the *SCL22A2* gene encoding the OCT2 transporter was associated with reduced cisplatin-induced nephrotoxicity [76]. Similarly, it was also found that *SLC22A2* gene polymorphism 808 G/T and cimetidine could attenuate cisplatin nephrotoxicity in Chinese cancer patients [77]. Indeed, in addition to cimetidine, many OCT2 inhibitors such as verapamil, imatinib, tropisetron, and carvedilol have demonstrated the potential protective effect from cisplatin-induced nephrotoxicity in studies carried out in animals as well as in human subjects [75, 78–81]. These results indicate that relatively selective OCT2 inhibitors may be used to attenuate nephrotoxicity of cisplatin.

However, despite the fact that many studies have suggested the nephroprotective role of cimetidine in cisplatin treatment due to its potent inhibition of OCT2, contradictory results have been reported with respect to its effectiveness in protection of cisplatin-induced nephrotoxicity [77, 82–85], which might be partially, if not completely, explained by the fact that cimetidine is a more potent inhibitor toward MATE1 than OCT2 at clinically relevant concentrations (Table 11.2) [29]. This is consistent with clinical studies reporting high doses of cimetidine could reduce cisplatin-induced nephrotoxicity due to inhibition of OCT2 [84, 85] with minimal effects on human pharmacokinetics and antitumor activity of cisplatin [85]. Indeed, among the platinum-based drugs, cisplatin, and oxaliplatin have been reported to exhibit more severe nephrotoxicity than other due to a higher affinity to OCT2. Interestingly, oxaliplatin induces less nephrotoxicity than cisplatin probably because it is a better substrate of MATE2K than MATE1 for excretion into the urine [78, 86, 87]. These data suggested that both OCT2 and MATE play important roles in the nephrotoxicity of cisplatin and the interplay is complex. Therefore, the risk of using transporter inhibitors as nephroprotective agents during cisplatin treatment should be carefully addressed given the opposing effect of OCT2 and MATEs in cisplatin related renal tubular accumulation and toxicity. Furthermore, the underlying mechanism and clinical implication of the interplay among these transporters related to nephrotoxicity deserve further investigation.

11.4 Biomarkers and Nephrotoxicity

11.4.1 Biomarkers for Detecting Glomerular Injury

Serum creatinine levels have been heavily relied on to monitor kidney function, particularly GFR. The basic premise for use of serum creatinine is that it is an endogenous metabolite that is readily filtered in the glomerulus, has little reabsorption in the kidney tubules and can be quantitatively measured rapidly and economically. Thus, increases in serum creatinine, and to a lesser extent BUN are often used diagnostically to identify AKI [88].

While serum creatinine can provide useful information pertaining to GFR and kidney function, reliance on this blood parameter alone for assessment of AKI has several confounding issues. First is the sensitivity of serum creatinine in picking up early stages of kidney decline [89, 90]. Case in point, many clinical trials set a 50% increase in serum creatinine as an acceptable upper limit for drugs that are considered safe for renal function. However a 50% increase in normal serum creatinine levels could represent as much as a 30% decrease in glomerular filtration [91]. Another disadvantage with reliance on serum creatinine levels to assess drug-induced kidney injury is the potential for false positive designations of toxicity where elevations in serum creatinine may reflect DDIs with kidney transporters involved in the tubular secretion of creatinine rather than actual reductions in GFR [40, 41]. For these reasons, other more sensitive blood biomarkers have been evaluated to mechanistically aid in identification of drug-induced changes in the filtration capacity of the kidneys (Table 11.5).

Cystatin C (CysC) is a cysteine proteinase inhibitor produced by all nucleated cells [92]. Intracellularly, CysC is found within the lysosomes where it modulates enzyme activity within this subcellular organelle. However, CysC is also an important extracellular inhibitor of cysteine proteases and can be detected in all body fluids. With a molecular weight of approximately 13 kD, CysC present in the blood is readily filtered by the glomerulus. In a healthy kidney, the filtered CysC would be almost completely reabsorbed through catabolic processes and uptake as the amino acid and di-and tri-peptide metabolites [93]. In contrast to creatinine, serum levels of CysC are not as affected by age and differences in muscle mass. Furthermore, as CysC only undergoes filtration with no tubular secretion [97], serum CysC levels are considered a more accurate reflection of GFR. For these reasons, serum CysC is increasingly used in place of serum creatinine to monitor kidney function during both preclinical and clinical drug development process [94, 95].

Recent studies by Ballew and colleagues [96] illustrate how the use of serum CysC for monitoring of kidney function could also provide advantages to patient care. The study examined kidney function in elderly patients using multiple determinants including serum CysC, serum creatinine, and urine albumin-creatinine ratio (ACR). These studies showed an age dependent reduction in kidney function with an even greater reduction in the subset of patients who were considered frail. When adjusted for covariates, serum creatinine assessment of GFR was only moderately associated with frailty, while serum CysC estimates of GFR and ACR were strongly associated with frailty [96]. Using an estimated GFR $< 60\,\text{ml}\,\text{min}^{-1}/1.73\,\text{m}^2$ as a measure of compromised kidney function, 45% of the frail patients were renal compromised using serum creatinine to estimate GFR compared to 77% with serum CysC [96]. As estimated GFR is commonly used to guide physicians on dosage adjustments for drugs with renal clearance these studies suggest an overestimation of kidney function using serum creatinine vs serum CysC. The clinical ramifications of this are increased risk of potential drug toxicity in this patient population.

Beta-2 microglobulin (B2M) is also used as a blood biomarker for glomerular function. This immunoglobulin is present in the blood and is actively removed from the basement membrane of the glomerulus by neonatal Fc receptor (FcRn) expressed in the podocytes that line the Bowman's capsule [97]. Studies from the Chronic Kidney Disease Epidemiology Collaboration (CKD-EPI) group compared serum B2M based GFR assessment with both CysC and creatinine in a cohort of 2380 patients. Primarily of Caucasian or African American classification [98]. The findings of this study were that serum B2M was inversely correlated with GFR and highly correlated with both serum CysC and creatinine [98]. The variability within serum B2M based on age, sex, and race were similar to that of serum CysC and considerably less than serum creatinine [98]. These findings suggest that for adult patients, both serum CysC and B2M are likely more reflective of GFR than serum creatinine and likely more useful biomarkers for assessing kidney function in clinical trials. Despite the more accurate reflection of GFR for both B2M and CysC, serum creatinine remains the more widely used blood biomarker in the general clinical setting due to ease and cost of assay.

11.4.2 Biomarkers for Drug-Induced Injury to Proximal and Distal Tubules

The assessment of injury to the kidney tubule epithelial cells primarily relies on protein-based urine biomarkers (Table 11.5). These urine biomarkers can be further categorized into the smaller molecular weight proteins that are readily filtered from the blood into the glomerulus, such as CysC, B2M, neutrophil gelatinase-associated lipocalin (NGAL), osteopontin (OPN), and interleukin-18 (IL-18). Under normal conditions these filtered proteins would either be efficiently reabsorbed in the tubules through various receptor-mediated process, as is the case for B2M and FcRn in the proximal tubules [97], or catabolized and reabsorbed as various amino acid and peptide constituents, as is the case for CysC [93]. Thus, following injury to the tubule epithelial cells in the kidney, urine levels of these proteins increase due to diminished capacity for reabsorption in the injured tubules. The remaining urine biomarkers reflect large proteins that do not undergo glomerular filtration such as *N*-acetyl-beta-D-glucosaminidase (NAG), alpha-glutathione-S-transferase (αGST), kidney injury molecule-1 (KIM-1), and clusterin (CLU). For these biomarkers, the kidney epithelial cells produce these proteins in response to injury, and their presence in the urine reflects increased excretion and/or cytotoxic release by the tubular epithelial cells [91]. CysC, B2M, KIM-1, CLU, and OPN are all recognized and accepted biomarkers by both FDA and EMEA for preclinical assessment of drug-induced nephrotoxicity [99–101] and show good correlation with histopathological findings [91]. A list of qualified biomarkers can also be found at the FDA webpage: https://www.fda.gov/drugs/cder-biomarker-qualification-program/list-qualified-biomarkers. In the case of NGAL and IL-18 there is emerging data suggesting usefulness of these biomarkers for assessing kidney injury [101]. While these biomarkers are briefly summarized below, additional information can be found in the comprehensive reviews of van Meer et al. [91] and Bonventre and colleagues [47].

Using a rodent model, Hoffman et al. [102] examined various urine biomarkers in response to treatment with three different kidney toxins. Following gentamicin exposure, the urine levels of CysC, NGAL, B2M, CLU, and KIM-1 were all significantly elevated [102]. In the case of NGAL, CysC, and Kim-1 the urine levels were more predictive of proximal tubule toxicity than BUN or serum creatinine, with CysC being the best of all the markers examined, and NGAL being the most sensitive for proximal tubule damage [102]. While, neither OPN nor α-GST was able to accurately capture gentamicin renal toxicity in this study, other reports have shown increased α-GST [103] and OPN [104] following gentamicin exposure. Increased α-GST, along with CLU and KIM-1 were detected in the urine following exposure to cisplatin, another drug known to produce proximal tubule damage [105]. In studies by Harpur et al. [106] using rodent models exposed to either proximal (gentamicin and cisplatin) or collecting duct (*N*-phenylanthranilic acid) toxins, urinary α-GST, and the other urine markers out performed serum creatinine in identifying proximal tubule toxicity. However, urine levels of α-GST were actually decreased in collecting duct injury and failed to correlate with histological examinations. Of the urine biomarkers examined, CLU was able to effectively identify both proximal and collecting duct injury [106]. In order to fully capture drug-induced kidney damage, it is recommended that these urine biomarkers be collectively used as a panel as opposed to single biomarker [99, 100].

Clinical use of urine biomarkers to assess drug-induced kidney damage is somewhat limited. In patients treated with amphotericin B, a drug that commonly produces distal tubule injury, urinary NGAL was able to detect acute kidney damage earlier than standard serum creatinine [107]. Urine levels of NGAL and Il-18 were also early indicators of renal toxicity in children receiving ifosfamide [108]. In contrast several clinical studies examining cisplatin treatment in cancer patients, found urine levels of KIM-1 were consistently correlated with AKI while NGAL was not [109, 110]. One of the more commonly used urine biomarkers in the clinic is B2M [101]. While increased concentrations of B2M have been observed in urine from patients treated with nephrotoxic agents such as cyclosporin A, gentamicin, and cisplatin, the correlation with development of AKI remains to be determined [101]. These studies speak strongly toward the utility of a panel of urine biomarkers that can be used collectively to identify drug-induced nephrotoxicity instead of a single biomarker that may be subject to considerable drug and timing variations.

11.5 Conclusion

It is commonly recognized that drugs interacting with renal transporters may have marked changes in the pharmacokinetics and clinically significant DDIs. Furthermore, inhibition of renal drug transporters may result in abnormal drug accumulations in renal tubular cells, leading to drug-induced nephrotoxicity. It has become increasingly important to identify sensitive biomarkers to be used in early drug discovery as well as in clinical studies to aid in identification and diagnosis of drug-induced changes in the kidney function.

From the perspective of DDIs, it is important to identify reliable victim drugs that are safe and well characterized and can be used in preclinical and clinical DDI studies involving renal transporters. Metformin is one of the most widely prescribed drugs in the United States with a wide therapeutic index. It was reported that among the top 200 prescribed drugs in the United States, metformin ranks number 4 [111]. Convenience of monitoring blood glucose levels and other related pharmacodynamic end points for metformin and easy to adjust doses makes it a good victim drug for renal transporter DDI study. It should be noted that there are limitations in using metformin as a victim drug for renal transporter mediated DDIs since it may not be representative of drugs that undergo metabolism by cytochrome P450s and involving other renal transporters. In addition, renal function is an important factor for controlling optimal metformin dosage. Caution should be exercised when metformin is combined with other OCT2 substrates as an increased exposure to such drugs may occur. In addition, patients that are elderly or with compromised renal function, may have higher risk of potential DDIs with metformin.

Although there are many DDI studies involving metformin, only a small number achieve clinical importance. In many cases such as with cimetidine, dolutegravir, pyrimethamine, and trimethoprim, although the magnitude of metformin pharmacokinetic changes were significant the pharmacodynamics remained unchanged [13, 19, 20, 34]. In contrast, given multiple transporters involved in the transport of metformin, the total exposure may change only to a minor extent when multiple transport mechanisms are inhibited. A number of DDI studies describing changes in the pharmacodynamic response to metformin without changes in metformin pharmacokinetic have been reported which may have potential involvement of both renal and hepatic metformin uptake transporters [112, 113]. This highlights the importance of including pharmacodynamic measures in DDI studies of metformin, as not all DDIs will be explained solely by changes in plasma concentrations.

Renal transporter mediated DDIs are less common compared to those involving the hepatic transporters even though many drugs and/or their metabolites are eliminated in the urine. A report from 2010 suggested that 32% of the 200 top prescribed drugs in 2010 are cleared by renal mechanisms with greater than 25% of the absorbed dose excreted unchanged in urine [3]. Despite renal elimination being a major excretion pathway, *in vivo* DDI studies involving MATE1 and OCT2 were much less common compared to those of hepatic transporters such as OATP1B1 and OATP1B3 based on an analysis of 103 drugs approved by the FDA from 2013 to 2016 [114]. This may be due in part to the ability of glomerular filtration to compensate for altered transporter-based secretion routes of drug elimination in the kidney. Therefore, DDIs involving inhibition of renal transporters would only account for a portion of the net renal excretion. Among the 103 drugs being approved by the FDA from 2013 to 2016, in vivo transporter mediated DDI studies which have demonstrated ≥1.25fold AUC changes were mainly related to OATP1B1 and OATP1B3 followed by P-gp and BCRP. No cases of DDIs resulting in plasma AUC changes greater than 1.25-fold was mediated by renal transporters [114]. That being said, it is important to include clear information regarding contraindications on the drug labels for those that are potent inhibitors of OCT2 given their potential for enhancing systemic exposures of substrate drugs that have narrow therapeutic windows, such as dofetilide. Drugs that are potent inhibitors of OCT2, including dolutegravir [115], bictegravir [27], and the most recently approved FGFR inhibitor erdafitinib [116], have included label warnings that dofetilide is contraindicated and suggest reducing the dose of OCT2 substrates or consider alternative agents.

Although DDIs involving renal transporters may not be as meaningful as the other transporters in terms of drug systemic exposure, interactions with renal transporters may have great impact on tissue accumulation, therefore, play a crucial role in kidney toxicity. As described in this Chapter drugs with different mechanisms of causing toxicity often affect different parts of the kidney, thereby demonstrating their primary sites of nephron toxicity for each drug or drug class (Table 11.5). This observation is likely explained by that different regions of the nephron are characterized by different expressions of transporters and metabolic enzymes, in addition to distinct blood flow and oxygen tensions. Furthermore, genetic and/or physical characteristics of the patients may enhance their risk for kidney injury. Under certain circumstances, these mechanisms and risk factors may present simultaneously and ultimately explain the variability and heterogeneity observed with drug-induced nephrotoxicity. Early efforts to understand the roles of renal transporters in kidney toxicity are important steps when considering the drug kidney accumulation or active renal secretion in animals or clinical studies. These assessments provide important parameters to help predict drug disposition and understand the potential effect of DDI on systemic exposure or tissue partitioning in order to develop a safe drug and reduce the attrition rate.

There is increasing interest in identifying sensitive biomarkers to more efficiently identify and predict nephrotoxicity not only in early drug development but also during clinical studies. As discussed, drugs-induced nephrotoxicity may affect different parts of the kidney and demonstrate their primary sites of nephron toxicity. Therefore, it is of great interest to identify biomarkers that are associated with injury to their primary sites in the kidney, such as the glomerulus or the proximal tubule, which together represent the major sites of toxicity related to >90% of drugs [47]. Some of these markers of glomerular, proximal tubule as well as the distal and collecting tubules injury are listed in Table 11.5. Many biomarkers have been identified and emerging studies suggest strongly toward the utility of a panel of biomarkers that can be used collectively to identify drug-induced nephrotoxicity as opposed to single biomarker for the consideration of time, resource, and budget efficiency. To this end, earlier identification of nephrotoxicity in preclinical studies using kidney injury biomarkers is beneficial for clinical studies with recommendation of translational biomarkers to facilitate the early management of patients when there is a concern about toxicity. Most importantly, identification of nephrotoxicity earlier and sooner will potentially prevent or reduce the entry of nephrotoxic drugs into the market.

References

1 Cheng Y, El-Kattan A, Zhang Y, Ray AS & Lai Y (2016) Involvement of drug transporters in organ toxicity: the fundamental basis of drug discovery and development. *Chem Res Toxicol* 29: 545–563. doi:https://doi.org/10.1021/acs.chemrestox.5b00511.

2 George B, You D, Joy MS & Aleksunes LM (2017) Xenobiotic transporters and kidney injury. *Adv Drug Deliv Rev* 116: 73–91. doi:https://doi.org/10.1016/j.addr.2017.01.005.

3 Morrissey KM, Stocker SL, Wittwer MB, Xu L & Giacomini KM (2013) Renal transporters in drug development. *Annu Rev Pharmacol Toxicol* 53: 503–529. doi:https://doi.org/10.1146/annurev-pharmtox-011112-140317.

4 Yin J & Wang J (2016) Renal drug transporters and their significance in drug-drug interactions. *Acta Pharm Sin B* 6: 363–373. doi:https://doi.org/10.1016/j.apsb.2016.07.013.

5 Zamek-Gliszczynski MJ, Taub ME, Chothe PP, Chu X, Giacomini KM, Kim RB, Ray AS, Stocker SL, Unadkat JD, Wittwer MB, Xia C, Yee SW, Zhang L, Zhang Y & International Transporter C (2018) Transporters in drug development: 2018 ITC recommendations for transporters of emerging clinical importance. *Clin Pharmacol Ther* 104: 890–899. doi:https://doi.org/10.1002/cpt.1112.

6 Giacomini, K.M. & Sugiyama, Yuichi (2005) Membrane transporters and drug response. In: *Goodman & Gilman's The Pharmacological Basis of Therapeutics* (Brunton, L.L., Lazo, J.S., and Parker, K.L. eds.). McGraw-Hill New York. 41–70

7 International Transporter Consortium, Giacomini KM, Huang SM, Tweedie DJ, Benet LZ, Brouwer KL, Chu X, Dahlin A, Evers R, Fischer V, Hillgren KM, Hoffmaster KA, Ishikawa T, Keppler D, Kim RB, Lee CA, Niemi M, Polli JW, Sugiyama Y, Swaan PW, Ware JA, Wright SH, Yee SW, Zamek-Gliszczynski MJ & Zhang L (2010) Membrane transporters in drug development. *Nat Rev Drug Discov* 9: 215–236. doi:https://doi.org/10.1038/nrd3028.

8 Zhang Y (2018) Overview of transporters in pharmacokinetics and drug discovery. *Curr Protoc Pharmacol* 82: e46. doi:https://doi.org/10.1002/cpph.46.

9 Otsuka M, Matsumoto T, Morimoto R, Arioka S, Omote H & Moriyama Y (2005) A human transporter protein that mediates the final excretion step for toxic organic cations. *Proc Natl Acad Sci U S A* 102: 17923–17928. doi:https://doi.org/10.1073/pnas.0506483102.

10 FDA (2020) in vitro drug interaction studies — Cytochrome P450 enzyme- and transporter-mediated drug interactions guidance for industry. https://www.fda.gov/regulatory-information/search-fda-guidance-documents/vitro-drug-interaction-studies-cytochrome-p450-enzyme-and-transporter-mediated-drug-interactions.

11 Pentikainen PJ, Neuvonen PJ & Penttila A (1979) Pharmacokinetics of metformin after intravenous and oral administration to man. *Eur J Clin Pharmacol* 16: 195–202. doi:https://doi.org/10.1007/bf00562061.

12 Scheen AJ (1996) Clinical pharmacokinetics of metformin. *Clin Pharmacokinet* 30: 359–371. doi:https://doi.org/10.2165/00003088-199630050-00003.

13 Song IH, Zong J, Borland J, Jerva F, Wynne B, Zamek-Gliszczynski MJ, Humphreys JE, Bowers GD & Choukour M (2016) The effect of dolutegravir on the pharmacokinetics of metformin in healthy subjects. *J Acquir Immune Defic Syndr* 72: 400–407. doi:https://doi.org/10.1097/QAI.0000000000000983.

14 Shu Y, Brown C, Castro RA, Shi RJ, Lin ET, Owen RP, Sheardown SA, Yue L, Burchard EG, Brett CM & Giacomini KM (2008) Effect of genetic variation in the organic cation transporter 1, OCT1, on metformin pharmacokinetics. *Clin Pharmacol Ther* 83: 273–280. doi:https://doi.org/10.1038/sj.clpt.6100275.

15 Chen Y, Li S, Brown C, Cheatham S, Castro RA, Leabman MK, Urban TJ, Chen L, Yee SW, Choi JH, Huang Y, Brett CM, Burchard EG & Giacomini KM (2009) Effect of genetic variation in the organic cation transporter 2 on the renal elimination of metformin. *Pharmacogenet Genomics* 19: 497–504. doi:https://doi.org/10.1097/FPC.0b013e32832cc7e9.

16 Christensen MM, Pedersen RS, Stage TB, Brasch-Andersen C, Nielsen F, Damkier P, Beck-Nielsen H & Brosen K (2013) A gene-gene interaction between polymorphisms in the OCT2 and MATE1 genes influences the renal clearance of metformin. *Pharmacogenet Genomics* 23: 526–534. doi:https://doi.org/10.1097/FPC.0b013e328364a57d.

17 Tanihara Y, Masuda S, Sato T, Katsura T, Ogawa O & Inui K (2007) Substrate specificity of MATE1 and MATE2-K, human multidrug and toxin extrusions/H(+)-organic cation antiporters. *Biochem Pharmacol* 74: 359–371. doi:https://doi.org/10.1016/j.bcp.2007.04.010.

18 Chu X, Bleasby K, Chan GH, Nunes I & Evers R (2016) The complexities of interpreting reversible elevated serum creatinine levels in drug development: does a correlation with inhibition of renal transporters exist? *Drug Metab Dispos* 44: 1498–1509. doi:https://doi.org/10.1124/dmd.115.067694.

18 Deeks ED (2018) Bictegravir/emtricitabine/tenofovir alafenamide: a review in HIV-1 infection. *Drugs* 78: 1817–1828. doi:https://doi.org/10.1007/s40265-018-1010-7.

19 Kusuhara H, Ito S, Kumagai Y, Jiang M, Shiroshita T, Moriyama Y, Inoue K, Yuasa H & Sugiyama Y (2011) Effects of a MATE protein inhibitor, pyrimethamine, on the renal elimination of metformin at oral microdose and at therapeutic dose in healthy subjects. *Clin Pharmacol Ther* 89: 837–844. doi:https://doi.org/10.1038/clpt.2011.36.

20 Somogyi A, Stockley C, Keal J, Rolan P & Bochner F (1987) Reduction of metformin renal tubular secretion by cimetidine in man. *Br J Clin Pharmacol* 23: 545–551. doi:https://doi.org/10.1111/j.1365-2125.1987.tb03090.x.

21 Tsuda M, Terada T, Ueba M, Sato T, Masuda S, Katsura T & Inui K (2009) Involvement of human multidrug and toxin extrusion 1 in the drug interaction between cimetidine and metformin in renal epithelial cells. *J Pharmacol Exp Ther* 329: 185–191. doi:https://doi.org/10.1124/jpet.108.147918.

21 Johansson S, Read J, Oliver S, Steinberg M, Li Y, Lisbon E, Mathews D, Leese PT & Martin P (2014) Pharmacokinetic evaluations of the co-administrations of vandetanib and metformin, digoxin, midazolam, omeprazole or ranitidine. *Clin Pharmacokinet* 53: 837–847. doi:https://doi.org/10.1007/s40262-014-0161-2.

22 Koteff J, Borland J, Chen S, Song I, Peppercorn A, Koshiba T, Cannon C, Muster H & Piscitelli SC (2013) A phase 1 study to evaluate the effect of dolutegravir on renal function via measurement of iohexol and para-aminohippurate clearance in healthy subjects. *Br J Clin Pharmacol* 75: 990–996. doi:https://doi.org/10.1111/j.1365-2125.2012.04440.x.

23 Thornton K, Kim G, Maher VE, Chattopadhyay S, Tang S, Moon YJ, Song P, Marathe A, Balakrishnan S, Zhu H, Garnett C, Liu Q, Booth B, Gehrke B, Dorsam R, Verbois L, Ghosh D, Wilson W, Duan J, Sarker H, Miksinski SP, Skarupa L, Ibrahim A, Justice R, Murgo A & Pazdur R (2012) Vandetanib for the treatment of symptomatic or progressive medullary thyroid cancer in patients with unresectable locally advanced or metastatic disease: U.S. food and drug administration drug approval summary. *Clin Cancer Res* 18: 3722–3730. doi:https://doi.org/10.1158/1078-0432.CCR-12-0411.

23 Urakami Y, Kimura N, Okuda M & Inui K (2004) Creatinine transport by basolateral organic cation transporter hOCT2 in the human kidney. *Pharm Res* 21: 976–981. doi:https://doi.org/10.1023/b:pham.0000029286.45788.ad.

24 Lepist EI, Zhang X, Hao J, Huang J, Kosaka A, Birkus G, Murray BP, Bannister R, Cihlar T, Huang Y & Ray AS (2014) Contribution of the organic anion transporter OAT2 to the renal active tubular secretion of creatinine and mechanism for serum creatinine elevations caused by cobicistat. *Kidney Int* 86: 350–357. doi:https://doi.org/10.1038/ki.2014.66.

25 Shen H, Liu T, Morse BL, Zhao Y, Zhang Y, Qiu X, Chen C, Lewin AC, Wang XT, Liu G, Christopher LJ, Marathe P & Lai Y (2015) Characterization of organic anion transporter 2 (SLC22A7): a highly efficient transporter for creatinine and species-dependent renal tubular expression. *Drug Metab Dispos* 43: 984–993. doi:https://doi.org/10.1124/dmd.114.062364.

26 Imamura Y, Murayama N, Okudaira N, Kurihara A, Okazaki O, Izumi T, Inoue K, Yuasa H, Kusuhara H & Sugiyama Y (2011) Prediction of fluoroquinolone-induced elevation in serum creatinine levels: a case of drug-endogenous substance interaction involving the inhibition of renal secretion. *Clin Pharmacol Ther* 89: 81–88. doi:https://doi.org/10.1038/clpt.2010.232.

27 Bictegravir label (2018) BIKTARVY® (bictegravir, emtricitabine, and tenofovir alafenamide) tablets, for oral use https://www.accessdata.fda.gov/drugsatfda_docs/label/2018/210251s000lbl.pdf.

28 Custodio J, West S, Yu A, Martin H, Graham H, Quirk E, & Kearney B (2017) Lack of clinically relevant effect of Bictegravir (BIC, B) on Metformin (MET) Pharmacokinetics (PK) and Pharmacodynamics (PD). *Open Forum Infect Dis* 4(Suppl 1): S429. doi: https://doi.org/10.1093/ofid/ofx163.1082.

29 Ito S, Kusuhara H, Yokochi M, Toyoshima J, Inoue K, Yuasa H & Sugiyama Y (2012) Competitive inhibition of the luminal efflux by multidrug and toxin extrusions, but not basolateral uptake by organic cation transporter 2, is the likely mechanism underlying the pharmacokinetic drug-drug interactions caused by cimetidine in the kidney. *J Pharmacol Exp Ther* 340: 393–403. doi:https://doi.org/10.1124/jpet.111.184986.

30 Somogyi A & Gugler R (1983) Clinical pharmacokinetics of cimetidine. *Clin Pharmacokinet* 8: 463–495. doi:https://doi.org/10.2165/00003088-198308060-00001.

31 Elsby R, Chidlaw S, Outteridge S, Pickering S, Radcliffe A, Sullivan R, Jones H & Butler P (2017) Mechanistic in vitro studies confirm that inhibition of the renal apical efflux transporter multidrug and toxin extrusion (MATE) 1, and not altered absorption, underlies the increased metformin exposure observed in clinical interactions with cimetidine, trimethoprim or pyrimethamine. *Pharmacol Res Perspect* 5. e00357 doi:https://doi.org/10.1002/prp2.357.

32 Zolk O, Solbach TF, Konig J & Fromm MF (2009) Functional characterization of the human organic cation transporter 2 variant p.270Ala>Ser. *Drug Metab Dispos* 37: 1312–1318. doi:https://doi.org/10.1124/dmd.108.023762.

33 Muller F, Konig J, Glaeser H, Schmidt I, Zolk O, Fromm MF & Maas R (2011) Molecular mechanism of renal tubular secretion of the antimalarial drug chloroquine. *Antimicrob Agents Chemother* 55: 3091–3098. doi:https://doi.org/10.1128/AAC.01835-10.

34 Grun B, Kiessling MK, Burhenne J, Riedel KD, Weiss J, Rauch G, Haefeli WE & Czock D (2013) Trimethoprim-metformin interaction and its genetic modulation by OCT2 and MATE1 transporters. *Br J Clin Pharmacol* 76: 787–796. doi:https://doi.org/10.1111/bcp.12079.

35 Hruska MW, Amico JA, Langaee TY, Ferrell RE, Fitzgerald SM & Frye RF (2005) The effect of trimethoprim on CYP2C8 mediated rosiglitazone metabolism in human liver microsomes and healthy subjects. *Br J Clin Pharmacol* 59: 70–79. doi:https://doi.org/10.1111/j.1365-2125.2005.02263.x.

36 Hibma JE, Zur AA, Castro RA, Wittwer MB, Keizer RJ, Yee SW, Goswami S, Stocker SL, Zhang X, Huang Y, Brett CM, Savic RM & Giacomini KM (2016) The effect of famotidine, a MATE1-selective inhibitor, on the pharmacokinetics and pharmacodynamics of metformin. *Clin Pharmacokinet* 55: 711–721. doi:https://doi.org/10.1007/s40262-015-0346-3.

37 Liang X & Giacomini KM (2017) Transporters involved in metformin pharmacokinetics and treatment response. *J Pharm Sci* 106: 2245–2250. doi:https://doi.org/10.1016/j.xphs.2017.04.078.

38 Levey AS, Perrone RD & Madias NE (1988) Serum creatinine and renal function. *Annu Rev Med* 39: 465–490. doi:https://doi.org/10.1146/annurev.me.39.020188.002341.

39 Naderer O, Nafziger AN & Bertino JS, Jr. (1997) Effects of moderate-dose versus high-dose trimethoprim on serum creatinine and creatinine clearance and adverse reactions. *Antimicrob Agents Chemother* 41: 2466–2470.

40 Zhang Y, Warren MS, Zhang X, Diamond S, Williams B, Punwani N, Huang J, Huang Y & Yeleswaram S (2015) Impact on creatinine renal clearance by the interplay of multiple renal transporters: a case study with INCB039110. *Drug Metab Dispos* 43: 485–489. doi:https://doi.org/10.1124/dmd.114.060673.

41 Chappell JC, Turner PK, Pak YA, Bacon J, Chiang AY, Royalty J, Hall SD, Kulanthaivel P & Bonventre JV (2019) Abemaciclib inhibits renal tubular secretion without changing glomerular filtration rate. *Clin Pharmacol Ther* 105: 1187–1195. doi:https://doi.org/10.1002/cpt.1296.

42 Kopec KT & Kowalski MJ (2013) Metformin-associated lactic acidosis (MALA): case files of the Einstein Medical Center medical toxicology fellowship. *J Med Toxicol* 9: 61–66. doi:https://doi.org/10.1007/s13181-012-0278-3.

43 Phillips PJ, Scicchitano R, Clarkson AR, Gilmore HR (1978) Metformin associated lactic acidosis *Aust N Z J Med* 8: 281–284. doi:https://doi.org/10.1111/j.1445-5994.1978.tb04524.x.

44 Abel S, Nichols DJ, Brearley CJ & Eve MD (2000) Effect of cimetidine and ranitidine on pharmacokinetics and pharmacodynamics of a single dose of dofetilide. *Br J Clin Pharmacol* 49: 64–71. doi:https://doi.org/10.1046/j.1365-2125.2000.00114.x.

45 Smith DA, Rasmussen HS, Stopher DA & Walker DK (1992) Pharmacokinetics and metabolism of dofetilide in mouse, rat, dog and man. *Xenobiotica* 22: 709–719. doi:https://doi.org/10.3109/00498259209053133.

46 Dofetilide label (2012) TIKOSYN®. https://labeling.pfizer.com/ShowLabeling.aspx?id=639.

47 Bonventre JV, Vaidya VS, Schmouder R, Feig P & Dieterle F (2010) Next-generation biomarkers for detecting kidney toxicity. *Nat Biotechnol* 28: 436–440. doi:https://doi.org/10.1038/nbt0510-436.

48 Naughton CA (2008) Drug-induced nephrotoxicity. *Am Fam Physician* 78: 743–750.

49 Perazella MA (2018) Pharmacology behind common drug nephrotoxicities. *Clin J Am Soc Nephrol* 13: 1897–1908. doi:https://doi.org/10.2215/CJN.00150118.

50 Saito H (2010) Pathophysiological regulation of renal SLC22A organic ion transporters in acute kidney injury: pharmacological and toxicological implications. *Pharmacol Ther* 125: 79–91. doi:https://doi.org/10.1016/j.pharmthera.2009.09.008.

51 Dixit M, Doan T, Kirschner R & Dixit N (2010) Significant acute kidney injury due to non-steroidal anti-inflammatory drugs: inpatient setting. *Pharmaceuticals (Basel)* 3: 1279–1285. doi:https://doi.org/10.3390/ph3041279.

52 Black HE (1986) Renal toxicity of non-steroidal anti-inflammatory drugs. *Toxicol Pathol* 14: 83–90. doi:https://doi.org/10.1177/019262338601400110.

53 Lucas GNC, Leitao ACC, Alencar RL, Xavier RMF, Daher EF & Silva Junior GBD (2019) Pathophysiological aspects of nephropathy caused by non-steroidal anti-inflammatory drugs. *J Bras Nefrol* 41: 124–130. doi:https://doi.org/10.1590/2175-8239-JBN-2018-0107.

54 Perazella MA (2005) Drug-induced nephropathy: an update. *Expert Opin Drug Saf* 4: 689–706. doi:https://doi.org/10.1517/14740338.4.4.689.

55 Leibbrandt ME, Wolfgang GH, Metz AL, Ozobia AA & Haskins JR (1995) Critical subcellular targets of cisplatin and related platinum analogs in rat renal proximal tubule cells. *Kidney Int* 48: 761–770. doi:https://doi.org/10.1038/ki.1995.348.

56 Matsushima H, Yonemura K, Ohishi K & Hishida A (1998) The role of oxygen free radicals in cisplatin-induced acute renal failure in rats. *J Lab Clin Med* 131: 518–526. doi:https://doi.org/10.1016/s0022-2143(98)90060-9.

57 Thakur V, Walker PD & Shah SV (1988) Evidence suggesting a role for hydroxyl radical in puromycin aminonucleoside-induced proteinuria. *Kidney Int* 34: 494–499. doi:https://doi.org/10.1038/ki.1988.208.

58 Walker PD & Shah SV (1988) Evidence suggesting a role for hydroxyl radical in gentamicin-induced acute renal failure in rats. *J Clin Invest* 81: 334–341. doi:https://doi.org/10.1172/JCI113325.

59 Kirsch LS, Arevalo JF, Chavez de la Paz E, Munguia D, de Clercq E & Freeman WR (1995) Intravitreal cidofovir (HPMPC) treatment of cytomegalovirus retinitis in patients with acquired immune deficiency syndrome. *Ophthalmology* 102: 533–542; discussion 542–533. doi:https://doi.org/10.1016/s0161-6420(95)30985-2.

60 Cundy KC, Petty BG, Flaherty J, Fisher PE, Polis MA, Wachsman M, Lietman PS, Lalezari JP, Hitchcock MJ & Jaffe HS (1995) Clinical pharmacokinetics of cidofovir in human immunodeficiency virus-infected patients. *Antimicrob Agents Chemother* 39: 1247–1252. doi:https://doi.org/10.1128/aac.39.6.1247.

61 Ho ES, Lin DC, Mendel DB & Cihlar T (2000) Cytotoxicity of antiviral nucleotides adefovir and cidofovir is induced by the expression of human renal organic anion transporter 1. *J Am Soc Nephrol* 11: 383–393.

62 Yao X, Panichpisal K, Kurtzman N & Nugent K (2007) Cisplatin nephrotoxicity: a review. *Am J Med Sci* 334: 115–124. doi:https://doi.org/10.1097/MAJ.0b013e31812dfe1e.

63 Ruggiero A, Rizzo D, Trombatore G, Maurizi P & Riccardi R (2016) The ability of mannitol to decrease cisplatin-induced nephrotoxicity in children: real or not? *Cancer Chemother Pharmacol* 77: 19–26. doi:https://doi.org/10.1007/s00280-015-2913-6.

64 Filipski KK, Loos WJ, Verweij J & Sparreboom A (2008) Interaction of cisplatin with the human organic cation transporter 2. *Clin Cancer Res* 14: 3875–3880. doi:https://doi.org/10.1158/1078-0432.CCR-07-4793.

65 Nakamura T, Yonezawa A, Hashimoto S, Katsura T & Inui K (2010) Disruption of multidrug and toxin extrusion MATE1 potentiates cisplatin-induced nephrotoxicity. *Biochem Pharmacol* 80: 1762–1767. doi:https://doi.org/10.1016/j.bcp.2010.08.019.

66 Kou W, Qin H, Hanif S & Wu X (2018) Nephrotoxicity evaluation on cisplatin combined with 5-HT3 receptor antagonists: a retrospective study. *Biomed Res Int* 2018: 1024324. doi:https://doi.org/10.1155/2018/1024324.

67 Li Q, Guo D, Dong Z, Zhang W, Zhang L, Huang SM, Polli JE & Shu Y (2013) Ondansetron can enhance cisplatin-induced nephrotoxicity via inhibition of multiple toxin and extrusion proteins (MATEs). *Toxicol Appl Pharmacol* 273: 100–109. doi:https://doi.org/10.1016/j.taap.2013.08.024.

68 Hesketh PJ, Grunberg SM, Gralla RJ, Warr DG, Roila F, de Wit R, Chawla SP, Carides AD, Ianus J, Elmer ME, Evans JK, Beck K, Reines S, Horgan KJ & Aprepitant Protocol 052 Study G (2003) The oral neurokinin-1 antagonist aprepitant for the prevention of chemotherapy-induced nausea and vomiting: a multinational, randomized, double-blind, placebo-controlled trial in patients receiving high-dose cisplatin--the Aprepitant Protocol 052 Study Group. *J Clin Oncol* 21: 4112–4119. doi:https://doi.org/10.1200/JCO.2003.01.095.

69 Shen H, Yang Z, Zhao W, Zhang Y & Rodrigues AD (2013) Assessment of vandetanib as an inhibitor of various human renal transporters: inhibition of multidrug and toxin extrusion as a possible mechanism leading to decreased cisplatin and creatinine clearance. *Drug Metab Dispos* 41: 2095–2103. doi:https://doi.org/10.1124/dmd.113.053215.

70 Blackhall FH, O'Brien M, Schmid P, Nicolson M, Taylor P, Milenkova T, Kennedy SJ & Thatcher N (2010) A phase I study of Vandetanib in combination with vinorelbine/cisplatin or gemcitabine/cisplatin as first-line treatment for advanced non-small cell lung cancer. *J Thorac Oncol* 5: 1285–1288. doi:https://doi.org/10.1097/JTO.0b013e3181e3a2d1.

71 Izzedine H, Hulot JS, Villard E, Goyenvalle C, Dominguez S, Ghosn J, Valantin MA, Lechat P & Deray AG (2006) Association between ABCC2 gene haplotypes and tenofovir-induced proximal tubulopathy. *J Infect Dis* 194: 1481–1491. doi:https://doi.org/10.1086/508546.

72 Minematsu T & Giacomini KM (2011) Interactions of tyrosine kinase inhibitors with organic cation transporters and multidrug and toxic compound extrusion proteins. *Mol Cancer Ther* 10: 531–539. doi:https://doi.org/10.1158/1535-7163.MCT-10-0731.

73 Sprowl JA, Mathijssen RH & Sparreboom A (2013) Can erlotinib ameliorate cisplatin-induced toxicities? *J Clin Oncol* 31: 3442–3443. doi:https://doi.org/10.1200/JCO.2013.50.8184.

74 Martins RG, Parvathaneni U, Bauman JE, Sharma AK, Raez LE, Papagikos MA, Yunus F, Kurland BF, Eaton KD, Liao JJ, Mendez E, Futran N, Wang DX, Chai X, Wallace SG, Austin M, Schmidt R & Hayes DN (2013) Cisplatin and radiotherapy with or without erlotinib in locally advanced squamous cell carcinoma of the head and neck: a randomized phase II trial. *J Clin Oncol* 31: 1415–1421. doi:https://doi.org/10.1200/JCO.2012.46.3299.

75 Ciarimboli G, Deuster D, Knief A, Sperling M, Holtkamp M, Edemir B, Pavenstadt H, Lanvers-Kaminsky C, am Zehnhoff-Dinnesen A, Schinkel AH, Koepsell H, Jurgens H & Schlatter E (2010) Organic cation transporter 2 mediates cisplatin-induced oto- and nephrotoxicity and is a target for protective interventions. *Am J Pathol* 176: 1169–1180. doi:https://doi.org/10.2353/ajpath.2010.090610.

76 Filipski KK, Mathijssen RH, Mikkelsen TS, Schinkel AH & Sparreboom A (2009) Contribution of organic cation transporter 2 (OCT2) to cisplatin-induced nephrotoxicity. *Clin Pharmacol Ther* 86: 396–402. doi:https://doi.org/10.1038/clpt.2009.139.

77 Zhang J & Zhou W (2012) Ameliorative effects of SLC22A2 gene polymorphism 808 G/T and cimetidine on cisplatin-induced nephrotoxicity in Chinese cancer patients. *Food Chem Toxicol* 50: 2289–2293. doi:https://doi.org/10.1016/j.fct.2012.03.077.

78 Guo D, Yang H, Li Q, Bae HJ, Obianom O, Zeng S, Su T, Polli JE & Shu Y (2018) Selective inhibition on organic cation transporters by carvedilol protects mice from cisplatin-induced nephrotoxicity. *Pharm Res* 35: 204. doi:https://doi.org/10.1007/s11095-018-2486-2.

79 Offerman JJ, Meijer S, Sleijfer DT, Mulder NH, Donker AJ, Schraffordt Koops H & van der Hem GK (1985) The influence of verapamil on renal function in patients treated with cisplatin. *Clin Nephrol* 24: 249–255.

80 Tanihara Y, Masuda S, Katsura T & Inui K (2009) Protective effect of concomitant administration of imatinib on cisplatin-induced nephrotoxicity focusing on renal organic cation transporter OCT2. *Biochem Pharmacol* 78: 1263–1271. doi:https://doi.org/10.1016/j.bcp.2009.06.014.

81 Zirak MR, Rahimian R, Ghazi-Khansari M, Abbasi A, Razmi A, Mehr SE, Mousavizadeh K & Dehpour AR (2014) Tropisetron attenuates cisplatin-induced nephrotoxicity in mice. *Eur J Pharmacol* 738: 222–229. doi:https://doi.org/10.1016/j.ejphar.2014.05.050.

82 Dorr RT & Soble MJ (1988) Cimetidine enhances cisplatin toxicity in mice. *J Cancer Res Clin Oncol* 114: 1–2. doi:https://doi.org/10.1007/bf00390477.

83 Katsuda H, Yamashita M, Katsura H, Yu J, Waki Y, Nagata N, Sai Y & Miyamoto K (2010) Protecting cisplatin-induced nephrotoxicity with cimetidine does not affect antitumor activity. *Biol Pharm Bull* 33: 1867–1871. doi:https://doi.org/10.1248/bpb.33.1867.

84 Sleijfer DT, Offerman JJ, Mulder NH, Verweij M, van der Hem GK, Schraffordt Koops HS & Meijer S (1987) The protective potential of the combination of verapamil and cimetidine on cisplatin-induced nephrotoxicity in man. *Cancer* 60: 2823–2828. doi:https://doi.org/10.1002/1097-0142(19871201)60:11<2823::aid-cncr2820601138>3.0.co;2-c.

85 Sprowl JA, van Doorn L, Hu S, van Gerven L, de Bruijn P, Li L, Gibson AA, Mathijssen RH & Sparreboom A (2013) Conjunctive therapy of cisplatin with the OCT2 inhibitor cimetidine: influence on antitumor efficacy and systemic clearance. *Clin Pharmacol Ther* 94: 585–592. doi:https://doi.org/10.1038/clpt.2013.145.

86 Yokoo S, Yonezawa A, Masuda S, Fukatsu A, Katsura T & Inui K (2007) Differential contribution of organic cation transporters, OCT2 and MATE1, in platinum agent-induced nephrotoxicity. *Biochem Pharmacol* 74: 477–487. doi:https://doi.org/10.1016/j.bcp.2007.03.004.

87 Yonezawa A, Masuda S, Yokoo S, Katsura T & Inui K (2006) Cisplatin and oxaliplatin, but not carboplatin and nedaplatin, are substrates for human organic cation transporters (SLC22A1-3 and multidrug and toxin extrusion family). *J Pharmacol Exp Ther* 319: 879–886. doi:https://doi.org/10.1124/jpet.106.110346.

88 Kellum JA, Levin N, Bouman C & Lameire N (2002) Developing a consensus classification system for acute renal failure. *Curr Opin Crit Care* 8: 509–514. doi:https://doi.org/10.1097/00075198-200212000-00005.

89 Botev R, Mallie JP, Wetzels JF, Couchoud C & Schuck O (2011) The clinician and estimation of glomerular filtration rate by creatinine-based formulas: current limitations and quo vadis. *Clin J Am Soc Nephrol* 6: 937–950. doi:https://doi.org/10.2215/CJN.09241010.

90 Murty MS, Sharma UK, Pandey VB & Kankare SB (2013) Serum cystatin C as a marker of renal function in detection of early acute kidney injury. *Indian J Nephrol* 23: 180–183. doi:https://doi.org/10.4103/0971-4065.111840.

91 van Meer L, Moerland M, Cohen AF & Burggraaf J (2014) Urinary kidney biomarkers for early detection of nephrotoxicity in clinical drug development. *Br J Clin Pharmacol* 77: 947–957. doi:https://doi.org/10.1111/bcp.12282.

92 Olafsson I (1995) The human cystatin C gene promoter: functional analysis and identification of heterogeneous mRNA. *Scand J Clin Lab Invest* 55: 597–607. doi:https://doi.org/10.3109/00365519509110259.

93 Tenstad O, Roald AB, Grubb A & Aukland K (1996) Renal handling of radiolabelled human cystatin C in the rat. *Scand J Clin Lab Invest* 56: 409–414. doi:https://doi.org/10.3109/00365519609088795.

94 Dieterle F, Marrer E, Suzuki E, Grenet O, Cordier A & Vonderscher J (2008) Monitoring kidney safety in drug development: emerging technologies and their implications. *Curr Opin Drug Discov Devel* 11: 60–71.

95 Kar S, Paglialunga S & Islam R (2018) Cystatin C is a more reliable biomarker for determining eGFR to support drug development studies. *J Clin Pharmacol* 58: 1239–1247. doi:https://doi.org/10.1002/jcph.1132.

96 Ballew SH, Chen Y, Daya NR, Godino JG, Windham BG, McAdams-DeMarco M, Coresh J, Selvin E & Grams ME (2017) Frailty, kidney function, and polypharmacy: the atherosclerosis risk in communities (aric) study. *Am J Kidney Dis* 69: 228–236. doi:https://doi.org/10.1053/j.ajkd.2016.08.034.

97 Akilesh S, Huber TB, Wu H, Wang G, Hartleben B, Kopp JB, Miner JH, Roopenian DC, Unanue ER & Shaw AS (2008) Podocytes use FcRn to clear IgG from the glomerular basement membrane. *Proc Natl Acad Sci U S A* 105: 967–972. doi:https://doi.org/10.1073/pnas.0711515105.

98 Argyropoulos CP, Chen SS, Ng YH, Roumelioti ME, Shaffi K, Singh PP & Tzamaloukas AH (2017) Rediscovering Beta-2 microglobulin as a biomarker across the spectrum of kidney diseases. *Front Med (Lausanne)* 4: 73. doi:https://doi.org/10.3389/fmed.2017.00073.

99 FDA (2009) Review of qualification data for biomarkers of nephrotoxicity submitted by the predictive safety testing consortium. https://www.fda.gov/media/87781/download.

100 FDA & EMEA (2009) Final conclusions on the pilot joint EMEA/FDA VXDS experience on qualification of nephrotoxicity biomarkers. https://www.ema.europa.eu/en/documents/regulatory-procedural-guideline/final-conclusions-pilot-joint-european-medicines-agency/food-drug-administration-vxds-experience-qualification-nephrotoxicity-biomarkers_en.pdf.

101 Griffin BR, Faubel S & Edelstein CL (2019) Biomarkers of drug-induced kidney toxicity. *Ther Drug Monit* 41: 213–226. doi:https://doi.org/10.1097/FTD.0000000000000589.

102 Hoffmann D, Fuchs TC, Henzler T, Matheis KA, Herget T, Dekant W, Hewitt P & Mally A (2010) Evaluation of a urinary kidney biomarker panel in rat models of acute and subchronic nephrotoxicity. *Toxicology* 277: 49–58. doi:https://doi.org/10.1016/j.tox.2010.08.013.

103 Feinfeld DA, Fleischner GM & Arias IM (1981) Urinary ligandin and glutathione-S-transferase in gentamicin-induced nephrotoxicity in the rat. *Clin Sci (Lond)* 61: 123–125. doi:https://doi.org/10.1042/cs0610123.

104 Phillips JA, Holder DJ, Ennulat D, Gautier JC, Sauer JM, Yang Y, McDuffie E, Sonee M, Gu YZ, Troth SP, Lynch K, Hamlin D, Peters DG, Brees D & Walker EG (2016) Rat urinary osteopontin and neutrophil gelatinase-associated lipocalin improve certainty of detecting drug-induced kidney injury. *Toxicol Sci* 151: 214–223. doi:https://doi.org/10.1093/toxsci/kfw038.

105 Sasaki D, Yamada A, Umeno H, Kurihara H, Nakatsuji S, Fujihira S, Tsubota K, Ono M, Moriguchi A, Watanabe K & Seki J (2011) Comparison of the course of biomarker changes and kidney injury in a rat model of drug-induced acute kidney injury. *Biomarkers* 16: 553–566. doi:https://doi.org/10.3109/1354750X.2011.613123.

106 Harpur E, Ennulat D, Hoffman D, Betton G, Gautier JC, Riefke B, Bounous D, Schuster K, Beushausen S, Guffroy M, Shaw M, Lock E, Pettit S & Nephrotoxicity HCoBo (2011) Biological qualification of biomarkers of chemical-induced renal toxicity in two strains of male rat. *Toxicol Sci* 122: 235–252. doi:https://doi.org/10.1093/toxsci/kfr112.

107 Rocha PN, Macedo MN, Kobayashi CD, Moreno L, Guimaraes LH, Machado PR, Badaro R, Carvalho EM & Glesby MJ (2015) Role of urine neutrophil gelatinase-associated lipocalin in the early diagnosis of amphotericin B-induced acute kidney injury. *Antimicrob Agents Chemother* 59: 6913–6921. doi:https://doi.org/10.1128/AAC.01079-15.

108 Sterling M, Al-Ismaili Z, McMahon KR, Piccioni M, Pizzi M, Mottes T, Lands LC, Abish S, Fleming AJ, Bennett MR, Palijan A, Devarajan P, Goldstein SL, O'Brien MM & Zappitelli M (2017) Urine biomarkers of acute kidney injury in noncritically ill, hospitalized children treated with chemotherapy. *Pediatr Blood Cancer* 64. e26538 doi:https://doi.org/10.1002/pbc.26538.

109 George B, Joy MS & Aleksunes LM (2018) Urinary protein biomarkers of kidney injury in patients receiving cisplatin chemotherapy. *Exp Biol Med (Maywood)* 243: 272–282. doi:https://doi.org/10.1177/1535370217745302.

110 Shinke H, Masuda S, Togashi Y, Ikemi Y, Ozawa A, Sato T, Kim YH, Mishima M, Ichimura T, Bonventre JV & Matsubara K (2015) Urinary kidney injury molecule-1 and monocyte chemotactic protein-1 are noninvasive biomarkers of cisplatin-induced nephrotoxicity in lung cancer patients. *Cancer Chemother Pharmacol* 76: 989–996. doi:https://doi.org/10.1007/s00280-015-2880-y.

111 Fuentes AV, Pineda MD & Venkata KCN (2018) Comprehension of top 200 prescribed drugs in the US as a resource for pharmacy teaching, training and practice. *Pharmacy (Basel)* 6. 43doi:https://doi.org/10.3390/pharmacy6020043.

112 Cho SK, Kim CO, Park ES & Chung JY (2014) Verapamil decreases the glucose-lowering effect of metformin in healthy volunteers. *Br J Clin Pharmacol* 78: 1426–1432. doi:https://doi.org/10.1111/bcp.12476.

113 Cho SK, Yoon JS, Lee MG, Lee DH, Lim LA, Park K, Park MS & Chung JY (2011) Rifampin enhances the glucose-lowering effect of metformin and increases OCT1 mRNA levels in healthy participants. *Clin Pharmacol Ther* 89: 416–421. doi:https://doi.org/10.1038/clpt.2010.266.

114 Yu J, Zhou Z, Tay-Sontheimer J, Levy RH & Ragueneau-Majlessi I (2018) Risk of clinically relevant pharmacokinetic-based drug-drug interactions with drugs approved by the U.S. food and drug administration between 2013 and 2016. *Drug Metab Dispos* 46: 835–845. doi:https://doi.org/10.1124/dmd.117.078691.

115 Dolutegravir label (2013) TIVICAY®, (dolutegravir tablets). for oral use https://www.accessdata.fda.gov/drugsatfda_docs/label/2013/204790lbl.pdf.

116 Erdafitinib label (2019) BALVERSA® tablets. for oral use https://www.accessdata.fda.gov/drugsatfda_docs/label/2019/212018s000lbl.pdf.

12

Blood–Brain Barrier Transporters and Central Nervous System Drug Response and Toxicity

Donald W. Miller[1], Stacey Line[1], Nur A. Safa[1], and Yan Zhang[2]

[1] *Department of Pharmacology and Therapeutics, University of Manitoba, Winnipeg, MB, Canada*
[2] *DMPK, Nuvation Bio, Inc., New York, NY, USA*

12.1 Over-View of the Brain Barriers

The brain is a complex organ tasked with carrying out sensory perception, coordinated movement, emotional responses, organized thought and both long- and short-term memory functions. These tasks rely on organized networks of neurons that participate in receiving and sending electrochemical signals. Within this context, there are two cellular barriers, the blood–brain barrier (BBB) and the blood–cerebral spinal fluid barrier (BCSFB) that together help maintain the proper extracellular environment required for normal brain function (Figure 12.1). While the BBB and BCSFB share many similar biochemical and anatomical features, they are distinct and separate barrier systems.

12.1.1 Blood–Brain Barrier (BBB)

The BBB is comprised of the brain capillary endothelial cells and surrounding astrocytic foot processes and scattered pericytes (see Figure 12.1). The continuous endothelial cells found within the brain microvasculature have well-established tight junctions and adherens junctions that limit paracellular diffusion of solutes from the blood to the brain [1, 2]. Additional features that differentiate brain capillary endothelial cells from those found within peripheral tissues are an absence of fenestrations, reduced vesicular transport activity and a wide variety of membrane proteins that facilitate the uptake and efflux of select solutes [1, 3–5]. Together these anatomical and biochemical features provide for a selectively impermeable cell interface to solutes, macromolecules, and drugs.

While it is the brain endothelial cells that provide the cellular barrier to solute permeability, the astrocytes and pericytes provide structural support and can modulate the barrier properties of the brain endothelium [6]. Astrocytes provide

Transporters and Drug-Metabolizing Enzymes in Drug Toxicity, First Edition. Edited by Albert P. Li.

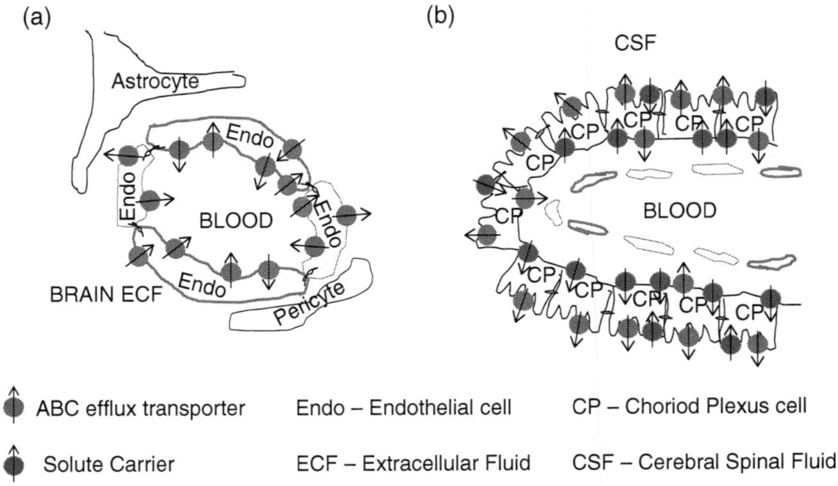

Figure 12.1 Schematic of blood–brain (a) and blood–cerebral spinal (b) barriers. Both cellular barriers are characterized by tight intercellular junctions that limit paracellular diffusion of solutes and the presence of a variety of ABC efflux transporter proteins and solute carriers (SLC) for directional passage of solutes across the cellular interfaces. A more detailed listing of specific ABC and SLC transport proteins can be found in Table 12.1.

for local control of cerebrovascular blood flow and BBB permeability through the release of various agents such as nitric oxide, prostaglandins, and bioactive phospholipids [7, 8]. Pericytes are contractile cells that are believed to have an important role in regulating blood flow in capillaries [9]. These cells are also critical for early formation of tight junctions during development of the BBB [10], as well as reduced transcytosis of proteins and the regulation of immune cell migration in the mature BBB [11, 12]. Their importance in maintaining BBB integrity is best illustrated in the correlation of pericyte loss with the leak of macromolecules and proteins into the brain that occurs in various neurodegenerative conditions [13].

The importance of the brain microenvironment for establishing and maintaining the barrier properties of the brain microvessel endothelial cells was demonstrated by the studies of Stewart and Wiley [14]. Using quail embryos, vascular tissue from the intestines were implanted into the brain. The transplanted vascular tissue adopted the same BBB phenotype as the surrounding cerebral vasculature [14]. Furthermore, cerebral vascular tissue implanted into the periphery soon lost its BBB phenotype properties and resembled the more leaky vasculature of the tissue it was implanted in [14]. Similar studies were performed in developing chick embryos confirming the role of the brain microenvironment in inducing a BBB phenotype [15].

12.1.2 Blood–Cerebrospinal Fluid Barrier (BCSFB)

The BCSFB is formed by the epithelial cells in the choroid plexus (CP) that line the ventricles in the brain (see schematic Figure 12.1). The capillaries that supply blood to the CP are discontinuous and it is the epithelial cells of the CP that provide the cellular interface between the blood and cerebral spinal fluid (CSF). The BSCFB has many analogies with the kidney tubules. The cerebral vasculature within the CP is anatomically similar to the glomerulus and the CSF has been referred to as the urine of the brain. In addition to secreting CSF into the ventricles, the CP epithelial cells regulate CSF composition through the selected transfer of solutes, growth factors, cytokines, and even immune cells into the CSF [16, 17]. Like the proximal tubule cells of the kidney, the epithelial cells of the CP express a variety of solute carriers (SLC) and efflux transporters that are involved in the secretion of solutes into the CSF [16–18]. While there is significant overlap in both SLC and ATP-binding cassette (ABC) transporters in the BBB and BSCF, less is understood about the specific transporters in the BSCF and their function in regulating drug levels in the brain. Proteomics comparing transporter expression in rat CP cells to human CP suggest that there could be species differences with regard to specific transporters expressed at the BCSFB [18]. In addition, there are significant species differences in CSF turnover rates in rodents versus humans that may impact drug concentration in the CSF.

12.1.3 CSF as Predictor of Drug Exposure in the Brain

From a drug delivery and central nervous system (CNS) drug development standpoint, the BBB is the primary brain barrier interface of interest and the focus of this chapter. However, the BCSFB has a critical role in determining the composition of the CSF, and to the extent that drug levels in the CSF are used clinically as a surrogate marker of drug levels in the brain, knowledge of BCSFB dynamics can help guide interpretation of these results. From an industry perspective, CSF levels of drugs have long been considered to be representative of brain levels [19, 20]. While this holds true for many drugs, there are some notable exceptions where unbound CSF drug levels are substantially greater than actual extracellular brain concentrations of unbound drug. In these cases, the discrepancies between unbound extracellular brain concentrations of drug and unbound CSF levels of drug reflect the transport activity of the drug with ABC efflux transporters, most notably P-glycoprotein (P-gp) and breast cancer resistance protein (BCRP). This was demonstrated in the studies by Kodaira et al. [21] examining the steady-state plasma, CSF and brain concentrations of a series of 25 drugs in rats. For many of the drugs, the CSF/brain ratio was considerably greater than 1, indicative of substantially higher drug accumulation in the CSF compared to the brain [21]. As the drugs with the highest CSF/brain ratios were also either P-gp and/or BCRP substrates, follow-up studies with selected drugs were performed in *Mdr1a* −/−, *Bcrp* −/−, and *Mdr1a, Bcrp* −/− mice. The resulting CSF/brain ratios for the drugs were significantly reduced in the transporter deficient mice as well as in the mice receiving pharmacological inhibitors of P-gp and BCRP [21]. These studies point out the potential over-estimation of CNS drug concentrations when CSF is used as a surrogate marker for drugs with P-gp or BCRP transporter activity. Kodaira et al. [22] proposed a pharmacokinetic model with scaling of drug efflux transporter interactions that allows for more accurate predictions of brain concentrations of drugs. The model performed well in preclinical studies but remains to be fully validated for use in clinical trials.

12.1.4 Solute Carriers in the BBB

The SLC superfamily represents a large and diverse group of membrane transport proteins. SLC family member nomenclature follows the general format, SLCnXm, where SLC stands for SLC, n is the number from 1 to 65 designating the family of proteins, X is a letter representing the subfamily and m is an integer number representing the protein isoform. The human SLCs include 65 different protein families and at least 298 different genes [23–25]. These transport proteins are responsible for the exchange of a multitude of nutrients, metabolites, counter ions, and drugs in a variety of tissues throughout the body. Of the 298 SLC genes that have been characterized, approximately 250 are expressed in the brain microvessels [26]. The most prominent SLCs expressed in the brain microvessel endothelial cells that facilitate solute entry into the brain are listed in Table 12.1.

While most of the SLCs in the BBB participate in the blood to brain transfer of solutes, a few such as SLC22 (Oat2), SLC21 (Oat1), and SCL15 (PepT) function to remove solutes from the brain. While most of the studies of brain to blood transport for the SLCs is preclinical there is increasing evidence that establishes their role as efflux transporters [27, 28]. It is reported that Zidovudine influx into brain is limited by the SLC22 efflux transporter present on the apical plasma membrane of the CP epithelial cells [29]. However, SLC22A8 is also a major efflux transporter present in the BBB [28].

12.1.5 Drug Efflux Transporters in the BBB

The BBB also has a wide array of active drug efflux transporters, belonging to the ABC superfamily of proteins [4, 30]. The ABC transporters are integral membrane proteins located predominantly on the luminal (apical) side of the polarized brain endothelial cells and function to remove solutes from the cell through an active ATP-driven process. The ABC transporters expressed in the brain endothelial cells

Table 12.1 ABC and SLC transporters in CNS fluid barriers.

		Blood–brain barrier (BBB)	Blood–cerebrospinal fluid barrier (BCSFB)				
Transporter	**Drug substrates**[a]	**Apical**	**Basolateral**	**Effect**	**Apical**	**Basolateral**	**Effect**
P-gp	Abacavir	X		↓ Brain	X		↑ CSF
	Cyclosporin						
	Imatinib						
	Fexofenadine						
	Indinavir						
	Cetirizine						
	Cerivastatin						
	Ivermectin						
	Loratadine						
	Loperamide						
	Vinblastine						
BCRP	Cimetidine	X		↓ Brain	X		↑ CSF
	Gefitinib						
	Imatinib						
	Mitoxantrone						
	Topotecan						

(Continued)

Table 12.1 (Continued)

		Blood–brain barrier (BBB)	Blood–cerebrospinal fluid barrier (BCSFB)				
Transporter	**Drug substrates**[a]	**Apical**	**Basolateral**	**Effect**	**Apical**	**Basolateral**	**Effect**
MRP1		X		↓ Brain		X	↓ CSF
MRP4	Topotecan	X	X	??		X	↓ CSF
	PMEA						
OAT3	6-Mercaptopurine	X	X	↓ Brain	X		↓ CSF
	6-Thioguanine						
OATP1A2	Fexofenadine	X		↑ Brain			
	Levofloxacin						
	Methotrexate						
	Darunavir						
OCT1	Thiamine, choline, lamotrigene	X		↑ Brain			
OCTN1	Carnitine	X		↑ Brain			
LAT1 (SLC7A5)	Gabapentin Pregabalin Levadopa Methyldopa	X	X	↑ Brain			
Mono-carboxylate Transporter1 (MCT1/SLC16A1)	GHB, simvastatin, lovastatin, beta-lactam antibiotics	X		↑ Brain			
Concentrative nucleoside transporter (CNT2/SLC28A2) and (CNT3/SLC28A3)	Cladribine, gemcitabine, zidovudine	X		↑ Brain	X		↓ CSF

[a] Representative drugs that are substrates for the indicated transporters in Table 1.

forming the BBB play a critical role in neuroprotection, drug penetration into the brain and pharmacoresistance [31]. Efflux transporters act as a general defense mechanism at the BBB, protecting the brain from toxins and potentially harmful agents that would otherwise readily enter the brain by simple diffusion without limitations [30]. Efflux transporters function to pump a variety of lipophilic, amphipathic substrates across the plasma membrane back into the blood [4, 32].

12.2 General Influence of BBB Transporters on Drug Entry into the Brain

While most drugs cross the BBB through a passive diffusion process, the chemical space occupied by those agents is rather confined. This concept is eloquently illustrated by the *B*rain *O*r *I*ntestina*L* *E*stimate*D* permeation (BOILED-Egg) model created by Daina and Zoete [33] at the Swiss Institute for Bioinformatics (Figure 12.2). In contrast to rule-based models for predicting small molecule drug permeability, the BOILED-Egg models intestinal absorption and BBB permeability based on two physicochemical parameters, lipophilicity, as assessed by Wildman and Crippen method for calculation of octanol water partitioning coefficients (WLOGP), and topological polar surface area (TPSA) based on 2-D fragmentation modeling [33]. Appling WLOGP and TPSA to large datasets of small molecule drugs with known oral absorption and brain penetration properties, the authors found oral absorption via passive diffusion was highly probable for those compounds with polar surface area less than 142 A and a log P between −2.3 and 6.8 [33]. The drugs having those physicochemical properties are represented graphically by the white elliptical area in Figure 12.2. Extending this analysis to predict brain penetration, the authors found that those compounds with high probability of passive diffusion across the BBB were more confined in terms of both size (polar surface area less than 80 A) and lipophilicity (Log P between 0.4 and 6.0) [33]. The drugs having those physicochemical properties fell within the shaded area on the graph in Figure 12.2, and together the oral absorption and brain penetration of molecules resemble a boiled egg, hence the name.

While this simplistic method of predicting the passive permeability of small molecules in the BBB resulted in 90% or better prediction of brain penetrating drugs, there were compounds that satisfied the physicochemical criteria but lacked adequate brain penetration [33]. Most of the cases of false positives for brain penetration are attributable to drug efflux transporter liabilities of the drugs. Indeed, studies by Broccatelli et al. [34] examining the BBB penetration of Class 1 compounds in the biopharmaceutical classification system (BCS) found BBB drug efflux transport by P-gp to be the most likely reason for failure of Class I compounds to enter into the brain. Compared to other cellular barriers, drug

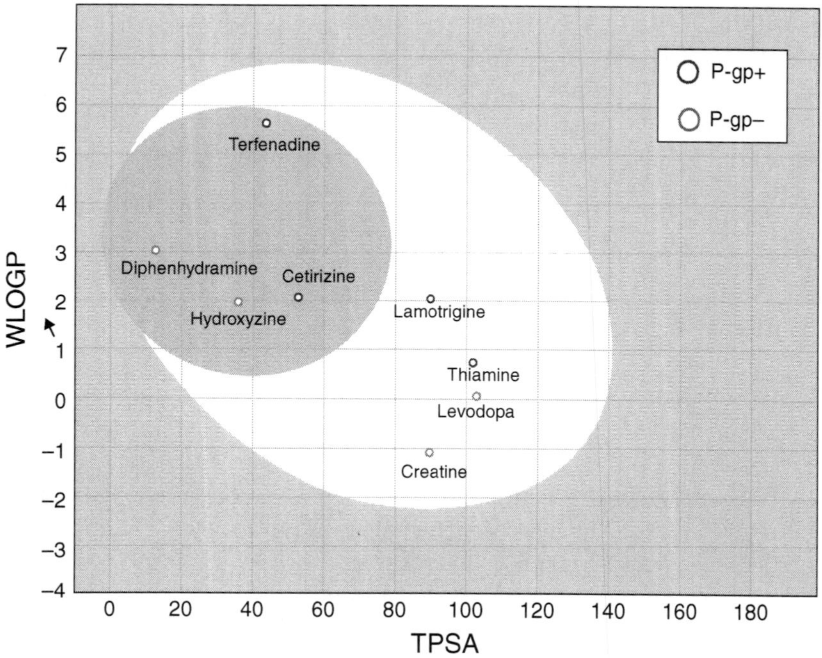

Figure 12.2 BOILED-Egg model depicting the lipophilicity (WLOGP) and topographical polar surface area (TPSA) properties occupied by drugs that are orally bioavailable (white area) and brain penetrable (shaded area) via passive diffusion. Within the shaded area are various antihistamines – diphenhydramine and hydroxyzine are CNS active, cetirizine and terfenadine have reduced CNS activity owing to P-gp transporter liabilities (filled symbol). Additional drugs (lamotrigine and levodopa) and nutrients (creatine and thiamine) that reside outside the shaded area enter the brain through SLC processes. BOILED-Egg model and coordinates for molecules listed were obtained from public database website http://www.swissadme.ch/index.php.

penetration in the BBB is heavily influenced by ABC efflux transporters. In this regard, the human multidrug resistance proteins (MRPs), P-gp, and BCRP are of particular clinical relevance, due to their high expression levels in human brain capillary endothelial cells, luminal distribution within the BBB, and broad range of drug substrates that are transported [4, 32]. Preclinical studies using knockout mice, pharmacological inhibitors and ABC transporter over-expressing cell lines have all contributed to a greater understanding of the functional role of active efflux transporter processes for drug permeability in the BBB [31]. The primary ABC transporters, their localization and primary impact on the barriers of the CNS can be found in Table 12.1.

While interactions with drug efflux transporters in the BBB can significantly limit the therapeutic effectiveness of a variety of agents, for non-CNS therapeutics, ABC transporter liability may have a desired beneficial role in preventing unwanted drug side effects. A prime example of this is the second and third generation antihistamines used by many for the treatment of seasonal allergies [35]. Histamine is an endogenous inflammatory mediator found in the mast cells and basophils. During an allergic reaction, histamine is released from mast cells and basophils triggering vasodilation and vascular leakage through activation of histamine H_1 receptor on the vascular smooth muscle and capillary endothelium [36]. Antihistamines effectively bind H_1 receptors reducing the symptoms of runny nose and watery eyes associated with allergic reactions [37]. However, as histamine is also an important neurotransmitter controlling wake and rest cycles, a major adverse effect of the antihistamines that can cross the BBB and interfere with CNS histaminergic neurotransmission is drowsiness. First generation antihistamines, while effective in treating allergic responses, produced a high incidence of sedation [36]. In contrast, the newer second-generation antihistamines, such as terfenadine, loratadine and cetirizine had fewer sedative side effects [36, 38].

Given the binding affinities and receptor occupancy profiles were comparable for the first and second-generation antihistamines [39], the available evidence suggests the reduced sedative properties were attributable in large part to reduced BBB penetration. As shown in Figure 12.2, both the first and second-generation antihistamines meet the criteria for passive diffusion across the BBB. However, examination of the brain penetration of cetirizine, loratadine, and fexofenadine in P-gp knockout mice showed significantly greater brain accumulation in the P-gp deficient mice compared to the wild-type controls expressing P-gp in the BBB [40]. Thus, despite adequate physicochemical structure requirements for brain penetration, the P-gp transporter liabilities of the second-generation antihistamines likely explains the reduced sedative effects with these agents (Figure 12.2) and provides an example of how ABC transporter interactions can reduce adverse effects for non-CNS drug candidates.

It should be noted that there are small molecule drugs and nutrients that achieve biologically relevant concentrations in the brain despite having physicochemical properties that would preclude adequate passive diffusion across the BBB (see Figure 12.2). In these instances, brain penetration is dependent on one of several SLC transporters expressed in the brain endothelial cells (Table 12.1). As is the case for some drugs and nutrients that are transported into the brain, the uptake process is sufficient to overcome efflux transporter interactions (see Figure 12.2). As discussed below, for those drugs and essential nutrients that rely on SLC mediated transport across the BBB, any changes in transporter activity or expression could have substantial impact on brain function (Table 12.2).

Table 12.2 Physicochemical and binding properties of selected anti-histamines.

Drugs	Molecular Wt ($g\,mol^{-1}$)[a]	Polar surface area[a]	Lipophilicity (LogP)[a]	P-gp substrate[b]	H1 receptor occupancy[c] (%)	CNS side effects[c]
First generation						
Hydroxyzine	348	36	3.4	NO	34	YES
Diphenhydramine	255	12	3.6	NO	56	YES
Triprolidine	278	16	4.1	NO	N/A	YES
Second generation						
Terfenadine	472	44	6.4	YES	17	LESS
Cetirizine	388	53	2.9	YES	84	LESS
Loratadine	381	42	4.8	YES	12	LESS

[a] Based on Chemspider (http://www.chemspider.com/).
[b] Based on information on SwissADME (http://www.swissadme.ch/index.php).
[c] Based on information on Drug Bank website (https://www.drugbank.ca/).

12.3 BBB-Transporter Effects on CNS Drug Response

Compared to other tissues such as the liver, kidney, and gastrointestinal tract, the impact of drug transporters within the BBB on pharmacokinetics of drugs is small [41, 42]. This is due to the relatively small tissue compartment of the brain in comparison to other distribution compartments such as skeletal muscle or skin [43]. However, as the examples below illustrate, alterations in solute and efflux transporter expression and activity in the BBB can have profound effects on both the therapeutic and neurotoxic effects of drugs in the CNS.

12.3.1 Influence of Efflux Transporters on Brain Disposition of Drugs

12.3.1.1 Anticancer Agents

The BBB is a major obstacle to chemotherapy delivery within the brain, severely limiting pharmacotherapy options in treating CNS tumors [41]. The ABC efflux transporters located on both the endothelial cells forming the BBB and tumors cells themselves significantly restrict exposure to cytotoxic chemotherapeutics by effectively preventing drug uptake into the tumor. Therefore, successful CNS anticancer drugs must pass both the BBB and blood–tumor barrier (BTB) to reach the desired target [41]. Chemotherapy is used either alone or in combination with surgical and radiation therapy to control cancer growth and proliferation.

Chemotherapeutics include a diverse array of chemical agents ranging from antimetabolites that are used to inhibit the synthesis of nucleotides required for DNA replication (purine, pyrimidine, and folate antagonists), genotoxic agents that damage DNA within the cellular nucleus (alkylating, intercalating agents, and topoisomerase inhibitors), mitotic spindle inhibitors that inhibit mitosis (vinca alkaloids and taxanes), and the newer kinase inhibitors that block specific signaling pathways in the tumor cell [44]. A variety of these chemotherapeutic agents are substrates for ABC efflux transporters expressed in drug resistance cancer cells. As these same ABC efflux transporters are also expressed within the brain endothelial cells of the BBB, many chemotherapeutic agents also have reduced brain penetration [4, 44].

As many anticancer therapeutics are substrates of P-gp, encoded by the multidrug resistance 1, (*MDR1*) gene, including doxorubicin, mitoxantrone, vinblastine, and vincristine [4, 44], modulating the efflux transporter activity for chemotherapeutics has been explored as a method to enhance efficacy of these drugs. A common approach for modulating efflux transporter activity consists of coadministration of pharmacological inhibitors of P-gp. Clinical reports have shown enhanced brain penetration of chemotherapeutics with coadministration of a Pgp inhibitor [44]. Therefore, proof-of-concept for potential clinical use of pharmacological efflux transporter inhibitors to increase brain delivery of drug therapies has been demonstrated. However, as coadministration of P-gp inhibitors can also alter drug processing in liver and kidney, reduced clearance and drug toxicity presenting a major drawback to the use of pharmacological drug efflux inhibitors [44]. Human trials reported serious adverse effects, low potency, and poor selectivity when using efflux transporter inhibitors [44].

Despite limited evidence for improved tumor response to chemotherapeutic agents, the development of ABC efflux transport modulators continues to be a pharmacological target of interest [44]. The newer second and third generation inhibitors for P-gp and BCRP have reduced toxicity and less off target effects [44]. However for applications directed at improving chemotherapeutic responses to CNS tumors, the studies with animal models suggest that improvements are achievable, but require reductions in dosing for the chemotherapeutic agents. An example is the studies by Fellner et al. [45] in which the P-gp inhibitor, valspodar, was used in conjunction with paclitaxel to treat brain tumors in mice. While a standard 8 $mg\,kg^{-1}$ dose of paclitaxel alone had no significant impact on brain tumor growth, doses as low as 4 $mg\,kg^{-1}$ when administered with valspodar showed tumor reduction and increased survival [45]. However the standard dose of paclitaxel when administered with the P-gp modulator or in P-gp knockout mice caused toxicity and death. Thus translation of these findings to the clinical setting may require either dose reductions for the chemotherapeutics to avoid CNS toxicity of the chemotherapeutic agents or less neurotoxic chemotherapeutics options.

12.3.1.2 Opioids

Opioid analgesics remain the standard choice for management of chronic and severe pain. According to the Canadian Institute for Health Information, in 2018 an estimated one in eight people were prescribed an opioid with one and five people living with chronic pain. The most studied and widely used opioids are morphine and codeine that is found in opium [46, 47]. However, semi-synthetic and synthetic opioids such as oxycodone, heroin, hydrocodone, and methadone are used both therapeutically and as recreational drugs of abuse [46]. Pharmacologically, opioids interact with G-protein coupled receptors (GPCRs) widely distributed throughout the CNS [47, 48]. Activation of the mu opioid receptors within the CNS produces the desired analgesic effect. In additional to central activity, opioid agonists can also bind to mu opioid receptors located in the gastrointestinal tract causing constipation [47, 48]. Thus while opioids are primarily used therapeutically for analgesia, agents such as loperamide, that have mu agonist activity but restricted brain penetration, have primary indications as antidiarrheal medications.

Many of the opioids, including morphine, methadone, fentanyl, and oxycodone, are substrates for drug efflux transporters expressed in the BBB [46]. In addition to being actively transported, several opiods are also capable of inducing drug efflux transporter expression [46] (Table 12.3). While P-gp is the most prevalent, some cross-over with MRP and BCRP transporters have also been reported for the opioid drugs [46] (Table 12.3). Aside from the transporter interactions of the parent drugs, active opioid metabolites may also have drug efflux transporter activity that impact on opioid levels in the brain. Notable examples are morphine-3-glucuronide and oxymorphone that undergo efflux transport out of the CNS through P-gp and MRP-mediated processes [46]. Given the range of potential interactions of the opioids and their respective metabolites with drug efflux transporters in the BBB, it is likely that such interactions play an important role in both interindividual variability and drug tolerance observed with the opioid analgesics.

Table 12.3 Selected opioids and drug transporter interactions at the BBB.

Opioids	Metabolites	Interaction with ABC efflux transporter	Inducer/ inhibitor
Morphine	Morphine-6-glucuronide, morphine-3-glucuronide	P-gp	Inducer
Loperamide	*N*-demethyl loperamide	P-gp	N/A
Fentanyl	Norfentanyl	P-gp	N/A
Methadone	Hydromorphone-3-glucuronide	P-gp	Inhibitor
Oxycodone	Oxymorphone	P-gp	Inducer

Using P-gp knockout mice, Schinkel and colleagues, reported a 1.7-fold higher brain accumulation of morphine compared to wild type mice, supporting the role of P-gp in restricting morphine penetration into the brain. Brain mirodialysis and *in situ* brain perfusion studies using P-gp knockout mice also supported morphine transport by P-gp in the BBB [49, 50]. Furthermore, as administration of a P-gp inhibitor such as verapamil resulted in significantly increased antinociceptive effects in mice exposed to morphine, these interactions with P-gp efflux transporters in the BBB influenced not only brain penetration but also therapeutic response to morphine [51]. Similar findings of increased antinociceptive response and increased brain penetration were observed with fentanyl following P-gp inhibition [52, 53]. Given the high potency of fentanyl and its illicit use, altered drug efflux transport at the BBB could cause serious adverse effects through opioid receptor-induced respiratory depression.

A common problem encountered with most opioid analgesics for long-term pain management is the development of tolerance resulting in the need for increased dosing to maintain the desired level of analgesia [46, 47]. Tolerance is described as a decrease in pharmacological response after repeated or prolonged exposure to drugs. Multiple molecular mechanisms have been proposed to account for opioid tolerance. Desensitization through internalization of receptors on the plasma membrane of the cell is a classical pathway for tolerance with GPCR [47]. However, tolerance through receptor internalization cannot account entirely for desensitization as not all opioid analgesics activate arrestin-mediated signaling pathways that promote receptor internalization [47]. Whistler et al. [54] examined ligand-induced signaling and endocytosis of an epitope tagged mu opioid receptor, to monitor real-time activation of receptor endocytosis through arrestin-dependent signaling pathways. Following addition of the opioid peptide [d-Ala^2,MePhe^4,Gly(ol)^5]encephalin (DAMGO), rapid endocytosis of μ opioid receptors was observed, consistent with the well-characterized desensitization process for GPCR. However, in these same studies, addition of morphine failed to promote detectable endocytosis following prolonged receptor activation [54], suggesting other mechanisms for tolerance to opioids were likely involved in opioid tolerance.

An additional theory of opioid tolerance involving altered efflux transporter activity at the BBB, specifically P-gp, is gaining experimental support [48]. As many of the opioids and/or their metabolites are transport substrates, these agents are particularly vulnerable to altered pharmacokinetic and pharmacodynamics response due to either transporter induction or inhibition. Evidence in support of this is the negative correlation between ATPase activity of P-gp and analgesic effects of morphine [46]. Given that many of the opioid analgesics are also inducers of P-gp transporter expression [46], the increased efflux transporter activity and expression resulting from long-term opioid exposure could reduce the amount of drug reaching the CNS and contribute to drug pharmacoresistance observed.

Evidence for a BBB transporter role in opioid tolerance has been demonstrated in preclinical rodent studies. Multiple studies from different research groups have reported increased expression of P-gp in cerebral blood vessels following repeated administration of morphine [43, 55–57]. While the focus of these studies was on P-gp expression within the brain, additional studies demonstrated repeated exposure of rodents to morphine caused 1.5-fold enhancement in P-gp and Bcrp expression in the cerebral vasculature [58]. In contrast, a single dose of morphine given to rats had no significant effect on *ABCB1* expression levels in the brain, indicating that prolonged exposure to morphine was essential for the induction of ABC efflux transporters [43].

In the studies by Yousif et al. [57, 58], the increased P-gp expression in the BBB in response to repeated morphine exposure did not appear to influence BBB integrity as measured by various permeability markers. However, the brain penetration and exposure of morphine and its metabolites were not examined. In contrast, studies by Aquilante et al. [43] and Kobori et al. [55] were able to correlate the alterations in P-gp expression in the cerebral vasculature observed following chronic morphine exposure and analgesic tolerance to decreased brain accumulation of the drug. Direct evidence that induction of P-gp in the BBB is mechanistically involved in the analgesic tolerance observed with chronic morphine exposure is the studies of Yanwei et al. [56]. In these studies, rats treated for five days with morphine showed a reduced analgesic response using the tail-flick pain model. Examining P-gp expression levels within the brain microvasculature indicated a threefold increase in P-gp in the morphine tolerant rats [56]. Using the alkaloid oxymatrine, together with morphine, reversed analgesic tolerance to morphine as well as restoring P-gp levels in the BBB to those observed in control rats [56]. Together these results support a contributing role for ABC efflux transporters in the BBB in opioid tolerance.

To further complicate the issue of tolerance, other analgesics commonly coadministered with opioids have been shown to regulate ABC efflux transporter expression. Yang et al. [59] demonstrated acetaminophen increased functional expression of P-gp at the BBB resulting in reducing brain uptake and analgesic effectiveness of morphine. Moreover, many highly prescribed drugs, dietary constituents, and nutraceuticals act as ligands for nuclear receptors (NRs) including pregnane X receptor (PXR) [60], glucocorticoid receptor (GR) [61], and constitutive androstane receptors (CARs) [62]. Activation of these NRs by coadministered drugs may also increase expression/activity of ABC transporters, further exacerbating the issue of tolerance for pain management. Overall, recent data suggests a role for morphine acting alone or together with other anti-inflammatory drugs to modulate ABC efflux transporter expression in the BBB and influence both CNS delivery therapeutic effectiveness of a variety of opioid drugs.

12.3.2 SLCs and BBB Transport of Drugs

Given the limited chemical space for small molecules undergoing passive diffusion across the BBB (see Figure 12.2), targeting SLCs for improved BBB delivery is an attractive approach for increasing drug levels in the brain [5, 63]. A clinically relevant example of this approach is Levadopa, a drug used in the clinical treatment of Parkinson's disease and dopamine-responsive dystonia. As shown in Figure 12.2, Levadopa is water soluble with insufficient lipophilicity and too large a size to passively diffuse across the BBB. However, it is able to penetrate the BBB and achieve therapeutic levels through carrier mediated transport (CMT) via the large neutral amino acid (LNAA) transporter type1 or LAT1 present on the apical and basolateral sides of BBB [64]. Gabapentin and melphalan are additional examples of water soluble drugs with limited ability to passively diffuse across the brain microvessels that are pharmacologically active in CNS due to CMT via LNAA at the BBB [65, 66].

There are several other types of SLCs present in the BBB that facilitate the transport of several types of endogenous and exogenous solutes into the brain. Additional SLCs expressed in the BBB and their role in drug transport can be found in Table 12.1.

12.4 Transporter Considerations Influencing CNS Drug Response

The utilization and modulation of transporter-mediated pathways for CNS drug delivery allows a wider and more diverse spectrum of small molecules to be considered for brain-related disorders. However, BBB transporter dependence and liabilities also introduces more variability within individuals or populations of individuals that may ultimately impact on CNS drug response. Altered expression of BBB transporters through genetic variation, normal development and aging process, pathological disease-related changes and drug interactions can all influence both the therapeutic and adverse responses to drugs. Examples of each of these are discussed below.

12.4.1 Transporter Polymorphisms

12.4.1.1 P-gp Polymorphism

P-gp is one of the vital efflux transporters present in the apical membrane of brain microvessel endothelial cells and has a major role in limiting the entry of a chemically diverse range of endogenous and exogenous agents [67, 68]. The neuroprotective role of P-gp in the BBB is best illustrated by the studies in *Mdr1a* knockout mice that are deficient in P-gp [69]. While the Mdr1a mice were viable and healthy, the absence of P-gp left the mice susceptible to neurotoxicity from exposure to agents that in control mice produced little if any adverse effects. The classic example of this was reported with ivermectin, a commonly used pesticide in animal care facilities. Exposure of mice to a room treated with ivermectin resulted in the loss of the entire *Mdr1* knockout mouse colony due to neurotoxicity while the wild-type mice with functional P-gp, housed in the same room, showed no toxic response [69]. Subsequent studies revealed several additional drugs that under normal conditions would have minimal brain penetration but in the *Mdr1* knockout mice had substantial and often toxic levels of accumulation in the brain [69].

In addition to preventing toxic agents from entering the brain, studies with the *Mdr1* knockout mouse have also identified the role of P-gp in CNS drug accumulation in the brain. Using the *Mdr1* gene knockout mice Uhr et al. [70] compared brain levels of several different antidepressant agents. The focus of these studies was on the tricyclic antidepressant amitriptyline and its active metabolite along with the newer agent, fluoxetine. Following single dose intraperitoneal injections of amitriptyline, the levels of both amitriptyline and its active metabolite were significantly higher in the brains of the Mdr1−/− compared to Mdr1+/+ mice [70]. In contrast, brain and plasma levels of fluoxetine were similar in both the Pgp knockout and wild-type mice [70]. Follow-up studies examined other antidepressant agents and found that the levels of citalopram, venlafaxine, and D-venlafaxine in brain tissue were 3.0, 1.7, and 4.1 times higher in the knockout mice compared to wildtype control mice [64]. Although these levels were much less than the 100-fold or more increases observed with ivermectin and other neurotoxic agents, they could certainly alter both therapeutic effectiveness and adverse affects of the CNS drugs.

Because of the dramatic changes observed in drug accumulation in the brain of P-gp deficient mice, there was much concern for potential neurotoxicity in humans having *MDR1* polymorphisms resulting in reduced activity and/or expression of P-gp. While several different single nucleotide polymorphisms (SNPs) have been identified in the human *MDR1* gene, the non-synonymous polymorphisms at positions 2677 (2677T>G/A) and 3435 (3435T>C) in the human *MDR1* gene appear to have the most clinical relevance [68, 71, 72]. Comparatively few incidences of CNS-related adverse responses with MDR1 polymorphism have been reported. However, the 3435T>C mutations resulting in either TT or CT genotypes were associated with reduced P-gp expression in brain endothelial cells [72]. Furthermore, patients with 3435 TT and CT genotype taking the antidepressant, nortriptyline were more likely to develop postural hypotension, presumably due to increased levels of nortriptyline in the brain [66].

The potential role of P-gp polymorphisms in drug response in schizophrenia has also been examined. Studies by Lin et al. [73] examined responses to olanzapine in patients with schizophrenia. They found that steady-state plasma levels of olanzapine were predictive of positive therapeutic response in patients with the TT or CT mutation in the C3435T allele. However those patients without the mutation showed no clear correlation between steady-state plasma levels of olanzapine and therapeutic response [73]. The primary explanation for these findings is that the reduced P-gp activity at the BBB in patients with the TT or CT mutations resulted in improved permeability of olanzapine into the brain. These studies demonstrate that polymorphisms in the *MDR* gene can influence CNS drug response.

12.4.1.2 BCRP Polymorphism

The ABC super-family G member 2 (ABCG2) gene is responsible for BCRP expression. Expression of BCRP in the BBB is responsible for preventing entry of different solutes and drugs into the brain. Indeed, in human brain microvessel endothelial cells, BCRP has higher expression than even P-gp [74]. Several SNPs for BCRP have been reported [75], however the importance of these mutations in CNS drug response is still relatively unknown. The ABCG2 ca.421C>A mutation is one of the more common polymorphisms observed. Studies by Tsuchiya [76] reported decreased cerebrospinal fluid concentrations of raltegravir, an antiviral medication used to treat patients with HIV who had the ca.421C>A allele. As BCRP is expressed in the epithelial cells of the chorioid plexus and actively effluxes solutes and drugs into the CSF, such findings fit the reduced BCRP activity associated with this polymorphism. A similar result has also been reported for the antibiotic, Ceftriaxone, in patients with CT/TT in the r513120400 BCRP allele [42]. Patients having the mutation had reduced levels of Cetriaxone in the CSF such that the CSF/plasma ratio for Cetraixone was reduced by approximately 50%. Whether similar reductions in the concentrations of these drugs in the brain extracellular fluid also occur is currently unknown.

12.4.1.3 SLC Polymorphism

SLC transporters are widely distributed throughout the body and have an important role in solute uptake and removal from the brain. They also represent attractive targets for small molecule drug delivery to the brain. Several polymorphisms have been reported within various members of this superfamily of proteins with important clinical implications [72, 77, 78]. However, most of the known SLC polymorphisms have not produced major brain-related toxicities due to altered BBB transport. From the standpoint of drug penetration into the brain, OAPT1A2 (SLC21A3) has intense immunostaining within the brain microvasculature [79] and its luminal distribution at the BBB coupled with the large number of drug transport substrates makes it a prime candidate for polymorphic changes in drug response in the brain. Multiple polymorphisms are present for OAPT1A2 with the 38T>C, 382A>T, 404A>T, 516A>C, 559 G>A, and 2003 C>G alleles all showing diminished transport function *in vitro* [80]. To date demonstration of clinical impact of OATP1A2 polymorphism is limited to the studies of Calcagno et al. [81] that reported CSF levels of the antiviral agent, darunavir, in patients with uncommon OATP1A2 polymorphisms were slightly higher. While OAPT1A2 polymorphisms may play a role in CNS drug toxicity and therapeutic response, more studies are required to definitively address this issue in larger patient populations.

From a BBB standpoint, mutations in SLC6A8, a gene responsible for creatine transport into cells, warrant discussion as mutations in this gene cause serious pathology. X-linked mental retardation includes over 150 genes that are associated with reduced mental capacity in males. Given the large number of X-linked gene mutations and the relative low frequency of occurrence in the general population, the involvement of any single gene mutation in mental retardation would be rare. The exception to this may reside in SLC6A8 mutations. Cecil et al. [82] were the first to report a mutation in SLC6A8 in a boy with mental retardation. The SLC6A8 mutation resulted in a substantial reduction in creatine content in the brain and an increased urinary creatine/creatinine ratio [82]. These findings were consistent with an inability to efficiently distribute creatine into the brain. Since this initial study, additional SLC6A8 mutations have been reported [83, 84]. It is now estimated that mutations within the SLC6A8 gene account for 1–2% of X-linked mental retardation. As illustrated in Figure 12.2, creatine falls well outside of the physicochemical space required for passive diffusion into the brain. Thus despite the increased identification of children with SLC6A8 mutations, treatments to date have been unsuccessful, although newer creatine analogs with improved lipophilicity show promise [85].

12.4.2 Age-Related Alterations in BBB Transporter Function and Drug Response

Both the SLCs and ABC efflux transporters are expressed early during BBB development. Notably, glucose transporter 1 and P-gp have been identified as two of the earliest markers of BBB in developing rodents [86]. In humans, Pgp, MRP1, and BCRP have been detected in postmortem brain tissue from neonates and infants [87, 88]. In the case of P-gp and BCRP immunohistochemistry detected expression in brain microvessels as early as 22 weeks of gestation [88]. While some MRP1 was detected in brain microvessel endothelial cells, immunohistochemical staining was much more intense for MRP1 within the CP in neonates [87, 88]. The early gestational expression of these transporters coincided with barrier function and suggested an important role in protection of the developing neonate from toxic insults to the brain. Interestingly, while the cell specific distribution of P-gp, MRP1, and BCRP in human adult brain tissue was similar to that observed in the newborn brain samples, the intensity of staining was substantially greater in the adult tissue [87], suggesting age-related differences in ABC transporter activity in both BBB and BCSFB.

A potential example of age-related differences in transporter expression in the BBB influencing drug response can be found with the opioid, loperamide, which is used as a antidiarrheal medication. Although considered a rather "safe" drug in adults, loperamide can produce central opiate effects such as respiratory depression in neonates and children [89, 90]. The differences in CNS response to loperamide in pediatric subjects may be due in part to the different expression levels of P-gp in this population. Although P-gp appears to be expressed to similar extents in the liver and kidney of young children and adults, expression in the BBB of infants and neonates has been reported to be approximately half of the levels observed in adults [91, 92]. A consequence of this reduced P-gp activity in the BBB during development may be elevated drug accumulation in the CNS and increased sensitivity to centrally acting P-gp substrates in neonates and young children (e.g. respiratory depression caused by loperamide) [89, 90].

Alterations in BBB transporters may also influence CNS drug responses on the other end of the age spectrum (i.e. elderly patient population). The effects of physiological aging on P-gp efflux transporter expression and activity in the BBB has been demonstrated in both preclinical and clinical studies [93]. Studies using aged rat models reported a significant decline in P-gp expression as a function of increasing age [94]. A similar finding has been reported in canines where postmortem analysis of brain tissue indicated an approximately 70% decrease in P-gp expression in the BBB of the old dogs compared to young adult dogs [95].

Much of the clinical evidence for age-related changes in P-gp activity in humans comes from positron emission tomography (PET) imaging studies using contrast agents that either bind to P-gp or are substrates for P-gp transport. Similar to the preclinical findings, experimental evidence using human subjects and PET studies found an 18% decrease in P-gp activity in the gray matter of elderly patients [96]. A decrease in P-gp function in the BBB in older patients could account for increased drug toxicity and side effects of drugs observed in the elderly [96]. As decreased P-gp function could lead to impairment of the natural protective barrier properties of the BBB, the elderly population may be more vulnerable to both exogenous and endogenous neurotoxins that are transported by P-gp and other ABC efflux transporters.

12.4.3 Disease-Dependent Modulation of BBB Transporters and Drug Response

12.4.3.1 Inflammation and Pain

Pathological conditions can also modulate transporter expression and activity. A well-known example of this is the alterations in ABC transporter expression in the BBB observed with pain and inflammation. Peripheral inflammatory pain (PIP) has been demonstrated to alter ABC efflux transport expression at the BBB. Cytokines such as TNF-a, IL-B, and IL-6 are inflammatory mediators released into blood circulation [59]. Prostaglandin released by nociceptors are triggered by cytokine release and signals are conveyed to the brain through neuronal signaling. Preclinical studies by Yang et al. [59] demonstrated increased P-gp expression in rat brain in response to carrageenan injection as a model of PIP. The increased P-gp expression observed in the PIP model was correlated with decreased morphine levels within the brain and reduced modulation of pain response [59] and indicative of influence altered BBB transporter function can have in treating pain.

An additional component in the PIP process that may influence analgesic response and delivery to the brain is a rapid translocation of P-gp in brain endothelial cells from intracellular organelles to the plasma membrane. Using the rodent carrageen PIP model, Seelbach et al. [97] reported altered morphine transport and reduced analgesia within three hours of carrageen injections. As the time course for such changes were relatively rapid for upregulation of protein translation, other mechanisms were examined. Follow-up studies confirmed that localized inflammatory responses to carrageen injection in the hind paw of rats were associated with a rapid redistribution of P-gp within the brain microvasculature [98, 99]. The redistribution of P-gp from intracellular stores to the plasma membrane following induction of peripheral inflammatory response was associated with increased transporter activity in the cerebral microvessels [98, 99]. As the cellular distribution of MRP4 in the brain microvessels showed no change the inflammatory effects appeared to be selective for P-gp efflux transporters within the BBB [98].

12.4.3.2 Epilepsy

Epilepsy is a common serious neurological disorder affecting millions of people worldwide. Epilepsy is characterized by recurrent seizures due to excitatory neuronal activity within localized foci in the brain [100]. A wide range of antiepileptic drugs (AEDs) are available for managing seizure episodes. The AEDs can be broadly defined as either working through inhibiting excitatory neuronal activity or through promoting inhibitory neuronal activity within the epileptic foci [100]. Carbamazepine, lamotrigine, and phenytoin exert their effects by inhibiting sodium channels preventing depolarization of the neurons, while clonazepam and phenobarbital enhance gamma-aminobutyric acid (GABA)-mediated chloride currents resulting in neuronal hyperpolarization [100]. Inhibition of excitatory neuron transmission through glutamate receptor and calcium flux has also emerged as a target for novel AEDs [100]. However, current antagonists of glutamate receptors have adverse effects that preclude their use as AEDs at this time.

Despite over 20 different drugs available for preventing seizure activity, it has been estimated that 30–40% of patients remain resistant to AED pharmacotherapy [101]. While multiple factors are likely to contribute to AED resistance, there is evidence, both from animal models as well as studies in epileptic patients, of an emerging role for altered ABC efflux transporter activity within the brain capillaries in and around the epileptic foci in the brain. Many of the AEDs have known ABC efflux transporter activity [101, 102] (see also Table 12.1). Pharmacological inhibition of P-gp significantly increased the brain penetration of several AEDs including lamotrigine, phenytoin, and carbamazepine [102]. Given the known P-gp interactions of several AEDs, potential drug efflux at the BBB could influence the ability of these agents to reach optimal therapeutic levels in the brain [101, 102]. While it has been argued that for many of the AEDs efflux transport interactions are relatively modest in magnitude, and unlikely to impact overall brain penetration of the AEDs, an increase in either expression or activity of ABC transporters in the cerebrovasculature within the lesion area could negatively impact drug distribution at the affected site. In support of this possibility it should be noted P-gp expression detected in excised brain tissue from epileptic patients was directly correlated with the degree of uncontrolled seizure activity [101]. While most of the increased P-gp was found within the microvessel endothelial cells, some P-gp was detected in the astrocytes and neurons of epileptic patients [101]. Such distribution was not observed in brain tissue from non-seizure sites and suggesting reduced accumulation within the cellular targets may also occur with many of the AEDs. Collectively these findings suggest that upregulation of P-gp within the brain capillary endothelial cells in and around the lesion site could result in subtherapeutic concentrations of AEDs in the brain.

Drug efflux transporters in the BBB may also have an important role in the pharmacoresistance observed with many AEDs. Using a rat model, induction of seizures where associated with increased ABC efflux transporter within the cerebral vasculature [103]. In this regard, the induction of ABC efflux transporters within epileptic foci in preclinical models resembles the findings reported in epileptic patients. However, the preclinical studies demonstrated that P-gp over expression could be induced through two mechanisms, one involving seizure activity itself and the other involving prolonged exposure to AEDs [103].

In the case of AEDs, alterations in ABC transporters within the BBB are likely attributable to the activation of ligand-dependent NR proteins that can transcriptionally regulate the expression of target genes including drug metabolizing enzymes and transporter proteins [104]. PXR and GR are key NRs involved in modifying drug metabolism through the CYP34A metabolizing enzyme and efflux transporter expression in brain micro-vessel endothelial cells [104]. Several of the AEDs are known to activate NRs [105]. Furthermore, examination of excised brain tissue from drug-resistant epileptic patients showed that PXR and GR were significantly enhanced in comparison to normal subjects, suggesting the role of NRs in epileptic drugs resistance [104]. Together these studies demonstrate the role that NRs play as key transcriptional regulatory factors modulating the expression of efflux transporters and metabolizing enzymes within the brain and contributing to the levels of AEDs in the brain and their respective therapeutic responses.

12.4.4 CNS Toxicity Caused by Drug Interactions at the BBB

Among all tissues, drug delivery to the brain is the most challenging due to the presence of the BBB and the BCSFB. In the BBB, in addition to the tight junctions limiting paracellular diffusion, many efflux transporters expressed at the BBB, as described in the previous sections, protect the brain from exposure to the foreign substances including toxins. Therefore, disrupting the protective barrier by either increasing the paracellular permeability or inhibiting the action of transporters in the BBB may lead to CNS toxicity. While CNS drug toxicity due to drug–drug interactions (DDIs) at the level of the BBB are not widespread, some clinically important examples involving transporter interactions in the BBB are discussed below.

It is widely accepted that P-gp expressed at the BBB protects the brain from toxic substances circulating in the blood and is one of the most important transporters in modulating the entry of drugs into the CNS. Inhibiting P-gp at the BBB may increase the brain delivery of certain therapeutic agents but also increase the propensity of DDI causing CNS toxicity. Among the clinically approved drugs that can inhibit P-gp, quinidine can produce significant inhibition at the human BBB [106]. Quinidine potently inhibits P-gp in vitro ($EC_{50} = 0.9\,\mu M$), with an unbound plasma concentration (C_u) of 1.3 μM, makes it a therapeutically relevant inhibitor at the BBB ($C_u/EC_{50} = 1.4$) [106]. Loperamide (Imodium) is a potent opiate that reduces gut motility by its action at opiate receptors in the gut. Although a potent opiate, at standard dosages used in adults (up to 16 mg in 24 hours), Loperamide does not produce opioid CNS effects, such as respiratory depression or euphoria. The lack of CNS effects are due to the low oral bioavailability, reduced BBB permeability, and Pgp transporter liability of Loperamide [107].

Loperamide has long been considered a safe over-the-counter (OTC) antidiarrheal medicine devoid of CNS opioid activity [108]. While this is likely true when taken alone, when in combination with drugs that inhibit P-gp in the BBB, there is potential for CNS toxicity. In a study carried out in eight healthy male volunteers given a 16 mg dose of loperamide either in the presence or the absence of 600 mg quinidine, the CNS effects of loperamide were measured by evaluation of the respiratory response to carbon dioxide rebreathing [107]. Opiate-induced respiratory depression occurred when loperamide was given with quinidine while no effect was observed in the placebo group who were treated with loperamide alone [107]. Therefore, coadministration of quinidine with loperamide resulted in significant CNS toxicity of loperamide demonstrating that a safe OTC drug such as loperamide can be reversed by a drug causing P-gp inhibition, resulting in serious CNS toxicity and abuse potential.

While there is clear evidence for quinidine and loperamide DDI, a literature review of adverse events in poison control centers across North America between year 2009 and 2015, suggested a clear trend toward increasing loperamide misuse and abuse, leading to significant neurological pathology, and even death [108]. Some of the neurotoxicity could be the result of other agents taken along with loperamide and many poisonings (intentional or accidental) involve multiple drug agents. This is of great importance given this trend may continue as long as loperamide is available without restrictions.

Inhibitors of P-gp, such as cyclosporine and itraconazole, may increase the efficacy of CNS drugs (e.g. AEDs) [109, 110] but also central toxicity of anticancer agents (e.g. vincristine). Case reports have suggested increased vincristine CNS penetration and neurotoxicity when coadministrated with the P-gp inhibitors itraconazole or cyclosporine in pediatric patients with acute lymphoblastic leukemia (ALL) [109, 111].

In addition to the efflux transporters such as P-gp, DDIs at the BBB involving uptake transporters are also possible. A well-known example of CNS toxicity caused by drug interactions with a BBB uptake transporter is the co-treatment of β-lactam antibiotics with probenecid. Probenecid is a well-characterized potent inhibitor of both OAT1 and OAT3. Its *Ki* values for human OAT1 and human OAT3 inhibition were reported to be 4.3–12.1 and 1.3–9.0 μM, respectively [112–114]. At clinical oral doses (0.5–2 g), unbound plasma concentrations of probenecid were reported to be 3–50 μM [115], suggesting that both human OAT1 and human OAT3 are likely to be the site of drug interactions with probenecid *in vivo*. In clinical practice, this DDI has been has been applied as a strategy to reduce the nephrotoxicity of β-lactam antibiotics by blocking the transport of these drugs into the proximal tubular cells from the basolateral side and increasing their plasma concentrations [116]. However, increasing plasma concentrations may ultimately increase the brain exposure of β-lactam antibiotics resulting in neurotoxicity due to the inhibition of brain OATs. Neurotoxic side effects of β-lactam antibiotics are well-known conditions for decades. β-lactam antibiotics can trigger epilepsy or seizures due to the binding of β-lactams to the GABA receptors in the brain. Some of the β-lactams are GABA receptor antagonists [117]. Not surprisingly, neurotoxicity induced by β-lactams can be a result of renal failure, which increases the amount of the antibiotic in the circulating blood. Hence, even in normal dosing ranges, β-lactam antibiotics pose CNS risks in case of renal failure [118].

CNS toxicity is not only associated with drug transporters such as OATs and P-gp as discussed herein, nutrient transporters such as thiamine transporters (THTRs) may also be important. THTR 1 (*SLC19A2*) and 2 (*SLC19A3*) are responsible for the oral absorption and tissue distribution of thiamine (vitamin B1). They are expressed in many organs and important physiologic barriers including the intestine, BBB, and the kidney proximal tubule [119, 120]. THTRs are important in thiamine intestinal absorption, brain penetration, and renal reabsorption. During phase III clinical studies of Fedratinib, a Janus Kinase 2 inhibitor, neurological adverse events consistent with Wernicke's encephalopathy (WE) were observed [121, 122]. The WE reported in the participants (how many?) from the fedratinib clinical trial was attributed to thiamine deficiency (TD), and were ameliorated, in at least one patient, upon thiamine supplementation [122]. Mechanistically, Fedratinib is reported to be a potent inhibitor of THTR2, the transporter responsible for the intestinal absorption of thiamine, and may have interfered with the oral absorption of thiamine resulting in TD [123]. As the mean steady-state plasma concentration of fedratinib in patients was 5–10 μM for a 400–500 mg daily dose [124], there is the possibility that fedratinib also inhibited THTR into the brain via THTR2, in addition to inhibition of thiamine uptake at the intestine. Considering that Fedratinib is also a substrate for THTR2 and the relatively high brain distribution of Fedratinib observed in preclinical studies [123, 125], the possibility that thiamine uptake by neurons and astrocytes might be also inhibited by fedratinib after this compound enters the brain may not be ruled out. During the clinical development of fedratinib, a TD neurological disorder, nausea, diarrhea, and vomiting were the major adverse events noted [121]. Therefore, unlike the common CNS DDIs, the neurological side effect of Fedratinib was not due to the excessive accumulation of the drug in the brain but rather attributable in preventing the brain accumulation of an essential nutrient required for CNS function. These findings observed with Fedratinib serve as an important reminder of the potential clinical implications of drug–nutrient transporter interactions [126].

12.5 Conclusions

The fluid barriers of the CNS are important cellular interfaces that are essential for normal brain function. Aside from the rare instances of direct injection or local polymer-based implantation of drug into the brain, most drugs must navigate the BBB to produce CNS effects. Given the restrictions in both size and lipophilicity required for passive diffusion in the BBB, both efflux and uptake transporters have substantial roles in determining drug distribution to the brain and therapeutic as well as adverse CNS responses. Fortunately, redundancy in terms of transporters and overlapping substrate activity has contributed to relatively limited impact of both pharmacogenomics and DDIs. However. as more research is performed additional examples of BBB transporter-related alterations in drug response in the brain are likely. Likewise our understanding of the physiological and pathophysiological alterations in both ABC and SLC transporters in the BBB are also at an early stage, and as more is learned our understanding of the effects of such changes on drug response in the brain is likely to result in improved therapeutics.

References

1 Abbott NJ, Patabendige AA, Dolman DE, Yusof SR & Begley DJ (2010) Structure and function of the blood-brain barrier. *Neurobiol Dis* 37:13–25. doi:https://doi.org/10.1016/j.nbd.2009.07.030.

2 Tietz S & Engelhardt B (2015) Brain barriers: crosstalk between complex tight junctions and adherens junctions. *J Cell Biol* 209:493–506. doi:https://doi.org/10.1083/jcb.201412147.

3 Liebner S, Dijkhuizen RM, Reiss Y, Plate KH, Agalliu D & Constantin G (2018) Functional morphology of the blood-brain barrier in health and disease. *Acta Neuropathol* 135:311–336. doi:https://doi.org/10.1007/s00401-018-1815-1.

4 On NH & Miller DW (2014) Transporter-based delivery of anticancer drugs to the brain: improving brain penetration by minimizing drug efflux at the blood-brain barrier. *Curr Pharm Des* 20:1499–1509. doi: https://doi.org/10.2174/13816128113199990458.

5 Tsuji A (2005) Small molecular drug transfer across the blood-brain barrier via carrier-mediated transport systems. *NeuroRx* 2:54–62. doi: https://doi.org/10.1602/neurorx.2.1.54.

6 Wolburg H, Noell S, Wolburg-Buchholz K, Mack A & Fallier-Becker P (2009) Agrin, aquaporin-4, and astrocyte polarity as an important feature of the blood-brain barrier. *Neurosci Rev J Bring Neurobiol, Neurol Psychiatry* 15:180–193. doi:https://doi.org/10.1177/1073858408329509.

7 Abbott NJ, Ronnback L & Hansson E (2006) Astrocyte-endothelial interactions at the blood-brain barrier. *Nat Rev. Neurosci* 7:41–53. doi:https://doi.org/10.1038/nrn1824 (2006).

8 Mehrabadi AR, Korolainen MA, Odero G, Miller DW & Kauppinen TM (2017) Poly(ADP-ribose) polymerase-1 regulates microglia mediated decrease of endothelial tight junction integrity. *Neurochem Int* 108:266–271. doi: https://doi.org/10.1016/j.neuint.2017.04.014

9 Sweeney MD, Ayyadurai S & Zlokovic BV (2016) Pericytes of the neurovascular unit: key functions and signaling pathways. *Nat Neurosci* 19:771–783. doi:https://doi.org/10.1038/nn.4288.

10 Daneman R & Prat A (2015) The blood-brain barrier. *Cold Spring Harbor Perspect Biol* 7:a020412. doi:https://doi.org/10.1101/cshperspect.a020412.

11 Armulik A, Genove G, Mae M, Nisancioglu MH, Wallgard E, Niaudet C, He L, Norlin J, Lindblom P, Strittmatter K, Johansson BR & Betsholtz C (2010) Pericytes regulate the blood-brain barrier. *Nature* 468:557–561. doi:https://doi.org/10.1038/nature09522.

12 Attwell D, Mishra A, Hall CN, O'Farrell FM & Dalkara T (2016) What is a pericyte? *J Cereb Blood Flow Metab* 36:451–455. doi: https://doi.org/10.1177/0271678X15610340.

13 Nikolakopoulou AM, Montagne A, Kisler K, Dai Z, Wang Y, Huuskonen MT, Sagare AP, Lazic D, Sweeney MD, Kong P, Wang M, Owens NC, Lawson EJ, Xie X, Zhao Z & Zlokovic BV (2019) Pericyte loss leads to circulatory failure and pleiotrophin depletion causing neuron loss. *Nat Neurosci* 22:1089–1098. doi: https://doi.org/10.1038/s41593-019-0434-z.

14 Stewart PA & Wiley M J (1981) Developing nervous tissue induces formation of blood-brain barrier characteristics in invading endothelial cells: a study using quail--chick transplantation chimeras. *Dev Biol* 84:183–192. doi: https://doi.org/10.1016/0012-1606(81)90382-1.

15 Ikeda E, Flamme I & Risau W (1996) Developing brain cells produce factors capable of inducing the HT7 antigen, a blood-brain barrier-specific molecule, in chick endothelial cells. *Neurosci Lett* 209:149–152. doi: https://doi.org/10.1016/0304-3940(96)12625-2.

16 Keep RF & Smith DE (2011) Choroid plexus transport: gene deletion studies. *Fluids Barriers CNS* 8:26. doi: https://doi.org/10.1186/2045-8118-8-26.

17 Spector R & Johanson CE (2010) Vectorial ligand transport thorugh mammalian choroid plexus. *Pharm Res* 27:2054–2062. doi: https://doi.org/10.1007/s11095-010-0162-2.

18 Uchica Y, Zhang Z, Tachikawa M & Terasaki T (2015) Quantitative targeted absolute proteomics of rat blood-cerebrospinal fluid barrier transporters: comparison with a human specimen. *J Neurochem* 134:1104–1115. doi:https://doi.org/10.1111/jnc.13147.

19 Lin JH (2008) CSF as a surrogate for assessing CNS exposure: an industrial perspective. *Curr Drug Metab* 9:46–59. https://doi.org/10.2174/138920008783331077.

20 Liu X, Van Natta K, Yeo H, Vilenski O, Weller PE, Worboys PD & Monshouwer M (2009) Unbound drug concentration in brain homogenate and cerebral spinal fluid at steady state as a surrogate for unbound concentration in brain interstitial fluid. *Drug Metab Dispos* 37:787–793. doi: https://doi.org/10.1124/dmd.108.024125.

21 Kodaira H, Kusuhara H, Fujita T, Ushiki J, Fuse E & Sugiyama Y (2011) Quantitative evaluation of the impact of active efflux by P-glycoprotein and breast cancer resistance protein at the blood-brain barrier on the predictability of the unbound concentrations of drugs in the brain using cerebrospinal fluid concentration as a surrogate. *J Pharmacol Exp Ther* 339:935–944. doi: https://doi.org/10.1124/jpet.111.180398.

22 Kodaira H, Kusuhara H, Fuse E, Ushiki J & Sugiyama Y (2014) Quantitative investigation of the brain-to-cerebrospinal fluid unbound drug concentration ratio under steady-state conditions in rats using a pharmacokinetic model and scaling factors for active efflux transporters. *Drug Metab Dispos* 42:983–989. doi: https://doi.org/10.1124/dmd.113.056606.

23 Girardin F (2006) Membrane transporter proteins: a challenge for CNS drug development. *Dialogues Clin Neurosci* 8:311–321.

24 Hediger MA, Romero MF, Peng J-B, Rolfs A, Takanaga H & Bruford EA (2004) The ABCs of solute carriers: physiological, pathological and therapeutic implications of human membrane transport proteins. Introduction. *Pflugers Arch Eur J Phys* 447:465–468. doi: https://doi.org/10.1007/s00424-003-1192-y.

25 Hu C, Tao L, Cao X & Chen L (2020) The solute carrier transporters and the brain: physiological and pharmacological implications. *Asian J Pharm Sci* 15:131–144. https://doi.org/10.1016/j.ajps.2019.09.002

26 Geier EG, Chen EC, Webb A, Papp AC, Yee SW, Sadee W & Giacomini KM (2013) Profiling solute carrier transporters in the human blood-brain barrier. *Clin Pharmacol Ther* 94:636–639. https://doi.org/10.1038/clpt.2013.175.

27 Strazielle N & Ghersi-Egea JF (2013) Physiology of blood-brain interfaces in relation to brain disposition of small compounds and macromolecules. *Mol Pharm* 10:1473–1491. doi: https://doi.org/10.1021/mp300518e.

28 Strazielle N & Ghersi-Egea JF (2015) Efflux transporters in blood-brain interfaces of the developing brain. *Front Neurosci* 9:21. doi:https://doi.org/10.3389/fnins.2015.00021.

29 Strazielle N, Belin MF & Ghersi-Egea JF (2003) Choroid plexus controls brain availability of anti-HIV nucleoside analogs via pharmacologically inhibitable organic anion transporters. *AIDS (London, England)* 17:1473–1485. https://doi.org/10.1097/00002030-200307040-00008.

30 Loscher W & Potschka H (2005a) Blood-brain barrier active efflux transporters: ATP-binding cassette gene family. *NeuroRx* 2:86–98. doi: doi: https://doi.org/10.1602/neurorx.2.1.86.

31 Loscher W & Potschka H (2005b) Drug resistance in brain diseases and the role of drug efflux transporters. *Nat Rev Neurosci* 6:591–602. doi: https://doi.org/10.1038/nrn1728.

32 Sun H, Dai H, Shaik N & Elmquist WF (2003) Drug efflux transporters in the CNS. *Adv Drug Del Rev* 55:83–105. doi: https://doi.org/10.1016/s0169-409x(02)00172-2.

33 Daina A & Zoete V (2016) A BOILED-Egg to predict gastrointestinal absorption and brain penetration of small molecules. *ChemMedChem* 11:1117–1121. doi: https://doi.org/10.1002/cmdc.201600182.

34 Broccatelli F, Larregieu CA, Cruciani G, Oprea TI & Benet LZ (2012). Improving the prediction of the brain disposition for orally administered drugs using BDDCS. *Adv Drug Del Rev* 64:95–109. doi:https://doi.org/10.1016/j.addr.2011.12.008.

35 Chishty M, Reichel A, Siva J, Abbott NJ & Begley DJ (2001) Affinity for the P-glycoprotein efflux pump at the blood-brain barrier may explain the lack of CNS side-effects of modern antihistamines. *J Drug Target* 9:223–228. doi: https://doi.org/10.3109/10611860108997930.

36 Obradovic T, Dobson GG, Shingaki T, Kungu T & Hidalgo IJ (2007) Assessment of the first and second generation antihistamines brain penetration and role of P-glycoprotein. *Pharm Res* 24:318–327. doi: https://doi.org/10.1007/s11095-006-9149-4.

37 Laak AM, Bijloo GJ, Fischer MJE, Donneop den Kelder GM, Wilting J & Timmerman H (1996) Serum protein binding of histamine H1-antagonists. A comparative study on the serum protein binding of a sedating ([3H]mepyramine) and a non-sedating H1-antagonist ([3H]loratadine). *Eur J Pharm Sci* 4:307–319. doi:https://doi.org/10.1016/0928-0987(96)00172-8.

38 Polli JW, Baughman TM, Humphreys JE, Jordan KH, Mote AL, Salisbury JA, Tippin TK & Serabjit-Singh CJ (2003) P-glycoprotein influences the brain concentrations of cetirizine (Zyrtec), a second-generation non-sedating antihistamine. *J Pharm Sci* 92:2082–2089. doi: https://doi.org/10.1002/jps.10453.

39 Ter Laak AM, Donne-Op den Kelder GM, Bast A & Timmerman H (1993) Is there a difference in the affinity of histamine H1 receptor antagonists for CNS and peripheral receptors? An in vitro study. *Eur J Pharmacol* 232:199–205. doi: https://doi.org/10.1016/0014-2999(93)90774-c.

40 Chen C, Hanson E, Watson JW & Lee JS (2003) P-glycoprotein limits the brain penetration of nonsedating but not sedating H1-antagonists. *Drug Metab Dispos* 31:312–318. doi: https://doi.org/10.1124/dmd.31.3.312.

41 Agarwal S, Sane R, Oberoi R, Ohlfest JR & Elmquist WF (2011) Delivery of molecularly targeted therapy to malignant glioma, a disease of the whole brain. *Expert Rev Mol Med* 13:e17. doi: https://doi.org/10.1017/S1462399411001888.

42 Allegra S, Cardellino CS, Fatiguso G, Cusato J, De Nicolo A, Avataneo V, Bonora S, D'Avolio A, Di Perri G & Calcagno A (2018) Effect of ABCC2 and ABCG3 gene polymorphisms and CSF-to-serum albumin ratio on ceftriaxone plasma and cerebrospinal fluid concentrations. *J Clin Pharmacol* 58:1550–1556. doi: https://doi.org/10.1992/jcph.1266.

43 Aquilante CL, Letrent SP, Pollack GM & Brouwer K L (2000) Increased brain P-glycoprotein in morphine tolerant rats. *Life Sci* 66:PL47–PL51. doi: https://doi.org/10.1010/s0024-3205(99)00599-8.

44 Choi YH & Yu AM (2014) ABC transporters in multidrug resistance and pharmacokinetics, and strategies for drug development. *Curr Pharm Des* 20:793–807. doi: https://doi.org/10.2174/138161282005140214165212.

45 Fellner S, Bauer B, Miller DS, Schaffrik M, Fankhanel M, Spruss T, Bernhardt G, Graeff C, Farber L, Gschaidmeier H, Buschauer A & Fricker G (2002) Transport of paclitaxel (Taxol) across the blood-brain barrier in vitro and in vivo. *J Clin Invest* 110:1309–1318. doi: https://doi.org/10.1172/JCI15451.

46 Chaves C, Remião F, Cisternino S & Declèves X (2017) Opioids and the blood-brain barrier: a dynamic interaction with consequences on drug disposition in brain. *Cur Neuropharm* 15:1156–1173. doi: https://doi.org/10.2174/1570159X15666170504095823.

47 Mercer SL & Coop A (2011) Opioid analgesics and P-glycoprotein efflux transporters: a potential systems-level contribution to analgesic tolerance. *Curr Top Med Chem* 11:1157–1164.

48 Schaefer CP, Tome ME & Davis TP (2017) The opioid epidemic: a central role for the blood brain barrier in opioid analgesia and abuse. *Fluids Barriers CNS* 14:32. doi: https://doi.org/10.1186/s12987-017-0080-3.

49 Xie R, Hammarlund-Udenaes M, de Boer AG & de Lange EC (1999) The role of P-glycoprotein in blood-brain barrier transport of morphine: transcortical microdialysis studies in mdr1a (−/−) and mdr1a (+/+) mice. *Br J Pharmacol* 128:563–568. doi: https://doi.org/10.1038/sj.bjp.0702804.

50 Cisterinino S, Rousselle C, Dagenais C & Scherrmann JM (2001) Screening of multidrug-resistance sensitive drugs by in situ brain perfusion in P-glycoprotein-deficient mice. *Pharm Res* 18:183–190. doi:https://doi.org/10.1023/a:1011080418027.

51 Letrent SP, Pollack GM, Brouwer KR & Brouwer KL (1999). Effects of a potent and specific P-glycoprotein inhibitor on the blood-brain barrier distribution and antinociceptive effect of morphine in the rat. *Drug Metab Dispos* 27:827–834.

52 Thompson SJ, Koszdin K & Bernards CM (2000). Opiate-induced analgesia is increased and prolonged in mice lacking P-glycoprotein. *Anesthesiology* 92:1392–1399. doi:https://doi.org/10.1097/00000542-200005000-0030.

53 Dagenais C, Graff CL & Pollack GM (2004). Variable modulation of opioid brain uptake by P-glycoprotein in mice. *Biochem Pharmacol* 67:269–276. doi:https://doi.org/10.1016/j.bcp.2003.08.027.

54 Whistler JL, Chuang H, Chu P, Jan LY & von Zastrow M (1999) Functional dissociation of μ opioid receptor signaling and endocytosis: implications for the biology of opiate tolerance and addiction. *Neuron* 23:737–746. doi: https://doi.org/10.1016/S0896-6273(01)80032-5.

55 Kobori T, Fujiwara S, Miyagi K, Harada S, Nakamoto K, Nakagawa T, Takahashi H, Narita M & Tokuyama S (2014) Involvement of moesin in the development of morphine analgesic tolerance through P-glycoprotein at the blood-brain barrier.

Drug Metab Pharmacokinet 29:482–489. doi: https://doi.org/10.2133/dmpk.DMPK-14-RG-042.

56 Yanwei L, Yue H & Xing Y (2010) Oxymatrine inhibits development of morphine-induced tolerance associated with decreased expression of P-glycoprotein in rats. *Integr Cancer Ther* 9:213–218. doi: https://doi.org/10.1177/1534735410369671.

57 Yousif S, Saubaméa B, Cisternino S, Marie-Claire C, Dauchy S, Scherrmann J-M & Declèves X (2008) Effect of chronic exposure to morphine on the rat blood-brain barrier: focus on the P-glycoprotein. *J Neurochem* 107:647–657. doi: https://doi.org/10.1111/j.1471-4159.2008.05647.

58 Yousif S, Chaves C, Potin S, Margaill I, Scherrmann J-M & Declèves X (2012). Induction of P-glycoprotein and Bcrp at the rat blood-brain barrier following a subchronic morphine treatment is mediated through NMDA/COX-2 activation. *J Neurochem* 123:491–503. doi: https://doi.org/10.1111/j.1471-4159.2012.07890.

59 Yang J, Reilly BG, Davis TP & Ronaldson PT (2018) Modulation of opioid transport at the blood-brain barrier by altered ATP-binding cassette (ABC) transporter expression and activity. *Pharmaceutics* 10:192. doi: https://doi.org/10.3390/pharmaceutics10040192.

60 Bauer B, Yang X, Hartz A M S, Olson E R, Zhao R, Kalvass JC, Pollack GM & Miller DS (2006) In vivo activation of human pregnane X receptor tightens the blood-brain barrier to methadone through P-glycoprotein up-regulation. *Mol Pharmacol* 70:1212–1219. doi: https://doi.org/10.1124/mol.106.023796.

61 Narang VS, Fraga C, Kumar N, Shen J, Throm S, Stewart CF & Waters CM (2008) Dexamethasone increases expression and activity of multidrug resistance transporters at the rat blood-brain barrier. *Am J Physiol Cell Physiol* 295:C440–C450 doi: https://doi.org/10.1152/ajpcell.00491.2007.

62 Wang X, Sykes DB & Miller DS (2010) Constitutive androstane receptor-mediated up-regulation of ATP-driven xenobiotic efflux transporters at the blood-brain barrier. *Mol Pharmacol* 78:376–383. doi: https://doi.org/10.1124/mol.110.063685.

63 Tanaka M, Dohgu S, Komabayashi G, Kiyohara H, Takata F, Kataoka Y, Nirasawa T, Maebuchi M & Matsui T (2019) Brain-transportable dipeptides across the blood-brain barrier in mice. *Sci Rep*, 9:5769. doi: https://doi.org/10.1038/s41598-019-42099-9.

64 Kageyama T, Nakamura M, Matsuo A, Yamasaki Y, Takakura Y, Hashida, M, Kanai Y, Naito M, Tsuruo T, Minato N & Shimohama S (2000) The 4F2hc/LAT1 complex transports L-DOPA across the blood-brain barrier. *Brain Res* 879:115–121. doi: https://doi.org/10.1016/s0006-8993(00)02758-x.

65 Cornford EM, Young D, Paxton JW, Finlay GJ, Wilson WR & Pardridge WM (1992) Melphalan penetration of the blood-brain barrier via the neutral amino acid transporter in tumor-bearing brain. *Cancer Res* 52(1), 138–143.

66 Roberts RL, Joyce PR, Mulder RT, Begg EJ & Kennedy MA (2002). A common P-glycoprotein polymorphism is associated with nortriptyline-induced postural hypotension in patients treated for major depression. *Pharmacogenomics J* 2:191–196. doi:https://doi.org/10.1038/sj.tpj.6500099.

67 Edwards, G. (2003). Ivermectin: does P-glycoprotein play a role in neurotoxicity? *Filaria J* 2:S8. doi: https://doi.org/10.1186/1475-2883-2-S1-S8.

68 Morris ME, Rodriguez-Cruz V & Felmlee MA (2017) SLC and ABC transporters: expression, localization, and species differences at the blood-brain and the blood-cerebrospinal fluid barriers. *AAPS J*, 19:1317–1331. doi: https://doi.org/10.1208/s12248-017-0110-8.

69 Schinkel AH, Wagenaar E, van Deemter L, Mol CA & Borst P (1995) Absence of the mdr1a P-glycoprotein in mice affects tissue distribution and pharmacokinetics of dexamethasone, digoxin, and cyclosporin A. *J Clin Invest* 96:1698–1705. doi: https://doi.org/10.1172/JCI118699.

70 Uhr M, Steckler T, Yassouridis A & Holsboer F (2000) Penetration of amitriptyline, but not of fluoxetine, into brain is enhanced in mice with blood-brain barrier deficiency due to mdr1a P-glycoprotein gene disruption. *Neuropsychopharmacology* 22:380–387. doi: https://doi.org/10.1016/S0893-133X(99)00095-0.

71 Hodges, L. M., Markova, S. M., Chinn, L. W., Gow, J. M., Kroetz, D. L., Klein, T. E., & Altman, R. B. (2011). Very important pharmacogene summary: ABCB1 (MDR1, P-glycoprotein). *Pharmacogenet Genom* 21:152–161. doi: https://doi.org/10.1097/FPC.0b013e3283385a1c.

72 Kerb R (2006) Implications of genetic polymorphisms in drug transporters for pharmacotherapy. *Cancer Lett* 234:4–33. doi: https://doi.org/10.1016/j.canlet.2005.06.051.

73 Lin Y-C, Ellingrod VL, Bishop JR & Miller DD (2006) The relationship between P-glycoprotein (PGP) polymorphisms and response to olanzapine treatment in schizophrenia. *Ther Drug Monit* 28:668–672. doi:https://doi.org/10.1097/01.ftd.0000246761.82377.a6.

74 Dauchy S, Dutheil F, Weaver RJ, Chassoux F, Daumas-Duport C, Couraud PO, Scherrmann JM, De Waziers I & Decleves X (2008) ABC transporters, cytochrome P450 and their main transcription factors: expression at the human blood-brain barrier. *J Neurochem* 107:1518–1528. doi: https://doi.org/10.1111/j.1471-4159.2008.05720.x.

75 Cascorbi I & Haenisch S (2010) Pharmacogenetics of ATP-binding cassette transporters and clinical implications. *Methods Mol Biol* 596:95–121. doi: https://doi.org/10.1007/978-1-60761-416-6_6.

76 Tsuchiya K, Hayashida T, Hamada A, Kato S, Oka S & Gatanaga H (2014) Low raltegravir concentration in cerebrospinal fluid in patients with ABCG2 genetic variants. *J AID Syndr* 66:484–486. doi: https://doi.org/10.1097/QAI.0000000000000222.

77 Pardridge WM (2015) Blood–brain barrier endogenous transporters as therapeutic targets: a new model for small molecule CNS drug discovery. *Expert Opin Ther Targets* 19:1059–1072. doi: https://doi.org/10.1517/14728222.2015.1042364.

78 Sanchez-Covarrubias L, Slosky LM, Thompson BJ, Davis TP & Ronaldson PT (2014) Transporters at CNS barrier sites: obstacles or opportunities for drug delivery? *Curr Pharm Des* 20:1422–1449. doi: https://doi.org/10.2174/13816128113199990463.

79 Lee W, Glaeser H, Smith H, Roberts RL, Moeckel GW, Gervasini G, Leake BF & Kim RB (2005) Polymorphisms in human organic anion-transporting polypeptide 1A2 (OATP1A2): implications for altered drug disposition and central nervous system drug entry. *J Biol Chem* 280:9610–9617. doi:https://doi.org/10.1074/jbc.M411092200.

80 Roth M, Obaidat A & Hagenbuch B (2012) OATPs, OATs and OCTs: the organic anion and cation transporters of the SLCO and SLC22A gene superfamilies. *Br J Pharmacol* 165:1260–1287. doi: https://doi.org/10.1111/j.1476-5381.2011.01724.x.

81 Calcagno A, Yilmaz A, Cusato J, Simiele M, Bertucci R, Siccardi M, Marinaro L, D'Avolio A, Di Perri G & Bonora S (2012) Determinants of darunavir cerebrospinal fluid concentrations: impact of once-daily dosing and pharmacogenetics. *AIDS* 26:1529–1533. doi: https://doi.org/10.1097/QAD.0b013e3283553619.

82 Cecil KM, Salomons GS, Ball WS Jr, Wong B, Chuck G, Verhoeven NM, Jakobs C & Degrauw TJ (2001) Irreversible brain creatine deficiency with elevated serum and urine creatine: a creatine transporter defect? *Ann Neurol* 49:401–404. doi: https://doi.org/10.1002/ana.79.

83 Clark AJ, Rosenberg EH, Almeida LS, Wood TC, Jakobs C, Stevenson RE, Schwartz CE & Salomons GS (2006) X-linked *creatine transporter (SLC6A8)* mutations in about 1% of males with mental retardation of unknown etiology. *Hum Genet* 119:604–610. doi:https://doi.org/10.1007/s00439-006-01262-9.

84 Rosenberg EH, Almeida LS, Kleefstra T, deGrauw RS, Yntema HG, Bahi N, Moraine C, Ropers HH, Fryus JP, Degrauw TJ, Jakobs C & Solomons GS (2004) High prevalence of SLC6A8 deficiency in X-linked mental retardation. *Am J Hum Genet* 75:97–105. doi: https://doi.org/10.1086/422102.

85 Trotier-Faurion A, Dezard S, Taran F, Valayannopoulos V, de Lonlay P & Mabondzo A (2013). Synthesis and biological evaluation of new creatine fatty esters revealed dodecyl creatine ester as a promising drug candidate for the treatment of the creatine transporter deficiency. *J Med Chem* 56:5173–5181. doi: https://doi.org/10.1021/jm400545n.

86 Qin Y & Sato TN (1995) Mouse multidrug resistance 1a/3 gene is the earliest known endothleial cell differentiation marker during blood-brain barrier development. *Dev Dyn* 202:172–180. doi: https://doi.org/10.1002/aja.1002020209.

87 Daood M, Tsai C, Ahdab-Barmada M & Watchko JF (2008) ABC transporter (P-gp/ABCB1, MRP1/ABCC1, BCRP/ABCG2) expression in the developing human CNS. *Neuropediatrics* 39:211–218. doi: https://doi.org/10.1055/s-0028-1103272.

88 Virgintino D, Errede M, Girolamo F, Capobianco C, Robertson D, Vimercati A, Serio G, Di Benedetto A, Yonekawa Y, Frei K & Roncali L (2008) Fetal blood-brain barrier P-glycoprotein contributes to brain potection during human development. *J Neuropathol Exp Neurol* 67:50–61. doi: https://doi.org/10.1097/nen.0b013e31815f65d9.

89 Bhutta TI & Tahir KI (1990) Loperamide poisoning in children. *Lancet* 335: 363. doi:https://doi.org/10.1016/0140-6736(90)90659-s.

90 McCowat LA, Cutting WA, Steinke D & MacDonald TM (1997) Treating diarrhoea. Children deserve special attention. *BMJ* 315: 1379–1380.

91 Brouwer KL, Aleksunes LM, Brandys B, Giacoia GP, Knipp G, Lukacova V, Meibohm B, Nigam SK, Rieder M, de Wildt SN & Pediatric Transporter Working Group (2015) Human ontogeny of drug transporters: review and recommendations of the pediatric transporter working group. *Clin Pharmacol Ther* 98: 266–287. doi:https://doi.org/10.1002/cpt.176.

92 Lam J, Baello S, Iqbal M, Kelly LE, Shannon PT, Chitayat D, Matthews SG & Koren G (2015) The ontogeny of P-glycoprotein in the developing human blood-brain barrier: implication for opioid toxicity in neonates. *Pediatr Res* 78: 417–421. doi:https://doi.org/10.1038/pr.2015.119.

93 Erdo F & Krajcsi P (2019) Age-related functional and expressional changes in efflux pathways at the blood-brain barrier. *Front Aging Neurosci* 11:196. doi: https://doi.org/10.3389/fnagi.2019.00196.

94 Silverberg GD, Messier AA, Miller MC, Machan JT, Majmudar SS, Stopa EG, Donahue JE & Johanson CE (2010) Amyloid efflux transporter expression at the blood-brain barrier declines in normal aging. *J Neuropathol Exp Neurol* 69:1034–1043. doi: https://doi.org/10.1097/NEN.0b013e3181f46e25.

95 Pekcec A, Schneider EL, Baumgartner W, Stein VM, Tipold A & Potschka H (2011) Age-dependent decline of blood-brain barrier P-glycoprotein expression in the canine brain. *Neurobiol Aging* 32:1477–1485. doi: https://doi.org/10.1016/j.neurobiolaging.2009.08.014.

96 van Assema DME, Lubberink M, Boellaard R, Schuit RC, Windhorst AD, Scheltens P, Lammertsma AA & van Berckel BNM (2012) P-glycoprotein function at the blood-brain barrier: effects of age and gender. *Mol Imaging Biol* 14:771–776. doi: https://doi.org/10.1007/s11307-012-0556-0.

97 Seelbach MJ, Brooks TA, Egleton RD & Davis TP (2007) Peripheral inflammatory hyperalgesia modulates morphine delivery to the brain: a role for P-glycoprotein. *J Neurochem* 102:1677–1690. doi: https://doi.org/10.1111/j.1471-4159.2007.04644.x.

98 McCaffrey G, Staatz WD, Sanchez-Covarrubias L, Finch JD, Demarco K, Laracuente ML, Ronaldson PT & Davis TP (2012) P-glycoprotein trafficking at the blood-brain barrier altered by peripheral inflammatory hyperalgesia. *J Neurochem* 122:962–975. doi: https://doi.org/10.1111/j.1471-4159.2012.07831.x.

99 Tome ME, Jarvis CK, Schaefer CP, Jacobs LM, Herndon JM, Hunn KC, Arkwright NB, Kellohen KL, Mierau PC & Davis TP (2018) Acute pain alters P-glycoprotein-containing protein complexes in rat cerebral microvessels: implications for P-glycoprotein trafficking. *J Cereb Blood Flow Metab* 38:2209–2222. doi: https://doi.org/10.1177/0271678X18803623.

100 Bromfield EB, Cavazos JE, Sirven JI (2006) An introduction to epilepsy [Internet]. West Hartford (CT): American Epilepsy Society. Chapter 3. *Neuropharmacology of Antiepileptic Drugs*. https://www.ncbi.nlm.nih.gov/books/NBK2513.

101 Zhang C, Kwan P, Zuo Z & Baum L (2012) The transport of antiepileptic drugs by P-glycoprotein. *Adv Drug Del Rev* 64:930–942. doi: https://doi.org/10.1016/j.addr.2011.12.003.

102 Abbott NJ, Khan EU, Rollinson CMS, Reichel A, Janigro D, Dombrowski SM, Dobbie MS & Begley DJ (2002) Drug resistance in epilepsy: the role of the blood-brain barrier. *Novatris Found Symp* 243:38–47.

103 Meldrum B S (1996) Update on the mechanism of action of antiepileptic drugs. *Epilepsia*, 37:S4–S11. https://doi.org/10.1111/j.1528-1157.1996.tb06038.x.

104 Xu D, Huang S, Wang H & Xie W (2018) Regulation of brain drug metabolizing enzymes and transporters by nuclear receptors. *Drug Metab Rev*50:407–414. https://doi.org/10.1080/03602532.2018.1554673.

105 Grewal GK, Kukal S, Kanojia N, Saso L, Kukreti S & Kukreti R (2017). Effect of oxidative stress on ABC transporters: contribution to epilepsy pharmacoresistance. *Molecules* 22:365. doi:https://doi.org/10.3390/molecules22030365.

106 Liu L, Collier AC, Link JM, Domino KB, Mankoff DA, Eary JF, Spiekerman CF, Hsiao P, Deo AK & Unadkat JD (2015) Modulation of P-glycoprotein at the human blood-brain barrier by quinidine or rifampin treatment: a positron emission tomography imaging study. *Drug Metab Dispos* 43: 1795–1804. doi:https://doi.org/10.1124/dmd.114.058685.

107 Sadeque AJ, Wandel C, He H, Shah S & Wood AJ (2000) Increased drug delivery to the brain by P-glycoprotein inhibition. *Clin Pharmacol Ther* 68: 231–237. doi:https://doi.org/10.1067/mcp.2000.109156.

108 Borron SW, Watts SH, Tull J, Baeza S, Diebold S & Barrow A (2017) Intentional misuse and abuse of loperamide: a new look at a drug with "Low Abuse Potential". *J Emerg Med* 53: 73–84. doi:https://doi.org/10.1016/j.jemermed.2017.03.018.

109 Bertrand Y, Capdeville R, Balduck N & Philippe N (1992) Cyclosporin A used to reverse drug rresistance increases vincristine neurotoxicity. *Am J Hematol* 40: 158–159. doi:https://doi.org/10.1002/ajh.2830400222.

110 Bisogno G, Cowie F, Boddy A, Thomas HD, Dick G & Pinkerton CR (1998) High-dose cyclosporin with etoposide--toxicity and pharmacokinetic interaction in children with solid tumours. *Br J Cancer* 77: 2304–2309. doi:https://doi.org/10.1038/bjc.1998.383.

111 Murphy JA, Ross LM & Gibson BE (1995) Vincristine toxicity in five children with acute lymphoblastic leukaemia. *Lancet* 346: 443. doi:https://doi.org/10.1016/s0140-6736(95)92816-2.

112 Maeda K, Tian Y, Fujita T, Ikeda Y, Kumagai Y, Kondo T, Tanabe K, Nakayama H, Horita S, Kusuhara H & Sugiyama Y (2014) Inhibitory effects of p-aminohippurate and probenecid on the renal clearance of adefovir and benzylpenicillin as probe drugs for organic anion transporter (OAT) 1 and OAT3 in humans. *Eur J Pharm Sci* 59: 94–103. doi:https://doi.org/10.1016/j.ejps.2014.04.004.

113 Tahara H, Kusuhara H, Endou H, Koepsell H, Imaoka T, Fuse E & Sugiyama Y (2005) A species difference in the transport activities of H2 receptor antagonists by rat and human renal organic anion and cation transporters. *J Pharmacol Exp Ther* 315: 337–345. doi:https://doi.org/10.1124/jpet.105.088104.

114 Takeda M, Narikawa S, Hosoyamada M, Cha SH, Sekine T & Endou H (2001) Characterization of organic anion transport inhibitors using cells stably expressing human organic anion transporters. *Eur J Pharmacol* 419: 113–120. doi:https://doi.org/10.1016/s0014-2999(01)00962-1.

115 Emanuelsson BM, Beermann B & Paalzow LK (1987). Non-linear elimination and protein binding of probenecid. *Eur J Clin Pharmacol* 32:395–401. doi:https://doi.org/10.1007/BF00543976.

116 Overbosch D, Van Gulpen C, Hermans J & Mattie H (1988) The effect of probenecid on the renal tubular excretion of benzylpenicillin. *Br J Clin Pharmacol* 25: 51–58. doi:https://doi.org/10.1111/j.1365-2125.1988.tb03281.x.

117 Chow KM, Hui AC & Szeto CC (2005) Neurotoxicity induced by beta-lactam antibiotics: from bench to bedside. *Eur J Clin Microbiol Infect Dis* 24: 649–653. doi:https://doi.org/10.1007/s10096-005-0021-y.

118 Chow KM, Szeto CC, Hui AC & Li PK (2004) Mechanisms of antibiotic neurotoxicity in renal failure. *Int J Antimicrob Agents* 23: 213–217. doi:https://doi.org/10.1016/j.ijantimicag.2003.11.004.

119 Boulware MJ, Subramanian VS, Said HM & Marchant JS (2003) Polarized expression of members of the solute carrier SLC19A gene family of water-soluble multivitamin transporters: implications for physiological function. *Biochem J* 376: 43–48. doi:https://doi.org/10.1042/BJ20031220.

120 Said HM, Balamurugan K, Subramanian VS & Marchant JS (2004) Expression and functional contribution of hTHTR-2 in thiamin absorption in human intestine. *Am J Physiol Gastrointest Liver Physiol* 286:G491–G498. doi:https://doi.org/10.1152/ajpgi.00361.2003.

121 Pardanani A, Harrison C, Cortes JE, Cervantes F, Mesa RA, Milligan D, Masszi T, Mishchenko E, Jourdan E, Vannucchi AM, Drummond MW, Jurgutis M,

Kuliczkowski K, Gheorghita E, Passamonti F, Neumann F, Patki A, Gao G & Tefferi A (2015a) Safety and efficacy of fedratinib in patients with primary or secondary myelofibrosis: a randomized clinical trial. *JAMA Oncol* 1: 643–651. doi:https://doi.org/10.1001/jamaoncol.2015.1590.

122 Pardanani A, Tefferi A, Jamieson C, Gabrail NY, Lebedinsky C, Gao G, Liu F, Xu C, Cao H & Talpaz M (2015b) A phase 2 randomized dose-ranging study of the JAK2-selective inhibitor fedratinib (SAR302503) in patients with myelofibrosis. *Blood Cancer J* 5: e335. doi:https://doi.org/10.1038/bcj.2015.63.

123 Zhang Q, Zhang Y, Diamond S, Boer J, Harris JJ, Li Y, Rupar M, Behshad E, Gardiner C, Collier P, Liu P, Burn T, Wynn R, Hollis G & Yeleswaram S (2014) The Janus kinase 2 inhibitor fedratinib inhibits thiamine uptake: a putative mechanism for the onset of Wernicke's encephalopathy. *Drug Metab Dispos* 42: 1656–1662. doi:https://doi.org/10.1124/dmd.114.058883.

124 Pardanani A, Vannucchi AM, Passamonti F, Cervantes F, Barbui T & Tefferi A (2011). JAK inhibitor therapy for myelofibrosis: critical assessment of value and limitations. *Leukemia* 25:218–225. doi:https://doi.org/10.1038/leu.2010.269.

125 Giacomini MM, Hao J, Liang X, Chandrasekhar J, Twelves J, Whitney JA, Lepist EI & Ray AS (2017) Interaction of 2,4-diaminopyrimidine-containing drugs including Fedratinib and Trimethoprim with thiamine transporters. *Drug Metab Dispos* 45: 76–85. doi:https://doi.org/10.1124/dmd.116.073338.

126 Zamek-Gliszczynski MJ, Taub ME, Chothe PP, Chu X, Giacomini KM, Kim RB, Ray AS, Stocker SL, Unadkat JD, Wittwer MB, Xia C, Yee SW, Zhang L, Zhang Y & International Transporter Consortium (2018) Transporters in drug development: 2018 ITC recommendations for transporters of emerging clinical importance. *Clin Pharmacol Ther* 104: 890–899. doi:https://doi.org/10.1002/cpt.1112.

13

Ototoxicity and Drug Transport in the Cochlea

Stefanie Kennon-McGill and Mitchell R. McGill

Department of Environmental and Occupational Health, Fay W. Boozman College of Public Health, University of Arkansas for Medical Sciences, Little Rock, AR, USA

13.1 Auditory System Anatomy

The auditory system is comprised of two divisions: the peripheral auditory system and the central auditory system. For the purposes of this chapter, we will focus mainly on the peripheral auditory system (Figure 13.1), but it is important to understand the general anatomy of the entire auditory system to better understand ototoxicity and why it is important.

13.1.1 External, Middle, and Inner Ear

The peripheral auditory system is comprised of the external, middle, and inner ear. The *external ear* is made up of the pinna, or what is colloquially referred to as the "ear lobe," as well as the external auditory canal. These components assist in delivering sound waves to the ear drum, or tympanic membrane, at the opening to the middle ear, although they do not alter the sound or transmit it. The *middle ear* is the area between the tympanic membrane and the oval window. This space consists of three small bones called the malleus, the incus, and the stapes. These three bones make contact with the tympanic membrane and transfer the vibrations from the sound waves from the membrane to the oval window, where the sound waves can then be transformed from compression waves in air to fluid-membrane waves in the inner ear. The *inner ear* is comprised of a bony labyrinth which is comprised of two parts: the cochlea and the vestibular system. This chapter will primarily focus on drug transport and toxicity in the cochlea, but it is

Transporters and Drug-Metabolizing Enzymes in Drug Toxicity, First Edition. Edited by Albert P. Li.

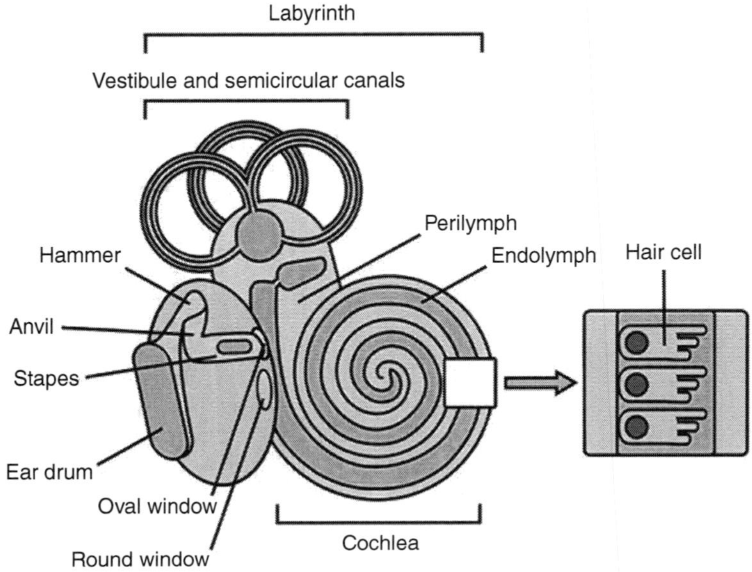

Figure 13.1 General anatomy of the inner ear.

important to note that the vestibular system, which is imperative for balance and positional processing, can be affected by certain drugs as well.

13.1.1.1 Anatomy of the Inner Ear

The word "cochlea" is derived from the Greek word *kokhlias*, which means spiral or snail shell. It was named this because the mammalian cochlea has a spiral shape and closely resembles a snail shell. Within that spiral is a complex system of ducts, membranes, and sensory cells that transduce the sound wave energy into neural signals that are communicated to the central auditory system. The physiological function of the cochlea will be discussed in the next section. The cochlea is divided into three compartments or ducts: the *scala vestibuli* (vestibular duct), the *scala tympani* (tympanic duct), and the *scala media* (cochlear duct), which are all fluid filled ducts that are lined with membranes and sensory cells. The *scala vestibuli* is superior to the *scala media* and touches the oval window. The *scala tympani* is inferior to the *scala media* and touches the round window. The *scala media*, which is most important for the purposes of this chapter, lies between the other two ducts and is the duct in which the stereocilia of the sensory cells project.

13.1.1.2 Hair Cell Anatomy

The sensory cells of the cochlea are also referred to as hair cells. These hair cells reside on the Organ of Corti, which is a sensory epithelium layer on the basilar membrane. They are the primary sensory receptors for the auditory system and their anatomical structure is delicate and intricate. The name "hair cells" was given to these sensory cells due to the appearance of the stereocilia that extend from the apical surface of the sensory cells in bundles that resemble tufts of hair. The basal ends of the hair cells are embedded in the basilar membrane, and the stereocilia-lined apical end of the hair cells protrude into the endolymph of the *scala media*. The ionic concentration of the endolymph is maintained by the stria vascularis, which is a highly vascularized epithelial tissue that is composed of multiple cell layers and tight junctions.

Throughout the length of the cochlear spiral, there are three rows of outer hair cells (OHCs) and one row of inner hair cells (IHCs), for a total of approximately 15 000 hair cells in the human cochlea. The functions of the two different types of hair cells will be discussed in the physiology section of this chapter. The hair cells of the inner ear synapse on the auditory nerve fibers. These nerve fibers project from the hair cells to the auditory brainstem and midbrain, which is comprised of multiple nuclei that process sound as it travels to the auditory cortex. This chapter will not focus on the anatomy of the central auditory system, as the unique structure and function of the hair cells and how they metabolize drugs are of primary interest, but it is important to note the various working parts of the entire auditory system that can be affected if drug metabolism and transport at the cochlear level are affected as well.

13.1.2 Blood–Labyrinth Barrier

One additional structure that is important to note in the context of drug toxicity and transport in the cochlea is the blood–labyrinth barrier (BLB). The BLB is similar in nature to the blood–brain barrier that lines the brain, except that the BLB separates the vasculature from the fluids within the inner ear. The BLB includes the blood–endolymph barrier, the blood–perilymph barrier, the cerebrospinal fluid–perilymph barrier, the endolymph–perilymph barrier, and the middle ear-labyrinth barrier. The presence of the BLB was first noted in the 1960s and was researched in the context of aminoglycoside ototoxicity [1, 2]. It was later characterized as having non-fenestrated capillaries accompanied by tight junctions that allowed for a decreasing rate of entry of compounds with increasing molecular weight into the perilymph from blood [3]. The BLB is an important biological barrier within the inner ear that serves to not only to protect the balance of fluids within the cochlea, but also to transport or extrude drugs into and out of the cochlear fluids. The BLB has been implicated in multiple clinically significant contexts, such as noise exposure induced hearing loss, infection, and ototoxicity. Various drugs, including ototoxic drugs which will be discussed later, have been shown to enter the inner ear via the BLB. The composition of the BLB, including the presence of drug transporters, will be covered in detail later in this chapter.

13.2 Auditory System Physiology

The peripheral auditory system relies on mechano-transduction as the main physiological mode of transforming sound waves into neural signals that are then relayed to the central auditory structures. When sound waves enter the ear via the external auditory canal, they travel to the tympanic membrane. The vibrations of the tympanic membrane then travel across the small bones in the middle ear, or the ossicles, and make contact with the oval window on the cochlea. Once these vibrations reach the oval window, they are transformed into waves in the fluid filled spaces of the cochlea. The waves in the perilymph cause the basilar membrane to move up and down in oscillations. When the basilar membrane moves, the hair cells that are embedded on it move as well, which causes the stereocilia at the tips of the hair cells to move side to side and bend. Because the stereocilia are connected to each other by tip links, this movement causes the tip links to stretch, which then causes ion channels to open [4]. Potassium ions from the endolymph then enter the hair cells, causing depolarization. IHCs synapse on type I auditory nerve fibers, and once the IHCs depolarize, they cause generation of action potentials in the nerve fibers, which sets off the chain of neural signals that eventually reach the auditory cortex in the central auditory system.

Because the hair cells of the cochlea are the primary sensory receptors of the auditory system, any effects on the function of the hair cells can have significant clinical effects. Thus, is it imperative to understand how drug transport occurs in the cochlea and how it can influence ototoxicity of certain drug classes. Approximately 95% of auditory nerve fiber projections form synapses with IHCs, which makes the IHCs the primary sensory receptors. However, the OHCs, of which there are three times more than the IHCs, are also important for hearing and in the context of drug transport and development. Synapses at the OHCs are primarily efferent inputs coming from the central auditory system. OHCs are thought to fine-tune the frequency resolving capabilities of the cochlea by actively contracting and relaxing, which changes the stiffness or laxity of the tectorial membrane. Therefore, damage to OHCs may have a significant effect on the ability of the hearer to distinguish certain sounds, especially in the context of background noise. It is important to consider drug toxicity and transport in the OHCs as well as the IHCs, though there does not seem to be much differentiation between the two hair cell types in the context of drug metabolism, transport, and toxicity.

13.3 Hearing Loss, Ototoxic Drugs, and Hair Cell Damage

The CDC estimates that upward of 15% of all adults 18 and older in the United States suffer from some degree of hearing loss, the vast majority of which is sensorineural hearing loss (i.e. hearing loss due to hair cell damage). Unfortunately, once mammalian hair cells are damaged, they do not regenerate. Interestingly, avian hair cells, which are similar in structure and function to mammalian hair cells, are able to regenerate which makes them a good model for studying hearing loss mechanisms. Avian hair cells have been shown to regenerate from adjacent supporting cells (i.e. non-sensory cells) through a process of direct transdifferentiation [5, 6]. Despite decades of research following this discovery, it is still unclear why mammalian hair cells do not share this feature, which makes further research into the mechanisms of ototoxicity and drug metabolism and transport in the inner ear that much more significant.

There are multiple methods by which hair cells can be damaged, including noise exposure, head and neck radiation therapy, and physical trauma, but for the purposes of this chapter, we will focus on drug-induced hair cell damage, or ototoxicity. Many classes of drugs have been found to cause hearing loss by hair cell damage, a partial list of which can be seen in Table 13.1. There are thought to be over 200 potentially ototoxic drugs on the market today, including both prescription and over the counter drugs, but three major classes of interest are aminoglycoside antibiotics such as gentamicin, platinum-based cancer therapeutics such as carboplatin or cisplatin, and salicylate pain relievers.

13.3.1 Aminoglycosides

Aminoglycoside antibiotics are broad-spectrum antibiotics that are primarily used in pediatric populations. One systematic review found a prevalence of aminoglycoside-induced ototoxicity in approximately 3% of neonates in low- and middle-income countries [7]. This class of drugs includes gentamicin, neomycin, kanamycin, and streptomycin and are typically used to treat gram-negative bacilli infections. Research into the toxic effects of aminoglycosides, specifically

Table 13.1 Partial list of ototoxic drugs.

Drug class	Examples	Proposed mechanisms of ototoxicity
Aminoglycosides	Gentamicin, neomycin, kanamycin, streptomycin, tobramycin	ROS, apoptosis
Platinum chemotherapeutics	Carboplatin, cisplatin	ROS, apoptosis
Salicylates	Sodium salicylate, aspirin	Increase of superoxide radicals in spiral ganglion cells, apoptosis
Loop diuretics	Furosemide, ethacrynic acid	Potentiates ototoxicity of other drugs; glutathione depletion, oxidative stress

streptomycin, in the 1940s precipitated the modern-day field of ototoxic research. This class of drugs is thought to cause permanent damage to hair cells and neurons by generating free radicals and can cause hearing loss as well as vestibular dysfunction. Interestingly, some aminoglycosides are selectively vestibulotoxic (gentamicin and tobramycin), while others are strictly cochleotoxic (neomycin, kanamycin, and amikacin) [8]. Damage to the vestibular system presents in up to 15% of patients after administration of aminoglycosides, and hearing loss presents in up to 25% of patients after administration [9–11].

While it is not completely clear how aminoglycosides enter hair cells, it is thought that their entry occurs either via endocytosis or through ion channels, and not drug transporters. Various aminoglycosides have been observed inside intracellular vesicles within hair cells after systemic injection in more than one animal model [12]. However, there is also evidence that aminoglycosides enter hair cells via mechanoelectrical transducers (METs) located at the tips of the stereocilia [13, 14]. Multiple factors point to MET channels as the primary route of entry of aminoglycosides into hair cells, including the exacerbation of ototoxicity by aminoglycosides when noise exposure is also present. Exposure to loud noises and acoustic stimuli causes the MET channel to open more frequently, which would allow for better entry of aminoglycosides, thus causing more significant ototoxic damage than if aminoglycosides were used in the absence of noise exposure [15]. There is also evidence of fluorescently labeled gentamicin in the tips of stereocilia before the fluorescent signal is seen further down in the cell body of the hair cells [16].

13.3.2 Platinum Chemotherapeutics

Platinum chemotherapeutics, such as carboplatin (cis-diammine 1,1-cyclobutane dicarboxylatoplatinum II) and cisplatin (cis-diamminedichloroplatinum II), are also major contributors to ototoxic hearing loss. The prevalence of cisplatin-related hearing loss is variable and appears to depend upon dose, method of administration, and duration of treatment. One systematic review found incidence of cisplatin-induced hearing loss ranged from 0% to 100% of patients, depending on the study [17]. Individuals who do suffer from ototoxic side-effects of cisplatin therapy commonly experience irreversible, severe hearing loss, as well as tinnitus. When administered systemically, platinum chemotherapeutics enter the cochlear space via the blood–endolymph barrier and are then transported into hair cells at the apical membrane. These chemotherapeutics are thought to cause hair cell death in a similar manner to aminoglycosides by creating reactive oxygen species and disrupting mitochondria, and eventually initiating apoptotic pathways [18]. However, unlike aminoglycosides, platinum therapeutics are thought to enter the hair cell via drug transporters. Evidence suggests that the organic cation transporter OCT2 and the influx copper transporter CTR1 could play significant roles in platinum chemotherapeutic transport across the apical membranes of hair cells. The role of these transporters will be discussed in greater detail later in this chapter.

13.3.3 Salicylate

Salicylate analgesics are another major class of drugs that can cause ototoxicity. While acute doses of salicylate cause only temporary hearing loss and tinnitus (ringing in the ears), chronic, high doses of sodium salicylate have been shown to cause permanent hair cell damage [19]. Unlike the previously mentioned ototoxic drug classes, salicylate appears to mainly target the OHCs, as opposed to the IHCs [20]. Salicylate has been shown to affect the electromotility of OHCs, which can suppress amplification of sound stimuli and result in hearing loss [21]. Interestingly, some studies have found low doses of salicylate to be otoprotective, especially when administered in conjunction with cisplatin. When Hyppolito et al. prophylactically administered sodium salicylate 90 minutes prior to administering cisplatin in a guinea pig model, they found all hair cells to be present (albeit with some disarrangement to stereocilia), compared damage of hair cells and stereocilia of OHCs when cisplatin was administered alone [22]. Conversely, when Deng et al. administered a high dose of salicylate, they found a large increase in the production of superoxide radicals in spiral ganglion cells, which triggered an apoptotic response [23, 24].

13.4 Drug Metabolism in the Ear

13.4.1 The Importance of Drug Metabolism in the Ear

There appears to be increasing interest in methods to deliver drugs directly to the ear, such as intratympanic injection and intracochlear infusion [25, 26–29, 52]. This has been precipitated both by (i) development of new approaches to treat hearing loss that require direct administration (e.g. gene therapy, hydrogel formulations) and (ii) the challenge of getting drugs across the BLB after systemic administration. However, very little is known about drug metabolism within ear cells, which could have a major impact on directly administered drugs. The lack of data is surprising, as there is strong evidence that drug-metabolizing enzymes (DMEs) and drug transporters play a role in damage to other organs caused by some ototoxic drugs. For example, we previously discussed cisplatin, which is a cancer drug that causes irreversible hearing loss. While the effects of DMEs on ototoxicity of cisplatin have not been investigated, multiple studies have demonstrated that the toxicity of cisplatin in the kidney and liver is partially dependent on cytochrome P450s [30–33]. Similarly, furosemide can cause transient hearing loss, and can potentiate the ototoxicity of other drugs like cisplatin, and it has been established that furosemide-induced liver injury requires P450s [34, 58]. Much of what is known about drug metabolism in the ear has been learned from indirect evidence from studies of drug ototoxicity, though a few systematic studies using more systematic approaches have been published. Here, we summarize the available literature on DME and transporter expression and activity in the inner ear.

13.4.2 Studies of Drug-Metabolizing Enzymes in Ototoxicity

Early studies of drug ototoxicity revealed that glutathione is protective in some cases. Hoffman et al. [35] reported that co-treatment with an inhibitor of glutathione synthesis, buthionine sulfoximine (BSO), dramatically potentiated hearing loss in chinchillas due to the combination of the furosemide-like drug ethacrynic acid and the aminoglycoside kanamycin. Together with the fact that ethacrynic acid and similar loop diuretics are known to form metabolites that can bind to sulfhydryl groups [36, 37, 58], those data are consistent with the hypothesis that these drugs form reactive metabolites in the ear, similar to other tissues. However, there is no direct evidence for that, and the enzymes that might form those reactive metabolites remain elusive. The DME that has been of greatest interest with regard to ototoxicity is Cyp2e1. It has been thoroughly demonstrated that the vestibular and ototoxicity of nitriles are not dependent upon Cyp2e1 in rodents, despite the fact that Cyp2e1 is necessary for the overall lethal effects of those compounds by releasing cyanide [38, 39, 57, 56]. Neither whole-body Cyp2e1 knockout nor specific pharmacologic Cyp2e1 inhibitors protect inner ear cells against nitrile toxicity. Consistent with that, no protein alkylation was observed in the cochlea from mice even after a large dose of the analgesic acetaminophen [40], which is known to react with proteins in the liver through a metabolite primarily generated by Cyp2e1 [41]. Finally, expression of *Cyp2e1* is very low in the cochlea compared to liver tissue [42]. Together, those data indicate that Cyp2e1 is unimportant for at least some cases of ototoxicity, and is likely altogether absent from the inner ear. Nevertheless, other DMEs may be expressed at relatively high levels. For example, a recent comparison of the cochlea and liver revealed that mRNA for some DMEs is similar between the two tissues [42]. In particular, Cyps 1a1, 1b1, and 2c66 mRNAs were the same, while Cyp2c65 expression was actually greater in the cochlea. Cyp2c is one of the most complex Cyp families, and the functions of Cyp2c65 are understudied. However, Cyp2c65 is homologous with human CYP2C9, which has several clinically important substrates such as nonsteroidal anti-inflammatory drugs (NSAIDs) and warfarin. Thus, the apparent presence of Cyp2c65 in the cochlea may be significant. However, those data need to be interpreted cautiously. Although Cyp1a1 mRNA values were similar in the cochlea and liver tissue, activity was undetectable in the cochlea using a substrate with at least moderate affinity for 1a1 [42]. Thus, mRNA may not accurately represent tissue levels. The same study revealed that Phase II enzymes (UDP-glucuronsyl transferases and sulfotransferases) have generally low expression in the cochlea vs. liver. Overall, DME expression seems to be low in the cochlea, with a few possible exceptions that may have major clinical implications. More studies are needed to determine the clinical importance of Cyps 1a1, 1b1, 2c65, and 2c66 in the ear, and to investigate expression of other Cyp isoforms that have not yet been measured in the cochlea.

13.4.3 Drug Transporters in the Ear

More definitive data are available from studies of transporter expression and function in the inner ear. It has been known for decades that certain ototoxic drugs (e.g. aminoglycoside antibiotics) enter the perilymph very slowly (well after the plasma C_{max} is achieved) in rodents, and that the drug concentrations achieved in that fluid are far lower than plasma concentrations [43]. That indicates that a BLB exists. Because the transporter multidrug resistance 1 (Mdr1)/P-glycoprotein (P-gp)/ATP-binding cassette b4 (Abcb4) (hereafter referred to as P-gp) is known to be important for maintenance of the blood–brain barrier, [55] hypothesized that it may be important for the BLB as well. To test that possibility, they measured P-gp expression in the guinea pig inner ear using immunohistochemistry and immunoblotting. Consistent with their hypothesis, they observed P-gp staining in capillary endothelial cells of the inner ear, and the staining intensity was similar to capillary endothelial cells in the brain [55]. More recent studies in mouse whole-cochlea extracts have confirmed those data, and revealed that P-gp expression is actually four-fold greater in cochlea than in liver tissue [42], where it is well known to be present in the bile canalicular membranes of hepatocytes [44, 55]. Together, those data indicate that P-gp may function as a critical efflux pump that keeps drugs out of the inner ear. In support of that hypothesis, it was later demonstrated that Mdr1a knockout (Mdr1a$^{-/-}$) mice have higher concentrations of doxorubicin (Dox) in the inner ear compared to Mdr1a$^{+/+}$ after treatment with the same Dox dose [29]. Importantly, the Mdr1a$^{-/-}$ mice also experienced greater hearing loss due to Dox ototoxicity, and co-administration of the P-gp inhibitor cyclosporin A with Dox in Mdr1a$^{+/+}$ mice had the same effect [29]. Together, these data demonstrate that P-gp is indeed important for drug extrusion out of the inner ear.

Other drug transporters that are present in the inner ear and have well-established clinical effects include organic cation transporter 1 and 2 (Oct1 and 2;

solute carrier (Slc) 22a1 and 22a2, respectively) and copper transporter 1 (Ctr1; Slc31a1). Ciarimboli et al. [45] reported expression of Oct1 and 2 in cochlea, and immunohistochemistry revealed the presence of Oct2 in hair cells, which would be expected if Oct2 is important for cisplatin-induced hair cell death and hearing loss. Importantly, Oct2 appeared to mediate uptake of cisplatin into cochlear fluids and tissue in that study. Cisplatin is a known substrate of Oct2 in other tissues, and Oct1/2 double knockout mice are resistant to the cisplatin otoxicity that occurs in wild-type mice [45]. Furthermore, the Oct2 inhibitor cimetidine was protective [45]. Together, those data indicate an important role for Oct2 in drug uptake in cochlea. However, a later study using parvalbumin staining to label hair cells revealed that Oct2 expression may be limited to other cell types within the supporting tissue [46]. Furthermore, it was reported that another Oct2 inhibitor, phenformin, had no effect on cochlear cisplatin concentration or ototoxicity [46]. While the immunohistochemistry results from that study are compelling, only one dose of phenformin was used and it is not clear if it was sufficient to inhibit Oct2. In addition, the authors used guinea pigs, which may be different in some way with respect to Oct2 function. Moreover, the specificity of the Oct2 antibody used in that study has been disputed [47]. And finally, more recent studies have provided support for the presence of Oct2 in hair cells [48]. Overall, there is strong evidence that Oct2 is an important drug uptake transporter in inner ear cells. Similarly, Ctr1 is expressed in multiple cell types throughout the inner ear and inhibition appears to decrease cisplatin ototoxicity [45, 49, 50]. More et al. [50] reported that competitive inhibition of Ctr1 with copper sulfate and knockdown using antisense RNA reduced cisplatin uptake by the hair cell-like cell line HEI-OC1 and cell death. They also observed reduced hearing loss in mice treated intratympanically with copper sulfate before intraperitoneal cisplatin injection compared to mice treated with cisplatin alone [50]. Similar results were obtained in cochlear explants. In the latter study, the researchers found that copper sulfate prevents loss of hair cells due to cisplatin, and the protection could be at least partially reversed by both copper and the Oct2 inhibitor cimetidine [49]. Interestingly, they also reported that the ATPase copper transporters ATP7A and ATP7B, which are known to pump cisplatin out of other tissues, are expressed in cochlear cells [49]. It is possible that those proteins are functional and reduce cisplatin ototoxicity, but that has not been tested.

Multidrug resistance-associated protein 1 (Mrp1)/Abcc1 has also been detected in the mouse and rat inner ear by immunohistochemistry and qPCR [42, 51, 53, 54]. Interestingly, Mrp1 variants have recently been linked to deafness in humans, and it is possible that that effect is related to the transport function of Mrp1 [51]. There is currently no direct physiological evidence that Mrp1 has a clinically important role in drug transport in the ear, but it seems highly likely.

A number of other transporters have been detected in cells of the inner ear, but their functional significance has not been tested. Manohar et al. [48] measured expression of a panel of 84 transporters in the stria vascularis, basilar membrane, and modiolus using qPCR, and confirmed several results by immunofluorescence staining. Their data partially confirmed results from the studies of Oct1, Oct2, and Ctr1 described above, but also identified other possible drug transporters that had not previously been measured in the inner ear, including Abcd3, voltage-dependent anion channel (VDAC) 1 and 2, and six other Slc superfamily members not mentioned above [48]. Unfortunately, the results are difficult to interpret because they did not measure those transporters in other tissues known to have qualitatively high or low expression for comparison, nor did they test the function of any of them. Nevertheless, the results are interesting and indicate that those transporters may be important for transport of xenobiotics and/or endogenous substrates in the inner ear.

13.5 Conclusion

Although there is relatively little known about drug metabolism and transport in the inner ear, the data discussed above point to a significant role for drug transporters in the auditory system. Further research into ototoxic mechanisms and the role of drug transporters in relation to those mechanisms could provide more insight into possible otoprotective therapeutics, as well as optimization of intratympanic drug delivery.

References

1 Gieldanowski J. The influence of drugs with various permeability through the vasculo-cerebral barrier on the microphonic potential of the organ of corti. II. Levels of the drugs in the cerebrospinal fluid. *Arch. Immunol. Ther. Exp.* 1963; 11:471–477

2 Misrahy GA, Spradley JF, Beran AV, Garwood VP. Permeability of cochlear partitions: comparison with blood-brain barrier. *Acta Otolaryngol.* 1960;52:525–534.

3 Juhn SK, Rybak LP, Prado S. Nature of blood-labyrinth barrier in experimental conditions. *Ann. Otol. Rhinol. Laryngol.* 1981;90(2 Pt 1):135–141.

4 Pickles JO, Comis SD, Osborne MP. Cross-links between stereocilia in the guinea pig organ of Corti, and their possible relation to sensory transduction. *Hear. Res.* 1984;15(2):103–112.

5 Girod DA, Rubel EW. Hair cell regeneration in the avian cochlea: if it works in birds, why not in man?. *Ear Nose Throat J.* 1991;70(6):343–354.

6 Stone JS, Cotanche DA. Hair cell regeneration in the avian auditory epithelium. *Int. J. Dev. Biol.* 2007;51(6–7):633–647.

7 Musiime GM, Seale AC, Moxon SG, Lawn JE. Risk of gentamicin toxicity in neonates treated for possible severe bacterial infection in low- and middle-income countries: systematic review. *Trop. Med. Int. Health* 2015;20(12):1593–1606.

8 Matz GJ. Aminoglycoside cochlear ototoxicity. *Otolaryngol. Clin. North Am.* 1993;26(5):705–712.

9 Fee WE Jr. Aminoglycoside ototoxicity in the human. *Laryngoscope* 1980;90(10 Pt 2 Suppl 24):1–19.

10 Lerner SA, Schmitt BA, Seligsohn R, Matz GJ. Comparative study of ototoxicity and nephrotoxicity in patients randomly assigned to treatment with amikacin or gentamicin. *Am. J. Med.* 1986;80(6B):98–104.

11 Moore RD, Smith CR, Lietman PS. Risk factors for the development of auditory toxicity in patients receiving aminoglycosides. *J. Infect. Dis.* 1984;149(1):23–30.

12 Hashino E, Shero M. Endocytosis of aminoglycoside antibiotics in sensory hair cells. *Brain Res.* 1995;704(1):135–140.

13 Imamura S, Adams JC. Distribution of gentamicin in the guinea pig inner ear after local or systemic application. *J. Assoc. Res. Otolaryngol.* 2003;4(2):176–195.

14 Marcotti W, van Netten SM, Kros CJ. The aminoglycoside antibiotic dihydrostreptomycin rapidly enters mouse outer hair cells through the mechano-electrical transducer channels. *J. Physiol.* 2005;567(Pt 2):505–521.

15 Hayashida T, Hiel H, Dulon D, Erre JP, Guilhaume A, Aran JM. Dynamic changes following combined treatment with gentamicin and ethacrynic acid with and without acoustic stimulation. Cellular uptake and functional correlates. *Acta Otolaryngol.* 1989;108(5–6):404–413.

16 Wang Q, Steyger PS. Trafficking of systemic fluorescent gentamicin into the cochlea and hair cells. *J. Assoc. Res. Otolaryngol.* 2009;10(2):205–219.

17 Paken J, Govender CD, Pillay M, Sewram V. Cisplatin-associated ototoxicity: a review for the health professional. *J. Toxicol.* 2016;2016:1809394.

18 Kopke RD, Liu W, Gabaizadeh R, *et al.* Use of organotypic cultures of Corti's organ to study the protective effects of antioxidant molecules on cisplatin-induced damage of auditory hair cells. *Am. J. Otolaryngol.* 1997;18(5):559–571.

19 Myers EN, Bernstein JM, Fostiropolous G. Salicylate ototoxicity: a clinical study *N. Engl. J. Med.* 1965; 273, 587–590.

20 Hakizimana P, Fridberger A. Effects of salicylate on sound-evoked outer hair cell stereocilia deflections. *Pflugers Arch.* 2015;467(9):2021–2029.

21 Kakehata S, Santos-Sacchi J. Effects of salicylate and lanthanides on outer hair cell motility and associated gating charge. *J. Neurosci.* 1996;16(16):4881–4889.

22 Hyppolito MA, de Oliveira JA, Rossato M. Cisplatin ototoxicity and otoprotection with sodium salicylate. *Eur. Arch. Otorhinolaryngol.* 2006; 263, 798–803.

23 Deng L, Ding D, Su J, Manohar S, Salvi R. Salicylate selectively kills cochlear spiral ganglion neurons by paradoxically up-regulating superoxide. *Neurotox. Res.* 2013; 24, 307–319.

24 Wei L, Ding D, Salvi R. Salicylate-induced degeneration of cochlea spiral ganglion neurons-apoptosis signaling. *Neuroscience*, 2010, 168, 288–299.

25 Peppi, M. *et al.* Intracochlear drug delivery systems: a novel approach whose time has come. *Expert Opin. Drug Deliv.* 2018; 15, 319–324.

26 Black, RJ, Cousins, VC, Chapman, P, Becvarovski, Z, Coates, HL, O'Leary, SJ, Perry, CF, Williams, BJ. Ototoxic ear drops with grommet and tympanic membrane perforations: a position statement, *Med. J. Aust.* 2007; 186, 605–606.

27 Rybak, LP, Mukherjea, D, Jajoo, S, Kaur, T, Ramkumar, V. siRNA-mediated knock-down of NOX3: therapy for hearing loss? 2012; *Cell Mol. Life Sci.* 69, 2429–2434.

28 Rua, F, Buffard, M, Sedo-Cabezon, L, Hernandez-Mir, G, de la Torre, A, Saldana-Ruiz, S, Chabbert, C, Bayona, JM, Messeguer, A, Llorens, J. Vestibulotoxic properties of potential metabolites of allylnitrile, *Toxicol. Sci.* 2013; 135, 182–192.

29 Zhang, C, Huang, W, Song, H. Expression of vascular cell adhesion molecule-1, alpha4-integrin and L-selectin during inner ear immunity reaction, *Acta Otolaryngol.* 2000; 120, 607–614.

30 Liu, H. *et al.* Effect of cytochrome P450 2E1 inhibitors on cisplatin-induced cytotoxicity to renal proximal tubular epithelial cells. *Anticancer Res.* 2002; 22, 863–868.

31 Liu, H, Baliga, R. Cytochrome P450 2E1 null mice provide novel protection against cisplatin-induced nephrotoxicity and apoptosis. *Kidney Int.* 2003; 63, 1687–1696.

32 Lu, Y, Cederbaum, AI. Cisplatin-induced hepatotoxicity is enhanced by elevated expression of cytochrome P450 2E1. *Toxicol. Sci.* 2006; 89, 515–523.

33 Quintanilha, JCF. *et al.* Involvement of cytochrome P450 in cisplatin treatment: implications for toxicity. *Cancer Chemother. Pharmacol.* 2017; 80, 223–233.

34 Mitchell, JR. *et al.* Metabolic activation of furosemide to a chemically reactive hepatotoxic metabolite. *J. Pharmacol. Exp. Ther.* 1976; 199, 41–52.

35 Hoffman, DW. *et al.* Potentiation of ototoxicity by glutathione depletion. *Ann. Otol. Rhinol. Laryngol.* 1988; 97, 36–41.

36 Cannon, PJ. *et al.* Ethacrynic acid: effectiveness and mode of diuretic action in man. *Circulation* 1965; 31, 5–18.

37 McGill, MR. *et al.* The role of the c-Jun N-terminal kinases 1/2 and receptor-interacting protein kinase 3 in furosemide-induced liver injury. *Xenobiotica* 2015; 45, 442–449.

38 Boadas-Vaello, P. *et al.* Allylnitrile metabolism by CYP2E1 and other CYPs leads to distinct lethal and vestibulotoxic effects in the mouse. *Toxicol. Sci.* 2009; 107, 461–472.

39 Boadas-Vaello, P. *et al.* Differential role of CYP2E1-mediated metabolism in the lethal and vestibulotoxic effects of cis-crotononitrile in the mouse. *Toxicol. Appl. Pharmacol.* 2007; 225, 310–317.

40 McGill, MR. *et al.* Hearing, reactive metabolite formation, and oxidative stress in cochleae after a single acute overdose of acetaminophen: an in vivo; study. *Toxicol. Mech. Methods* 2016; 26, 104–111.

41 McGill, MR, Jaeschke, H. Metabolism and disposition of acetaminophen: recent advances in relation to hepatotoxicity and diagnosis. *Pharm. Res.* 2013; 30.

42 Kennon-McGill, S. *et al.* Expression of drug metabolizing enzymes and transporters in the cochlea: Implications for drug delivery and ototoxicity. *Hear. Res.* 2019; 379, 98–102.

43 Dulon, D. *et al.* Comparative uptake of gentamicin, netilmicin, and amikacin in the guinea pig cochlea and vestibule. *Antimicrob. Agents Chemother.* 1986; 30, 96–100.

44 Nakatsukasa, H. *et al.* Expression of multidrug resistance genes in rat liver during regeneration and after carbon tetrachloride intoxication. *Hepatology* 1993; 18, 1202–1207.

45 Ciarimboli, G. *et al.* Organic cation transporter 2 mediates cisplatin-induced oto- and nephrotoxicity and is a target for protective interventions. *Am. J. Pathol.* 2010; 176, 1169–1180.

46 Hellberg, V. *et al.* Immunohistochemical localization of OCT2 in the cochlea of various species. *Laryngoscope* 2015; 125, E320–E325.

47 Ciarimboli, G. In reference to Immunohistochemical localization of OCT2 in the cochlea of various species. *Laryngoscope* 2016; 126, E231.

48 Manohar, S. *et al.* Quantitative PCR analysis and protein distribution of drug transporter genes in the rat cochlea. *Hear. Res.* 2016; 332, 46–54.

49 Ding, D. *et al.* Cisplatin ototoxicity in rat cochlear organotypic cultures. *Hear. Res.* 2011; 282, 196–203.

50 More, SS. *et al.* Role of the copper transporter, CTR1, in platinum-induced ototoxicity. *J. Neurosci.* 2010; 30, 9500–9509.

51 Li, M. *et al.* Extrusion pump ABCC1 was first linked with nonsyndromic hearing loss in humans by stepwise genetic analysis. *Genet. Med.* 2019; 21, 2744–2754.

52 Rybak LP, Dhukhwa A, Mukherjea D and Ramkumar V Local drug delivery for prevention of hearing loss. *Front. Cell Neurosci.* 2019;13:300.

53 Saito T, Zhang ZJ, Tokuriki M, Ohtsubo T, Noda I, Shibamori Y, Yamamoto T and Saito H. Expression of multidrug resistance protein 1 (MRP1) in the rat cochlea with special reference to the blood-inner ear barrier. *Brain Res.* 2001a;895(1–2):253–257.

54 Saito T, Zhang ZJ, Tokuriki M, Ohtsubo T, Noda I, Shibamori Y, Yamamoto T and Saito H. Expression of p-glycoprotein is associated with that of multidrug resistance protein 1 (MRP1) in the vestibular labyrinth and endolymphatic sac of the guinea pig. *Neurosci. Lett.* 2001b;303(3):189–192.

55 Saito T, Zhang ZJ, Tsuzuki H, Ohtsubo T, Yamada T, Yamamoto T and Saito H. Expression of P-glycoprotein in inner ear capillary endothelial cells of the guinea pig with special reference to blood-inner ear barrier. *Brain Res.* (1997;767(2):388–392.

56 Saldana-Ruiz S, Boadas-Vaello P, Sedo-Cabezon L and Llorens J. Reduced systemic toxicity and preserved vestibular toxicity following co-treatment with nitriles and CYP2E1 inhibitors: a mouse model for hair cell loss. *J. Assoc. Res. Otolaryngol.* 2013;14(5):661–671.

57 Saldana-Ruiz S, Soler-Martin C and Llorens J. Role of CYP2E1-mediated metabolism in the acute and vestibular toxicities of nineteen nitriles in the mouse. *Toxicol. Lett.* 2012;208(2):125–132.

58 Williams DP, Antoine DJ, Butler PJ, Jones R, Randle L, Payne A, Howard M, Gardner I, Blagg J and Park BK. The metabolism and toxicity of furosemide in the Wistar rat and CD-1 mouse: a chemical and biochemical definition of the toxicophore. *J. Pharmacol. Exp. Ther.* 2007;322(3): 1208–1220.

Part IV

Modeling Drug Metabolizing Enzymes-Transporters Interplay for The Prediction of Drug Toxicity

14

Application of a PBPK Model Incorporating the Interplay Between Transporters and Drug-Metabolizing Enzymes for the Precise Prediction of Drug Toxicity

Kazuya Maeda

Laboratory of Molecular Pharmacokinetics, Graduate School of Pharmaceutical Sciences, The University of Tokyo, 7-3-1 Hongo, Bunkyo-ku Tokyo, 113-0033, Japan

14.1 Importance of the Consideration of Intracellular Concentration of Drugs in the Tissue for Estimation of Pharmacological/Toxicological Effects of Drugs

The location of the target molecule is an important factor in predicting the pharmacological/toxicological effects of drugs. If the target molecule faces the blood circulation (e.g. a membrane-spanning receptor), the protein-unbound concentration of drug in the blood is a critical variable in the equation of an E_{max} model or a sigmoid E_{max} model for quantification of its pharmacological/toxicological effects. By contrast, if the target molecule is located inside cells (e.g. transcription factors, kinases), the intracellular unbound concentration of drugs will be important. Traditional pharmacokinetic (PK) analysis assumes that the unbound drug concentration in the cells is equal to that in the blood and that these rapidly equilibrate, meaning that the unbound blood concentration can be used for the estimation of a pharmacological/toxicological effect. However, if drugs cannot rapidly penetrate the plasma membrane and/or if uptake/efflux transporters are involved in drug movement across the plasma membrane, the aforementioned assumption is not applicable. Thus, the shape of the time profile of the drug concentration in blood will be different from that in the tissue, and

Transporters and Drug-Metabolizing Enzymes in Drug Toxicity, First Edition. Edited by Albert P. Li.

unbound drug concentrations in the blood and tissue are not equal even during the steady state. In that situation, the use of the unbound blood concentration as an indicator of pharmacological/toxicological effects is not appropriate. Although it is almost impossible to estimate the unbound tissue concentration of drugs *in vivo*, there are some clinical evidences indicating that the unbound tissue concentration of drugs must be considered when predicting the pharmacological/toxicological effects of drugs.

Sadeque et al. investigated the effect of quinidine administration on the loperamide side effect, an opiate-induced respiratory depression [1]. Loperamide works as an antidiarrheal drug via activation of the peripheral opioid receptor in the intestine. Because loperamide is a substrate of *P*-glycoprotein (*P*-gp), the brain distribution of loperamide is usually suppressed by *P*-gp at the blood–brain barrier (BBB), and under normal conditions the central nervous system (CNS) opioid receptor is not activated by loperamide. However, when loperamide was administered with quinidine, a potent *P*-gp inhibitor, the respiratory response to CO_2 was impaired, indicating that loperamide had activated the CNS opioid receptor. Quinidine coadministration also increased the plasma concentration of loperamide, but interestingly, the plasma concentration of loperamide was identical in the presence or absence of quinidine for up to 60 minutes after its administration, whereas respiratory depression in the presence of quinidine was rapidly initiated from 30 minutes. This discrepancy between the time profile of the plasma concentration of loperamide and that of its CNS side effect was thought to be explained by the difference in the loperamide concentrations in the blood and brain. Watson et al. investigated in rats whether the brain unbound concentration of drugs is a better biomarker of dopamine D_2 receptor occupancy than drug concentrations in other compartments, including the blood unbound concentration [2]. They measured the receptor occupancy, systemic PKs, unbound fraction in brain homogenate, and brain distribution of six commercial antipsychotics at different doses, and receptor occupancy (RO%) was plotted against the unbound concentration of these drugs in the blood ($C_{u,blood}$) or brain ($C_{u,brain}$) over the inhibition constant (K_i) for the D_2 receptor. A good relationship was observed between $C_{u,brain}/K_i$ and RO%, which followed the theoretical sigmoidal curve. However, plots of $C_{u,blood}/K_i$ vs. RO% were scattered, although the dose-dependent RO% change for each drug followed a sigmoidal curve. This clearly demonstrates that for each individual drug, the ratio of unbound drug concentration in the brain to that in the blood depends on the difference in their transport by drug transporters in the BBB. Such discrepancies between blood and local tissue concentrations of drugs are also observed in tissues such as the liver and kidney. Recently, Guo et al., on behalf of the International Transporter Consortium, published an excellent review article summarizing how to handle the prediction of intracellular drug concentration using various approaches, and how to incorporate into the physiologically based pharmacokinetic (PBPK) model the multiple factors that cause an imbalance between blood and tissue drug concentration [3]. This review includes various practical examples of the application of PBPK modeling to predict the pharmacological/toxicological effects of drugs at the stage of drug development. Different from that, the current review places more stress on the theoretical background of the imbalance of blood and tissue drug concentrations and the introduction and published examples of virtual clinical trials (VCTs) to facilitate understanding of the histogram of target clinical outcomes in a specific population using simulation by PBPK models incorporating different sets of model parameters produced randomly based on the initial distribution of each parameter.

14.2 Extended Clearance Concept as a Tool to Explain Theoretically Transporter and Drug-Metabolizing Enzyme Interplay

Conventional PK theory holds that if a drug is predominantly cleared by hepatic cytochrome P450 (CYP)-mediated metabolism, the intrinsic hepatic clearance is equal to the intrinsic metabolic clearance, which can be estimated from *in vitro* metabolism assays using human liver microsomes. Previous reports showed examples of the successful *in vitro–in vivo* correlation of the metabolic activities of drugs [4, 5]. Recently, however, many drugs thought to be eliminated from the blood circulation by extensive metabolism have also been recognized to be transporter substrates [6, 7]. In such cases, the "apparent" intrinsic hepatic clearance cannot be simply described using only the intrinsic metabolic clearance. The extended clearance concept suggests that the apparent intrinsic hepatic clearance comprises multiple primary processes including hepatic uptake from the blood to hepatocytes, backflux from hepatocytes to the blood, intracellular metabolism, and biliary excretion in an unchanged form (Figure 14.1) [6, 7]. Overall intrinsic hepatic clearance ($CL_{int,all}$) can be described by the following equation:

$$CL_{int,all} = PS_{inf} \cdot \frac{CL_{met} + PS_{bile}}{PS_{eff} + CL_{met} + PS_{bile}} \tag{14.1}$$

where PS_{inf}, PS_{eff}, CL_{met}, and PS_{bile} represent the intrinsic clearances for hepatic uptake, backflux to the blood, metabolism, and biliary excretion, respectively. In this equation, if the sum of CL_{met} and PS_{bile} is far larger than PS_{eff}, Eq. (14.1) can approximate PS_{inf}.

$$CL_{int,all} \sim PS_{inf} \tag{14.2}$$

By contrast, if the sum of CL_{met} and PS_{bile} is far smaller than PS_{eff}, Eq. (14.1) can be approximated by the following:

$$CL_{int,all} = PS_{inf} \cdot \frac{CL_{met} + PS_{bile}}{PS_{eff}} \tag{14.3}$$

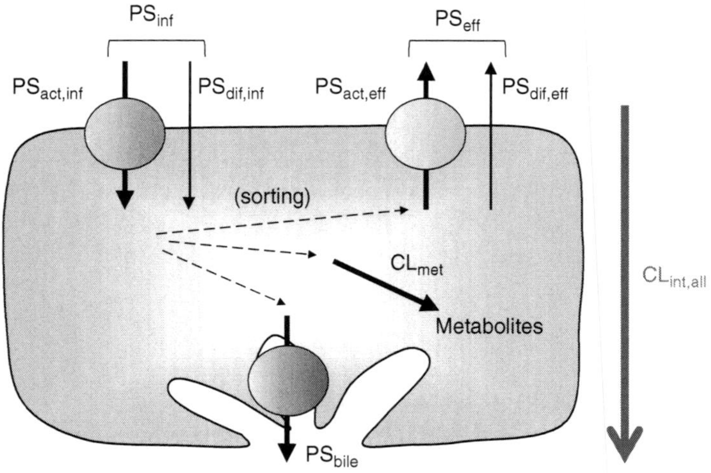

Figure 14.1 Schematic diagram of the hepatobiliary transport and metabolism of drugs.

Under the former situation (Eq. (14.2)), even if drugs are eliminated from the blood by metabolism, the observed intrinsic hepatic clearance is solely determined by the intrinsic uptake clearance and is not affected by the intrinsic metabolic clearance. This is usually termed "uptake-limited clearance". In this case, the intrinsic hepatic clearance cannot be estimated from *in vitro* metabolic assays using human liver microsomes. However, if the following conditions are fulfilled: (i) PS_{inf} is equal to PS_{eff}, (ii) PS_{bile} is negligible, and (iii) CL_{met} is far smaller than PS_{eff}, conventional theory can be applied to predict hepatic intrinsic clearance from an *in vitro* metabolism assay ($CL_{int,all} \approx CL_{met}$).

Watanabe et al. have demonstrated that the intrinsic hepatic clearance of four kinds of 3-hydroxy-3-methyl-glutaryl-coenzyme A (HMG-CoA) reductase inhibitors (statins), including two metabolized type of statins, fluvastatin (CYP2C9) and atorvastatin (CYP3A4), are much better predicted from *in vitro* uptake clearance in human cryopreserved hepatocytes than from *in vitro* metabolic clearance in human liver microsomes [8]. This indicated that the rate-determining process for these statins is likely to be the hepatic uptake process mediated by organic anion transporting polypeptide (OATP) 1Bs. We have confirmed this hypothesis in a clinical study investigating the effects of selective inhibitors of hepatic OATP1Bs (rifampicin) and CYP3A (itraconazole [intravenous dose]) on the plasma concentration of atorvastatin, a dual substrate of OATP1Bs and CYP3A, together with pravastatin (selective substrate for OATP1Bs) and midazolam (selective substrate for CYP3A) [9]. Coadministration of rifampicin, which causes a decrease in PS_{inf}, significantly increased the plasma concentration of atorvastatin and pravastatin. Conversely, intravenous dosing with itraconazole, which causes a decrease in

CL_{met}, did not alter the plasma concentration of atorvastatin but dramatically decreased the production of a major metabolite of atorvastatin mediated by CYP3A (2-hydroxy atorvastatin) and significantly increased the plasma concentration of midazolam. This confirmed the inhibition of hepatic CYP3A by itraconazole under these conditions. Therefore, we concluded that the rate-determining process of the hepatic clearance of atorvastatin *in vivo* in humans was the OATP1B-mediated hepatic uptake process. A similar clinical study reported by Yoshikado et al. has also demonstrated that rifampicin significantly increased the plasma concentration of bosentan, whereas itraconazole did not change it, indicating that the rate-determining process of hepatic clearance of bosentan is OATP1B-mediated hepatic uptake [10, 11].

14.3 Theoretical Consideration of the Intracellular Concentration of Drugs in the Tissue

To estimate accurately the pharmacological/toxicological effects of drugs when the drug target is located intracellularly, a precise estimation of the intracellular concentration of drugs in the target tissue is essential. Using the traditional assumption of rapid equilibrium of unbound drug concentrations between blood circulation and tissue without any active membrane transport, the tissue-to-blood concentration ratio (K_p value) can be described as follows:

$$K_p = \frac{C_{\text{tissue}}}{C_{\text{blood}}} = \frac{f_{\text{blood}}}{f_{\text{tissue}}} \tag{14.4}$$

where f_{blood} and f_{tissue} represent the protein-unbound fraction of drugs in blood and tissue, respectively. However, if an active transport system (uptake/efflux) is involved in the plasma membrane permeation of drugs, Eq. (14.4 should be modified as follows:

$$K_p = \frac{f_{\text{blood}}}{f_{\text{tissue}}} \cdot \frac{C_{\text{tissue,u}}}{C_{\text{blood,u}}} = \frac{f_{\text{blood}}}{f_{\text{tissue}}} \cdot K_{p,\text{uu}} \tag{14.5}$$

where $C_{\text{blood,u}}$ and $C_{\text{tissue,u}}$ represent the unbound drug concentrations in blood and tissue, respectively, and $K_{p,\text{uu}}$ is defined as the ratio of $C_{\text{tissue,u}}$ to $C_{\text{blood,u}}$. Efflux transporters on the plasma membrane contribute to the decrease in the $K_{p,\text{uu}}$ value and tissue distribution, while influx transporters contribute to their increase. In such a case, when the pharmacological/toxicological target is located inside the cells, the pharmacological/toxicological potency of drugs cannot be correctly estimated by the time course of unbound drug concentration in the blood, which can be easily measured, and estimation of the $K_{p,\text{uu}}$ value becomes important. In a traditional PBPK model, a single compartment is allocated to each organ and the tissue concentration of drugs is always instantly determined by multiplying K_p by the drug concentration in the blood vessels in the tissue. However, if the membrane permeation process becomes the rate-limiting step in tissue distribution (e.g. because of the involvement of drug transporters), each tissue compartment should be divided into two, which represent blood vessels and cells in the tissue, to reflect the discordance of the time-dependent unbound drug concentration in the blood and cells in the tissue. Considering the situation shown in Figure 14.1 and assuming that active membrane transporters on the basal side are involved in hepatic uptake, but not backflux ($PS_{act,eff} = 0$), the value of $K_{p,uu}$ can be described as shown below, using the intrinsic clearances of membrane transport and metabolism based on the extended clearance concept discussed above:

$$K_{p,uu} = \frac{C_{tissue,u}}{C_{blood,u}} = \frac{PS_{act,inf} + PS_{dif,inf}}{PS_{dif,eff} + PS_{bile} + CL_{met}} \tag{14.6}$$

where "act" and "dif" subscripts indicate clearance by active transport and passive diffusion, respectively. Yoshikado et al. previously reported the theoretical considerations for *in vitro* $K_{p,uu}$ estimation [12]. The traditional assumptions in many cases consider the passive diffusion clearances for influx and backflux on the same membrane to be the same. However, for ionized compounds at neutral pH, asymmetric distribution of unbound compounds in the blood and hepatocytes was identified during steady state because of the inside-negative membrane potential. Thus, the authors applied a new parameter, γ, which is defined as $PS_{dif,inf}/PS_{dif,eff}$. $PS_{dif,inf}$ is expressed by the following equation with separate consideration of the passive diffusion clearance of ionized and unionized forms of compounds inside and outside hepatocytes:

$$PS_{dif,inf} = f_{o,ion} \cdot PS_{dif,inf,ion} + f_{o,nonion} \cdot PS_{dif,inf,nonion} \tag{14.7}$$

where $f_{o,ion}$ and $f_{o,nonion}$ represent the fraction of ionized and nonionized form of compounds outside the hepatocytes and $PS_{dif,inf,ion}$ and $PS_{dif,inf,nonion}$ represent the intrinsic passive diffusion clearance for influx of ionized and nonionized form of compounds, respectively. When λ is defined as $PS_{dif,inf,ion}/PS_{dif,inf,nonion}$, Eq. (14.7 can be converted to the following equation based on the Henderson–Hasselbalch equation:

$$\begin{aligned} PS_{dif,inf} &= \left(1 - f_{o,nonion}\right) \cdot PS_{dif,inf,ion} + f_{o,nonion} \cdot PS_{dif,inf,nonion} \\ &= \left(\lambda \cdot \left(1 - f_{o,nonion}\right) + f_{o,nonion}\right) \cdot PS_{dif,inf,nonion} \\ &= \left(\lambda + \frac{1-\lambda}{10^{pH-pK_a} + 1}\right) \cdot PS_{dif,inf,nonion} \end{aligned} \tag{14.8}$$

Then, for statins, $PS_{dif,inf,nonion}$ and λ were optimized by fitting Eq. (14.8 to the pH-dependent apical-to-basal membrane permeation of compounds in Caco-2 cell monolayer in the presence of transporter inhibitors, which assumes no active transport. In the steady state, the passive permeation of nonionized compounds in both directions is equivalent regardless of the membrane potential ($PS_{dif,\,inf,\,nonion} = PS_{dif,\,eff,\,nonion}$), whereas that of ionized compounds is described as follows based on the Nernst equation:

$$PS_{dif,eff,ion} = \exp\left(\frac{z \cdot F \cdot \Delta\psi}{RT}\right) PS_{dif,inf,ion} = \Phi \cdot PS_{dif,inf,ion} \tag{14.9}$$

where z, F, $\Delta\psi$, R, and T represent the valence of the ion, the Faraday constant, the plasma membrane potential, the gas constant, and absolute temperature, respectively. Thus, γ can be described as follows:

$$\gamma = \frac{PS_{dif,inf}}{PS_{dif,eff}} = \frac{f_{o,nonion} + \lambda \cdot f_{o,ion}}{f_{i,nonion} + \Phi \cdot \lambda \cdot f_{i,ion}} \tag{14.10}$$

where $f_{i,ion}$ and $f_{i,nonion}$ represent the fractions of ionized and nonionized forms of compounds in the hepatocytes.

For example, in the case of pitavastatin, which has one carboxylic acid moiety and becomes a monovalent anion at neutral pH, $f_{o,ion}$, $f_{o,nonion}$, $f_{i,ion}$, and $f_{i,nonion}$ were estimated to be 0.998, 0.002, 0.998, and 0.002, respectively, based on the Henderson–Hasselbalch equation, with a pKa of 4.46 and intracellular/extracellular pH of 7.2/7.4. Lambda was estimated to be 0.0282 based on pH-dependent Caco-2 permeability and Φ was calculated to be 4.15 because the membrane potential in human hepatocytes was reported to be −38 mV at 37 °C. Thus, the γ of pitavastatin was calculated to be 0.247.

The accurate estimation of hepatic $K_{p,uu}$ value is important for quantitative prediction of the pharmacological/toxicological effects or the extent of drug–drug interaction when target molecules are located inside hepatocytes. However, researchers disagree about which *in vitro* methods are suitable for the estimation of $K_{p,uu}$ values in hepatocytes. Yoshikado et al. have demonstrated the relationship between experimentally obtained and true $K_{p,uu}$ values with the use of theoretical equations based on the abovementioned concept [12]. They found that $K_{p,uu,ss}$ estimated from the cellular accumulation of OATP1B substrates at 37 °C and on ice, which allows arrest of the active transport system, and $K_{p,uu,v0}$ estimated from the kinetic parameters for the concentration-dependent initial uptake of substrates at 37 °C are expressed differently by equations. They proved that the discrepancy between the experimental and true $K_{p,uu}$ depends on the λ value, and that $K_{p,uu,ss}$ of anionic compounds are closer to the true $K_{p,uu}$ values than $K_{p,uu,v0}$. This concept was incorporated into their PBPK models of hepatic OATP1B substrates to characterize accurately the risk of drug–drug interactions with OATP1B inhibitors, though it is difficult to directly show its validity because of a lack of real-time profiles of the intracellular concentrations of drugs in humans. One strategy to estimate the tissue concentration of compounds in humans is to utilize molecular imaging techniques such as positron emission tomography (PET) or single-photon emission computed tomography (SPECT). Previous studies have estimated the kinetic parameters of the radioactivity from OATP1B-substrate PET ligands as a measure of their hepatobiliary transport *in vivo* in humans [13]. However, strictly speaking, caution is needed concerning whether the radioactivity mainly comes from parent compounds, because the radioactivity from metabolites cannot be distinguished from that of the parent compound. This can lead to the miscalculation of the tissue distribution of the parent compound.

Nishiyama et al. have applied this concept to PBPK modeling of cationic metformin to reproduce quantitively the clinical drug–drug interaction between metformin and cimetidine, using *in vitro* K_i values of cimetidine for organic cation transporter 2 (OCT2) and multidrug and toxin extrusion proteins (MATEs) in the kidney [14]. For cationic compounds, the value of γ should be greater than 1 ($PS_{dif,inf} > PS_{dif,eff}$). Considering that the membrane potential of renal epithelial cells is approximately −70 mV, the γ_r, and γ_{urine} of metformin, which are defined as the passive influx-to-efflux ratio on the basolateral and apical sides, are estimated to be 13.7 and 12.3, respectively, calculated using $1/\Phi$ because metformin ($pK_a = 12.3$) is totally ionized at neutral pH. When Nishiyama et al. applied these parameters to a mechanistic kidney model, they found that their PBPK model could reproduce successfully the change in the plasma concentration of metformin in coadministration of cimetidine, using an uncorrected *in vitro* K_i value for MATEs. Thus, they showed that consideration of the constant membrane potential that drives the transporter-mediated membrane transport of metformin is necessary to capture drug–drug interactions involving renal transporters.

14.4 The Benefits of Using a PBPK Model for the Accurate Prediction of Pharmacological/Toxicological Effects of Drugs

The real-time simulation of drug PKs was originally initiated using a simple compartment model. In this approach, the whole body is divided into a small number of compartments (typically, one, two, or three), in each of which drugs distributed to various organs as well as blood circulation are combined, the shape of the drug concentration–time curve in each part is similar, and rapid equilibrium between parts can be assumed. The time profile of drug concentration in blood can then be apparently reproduced by optimizing the kinetic parameters in the compartment model. As discussed above, such a simple model is still useful to simulate the time-dependent pharmacological/toxicological potency of drugs when the drug binding site of pharmacological/toxicological target faces the blood circulation.

However, the drug concentration in a specific tissue cannot be estimated and no flow ever corresponds directly to the function of each isoform of the enzyme/transporter.

PBPK modeling is a powerful tool to simulate the time profile of drug concentrations not only in the blood circulation but also in various organs, including the pharmacological/toxicological target. Because each model parameter corresponds to physiological/anatomical properties such as blood flow and volume in each organ and biochemical properties such as expression and function of metabolic enzymes and transporters and concentration of drug binding proteins (albumin, α1-acid glycoprotein, etc.) in the blood and tissues throughout a human body, it is easy to estimate the change in the PKs and the subsequent clinical outcomes when the expression or function of multiple PK-related molecules is modified by several factors such as genetic polymorphisms and drug–drug interactions. Some of the model parameters such as intrinsic clearances can be theoretically extrapolated from *in vitro* experimental results by a simple scale-up method using cell counts per unit weight of tissue and tissue weight per unit of body weight. Moreover, a PBPK model can identify the critical parameters that have major effects on the pharmacological/toxicological potencies just by repetitive simulation systematically changing each model parameter (sensitivity analysis).

In particular, time profiles of both victim and perpetrator drugs are required to predict the risk of drug–drug interactions in order to consider accurately the time-dependent alteration in inhibition potencies of target proteins. In the early stages of the drug development process, a static model is often used to provide a rough estimation of the risk of drug–drug interactions involving a new drug candidate as a perpetrator. The US, EU, and Japanese regulatory guidance/guidelines for the evaluation of drug–drug interactions require that to avoid false-negative predictions (prediction: negative, reality: positive), the maximum inhibition potency is estimated by assuming that the theoretical maximum inhibitor concentration in the blood is maintained constantly. Thus, because the blood concentration of the inhibitor is always overestimated compared with the actual concentration, many false-positive predictions (prediction: positive, reality: negative) will be produced by such static model analyses. Conversely, in the static model, unbound inhibitor concentration in the blood is often used with the assumption that the unbound concentrations in the blood and tissue are the same, but the intracellular unbound concentration is sometimes higher than the unbound concentration in the blood because of the function of active uptake transporters such as hepatic OATPs. Therefore, the risk of drug–drug interactions involving the inhibition of an intracellular target (e.g. metabolic enzymes, efflux transporters), might be underestimated even by a static model using the maximum unbound concentration in the blood. By contrast, once PBPK models for both victim and perpetrator drugs are established, more accurate prediction of the extent of the change in blood exposure of victim drugs can be achieved by connecting these models at the site of inhibition, and the change in the tissue concentration, which is almost impossible to measure in humans, can also be simulated [15].

In the usual clinical settings, only blood and urine concentrations of drugs can be measured. Thus, traditionally, the pharmacological/toxicological effects of drugs are often expressed as a function of plasma unbound drug concentration or exposure (area under the curve [AUC]). However, as discussed above, discordance between drug concentrations in the blood and tissue sometimes leads to difficulty in estimation of pharmacological/toxicological effects based only on the systemic exposure of drugs.

Watanabe et al. have investigated the impact of the functional changes in hepatic uptake (OATP1Bs) and efflux transporters (multidrug resistance-associated protein [MRP] 2) on the pharmacological (decrease in low density lipoprotein [LDL]-cholesterol level) and toxicological (myopathy) effects of pravastatin using a PBPK model incorporating hepatic transporter functions [16]. The pharmacological target of statins is the HMG-CoA reductase expressed in hepatocytes; thus, their pharmacological effect is determined by the unbound intrahepatic concentration of statins. On the other hand, the muscle exposure of statins is thought to be a key factor in the risk of myopathy. Watanabe et al. first confirmed that their PBPK model of pravastatin accurately reproduced the time profiles of the plasma concentration of pravastatin at different doses and administration routes in both rats and humans (Figure 14.2a). Then, they performed sensitivity analyses to analyze the importance of hepatic uptake and biliary excretion of pravastatin in its pharmacological/toxicological effects (Figure 14.2b). In the simulation using the PBPK model, when the intrinsic hepatic uptake clearance, which corresponds to the function of OATP1Bs, was decreased, the plasma concentration of pravastatin was greatly increased, while its intrahepatic concentration was barely changed. This suggested that decreased hepatic uptake of pravastatin would result in an increased risk of myopathy, but no change in its pharmacological effect. By contrast, when intrinsic biliary excretion clearance, which corresponds to the function of MRP2, was decreased, the plasma concentration of pravastatin was not changed, whereas its intrahepatic concentration was greatly increased. This suggested that decreased biliary excretion of pravastatin would result in an increased pharmacological effect, but no change in the muscle exposure and consequent myopathy risk. These simulation results were indirectly supported by those of previous clinical studies. As for the cholesterol-lowering effect of statins, previous reports have shown that reduction of LDL-cholesterol by statin treatment in subjects with mutated alleles (c.521T>C) of *SLCO1B1*, which is known to decrease its transport function, is

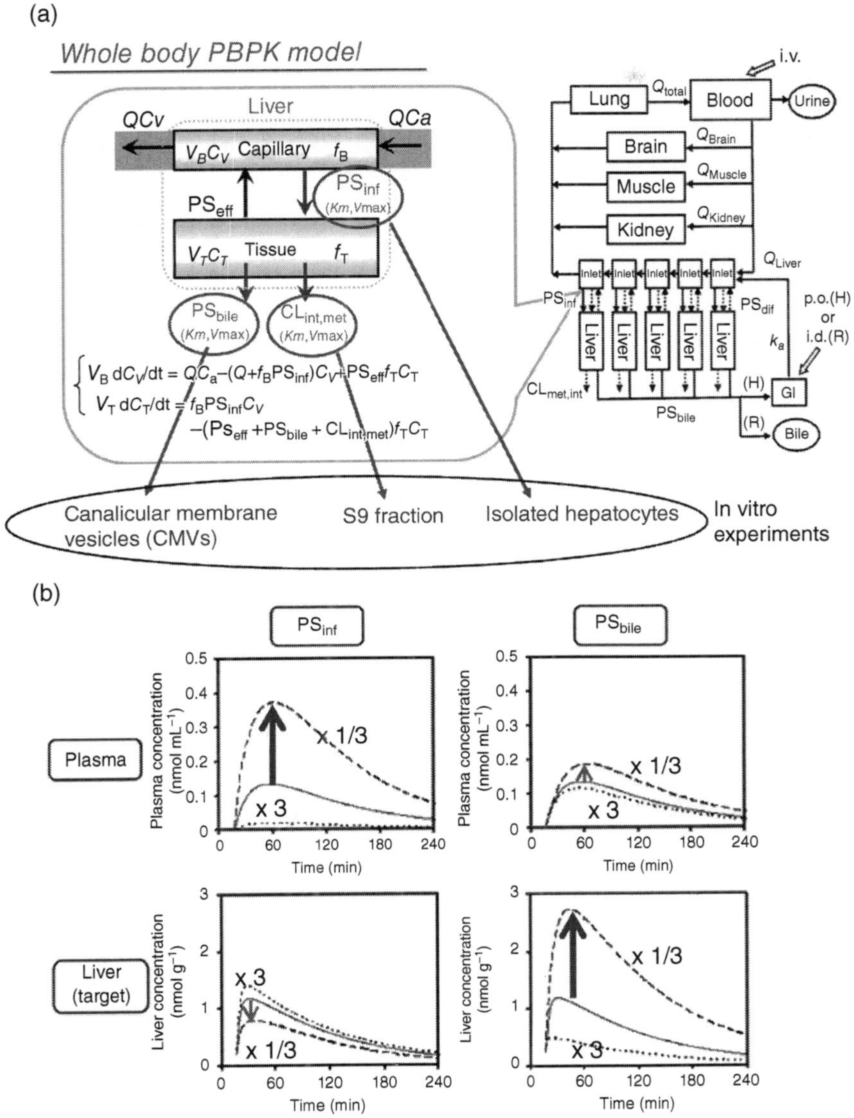

Figure 14.2 PBPK model analysis of pravastatin [16]. (a) Diagram of the structure of the PBPK model of pravastatin. As indicated, some of the model parameters were set based on in vitro experimental results. (b) Impact of the change in the function of hepatic uptake/biliary excretion transporters on the pravastatin concentration in the plasma and liver. The solid line represents the time profile of the plasma and liver concentrations of pravastatin under basal conditions. The broken and dotted lines represent the plasma and liver concentrations of pravastatin if transport function was decreased to one-third or increased three times, respectively.

comparable to that in subjects with wild-type alleles [17, 18]. Conversely, a genome-wide association study (GWAS) revealed that *SLCO1B1* c.521T>C is the allele most tightly linked to an increased risk of myopathy induced by simvastatin treatment [19]. Thus, PBPK modeling greatly assisted the estimation of local concentrations of drugs, which are impossible to measure in humans, although it is difficult to estimate accurately all the kinetic parameters related to metabolism and transport of drugs in the local tissue because of the lack of appropriate experimental systems.

14.5 VCT to Simulate the Distribution of Clinical Outcomes in a Specific Population with Defined Mean and Variability of Parameters in a PBPK Model

In general, only a small fraction of subjects exhibit the severe side effects of drugs, partly because they have abnormal function of metabolic enzymes and/or transporters with rare variants. Therefore, it is very difficult to detect a signal of toxicological effects of drugs in a regular clinical trial that includes a limited number of subjects. This sometimes leads to the unexpected expression of lethal side effects after new drugs enter the market and are prescribed to many patients, leading to the subsequent withdrawal of the drug from the market. Thus, pharmaceutical companies would like to have a package of *in vitro* methods and evaluation systems for the risk assessment of drugs. A VCT is one approach to simulate the PKs and pharmacological/toxicological effects of drugs in virtual patients by PBPK modeling incorporating intrinsic physiological/anatomical (e.g. blood flow, organ volume) and biochemical (e.g. expression and function of metabolic enzymes and transporters, albumin concentration) parameters allocated to each subject *in silico*, to analyze the whole distribution of clinical outcomes in a specific population (Figure 14.3) [20]. To perform a VCT, the average, variation, and distribution pattern (normal distribution, log-normal distribution, etc.) of each intrinsic parameter are first fixed or assumed based on reported information such as the clinical PK data of typical probe substrates for specific CYP and/or transporter isoforms or protein expression levels of metabolic enzymes/transporters in human-derived tissue samples. Once the information about the distribution of all the parameters used in the PBPK model and the covariances of each pair of different parameters is collated, many sets of PK parameters can be produced for a large number of "virtual" human subjects by randomly selecting each parameter based on its distribution within the constraint of covariances. Then, an appropriate PBPK model incorporating each parameter set is simulated to obtain the targeted output (e.g. pharmacological/

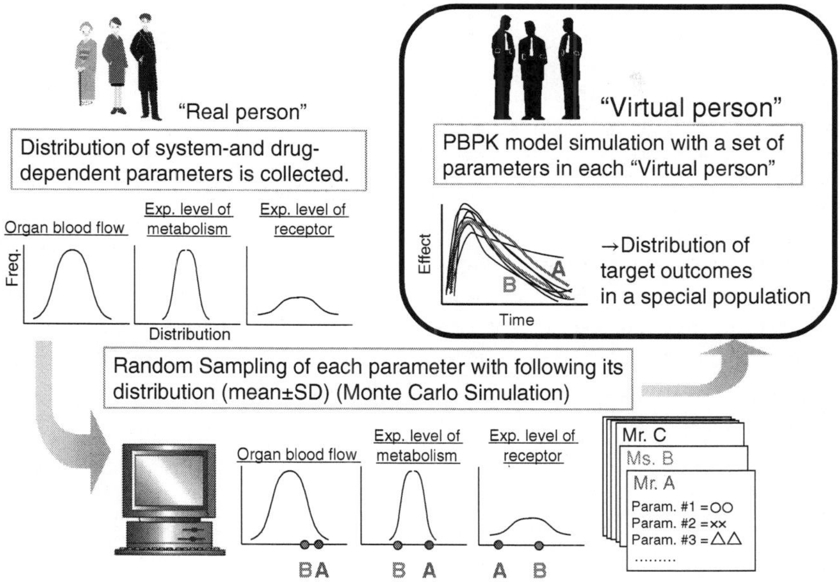

Figure 14.3 Basic concept of a virtual clinical trial (VCT).

toxicological effects, tissue concentration) in each virtual subject one by one. This analysis can produce a histogram of the outputs including outliers in a preset specific population. When considering the potential toxicity of drugs, it is particularly necessary to estimate the fraction of outliers who show excessive pharmacological/toxicological effects of drugs and the severity of the maximum impact of drug effects on outliers among a large number (~millions) of virtual subjects. Thanks to the rapid advances in computer technology, a large VCT (multiple trials with millions of subjects) can be performed using a regular laptop computer. This concept has also been introduced within several commercially available software packages that include a database of the distribution in a specific population of each parameter included in the PBPK model [21, 22]. Although the estimation of the inter-individual variability of each PK parameter *in vivo* in humans is difficult, Kato et al. estimated the coefficient of variation (CV) for hepatic CYP3A4 content at 33%, which best fits the reported CVs of dose-normalized AUCs of 40 CYP3A4 substrate drugs in humans [20]. A similar procedure was also applied to the determination of CVs for the content of other CYP isoforms [23–27].

Here, we show three examples of VCTs of drugs used to characterize the toxicity of transporter substrate drugs when the function of the transporters is modified by genetic polymorphisms.

14.5.1 VCT of Docetaxel to Estimate the Effects on the Risk of Neutropenia of Genetic Polymorphisms in OATP1B3 and MRP2

Docetaxel is used as first-line therapy in the treatment of several types of cancer. It is well known that neutropenia is a lethal dose-limiting toxicity of docetaxel, and thus, the identification of the factors that determine the risk of neutropenia induced by docetaxel is desirable. Kiyotani et al. attempted to determine the relationship between docetaxel-induced severe (Grade 3 and 4) neutropenia and the genetic background of cancer patients using samples deposited at BioBank Japan, and demonstrated that the risk of severe neutropenia induced by docetaxel was significantly increased in subjects with rs11045585 (A>G) in *SLCO1B3* encoding OATP1B3 and rs12762549 (G>C) in *ABCC2* encoding MRP2 [28]. However, because these mutations are located outside the exon region of these genes, it is not easy to identify the functional changes induced in transporters by these mutations using *in vitro* transporter gene expression systems. Therefore, Yamada et al. tried to estimate the quantitative impact of these mutations in *SLCO1B3* and *ABCC2* on docetaxel-induced neutropenia using a VCT that best reproduced the aforementioned clinical data [29]. They hypothesized that hepatic uptake of docetaxel is mainly mediated by OATP1B3, whereas MRP2 is involved in the efflux of intracellular docetaxel from bone marrow cells, and that these mutations decreased the functions of these transporters. Hence, decreased hepatic clearance and increased cellular accumulation of docetaxel in bone marrow cells led to a decreased number of neutrophils. Yamada et al. constructed a relatively simple PK/pharmacodynamic (PD) model for use in the VCT by combining the previously reported three-compartment model for the PKs of docetaxel [30] and the semi-mechanistic PD model to represent the proliferation and maturation of immature blood cells as proposed by Friberg et al. [31] (Figure 14.4). We propose that OATP1B3 function modified the elimination rate constant of docetaxel (K_{10}), while MRP2 function modified the apparent EC_{50} value in the E_{max} model (E_{drug}) for the suppression of the proliferation of immature cells. To conduct a VCT of docetaxel, we produced the parameters required in this PK/PD model for 500 virtual subjects using a Monte Carlo simulation approach, based on the published information about the mean and CV of each parameter estimated by nonlinear mixed-effects analysis, and simulated the time course of neutrophil concentration in the blood circulation after docetaxel administration in each virtual subject. Then, the grade of neutropenia was judged from the simulated nadir of neutrophil concentration in each subject, with Grade 3 and 4 defined as severe neutropenia (neutrophil concentration $< 1.0 \times 10^9\ l^{-1}$). To estimate the decrease in the MRP2 function in subjects with a mutated *ABCC2* gene, the change in the intrinsic hepatic clearance of docetaxel caused by this mutation was assumed to be the same as that for telmisartan estimated in our previous pharmacogenetic study

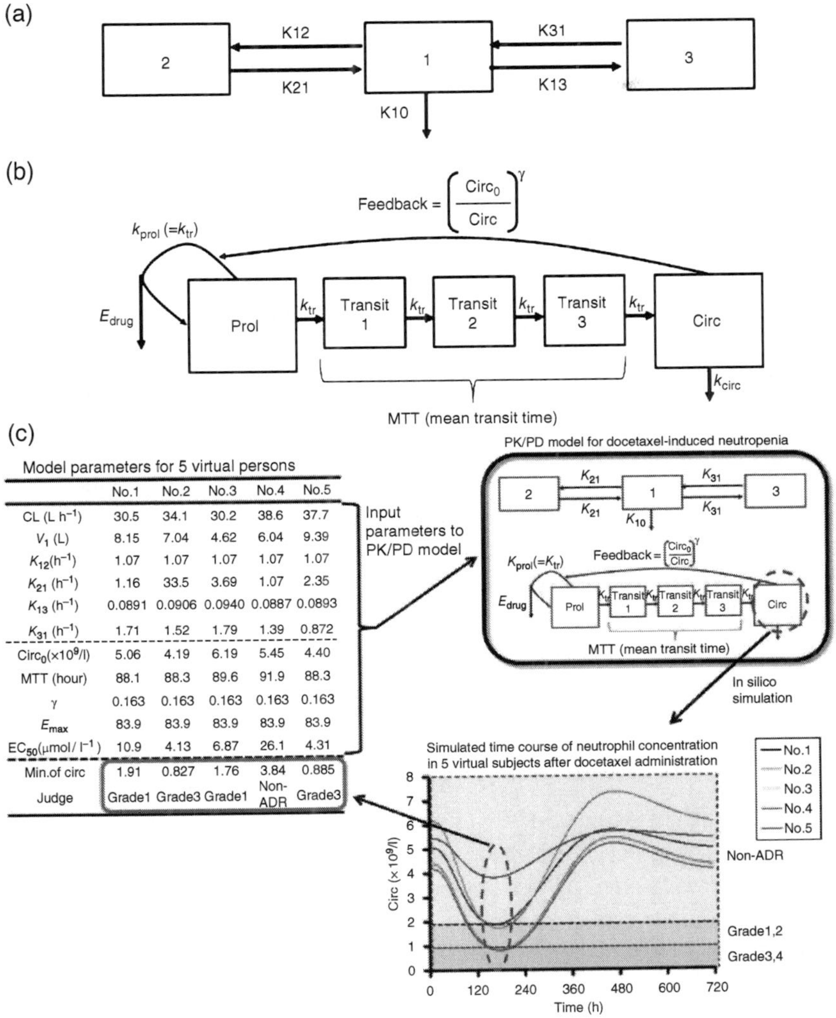

Model parameters for 5 virtual persons

	No.1	No.2	No.3	No.4	No.5
CL (L h^{-1})	30.5	34.1	30.2	38.6	37.7
V_1 (L)	8.15	7.04	4.62	6.04	9.39
K_{12}(h^{-1})	1.07	1.07	1.07	1.07	1.07
K_{21} (h^{-1})	1.16	33.5	3.69	1.07	2.35
K_{13} (h^{-1})	0.0891	0.0906	0.0940	0.0887	0.0893
K_{31} (h^{-1})	1.71	1.52	1.79	1.39	0.872
Circ$_0$(×10^9/l)	5.06	4.19	6.19	5.45	4.40
MTT (hour)	88.1	88.3	89.6	91.9	88.3
γ	0.163	0.163	0.163	0.163	0.163
E_{max}	83.9	83.9	83.9	83.9	83.9
EC$_{50}$(μmol/ l^{-1})	10.9	4.13	6.87	26.1	4.31
Min.of circ	1.91	0.827	1.76	3.84	0.885
Judge	Grade1	Grade3	Grade1	Non-ADR	Grade3

Figure 14.4 VCT of docetaxel to estimate the impact of OATP1B3 and MRP2 on the risk of neutropenia induced by docetaxel [29]. (a) Diagram of the pharmacokinetic model structure (three-compartment model). (b) Diagram of the pharmacodynamic model structure to simulate the cell count of immature and mature neutrophils. (c) Schematic diagram of the procedure for the VCT of docetaxel. First, a set of model parameters was produced for each virtual subject by Monte Carlo simulation following the distribution of each parameter in a patient population. Then, these parameter sets were input to the pharmacokinetic/pharmacodynamic model one by one, and the time course of neutrophil concentration in each virtual subject after docetaxel administration was simulated. Finally, based on the simulation results (nadir of neutrophil concentration), we judged the severity of neutropenia in each subject.

(41%) [32], and VCT was performed with different EC_{50} values to identify the optimized EC_{50} that best reproduced the clinically observed ratio of the number of patients with severe neutropenia to those without neutropenia (0.0968). From the ratio of EC_{50} in subjects with wild-type and mutated *ABCC2* (rs12762549), MRP2 function was expected to be decreased to 32% of wild type. Once we determined the functional changes in OATP1B3 and MRP2 caused by the genetic mutations, we could estimate the odds ratio for docetaxel-induced neutropenia in each combination of genotypes of OATP1B3 and MRP2. For example, a previous clinical study could not calculate the odds ratio for homozygotes of mutated alleles of *SLCO1B3* because of a lack of clinical samples due to the low allele frequency of rs11045585 in Japanese people (East Asian: 0.124 (dbSNP)). Our VCT model enabled easy calculation of this odds ratio with a large number of virtual subjects (e.g. 500) of each genotype, though in reality it is too difficult to recruit this number of subjects with minor alleles. The odds ratio for docetaxel-induced neutropenia in homozygotes with both mutated *SLCO1B3* and *ABCC2* alleles was estimated to be 88.4. Moreover, by comparing the effects on that odds ratio of mutations in *SLCO1B3* and *ABCC2*, we showed that *ABCC2* rs12762549 (13.2) is more critical in increasing the risk of neutropenia induced by docetaxel compared with *SLCO1B3* rs11045585 (5.11).

14.5.2 VCT of Oseltamivir and Its Active Metabolite (Ro 64-0802) to Estimate the Effects on Their Brain Exposure of Genetic Polymorphisms in Multiple Uptake/Efflux Transporters

Oseltamivir is a prodrug of Ro 64-0802, which is a selective inhibitor of influenza virus neuraminidase. Several years ago, abnormal behavior induced by oseltamivir administration was suspected in a small fraction of influenza patients. Although its exact cause has not been clarified, some researchers detected a CNS disorder induced by oseltamivir in animal experiments [33, 34]. A possible explanation of why the CNS side effect is expressed in only a limited number of subjects is interindividual variation in the brain distribution of Ro 64-0802, caused by rare variants of multiple enzymes and transporters. Considering the PKs of oseltamivir, the following four processes were considered to have a major influence on the regulation of the brain exposure of Ro 64-0802:

1) conversion of oseltamivir to Ro 64-0802 by hepatic carboxylesterase (CES)1A1;
2) renal secretion of Ro 64-0802 by OAT3;
3) suppression of brain penetration of oseltamivir by *P*-gp;
4) suppression of brain penetration of Ro 64-0802 by OAT3 and MRP4.

The advantage of the VCT approach is that it can capture the extreme cases and their probability of occurrence in a large population, which cannot be achieved easily in real clinical trials. Ito et al. performed a VCT to quantitatively predict the interindividual difference in oseltamivir and Ro 64-0802 exposure in the blood and brain [35]. They assumed a simple scenario in which oseltamivir in blood was eliminated by CES1A1-mediated hepatic clearance and renal clearance including OAT3-mediated tubular secretion, hepatically cleared oseltamivir was totally converted to Ro 64-0802 in the liver, and Ro 64-0802 was eliminated only from the kidney. Using these assumptions, they constructed a simple PBPK model consisting of blood compartments for oseltamivir and Ro 64-0802 and a liver compartment for Ro 64-0802 with first-order intestinal absorption of orally administered oseltamivir (Figure 14.5a, left). Then, information about the blood concentrations of oseltamivir and Ro 64-0802 was connected to a local-compartment model for the brain (Figure 14.5a, middle and right). In this brain model, conversion of oseltamivir to Ro 64-0802 in the brain by CES1A1 was considered. In addition, the brain distribution of oseltamivir and Ro 64-0802, *P*-gp-mediated efflux of oseltamivir, OAT3-mediated efflux of Ro 64-0802 from the brain to the BBB, and MRP4-mediated efflux of Ro 64-0802 from the BBB to the blood were incorporated into the local model based on the experimental results obtained in animals. The theoretical equations for blood and brain AUCs of oseltamivir and Ro 64-0802 were then derived from the differential equations for the abovementioned model. To consider the effects of genetic mutations of the target enzymes and transporters on the blood and brain exposure of oseltamivir and Ro 64-0802, allele frequencies of each mutation, and their relative activities compared with wild type were collected. In this VCT study, because of the rarity of the mutations, 100 000 virtual subjects were generated in each trial and the same trial was repeated 20 times. The predicted means and variabilities of blood AUCs of oseltamivir and Ro 64-0802 were similar to the clinical data [36]. Even in the highest case, the blood AUCs were at most three times higher than the mean values (Figure 14.5b, left). However, in the most extreme case, the brain AUCs of oseltamivir and Ro 64-0802 were increased by up to 7- and 40-fold, respectively (Figure 14.5b, right) compared with their mean values, because the virtual subject had multiple mutations of key enzymes/transporters. Under this condition, the highest unbound concentration of Ro 64-0802 exceeded the *in vitro* EC_{50} for the synchronized burst in cultured hippocampus slices, which may have led to the very rare but severe CNS side effect of oseltamivir. Although it is still difficult to establish accurately the variation in each intrinsic parameter (e.g. CV% of intrinsic renal secretion clearance, hepatic clearance, and glomerular filtration rate were assumed to be 20% in this study) and the alteration in the functions of key enzymes/transporters caused by genetic mutations, VCT is suitable for crudely capturing the maximum severity and the probability of the side effects of drugs in the worst-case scenario based on the population data from a large number of virtual patients. This information cannot be obtained in a real clinical study.

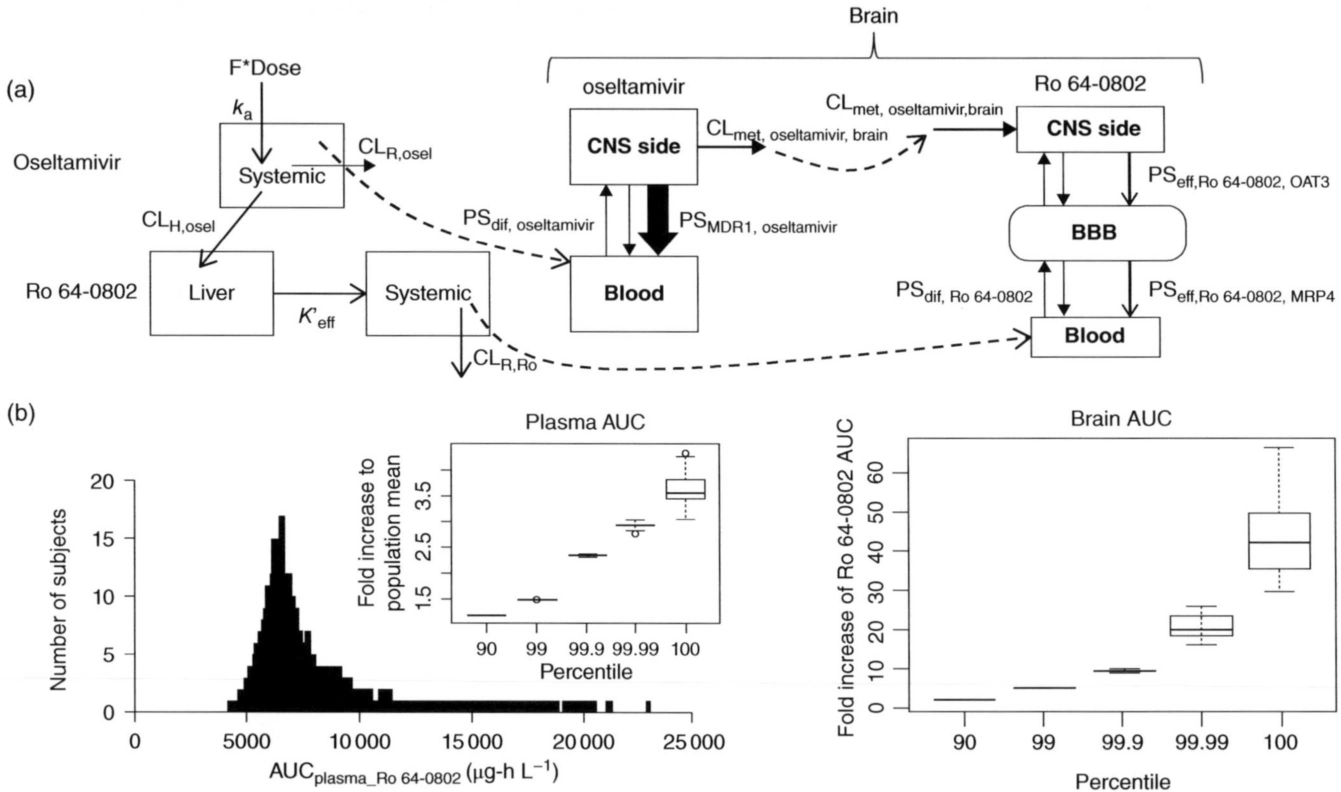

Figure 14.5 VCT of oseltamivir and its active metabolite, Ro 64-0802, to evaluate the distribution of the exposure of Ro 64-0802 in the plasma and brain in a large population [35]. (a) Diagram of the structure of the pharmacokinetic model of oseltamivir and Ro 64-0802 in the body (left) and brain (middle and right). Dotted arrows represent a direct link between compartments in different models. (b) Distribution of predicted plasma AUC of Ro 64-0802 in 100 000 virtual subjects after oral administration of oseltamivir (150 mg). The inset represents a box plot of the 90th, 99th, 99.9th, 99.99th, and 100th percentiles of fold increases relative to the population mean for 20 simulations, assuming that the efflux clearance of oseltamivir is much higher than the metabolic clearance of oseltamivir in the brain. (c) Distribution of predicted brain AUC of Ro 64-0802 after oral administration of oseltamivir (150 mg). The box plot shows the 90th, 99th, 99.9th, 99.99th, and 100th percentiles of fold increases relative to the population mean for 20 simulations, assuming that the

14.5.3 VCT of Irinotecan and Its Metabolites to Estimate the Effects of Genetic Polymorphisms in Multiple Uptake/Efflux Transporters on Irinotecan-Induced Side Effects (Neutropenia, Diarrhea)

Irinotecan (CPT-11) is a prodrug of SN-38, which exerts anticancer effects by the inhibition of topoisomerase I. The PKs of irinotecan are very complex because it forms various metabolites such as SN-38, SN-38 glucuronide, 7-ethyl-10-[4-*N*-(5-aminopentanoic acid)-1-piperidino] carbonyloxycamptothecin (APC), and 7-ethyl-10-(4-amino-1-piperidino) carbonyloxycamptothecin (NPC). Multiple enzymes and transporters in the gut and liver are involved in the regulation of the PKs of irinotecan and its metabolites. Irinotecan sometimes produces severe side effects such as neutropenia and delayed diarrhea, which are thought to be caused by high exposure of SN-38 in the blood and intestinal lumen, respectively. Previous pharmacogenetic studies have demonstrated the effects of genetic polymorphisms of enzymes and transporters on the expression of these side effects induced by irinotecan, but previous studies with a limited number of cancer patients have produced conflicting results [37]. Therefore, Toshimoto et al. attempted a VCT study of irinotecan and its metabolites to show the impact of genetic polymorphisms in each enzyme/transporter on irinotecan-induced neutropenia and diarrhea [38]. Figure 14.6a shows the whole structure of a PBPK model of irinotecan. A standard module PBPK model was used for each molecule and these modules were interconnected at the sites of metabolic conversion. Although this PBPK model precisely reflected the movement of drugs mediated by enzymes/transporters in the whole body, one of the major problems is the difficulty in determining accurately all the model parameters. Some of the parameters cannot be easily estimated from *in vitro*/animal experiments and it is also still debated whether *in vitro* parameters can be used directly in PBPK modeling using a simple scale-up approach without any empirical scaling factors. A small number of unknown parameters can be optimized to best explain the clinical PK data, but conventional fitting algorithms (e.g. Gauss–Newton method, Levenberg–Marquardt method) cannot identify optimum values for a large number of unknown parameters (e.g. 46 unknown parameters in this study) and require reliable initial values for all the parameters, which are often difficult to estimate. Thus, Toshimoto et al. introduced a new fitting algorithm, the Cluster Newton method (CNM), for the estimation of parameters in such a complex PBPK model [39, 40]. The advantage of CNM is that only a broad range of values is needed as an initial condition rather than fixed values at the first step of parameter optimization, and that multiple sets of optimized parameters that reproduce the observed data can be obtained eventually. CNM has been successfully utilized for the optimization of multiple parameters in various types of PBPK model. By introducing reasonable assumptions in

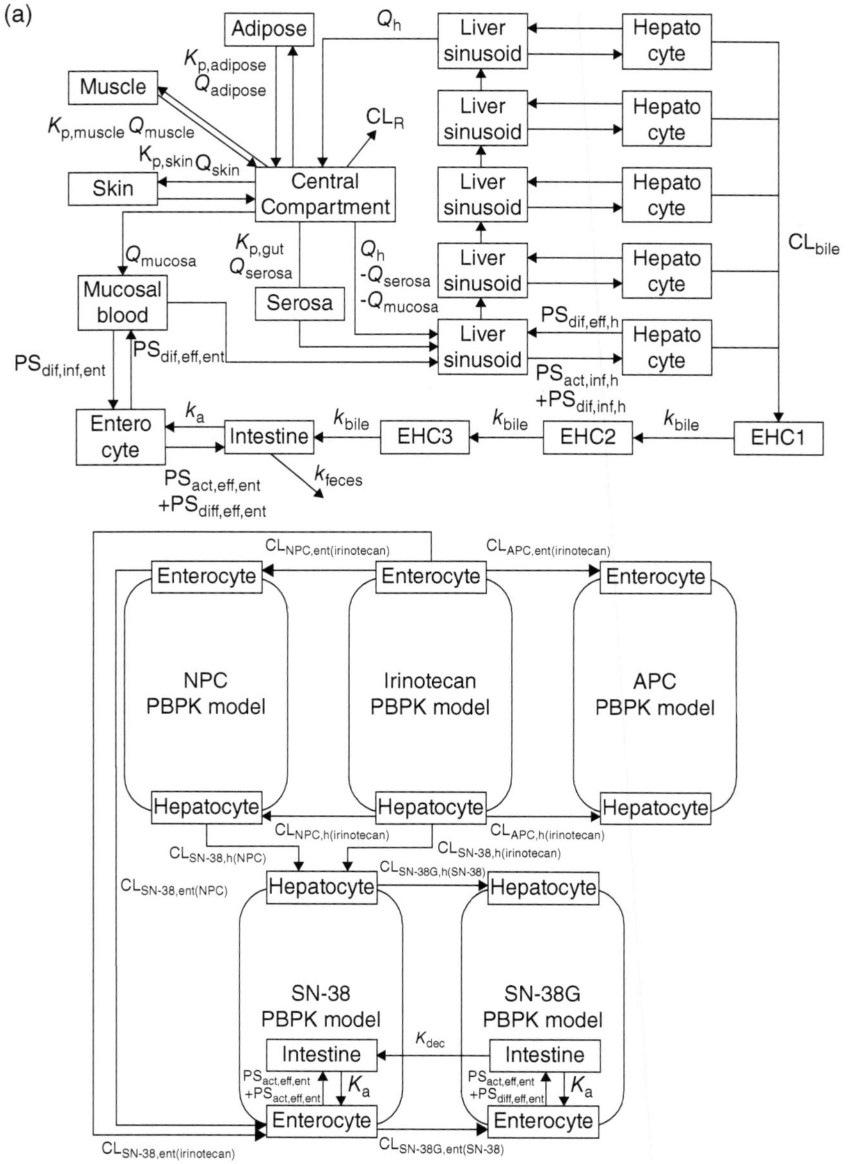

Figure 14.6 VCT of irinotecan to analyze the impact of genetic polymorphisms in enzymes and transporters on the risk of severe side effects of irinotecan (neutropenia, diarrhea) [38]. (a) Diagram of the structure of the PBPK model of irinotecan and its metabolites. (Left) A module of the PBPK model for each compound. (Right) The whole picture of the structure of the PBPK model of irinotecan and its metabolites. Each module for irinotecan and its metabolites is connected at the liver, enterocyte, and intestine compartments as shown. (b) Frequency of significance match of the 100 VCTs to a target clinical study for each of the 30 sets of parameters determined by CNM. Frequency of significance represents how many times statistical significance was achieved for each association between genotypes and SN-38 plasma concentration/neutropenia/diarrhea in 100 VCTs. Each dot represents the frequency of significance in 100 VCTs using a specific set of parameters

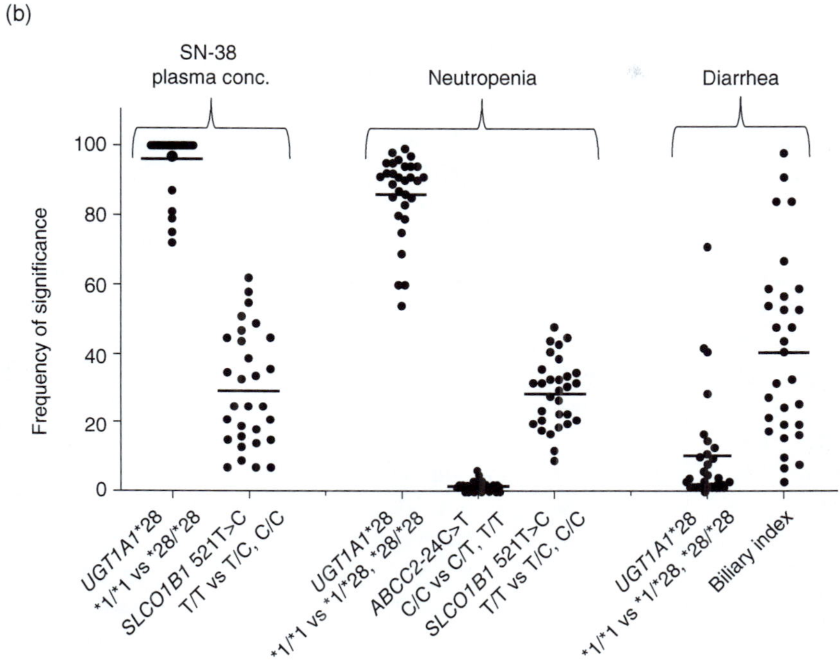

Figure 14.6 (Continued)

the determination of some model parameters and their variations, a VCT of irinotecan and its metabolites was performed using a PBPK model with 30 different sets of model parameters optimized by CNM. In this study, 127 subjects (the same number of patients as in the previous target study [41]) were enrolled in each in silico trial and these trials were repeated 100 times. Then, the effects on the PKs of SN-38 and subsequent expression of side effects of six genetic polymorphisms in enzymes and transporters (*UGT1A1**28, *SLCO1B1* c.521T>C and c. 388A>G, *ABCG2* c.421C>A, *ABCB1* c.3435C>T, and *ABCC2* c.-24C>T) were investigated. For the definition of the exposure of SN-38 that induces side effects, it was assumed that the virtual patients who showed the top 21 highest unbound plasma AUCs of SN-38 developed neutropenia and that those who showed the top 8 highest unbound enterocyte AUCs of SN-38 developed severe diarrhea, which mimics the situation in the previous study [41]. The results using different sets of model parameters in the PBPK model showed that in most cases, the plasma concentration of SN-38 and subsequent neutropenia risk was significantly higher in subjects with *UGT1A1**28 or *SLCO1B1* c.521T>C (Figure 14.6b). However, although the previous target study indicated a significant association between *ABCC2* c.-24C>T and neutropenia risk, the results of the VCT could not reproduce that clinical study outcome with any set of model parameters. As for diarrhea, previous reports had indicated that the "biliary index" ($=AUC_{irinotecan}*AUC_{SN\text{-}38}/AUC_{SN\text{-}38G}$) proposed by Gupta et al. [42] together with the *UGT1A1**28 genotype were significantly linked to the severe diarrhea induced by irinotecan. In this VCT, the frequency of significant association between the biliary index and diarrhea was higher than that of the association between *UGT1A1**28 and diarrhea for all sets of model parameters (Figure 14.6b). Thus, VCT suggested that the biliary index is a better biomarker than *UGT1A1**28 for the expression of severe diarrhea induced by irinotecan, although *UGT1A1**28 tended to increase the biliary index. VCT also provides a rationale for setting the number of subjects in a clinical study. Interindividual variability in the physiological/biochemical parameters of each patient sometimes causes conflicting outcomes in different clinical studies including small numbers of patients. For example, the target clinical study used for this VCT failed to show a positive association between *SLCO1B1* c.521T>C and irinotecan-induced neutropenia [41], whereas another clinical study demonstrated a significant association [37]. In this study, 100 VCTs were performed with different numbers of virtual subjects in a single trial and the frequency of a significant effect of *UGT1A1**28 or *SLCO1B1* c.521T>C on the plasma concentration of SN-38 (reproduction frequency) was calculated. As expected, to achieve a reproduction frequency >0.8, a smaller number of subjects is required to clarify the effect of *UGT1A1**28 than that of *SLCO1B1* c.521T>C, and at least 192 subjects are needed to obtain a reproducible effect of *SLCO1B1* c.521T>C on the plasma concentration of SN-38 in a single study. Thus, VCT is useful for rational design of a clinical study to convincingly assess a clinical outcome based on the effect size and variability of the modified functions in mutated alleles and the interindividual variability of other physiological/biochemical model parameters incorporated in a PBPK model.

14.6 Conclusions and Future Perspectives

We have briefly overviewed the theoretical considerations for the factors to determine the intracellular concentration of drugs, PBPK modeling analyses incorporating multiple metabolism and transport processes for predicting the different time profiles of drug concentrations in the local tissue and blood circulation, and VCTs to obtain the distribution of clinical outcomes in a specific population. Thanks to the rapid progress in PBPK modeling techniques and environments, many papers on PBPK models have been published. However, no unified procedure for constructing PBPK models and model parameters for new drug candidates has been established, especially when clinical PK data for drug candidates are not available. Further validation and improvements are needed to use PBPK models to predict *in vivo* PK profiles from *in vitro* experimental results (the so-called "bottom-up approach"). PBPK modeling is also expected to clarify the mean and deviation of the target outcomes in a specific population such as infants, older persons, and cancer patients, for whom participation in clinical trials is highly problematic. For this purpose, constructing a shared database of the distribution of model parameters in a wide variety of specific populations is of great importance. By using such data, optimized dose regimens can be proposed for specific populations, based on the PK data for healthy adults enrolled in phase I trials. For quantitative analysis of the pharmacological/toxicological effects of drugs, a molecular-based PD model or a quantitative systems pharmacology model should be connected to a PBPK model. Although there is no best answer for the method to establish a large number of optimum model parameters in a PD model, a PK/PD model provides directly the inter-individual differences in the time profiles of pharmacological/toxicological drug effects in a specific population while considering the change in the expression/function of multiple molecules involved in the pathways of pharmacological/toxicological drug action. Future advances in this field deserve continued attention.

References

1 Sadeque, A.J., Wandel, C., He, H. et al. (2000). Increased drug delivery to the brain by *P*-glycoprotein inhibition. *Clin Pharmacol Ther* 68 (3): 231–237.

2 Watson, J., Wright, S., Lucas, A. et al. (2009). Receptor occupancy and brain free fraction. *Drug Metab Dispos* 37 (4): 753–760.

3 Guo, Y., Chu, X., Parrott, N.J. et al. (2018). Advancing predictions of tissue and intracellular drug concentrations using in vitro, imaging and physiologically based pharmacokinetic modeling approaches. *Clin Pharmacol Ther* 104 (5): 865–889.

4 Chiba, M., Ishii, Y., and Sugiyama, Y. (2009). Prediction of hepatic clearance in human from in vitro data for successful drug development. *AAPS J* 11 (2): 262–276.

5 Iwatsubo, T., Hirota, N., Ooie, T. et al. (1997). Prediction of in vivo drug metabolism in the human liver from in vitro metabolism data. *Pharmacol Ther* 73 (2): 147–171.

6 Maeda, K. (2015). Organic anion transporting polypeptide (OATP)1B1 and OATP1B3 as important regulators of the pharmacokinetics of substrate drugs. *Biol Pharm Bull* 38 (2): 155–168.

7 Shitara, Y., Horie, T., and Sugiyama, Y. (2006). Transporters as a determinant of drug clearance and tissue distribution. *Eur J Pharm Sci* 27 (5): 425–446.

8 Watanabe, T., Kusuhara, H., Maeda, K. et al. (2010). Investigation of the rate-determining process in the hepatic elimination of HMG-CoA reductase inhibitors in rats and humans. *Drug Metab Dispos* 38 (2): 215–222.

9 Maeda, K., Ikeda, Y., Fujita, T. et al. (2011). Identification of the rate-determining process in the hepatic clearance of atorvastatin in a clinical cassette microdosing study. *Clin Pharmacol Ther* 90 (4): 575–581.

10 Yoshikado, T., Maeda, K., Furihata, S. et al. (2017a). A clinical cassette dosing study for evaluating the contribution of hepatic OATPs and CYP3A to drug–drug interactions. *Pharm Res* 34 (8): 1570–1583.

11 Yoshikado, T., Maeda, K., Kusuhara, H. et al. (2017b). Quantitative analyses of the influence of parameters governing rate-determining process of hepatic elimination of drugs on the magnitudes of drug–drug interactions via hepatic OATPs and CYP3A using physiologically based pharmacokinetic models. *J Pharm Sci* 106 (9): 2739–2750.

12 Yoshikado, T., Toshimoto, K., Nakada, T. et al. (2017c). Comparison of methods for estimating unbound intracellular-to-medium concentration ratios in rat and human hepatocytes using statins. *Drug Metab Dispos* 45 (7): 779–789.

13 Maeda, K. (2020). Recent progress in in vivo phenotyping technologies for better prediction of transporter-mediated drug–drug interactions. *Drug Metab Pharmacokinet* 35 (1): 76–88.

14 Nishiyama, K., Toshimoto, K., Lee, W. et al. (2019). Physiologically-based pharmacokinetic modeling analysis for quantitative prediction of renal transporter-mediated interactions between metformin and cimetidine. *CPT Pharmacometrics Syst Pharmacol* 8 (6): 396–406.

15 Taskar, K.S., Pilla Reddy, V., Burt, H. et al. (2020). Physiologically-based pharmacokinetic models for evaluating membrane transporter mediated drug–drug interactions: current capabilities, case studies, future opportunities, and recommendations. *Clin Pharmacol Ther* 107 (5): 1082–1115.

16 Watanabe, T., Kusuhara, H., Maeda, K. et al. (2009). Physiologically based pharmacokinetic modeling to predict transporter-mediated clearance and distribution of pravastatin in humans. *J Pharmacol Exp Ther* 328 (2): 652–662.

17 Igel, M., Arnold, K.A., Niemi, M. et al. (2006). Impact of the SLCO1B1 polymorphism on the pharmacokinetics and lipid-lowering efficacy of multiple-dose pravastatin. *Clin Pharmacol Ther* 79 (5): 419–426.

18 Yang, G.P., Yuan, H., Tang, B. et al. (2010). Lack of effect of genetic polymorphisms of SLCO1B1 on the lipid-lowering response to pitavastatin in Chinese patients. *Acta Pharmacol Sin* 31 (3): 382–386.

19 Group, S.C., Link, E., Parish, S. et al. (2008). SLCO1B1 variants and statin-induced myopathy--a genomewide study. *N Engl J Med* 359 (8): 789–799.

20 Kato, M., Chiba, K., Ito, T. et al. (2010). Prediction of interindividual variability in pharmacokinetics for CYP3A4 substrates in humans. *Drug Metab Pharmacokinet* 25 (4): 367–378.

21 Miller, N.A., Reddy, M.B., Heikkinen, A.T. et al. (2019). Physiologically based pharmacokinetic modelling for first-in-human predictions: an updated model

building strategy illustrated with challenging industry case studies. *Clin Pharmacokinet* 58 (6): 727–746.

22 Tsamandouras, N., Rostami-Hodjegan, A., and Aarons, L. (2015). Combining the 'bottom up' and 'top down' approaches in pharmacokinetic modelling: fitting PBPK models to observed clinical data. *Br J Clin Pharmacol* 79 (1): 48–55.

23 Chiba, K., Kato, M., Ito, T. et al. (2012). Inter-individual variability of in vivo CYP2D6 activity in different genotypes. *Drug Metab Pharmacokinet* 27 (4): 405–413.

24 Chiba, K., Shimizu, K., Kato, M. et al. (2017). Estimation of interindividual variability of pharmacokinetics of CYP2C9 substrates in humans. *J Pharm Sci* 106 (9): 2695–2703.

25 Chiba, K., Shimizu, K., Kato, M. et al. (2014). Prediction of inter-individual variability in the pharmacokinetics of CYP2C19 substrates in humans. *Drug Metab Pharmacokinet* 29 (5): 379–386.

26 Haraya, K., Kato, M., Chiba, K., and Sugiyama, Y. (2016). Prediction of inter-individual variability on the pharmacokinetics of CYP1A2 substrates in non-smoking healthy volunteers. *Drug Metab Pharmacokinet* 31 (4): 276–284.

27 Haraya, K., Kato, M., Chiba, K., and Sugiyama, Y. (2017). Prediction of inter-individual variability on the pharmacokinetics of CYP2C8 substrates in human. *Drug Metab Pharmacokinet* 32 (6): 277–285.

28 Kiyotani, K., Mushiroda, T., Kubo, M. et al. (2008). Association of genetic polymorphisms in SLCO1B3 and ABCC2 with docetaxel-induced leukopenia. *Cancer Sci* 99 (5): 967–972.

29 Yamada, A., Maeda, K., Kiyotani, K. et al. (2014). Kinetic interpretation of the importance of OATP1B3 and MRP2 in docetaxel-induced hematopoietic toxicity. *CPT Pharmacometrics Syst Pharmacol* 3: e126.

30 Bruno, R., Vivier, N., Vergniol, J.C. et al. (1996). A population pharmacokinetic model for docetaxel (Taxotere): model building and validation. *J Pharmacokinet Biopharm* 24 (2): 153–172.

31 Friberg, L.E., Henningsson, A., Maas, H. et al. (2002). Model of chemotherapy-induced myelosuppression with parameter consistency across drugs. *J Clin Oncol* 20 (24): 4713–4721.

32 Yamada, A., Maeda, K., Ishiguro, N. et al. (2011). The impact of pharmacogenetics of metabolic enzymes and transporters on the pharmacokinetics of telmisartan in healthy volunteers. *Pharmacogenet Genomics* 21 (9): 523–530.

33 Izumi, Y., Tokuda, K., O'Dell, K.A. et al. (2007). Neuroexcitatory actions of Tamiflu and its carboxylate metabolite. *Neurosci Lett* 426 (1): 54–58.

34 Usami, A., Sasaki, T., Satoh, N. et al. (2008). Oseltamivir enhances hippocampal network synchronization. *J Pharmacol Sci* 106 (4): 659–662.

35 Ito, M., Kusuhara, H., Ose, A. et al. (2017). Pharmacokinetic modeling and Monte Carlo simulation to predict interindividual variability in human exposure to oseltamivir and its active metabolite, Ro 64-0802. *AAPS J* 19 (1): 286–297.

36 He, G., Massarella, J., and Ward, P. (1999). Clinical pharmacokinetics of the prodrug oseltamivir and its active metabolite Ro 64-0802. *Clin Pharmacokinet* 37 (6): 471–484.

37 Fujiwara, Y. and Minami, H. (2010). An overview of the recent progress in irinotecan pharmacogenetics. *Pharmacogenomics* 11 (3): 391–406.

38 Toshimoto, K., Tomaru, A., Hosokawa, M., and Sugiyama, Y. (2017). Virtual clinical studies to examine the probability distribution of the AUC at target tissues using physiologically-based pharmacokinetic modeling: application to analyses of the effect of genetic polymorphism of enzymes and transporters on irinotecan induced side effects. *Pharm Res* 34 (8): 1584–1600.

39 Aoki, Y., Hayami, K., Sterck, H.D., and Konagaya, A. (2014). Cluster Newton method for sampling multiple solutions of underdetermined inverse problems: application to a parameter identification problem in pharmacokinetics. *SIAM J Sci Comput* 36: B14–B44.

40 Yoshida, K., Maeda, K., Kusuhara, H., and Konagaya, A. (2013). Estimation of feasible solution space using Cluster Newton Method: application to pharmacokinetic analysis of irinotecan with physiologically-based pharmacokinetic models. *BMC Syst Biol* 7 (Suppl 3): S3.

41 Teft, W.A., Welch, S., Lenehan, J. et al. (2015). OATP1B1 and tumour OATP1B3 modulate exposure, toxicity, and survival after irinotecan-based chemotherapy. *Br J Cancer* 112 (5): 857–865.

42 Gupta, E., Lestingi, T.M., Mick, R. et al. (1994). Metabolic fate of irinotecan in humans: correlation of glucuronidation with diarrhea. *Cancer Res* 54 (14): 3723–3725. *Application of a PBPK Model Incorporating the Interplay Between Transporters*

15

The Extended Clearance Model

A Valuable Tool For Drug-Induced Liver Injury Risk Prediction

Birk Poller[1], Felix Huth[1], Vlasia Kastrinou-Lampou[1,2], Gerd A. Kullak-Ublick[3,4], Michael Arand[2], and Gian Camenisch[1]

[1] *ADME Department of PK Sciences, Translational Medicine, Novartis Institutes for BioMedical Research, Basel, Switzerland*
[2] *Institute of Pharmacology and Toxicology, University of Zürich, Zürich, Switzerland*
[3] *Department of Clinical Pharmacology and Toxicology, University Hospital Zürich, University of Zurich, Zürich, Switzerland*
[4] *Mechanistic Safety, CMO & Patient Safety, Global Drug Development, Novartis, Basel, Switzerland*

15.1 Introduction

As the major detoxification organ, the liver is often affected by adverse drug reactions (ADRs), and drug-induced liver injury (DILI) is the most frequent reason for drug withdrawal from the market [1–3]. DILI is a generic term summarizing a number of mechanistically different drug-induced liver injuries (e.g. cholestasis or hepatocellular injury) [4]. The severity of DILI ranges from asymptomatic hepatic biomarker elevations (such as liver enzymes, (un)conjugated bilirubin, or bile acids) to acute liver failure [5].

Drug-induced cholestasis (DIC) is reported to account for up to 26% of all hepatic ADRs [6, 7]. Cholestasis describes the accumulation of bile acids in the hepatocytes, arising from inadequate clearance of bile acids in the liver, which leads to consecutive liver damage [8]. The elimination of unconjugated and conjugated bile acids is a multifaceted process (Figure 15.1a). Most bile acids are actively taken up into the hepatocytes facilitated by solute-carriers such as the organic anion-transporting polypeptides (OATP1B1, OATP1B3) or the Na^+ taurocholate cotransporting polypeptide (NTCP). Once in hepatocytes, bile acid metabolizing enzymes play an important role in the depletion of intracellular levels. While cytochrome P450 7A1 (CYP7A1) is responsible for the synthesis of bile acids, CYP3A4 is involved in their oxidative catabolism (e.g. transformation of chenodeoxycholic acid into 7α-dihydroxy-3-oxo-5β-cholanoic acid) [9–11]. For the subsequent conjugation of the hydroxylated bile acids, several UDP-glucuronosyltransferases (UGTs), and sulfotransferases (SULTs) are responsible [12, 13]. Glucuronidation at the 6α-hydroxy position is catalyzed by UGT2B4,

Transporters and Drug-Metabolizing Enzymes in Drug Toxicity, First Edition. Edited by Albert P. Li.

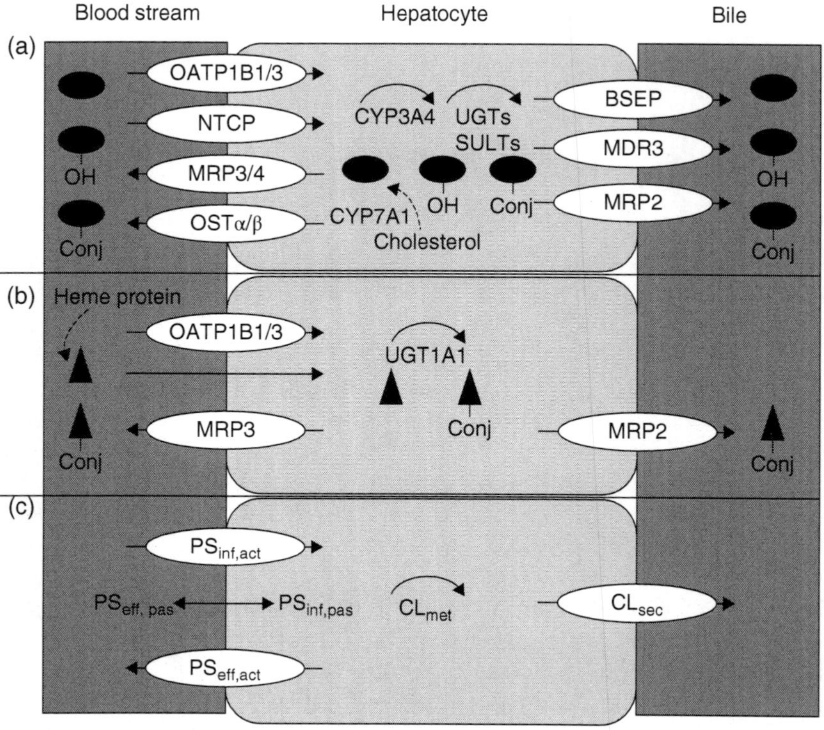

Figure 15.1 Cell stress perturbations in the liver are frequently the result of inhibitory insults on one or several of the active enzyme- or transporter-mediated hepatic clearance processes involved in the disposition of bile acids (circles in (a)) and/or bilirubin (triangles in (b)). The interplay of all hepatic clearance processes can be described by the Extended Clearance Model (ECM, Eq. (15.1), (c)) where unbound (drug) compound in the blood stream is taken up into hepatocytes (PS_{upt}) by transporters ($PS_{upt,act}$) and/or by passive diffusion ($PS_{upt,pas}$). Intrinsic elimination of drug from the hepatocyte occurs via hepatic metabolism (CL_{met}), by active secretion into bile (CL_{sec}), and by sinusoidal efflux (PS_{eff}) via active transport ($PS_{eff,act}$) and/or passive diffusion ($PS_{eff,pas}$).

whereas UGT2B7 generates 3α- and 6α-hydroxy glucuronides. UGT1A3 was identified to almost exclusively conjugate the C24-carboxyl group. Moreover, sulfatation at the 3α-hydroxy position is described to be catalyzed by SULT2A1. Eventually, unchanged bile acids and/or their conjugate metabolites can be excreted into the bile via canalicular efflux transporters such as the multidrug resistance-associated protein 2 (MRP2), the multidrug resistance protein 3 (MDR3), or the bile salt efflux pump (BSEP) [14–17]. Alternatively, bile acids can be transported back into blood across the sinusoidal membrane via MRP3, MRP4, or organic solute transporters (OSTs-alpha and beta) [18].

Besides the effects on hepatic bile acid disposition, impaired function of hepatic metabolic and transport processes can also result in impaired elimination of bilirubin and/or its glucuronide metabolites, clinically manifesting as hyperbilirubinemia. As Figure 15.1b illustrates, unconjugated bilirubin enters the liver likely through a passive diffusion process. The uptake of the glucuronide metabolites into the liver, however, depends on active transport mediated by OATP1B1 and OATP1B3 [19]. Once it enters the liver, bilirubin is extensively metabolized by UGT1A1 to form mono- and diglucuronide conjugates. The glucuronide metabolites are recognized by several efflux pumps from the MRP family. The primary route of excretion is into bile via MRP2 expressed at the apical membrane facing the bile canaliculi but glucuronide metabolites are also excreted into the blood by sinusoidally expressed MRP3 [20, 21].

Disturbance of the balance between transport and metabolism results in altered endobiotics concentrations (such as hepatocellular bile acids), which may ultimately induce liver injury [22]. Therefore, modeling the interplay of all the contributing enzymes and transporters may lead to better DILI predictions [23, 24]. The aim of the present article was to provide an overview about our in-house DILI anticipation and quantification attempts using the integrative hepatic Extended Clearance Model (ECM) as a hypothetical basis (Figure 15.1c). Potential follow-up needs and requirements are elaborated and discussed.

15.2 Application of the ECM to Estimate Kp_{uu} Liver

15.2.1 Introduction to the ECM: Concepts and Application for the Prediction of Hepatic Clearance and Drug–Drug Interactions

The disposition, metabolism, and excretion of endogenous and xenobiotic compounds by the liver is determined by a complex interplay of metabolism, passive diffusion, and active transport processes. To quantitatively describe the overall hepatic elimination of any compound, the ECM has been developed, which incorporates the individual clearance processes according to Eq. (15.1) (Figure 15.1c) [25, 26]:

$$CL_{int,all} = \frac{\left(PS_{upt,pas} + PS_{upt,act}\right)\cdot\left(CL_{sec} + CL_{met}\right)}{PS_{eff,pas} + PS_{eff,act} + CL_{sec} + CL_{met}} = \frac{PS_{upt}\cdot\left(CL_{sec} + CL_{met}\right)}{PS_{eff} + CL_{sec} + CL_{met}} \quad (15.1)$$

Where, $CL_{int,all}$ represents the overall apparent intrinsic hepatic clearance, $PS_{upt,act}$ is the intrinsic membrane clearance for basolateral (sinusoidal) transporter-mediated (active) influx, $PS_{upt,pas}$ is the intrinsic membrane clearances for basolateral influx via passive diffusion, PS_{upt} is the total intrinsic membrane clearances for basolateral influx, $PS_{eff,act}$ is the intrinsic clearance for active efflux back into the blood across the sinusoidal membrane, $PS_{eff,pas}$ is the intrinsic clearance for the passive back diffusion into the blood at the sinusoidal membrane, PS_{eff} is the total intrinsic membrane clearances for basolateral efflux, CL_{met} is the metabolic intrinsic hepatic clearance, and CL_{sec} is the intrinsic membrane clearance for biliary secretion at the canalicular side of hepatocytes (generally assumed to be solely a transporter-mediated process).

During the past years, our group developed a bottom-up approach for the calculation of $CL_{int,all}$ by applying data from different *in vitro* systems for the individual clearance processes in Eq. (15.1) [27, 28]. Passive and active hepatic uptake ($PS_{upt,pas}$, $PS_{upt,act}$) are determined in primary hepatocytes using the oil-spin method. Metabolic clearance (CL_{met}) is determined in liver microsomes supplied with cofactors for either CYP or UGT reactions. Alternatively, hepatocytes or S9 fractions can be used to determine CL_{met} for compounds that are metabolized by cytosolic enzymes. Sandwich-cultured hepatocytes are applied to measure the transporter-mediated biliary excretion (CL_{sec}) following the B-Clear method. Sinusoidal back-flux from hepatocytes into the systemic circulation ($PS_{eff,act}$) is most challenging to assess *in vitro*. Different approaches have been developed including a calculation method from hepatocyte excretion data and mathematical modeling of *in vitro* uptake and efflux measurements in sandwich-cultured hepatocytes [29, 30]. Quantitative knowledge about the contribution of the individual process clearances allows prediction of the rate-limiting hepatic elimination step and, as a result, assignment of compounds into four distinct ECM classes as depicted in Figure 15.2 [28, 31, 32].

Applying $CL_{int,all}$, the unbound fraction in blood (fu_b) and the hepatic blood flow (Q_h) in the well-stirred liver model allows the prediction of the hepatic clearance using:

$$CL_h = \frac{Q_h \cdot fu_b \cdot CL_{int,all}}{Q_h + fu_b \cdot CL_{int,all}} \tag{15.2}$$

The ECM has been widely applied for the prediction of hepatic clearance and the drug–drug interaction (DDI) potential upon inhibition of hepatic transporters and enzymes [26, 27, 33]. Traditionally, *in vitro-in vivo extrapolation* (IVIVE) has been conducted using liver microsomal or hepatocyte clearance data, accounting only for metabolic processes. Since the ECM-based IVIVE approach integrates active transport processes in the hepatic disposition and clearance of drugs, significantly improved predictions of rat and human hepatic clearance have been obtained by this method, in particular for low permeable compounds (ECM classes 3 and 4) [27, 29]. For compounds in ECM class 3, hepatic uptake represents the rate-limiting clearance and often involves uptake transporters (e.g. OATPs). Hepatic clearance of class 4 compounds is mediated by the interplay of hepatic

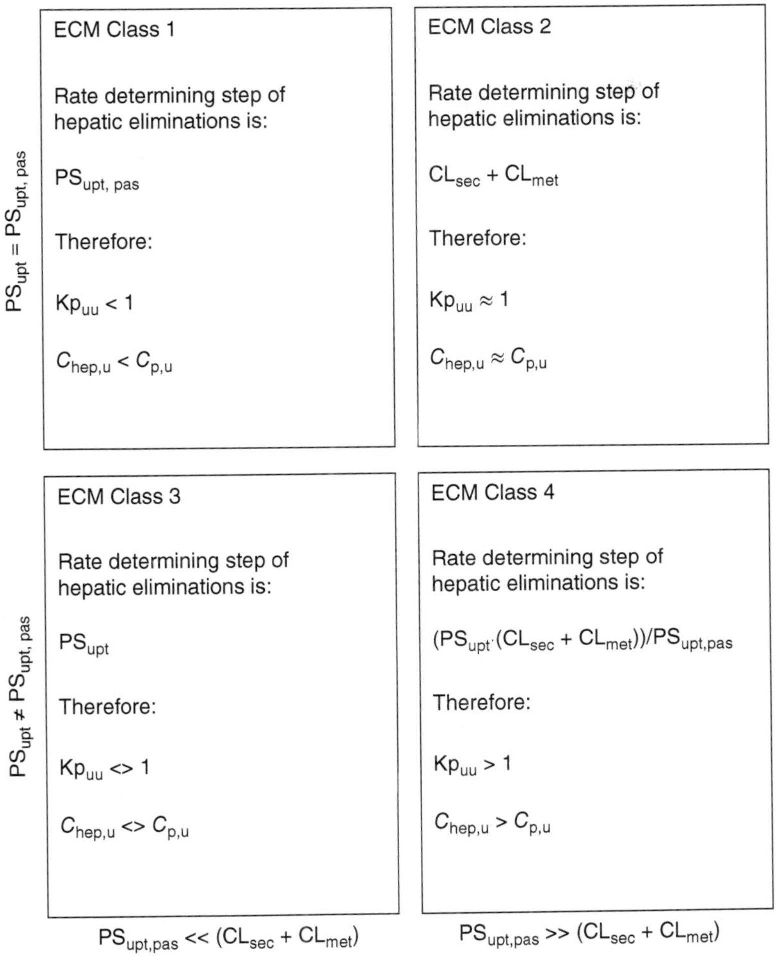

Figure 15.2 Drug classification according to ECM (Eq. (15.1)) and expected impact on the liver-to-blood partition coefficient for unbound drug Kp_{uu} and, as a result, on the unbound intracellular hepatic concentration $C_{hep,u}$ according to Eq. (15.6) assuming PS_{eff} to occur only by passive diffusion and to be equal to $PS_{upt,pas}$ (i.e. $PS_{eff,act} = 0$). ECM class 1 or 3 is applicable if $PS_{eff} = PS_{upt,pas} << (CL_{sec} + CL_{met})$ whereas ECM class 2 and 4 applies if $PS_{eff} = PS_{upt,pas} >> (CL_{sec} + CL_{met})$. For class 1 and 2 $PS_{upt} = PS_{eff} = PS_{upt,pas}$ while for class 3 and 4 $PS_{upt} \neq PS_{eff} = PS_{upt,pas}$.

uptake, metabolism and/or biliary excretion. Accordingly, incorporation of *in vitro* measurements for these processes enabled the clearance prediction for these drugs by a more elaborate IVIVE approach.

Referring to Eq. (15.1), the overall apparent intrinsic hepatic clearance in the presence of a perpetrator compound ($CL_{int,all,i}$) can be expressed as follows [27, 33]:

$$\mathrm{CL}_{\mathrm{int,all,i}} = \frac{\left(\mathrm{PS}_{\mathrm{upt,pas}} + \left(1 - f_{\mathrm{i,upt}}\right) \cdot \mathrm{PS}_{\mathrm{upt,act}}\right) \cdot \left(\left(1 - f_{\mathrm{i,sec}}\right) \cdot \mathrm{CL}_{\mathrm{sec}} + \left(1 - f_{\mathrm{i,met}}\right) \cdot \mathrm{CL}_{\mathrm{met}}\right)}{\mathrm{PS}_{\mathrm{eff,pas}} + \left(1 - f_{\mathrm{i,eff}}\right) \cdot \mathrm{PS}_{\mathrm{eff,act}} + \left(1 - f_{\mathrm{i,sec}}\right) \cdot \mathrm{CL}_{\mathrm{sec}} + \left(1 - f_{\mathrm{i,met}}\right) \cdot \mathrm{CL}_{\mathrm{met}}} \tag{15.3}$$

Where, $f_{\mathrm{i,upt}}$, $f_{\mathrm{i,eff}}$, $f_{\mathrm{i,sec}}$, and $f_{\mathrm{i,met}}$ denote the inhibited fractions of active influx, sinusoidal efflux, canalicular secretion, or metabolism, respectively. A value of zero thereby indicates no inhibition whereas a value of one refers to complete inhibition.

Consequently, following oral (po) administration of a drug and its perpetrator and assuming the liver being the only elimination pathway, the resulting exposure (area under the curve [AUC]) fold-change (expressed as AUCpo, i/AUCpo) can be described as follows:

$$\frac{\mathrm{AUC}_{\mathrm{po,i}}}{\mathrm{AUC}_{\mathrm{po}}} = \frac{\mathrm{CL}_{\mathrm{int,all}}}{\mathrm{CL}_{\mathrm{int,all,i}}} \tag{15.4}$$

Following this concept, the potential inhibition of single or multiple pathways and the resulting change in AUC can be calculated by assuming partial or complete inhibition of the respective mechanisms. We used this static DDI model to predict observed clinical DDI for statins coadministered with different perpetrator drugs. Depending on the inhibition potential of the perpetrator drug, different outcomes were predicted which correlated well with the observed AUC fold changes. For example, a 15-fold AUC change was observed for atorvastatin when co-administered with cyclosporine A, and our model predicted a 13.5-fold AUC change [33]. In line with clinically observed interaction data for the statins, the ECM approach predicted that inhibition of single individual processes (e.g. inhibition of hepatic uptake or metabolism only) resulted in maximum threefold AUC changes. Simultaneous inhibition of several pathways, however, predicted much higher AUC fold changes. In accordance with the ECM drug classification idea and the consequent class-dependent rate-limiting clearance processes (Figure 15.2), this was predominantly true for class 4 compounds.

15.2.2 Concept of $\mathrm{Kp}_{\mathrm{uu}}$ Liver

In pharmacokinetic and pharmacodynamic models the "free-drug hypothesis" is a commonly used concept, which assumes that the unbound drug concentration in blood and tissues are in equilibrium under steady-state conditions. This concept, however, does not generally apply for the liver and for other organs where active transport and metabolic processes shift the equilibrium of the free drug. As a result, unbound concentrations in plasma ($C_{\mathrm{p,u}}$) and the liver ($C_{\mathrm{hep,u}}$) are not necessarily identical and the equilibrium is described by the tissue-to-blood partition coefficient for unbound drug at steady-state (Kp_{uu}) [34, 35]:

$$Kp_{uu} = \frac{C_{hep,u}}{C_{p,u}} \tag{15.5}$$

The concept of Kp_{uu} has widely been accepted and several methods to determine Kp_{uu} from *in vitro* and *in vivo* data have recently been described in detail [36]. A common approach to determine Kp_{uu} in pre-clinical species is based on measurements of total drug concentrations in blood and liver under steady-state conditions, e.g. under constant infusion, to determine the tissue-to-blood partition coefficient for total drug (Kp). The unbound fractions in plasma (fu_p) and in liver tissue (fu_{hep}) are measured in independent experiments typically using equilibrium dialysis. Multiplication of Kp with the ratio of fu_{hep}/fu_p provides Kp_{uu}. While this method provides good estimates for Kp_{uu} in animals, it cannot easily be applied to human due to limitations in obtaining liver samples for concentration measurements. Imaging techniques, using positron emission tomography (PET) or single-photon emission computed tomography (SPECT) tracers, provide noninvasive measurements of human tissue concentrations. However, such studies are complex and typically only conducted at a late stage of the drug development. Therefore, to estimate human Kp_{uu} during early drug development, a method was derived from the ECM concept using only *in vitro* data, as described in the following section.

15.2.3 Estimation of Kp_{uu} Liver from *In Vitro* Data Using the ECM

While the original equation of the hepatic ECM was developed to determine the overall elimination of a drug by the liver, the ECM equation (Eq. (15.1)) can be rearranged to derive Kp_{uu} [28, 35]:

$$Kp_{uu} = \frac{CL_{int,all}}{CL_{sec} + CL_{met}} = \frac{PS_{upt}}{PS_{eff} + CL_{sec} + CL_{met}} \tag{15.6}$$

From this equation it is evident that the balance of hepatic uptake and intrinsic clearance of a drug determines Kp_{uu}. Same as for $CL_{int,all}$ (Section 15.2.1), Kp_{uu} is dependent on the ECM class of a drug as depicted in Figure 15.2. For instance, $C_{hep,u}$ of hepatic uptake transporters substrates (e.g. statins) may become greater than $C_{p,u}$ if the sum of the intrinsic clearance processes (PS_{eff}, CL_{sec}, CL_{met}) is slower than the uptake clearance ($Kp_{uu} > 1$ for ECM class 3 or 4). In contrast, Kp_{uu} is <1 for compounds with high metabolic or biliary clearance, for which uptake is the rate-limiting clearance process (ECM class 1 or 3). For ECM class 2 compounds, Kp_{uu} is close to unity since metabolic and/or biliary elimination are the rate-limiting clearance steps and high passive diffusion allows a rapid equilibration of the drug between plasma and the cellular compartments ($PS_{upt} = PS_{upt,pas}$).

In a recent study by our group we aimed to estimate the risk for DIC by correlating BSEP inhibition potency (IC_{50}) with unbound intrahepatic drug concentrations by using Kp_{uu} liver determined from the *in vitro*-based ECM approach [31]. For 18 compounds, hepatic uptake (PS_{upt}) was measured in human hepatocytes using the oil-spin method, metabolic clearance (CL_{met}) was determined in human liver microsomes and biliary secretion (CL_{sec}) was derived from sandwich-cultured hepatocytes. The *in vitro* Kp_{uu}, obtained from Eq. (15.6), are shown in Table 15.1. In line with the ECM theory, Kp_{uu} was below unity for ketoconazole, the only class 1 compound. The class 2 compounds verapamil and imatinib had Kp_{uu} values close to unity. Two of the seven tested compounds in ECM class 3 had $Kp_{uu} > 3$, whereas Kp_{uu} of the remaining class 3 compounds was below unity. Atazanavir was the compound with the lowest Kp_{uu} (0.07), indicating that $C_{hep,u}$ is 14-fold lower than $C_{p,u}$. All seven ECM class 4 compounds had $C_{hep,u}$ values greater than 1, and the highest Kp_{uu} (2.41) was obtained for pravastatin.

To assess whether Kp_{uu} from the *in vitro*-based ECM approach provides relevant estimates for *in vivo* situation, we conducted a validation for Kp_{uu} liver in the rat [31]. The ECM approach using rat *in vitro* data was compared to a conventional method based on *in vivo* Kp data (total liver tissue to blood concentration ratio) and the unbound fractions in plasma and liver tissue determined by equilibrium dialysis as described above. The obtained Kp_{uu} data from both methods were in very good agreement with values ranging from 0.28 to 7.9. The successful *in vitro-in vivo* correlation in the rat (IVIVC) together with the previously established extended human IVIVE for hepatic clearance (Section 15.2.1) provided confidence that relevant estimates of the human liver Kp_{uu} can be obtained by the ECM-based approach.

15.3 Relevant Concentrations for the DILI Risk Assessment

In order to predict whether the *in vitro* inhibition of a target enzyme results in clinically relevant toxicity or DDI, the *in vitro* potency (IC_{50}, K_i) of a compound is compared to its exposure in humans [38]. This approach is widely accepted and used by regulatory agencies, to assess the clinical DDI risk of a new drug candidate [39, 40]. Depending on the organ expression of the target, relevant *in vivo* concentrations such as the systemic plasma concentration, the hepatic inlet concentration in the liver portal vein, or the intracellular concentration should be used for the risk assessment. For all these matrices total or unbound concentrations can be considered. The selection of the appropriate matrix is controversially

Table 15.1 *In vitro* enzyme/transporter inhibition constants and human drug exposure.

Compound name	DILI or DIC concern[a]	$I_{sys,u}$ (µM)[b]	$I_{inlet,u}$ (µM)[c]	Kp_{uu} [d]	CYP3A4 k_{inact} (min^{-1})	CYP3A4	UGT1A1	BSEP	MRP2	MDR3	OATP1B1	OATP1B3	NTCP	1/*R* value according to:[f] Eq. (15.12) with $I_{sys,u}$	Eq. (15.12) with $I_{inlet,u}$	Eq. (15.12) with $I_{hep,inlet,u}$	Eq. (15.13) with $I_{inlet,u}$
						$K_{i,u}$ (µM)[e]											
Rosiglitazone	No	0.004	0.005	0.64	0.011	4.4	ni	0.95	13.5	ni	2.1	5.5	ni	0.99	0.98	0.99	0.99
Fluvastatin	No	0.005	0.013	1.15	ni	ni	ni	18.1	14.0	ni	nr	nr	20.0	1.00	1.00	1.00	1.00
Lovastatin acid	No	0.005	0.011	0.51	ni	ni	ni	9.7	119	ni	3.1	46.3	ni	1.00	1.00	1.00	1.00
Simvastatin acid	No	0.001	0.092	0.39	0.006	4.6	42.3	10.5	28.0	ni	2.5	nr	ni	0.98	0.71	0.77	0.72
Cerivastatin	No	0.000	0.001	1.60	ni	ni	ni	9.4	ni	ni	0.9	nr	ni	1.00	1.00	1.00	1.00
Rosuvastatin	No	0.010	0.067	1.62	ni	ni	ni	98.8	7.5	ni	0.03	1.8	250.0	0.86	0.63	0.59	0.64
Verapamil	Less	0.061	0.242	0.66	0.07	2.4	nr	10.3	nr	3.5	7.7	43.7	ni	0.69	0.63	0.65	0.66
Pravastatin	Less	0.073	0.801	2.41	nr	nr	ni	134.2	42.0	ni	1.8	31.0	nr	0.97	0.83	0.72	0.84
Ibuprofen	Less	3.929	5.188	0.66	ni	ni	nr	299.3	206.0	ni	nr	nr	ni	0.99	0.99	0.99	1.00
Glibenclamide	Less	0.009	0.013	1.12	ni	ni	ni	0.8	28.5	ni	0.6	1.4	0.3	0.97	0.96	0.95	0.98
Pitavastatin	Less	0.011	0.039	2.25	ni	ni	ni	21.1	ni	ni	nr	nr	80.0	1.00	1.00	1.00	1.00
Erythromycin	Most	1.008	1.134	0.24	0.12	8.8	ni	2.1	nr	ni	2.5	0.8	ni	0.35	0.34	0.55	0.42
Atorvastatin	Most	0.023	0.082	1.48	nr	nr	ni	7.5	60.0	ni	0.02	1.3	ni	0.73	0.59	0.56	0.59
Imatinib	Most	0.478	1.154	0.87	0.028	4.4	5.5	5.0	nr	ni	nr	nr	ni	0.65	0.59	0.60	0.59
Ketoconazole	Most	0.166	0.294	0.32	ni	ni	1.7	1.5	nr	2.3	0.9	2.0	231.0	0.76	0.64	0.84	0.84
Atazanavir	Most	1.162	2.351	0.07	0.05	0.8	1.9	1.6	nr	ni	0.5	1.9	ni	0.28	0.19	0.54	0.20
Bosentan	Most	0.464	0.931	1.17	nr	nr	ni	11.0	ni	ni	2.5	2.6	12.0	0.84	0.75	0.73	0.84
Cyclosporine A	Most	0.051	0.137	1.21	0.033	3.3	48.0	0.2	2.7	ni	0.3	0.1	1.2	0.56	0.36	0.32	0.46

nr: negative or irrelevant experimental outcome.
ni: not investigated (i.e. absence of experimental data).
[a] DILI classification (less concern vs most concern) based on FDA-approved drug labels as described in [5].
[b] Unbound steady-state drug concentration in plasma as determined upon oral administration of the maximum recommended dose in healthy subjects according to [31].
[c] Unbound drug concentration at the hepatic inlet equal to the sum of drug in the systemic calculation reaching the liver via the hepatic artery (i.e. $I_{sys,u}$) and drug that is delivered by the portal vein upon intestinal absorption according to [31, 37].
[d] Liver-to-blood partition coefficient for unbound drug at steady-state according to Eq. (15.6) [31].
[e] In the absence of experimental confirmation calculated as $K_{I,u} = fu_{mic} {}^* K_I$, $K_{i,u} = fu_{mic} {}^* IC_{50}/2$ (enzymes) or $K_{i,u} = IC_{50}/2$ (transporters).
[f] 1/*R* value risk assessment as described in Section 15.5; with $k_{deg,CYP3A4}$ equal to 0.000321 min^{-1} [38]. For any missing inhibition information the $CL_{int,i}/CL_{int}$ ratio calculation according to Eq. (15.10) was fixed, in accordance with lack of process inhibition, at a value of 1 (compare also Table 15.2). The gray highlighted 1/*R*-values refer to the most significant changes (i.e. >0.1) as compared to the "best prediction" reference illustrated in Figure 15.4 (i.e. Eq. (15.12) with $I_{inlet,u}$ as input matrix).

discussed and is critical for the outcome of any risk assessment. In the following sections, the different matrices are presented including a discussion of the associated advantages and limitations.

15.3.1 Maximum Plasma Concentrations

The maximum plasma concentration of an inhibitory drug (I) is a directly measured parameter. Determination of the unbound fraction in plasma (fu_p) allows to calculate the free plasma concentration ($I_u = I \cdot fu_p$), assuming that only the unbound drug is interacting with DDI or safety target. For compounds with long half-lives multiple dosing results in accumulation in the body, therefore the unbound systemic steady-state plasma concentration at the highest efficacious daily dose ($I_{sys,u}$) is considered to represent the most relevant systemic parameter for safety risk assessments.

15.3.2 Maximum Hepatic Inlet Concentrations

Upon oral administration, drugs enter the liver via the portal vein blood, prior to being available in the systemic circulation. The concentrations in the hepatic inlet often exceed systemic concentrations, especially for drugs with high oral doses, rapid and complete absorption and limited intestinal metabolism. In humans, hepatic inlet concentrations are not experimentally accessible, therefore the unbound hepatic inlet concentration ($I_{inlet,u}$) is calculated as the sum of drug in the systemic circulation reaching the liver via the hepatic artery (i.e. $I_{sys,u}$) and drug that is delivered by the portal vein upon intestinal absorption [37]:

$$I_{inlet,u} = I_{sys,u} + \frac{f_{u,p} \cdot k_a \cdot F_a \cdot F_g \cdot D}{Q_h \cdot R_b} \tag{15.7}$$

Where, k_a is the absorption rate constant, F_a is the fraction absorbed, F_g is the fraction escaping gut metabolism, D is the (oral) dose, Q_h is the hepatic blood flow, and R_b is the blood-to-plasma partition coefficient.

Compared to systemic concentrations, $I_{inlet,u}$ may represent the more relevant concentrations related to safety assessment in the liver. In particular for drugs with high hepatic extraction and extensive first-pass metabolism, $I_{inlet,u}$ is expected to represent a more relevant matrix than systemic concentrations. Accordingly, health authority guidelines recommend the usage of $I_{inlet,u}$ for DDI risk assessments of transporters and enzymes in the liver, whereas DDI risk assessments for extra-hepatic elimination mechanisms should be conducted based on systemic $I_{sys,u}$ [39, 40].

15.3.3 Maximum Intracellular Hepatocyte Concentrations

Both $I_{sys,u}$ as well as $I_{inlet,u}$ refer to plasma as the reference matrix. This assessment presumes complete distribution equilibrium of the unbound drug between plasma and liver at steady-state following the "free-drug hypothesis". However, as described in Section 15.2, the assumption is unlikely to apply for the liver as active cellular transport and metabolic processes shift the distribution equilibrium of a drug. Application of the *in vitro* derived Kp_{uu} allows to account for these processes and to estimate unbound intracellular hepatocyte concentration on the basis of either $I_{sys,u}$ or $I_{inlet,u}$ [31]:

$$I_{hep,sys,u} = Kp_{uu} \cdot I_{sys,u} \tag{15.8}$$

$$I_{hep,inlet,u} = Kp_{uu} \cdot I_{inlet,u} \tag{15.9}$$

15.4 Assessing the DIC Risk Using ECM-Based Unbound Intrahepatic Concentrations and Accounting for BSEP Inhibition as a Single Mechanism

One of the first explored factors contributing to DILI (or more precisely DIC) is bile acid accumulation upon inhibition of BSEP [41–45]. BSEP is a member of the adenosine triphosphate (ATP)-binding cassette transporter family, encoded by the gene ABCB11, expressed at the canalicular membrane of hepatocytes (Figure 15.1). It is the most important transporter pumping unconjugated, major bile acids into the bile duct. A number of investigations were performed trying to correlate the severity of DIC with the BSEP inhibition potential of drugs. Morgan et al. [46] demonstrated a strong correlation between the potency of BSEP inhibition and DIC for 200 marketed and withdrawn drugs. Accordingly, initial studies aimed to estimate the risk for DIC by comparing the *in vitro* BSEP inhibition potency (IC_{50} or Ki_u) to the measured clinical exposure (I_{sys} or $I_{sys,u}$). In order to inhibit BSEP, drugs need to achieve sufficiently high local concentrations, i.e. unbound cytosol concentrations. These previous risk assessments, however, were conducted based on the "free-drug hypothesis", assuming that drug concentrations in plasma are representative surrogates of unbound intrahepatic concentrations. Measurements of Kp_{uu} *in vitro* and *in vivo*, however, revealed that the "free-drug hypothesis" only applies for a limited fraction of all drugs, namely uptake-limited drugs in ECM class 2 as discussed in Section 15.2.3.

To overcome this limitation, our group conducted a study to predict the DIC risk for the 18 drugs shown in Table 15.1 using unbound intrahepatic concentrations and BSEP inhibition potency [31]. Thereby, $I_{hep,sys,u}$ and $I_{hep,inlet,u}$ were estimated from the corresponding systemic or hepatic inlet concentrations ($I_{sys,u}$ and $I_{inlet,u}$) and corrected for the ECM-based *in vitro* Kp_{uu} (Eqs. (15.8) and (15.9)). Consequently, the ratio of BSEP Ki_u divided by one of the different drug concentration matrices was calculated (Ki_u/I ratios). These ratios were correlated with the clinical DIC frequency of reported cholestasis events for the tested drugs as outlined in Table 15.1.

When applying unbound systemic steady-state concentrations ($I_{sys,u}$), Ki_u/I ratios across several magnitudes were obtained (Figure 15.3a). The assessment allowed separating "no" from "most" DIC classes. Ratios for compounds with "less" DIC risk, however, overlapped with the other classes. Replacing $I_{sys,u}$ with $I_{inlet,u}$ provided a comparable performance, although Ki_u/I ratios were lower (Figure 15.3b). Application of the unbound intrahepatic concentrations ($I_{hep,sys,u}$, $I_{hep,inlet,u}$), resulted in significantly improved risk assessments. In particular, using $I_{hep,inlet,u}$ enabled an almost complete separation of the three frequency classes, and allowed establishing Ki_u/I thresholds between the classes (Figure 15.3c). For Ki_u/I ratios greater than ~500 "no" DIC is expected, whereas a ratio below ~30 indicates "most" DIC risk. For drugs with Ki_u/I ratios in between, a "less"

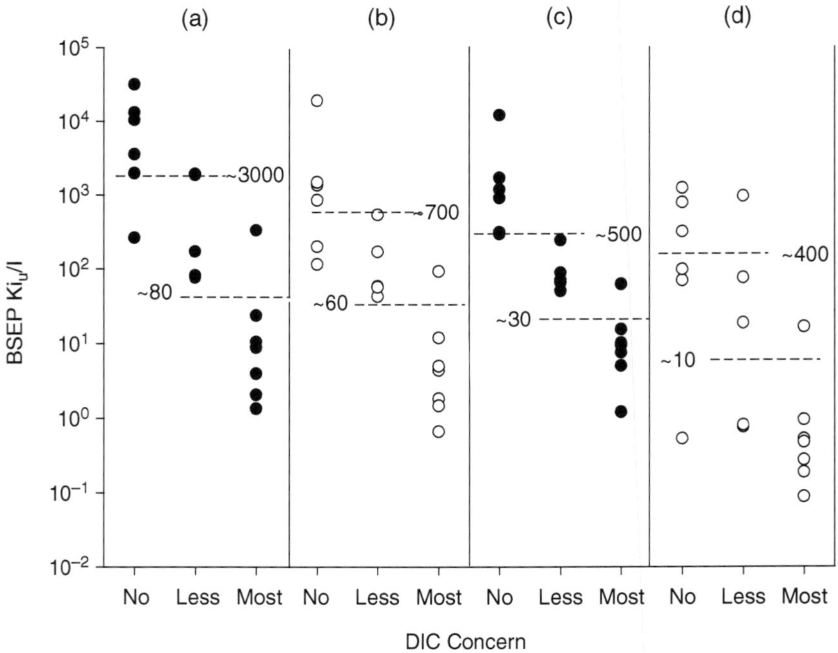

Figure 15.3 BSEP Ki_u/I ratios vs DIC relationships for the dataset summarized in Table 15.1. (a)–(d) refer to $I_{sys,u}$, $I_{inlet,u}$, $I_{hep,inlet,u}$, and I_{sys} as reference matrices, respectively. Estimated Safety Margin (SM) thresholds between the DIC classes "no-less" and "less-most" are shown next to the horizontal dashed lines.

DIC frequency is predicted. Figure 15.3d shows the Ki_u/I ratios for total systemic concentrations (I_{sys}). Similar as for $I_{sys,u}$ a separation of "no" from "most" DIC concern is achieved with one exception, but ratios for drugs with "less" DIC concern are highly overlapping with the other classes.

The outcome of this study revealed that unbound intrahepatic drug concentrations provide the best predictions for the interaction with an intracellular enzyme or transporter such as BSEP. The results further supported the validity of ECM-based Kp_{uu} liver method relying only on *in vitro* data. Due to the limited size of the test set, future research is required to corroborate the risk assessment and to refine the thresholds for the DIC frequency predictions.

15.5 Assessing the DILI Risk Using the "1/*R*-Value Model" to Account for the Inhibition of Multiple Pathways

DILI may arise due to inadequate clearance of bile acids and/or potentially toxic metabolites from the liver. As described in Section 15.1, various enzymes and transporters contribute to the hepatic disposition and elimination of the individual endobiotics. Although reduced BSEP activity is clearly associated with clinical manifestations of cholestasis, it may therefore not be sufficient to estimate the overall risk for DIC or even DILI. In order to integrate simultaneous interactions of a drug with multiple disposition mechanisms, the so-called "1/*R*-value model" has been developed based on the principles of the ECM [23].

15.5.1 ECM-Based 1/*R*-Value Model

In the absence of induction, the change of the risk factor R of any intrinsic enzyme and/or transporter clearance process following reversible and/or irreversible inhibition by a perpetrator (inhibitor) drug can be expressed as [38]:

$$R=\frac{CL_{int}}{CL_{int,i}}=\left(1+\Sigma\frac{I_u}{K_{i,u}}\right)\cdot\left(1+\Sigma\frac{k_{inact}\cdot I_u}{k_{deg}\cdot\left(K_{I,u}+I_u\right)}\right) \quad (15.10)$$

Where, CL_{int} and $CL_{int,i}$ refer to any intrinsic clearance in the absence and presence of process inhibitors, respectively. I_u is the unbound (cellular) concentration of the inhibitor, $K_{i,u}$ is its associated reversible unbound inhibition constant, $K_{I,u}$ is the irreversible unbound inhibition constant, k_{inact} is the maximal inactivation rate constant, and k_{deg} represents the degradation rate constant for the active process of interest.

Assuming passive diffusion is identical for efflux and uptake (i.e. $PS_{eff,pas}$ = $PS_{upt,pas} = PS_{pas}$), efflux across the sinusoidal membrane to be unaffected by the inhibitor compound (i.e. $PS_{eff,act} = PS_{eff,act,i}$) and canalicular secretion to be a solely active transport process ($PS_{sec,act} = PS_{sec}$), a combination of Eqs. (15.1) and (15.10) provides a new approximation generally referred to as the 1/*R*-value DILI model [23]:

$$\frac{\mathrm{CL}_{\mathrm{int,all,i}}}{\mathrm{CL}_{\mathrm{int,all}}}=\frac{1}{R}\propto\frac{\left(1+\frac{\mathrm{PS}_{\mathrm{upt,act,i}}}{\mathrm{PS}_{\mathrm{upt,act}}}\right)\cdot\left(\frac{\mathrm{PS}_{\mathrm{sec,i}}}{\mathrm{PS}_{\mathrm{sec,i}}}+\frac{\mathrm{PS}_{\mathrm{met,i}}}{\mathrm{PS}_{\mathrm{met}}}\right)}{2+\left(\frac{\mathrm{PS}_{\mathrm{sec,i}}}{\mathrm{PS}_{\mathrm{sec,i}}}+\frac{\mathrm{PS}_{\mathrm{met,i}}}{\mathrm{PS}_{\mathrm{met}}}\right)}=\frac{\left(1+\frac{1}{1+\sum\frac{I_{\mathrm{u}}}{\mathrm{Ki,u_{upt,act}}}}\right)\cdot\left[\left(\frac{1}{1+\sum\frac{I_{\mathrm{u}}}{\mathrm{Ki,u_{sec}}}}\right)+\left(\frac{1}{1+\sum\frac{I_{\mathrm{u}}}{\mathrm{Ki,u_{met}}}}\right)\cdot\left(\frac{1}{1+\sum\frac{k_{\mathrm{inact}}\cdot I_{\mathrm{u}}}{k_{\mathrm{deg}}\cdot\left(\mathrm{Ki,u_{met}}+I_{\mathrm{u}}\right)}}\right)\right]}{2+\left[\left(\frac{1}{1+\sum\frac{I_{\mathrm{u}}}{\mathrm{Ki,u_{sec}}}}\right)+\left(\frac{1}{1+\sum\frac{I_{\mathrm{u}}}{\mathrm{Ki,u_{met}}}}\right)\cdot\left(\frac{1}{1+\sum\frac{k_{\mathrm{inact}}\cdot I_{\mathrm{u}}}{k_{\mathrm{deg}}\cdot\left(\mathrm{Ki,u_{met}}+I_{\mathrm{u}}\right)}}\right)\right]} \tag{15.11}$$

Assuming DILI to be a consequence of reversible (OATP1B1, OATP1B3, NTCP, BSEP, MRP2, MDR3, and UGT1A1) and irreversible (CYP3A4) enzyme/transporter inhibition, Eq. (15.11) can be rewritten as follows:

$$\frac{\mathrm{CL}_{\mathrm{int,all,i}}}{\mathrm{CL}_{\mathrm{int,all}}}=\frac{1}{R}\propto$$

$$\frac{\left(\frac{1}{1+\frac{I_{\mathrm{u}}}{K_{\mathrm{i,u,OATP1B1}}}+\frac{I_{\mathrm{u}}}{K_{\mathrm{i,u,OATP1B3}}}}+1\right)\cdot\left[\left(\frac{1}{1+\frac{I_{\mathrm{u}}}{K_{\mathrm{i,u,BSEP}}}+\frac{I_{\mathrm{u}}}{K_{\mathrm{i,u,MRP2}}}+\frac{I_{\mathrm{u}}}{K_{\mathrm{i,u,MDR3}}}}\right)+\left(\frac{1}{1+\frac{I_{\mathrm{u}}}{K_{\mathrm{i,u,UGT1A1}}}}\right)\cdot\left(\frac{1}{1+\frac{k_{\mathrm{inact,CYP3A4}}\cdot I_{\mathrm{u}}}{k_{\mathrm{deg,CYP3A4}}\cdot\left(K_{\mathrm{I,u,CYP3A4}}+I_{\mathrm{u}}\right)}}\right)\right]}{2+\left(\frac{1}{1+\frac{I_{\mathrm{u}}}{K_{\mathrm{i,u,BSEP}}}+\frac{I_{\mathrm{u}}}{K_{\mathrm{i,u,MRP2}}}+\frac{I_{\mathrm{u}}}{K_{\mathrm{i,u,MDR3}}}}\right)+\left(\frac{1}{1+\frac{I_{\mathrm{u}}}{K_{\mathrm{i,u,UGT1A1}}}}\right)\cdot\left(\frac{1}{1+\frac{k_{\mathrm{inact,CYP3A4}}\cdot I_{\mathrm{u}}}{k_{\mathrm{deg,CYP3A4}}\cdot\left(K_{\mathrm{I,u,CYP3A4}}+I_{\mathrm{u}}\right)}}\right)} \tag{15.12}$$

Based on the dataset given in Table 15.1 and using the integrated $1/R$ ratios between systemic drug concentrations and inhibition constants according to Eq. (15.12) (with I_u equal to $I_{sys,u}$), a distinct separation between drug classes tagged with "no-less" and "most" DILI concerns could be established (Figure 15.4a). With reference to the apparent $1/R$-value threshold at 0.7 (horizontal solid line in Figure 15.4), the number of false positive compounds (i.e. number of compounds below the line) in the "no-less" DILI concern category was one (equivalent to about 9% for all compounds assigned to this group). Accordingly, the number of false negative compounds (i.e. number of compounds above the line) in the "most" DILI concern category was three (equivalent to about 42% for all compounds assigned to this group). The same breakup, with a significantly better separation power (two compounds wrongly assigned to the "no-less" and only one compound incorrectly assigned to the "most" DILI concern categories), is also seen for I_u equal to $I_{inlet,u}$ (Figure 15.4b). Surprisingly, both DILI prediction efforts could not be improved using unbound intracellular concentrations according to Eq. (15.8) (data not shown) and Eq. (15.9) (Figure 15.4c). For the latter, the

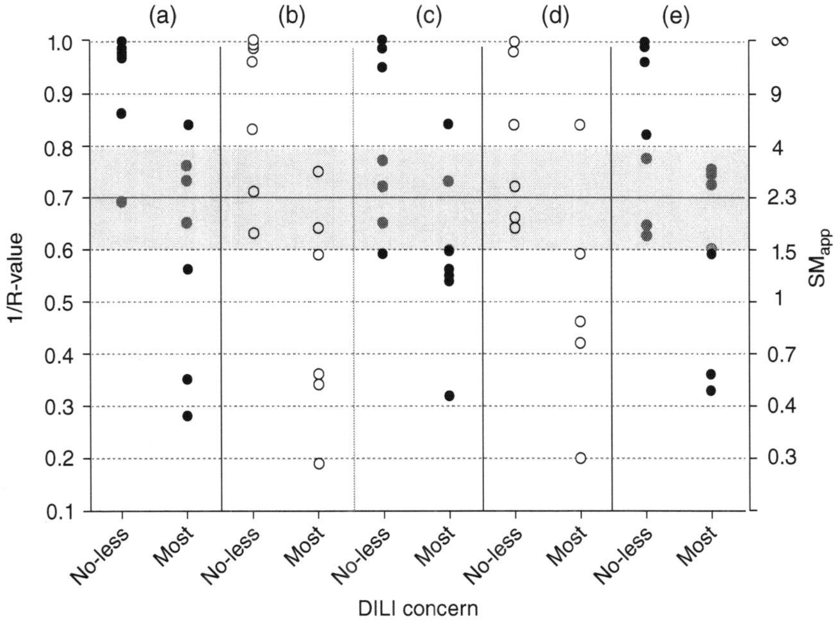

Figure 15.4 $1/R$- and SM_{app}-value (according to Eq. (15.14)) vs DILI relationships for the dataset summarized in Table 15.1. (a)–(c), (e) refer to Eq. (15.12) using $I_{sys,u}$, $I_{inlet,u}$, $I_{hep,inlet,u}$, and combination of $I_{inlet,u}$ (for sinusoidal uptake transporter inhibition) and $I_{hep,inlet,u}$ (for intracellular metabolism as well as canalicular efflux inhibition) as reference matrices, respectively. (d) refers to Eq. (15.13) with $I_{inlet,u}$ as input matrix. The gray highlighted area with a mean $1/R$-value of 0.7 (horizontal solid line) defines the transfer between "no-less" and "most" DILI concerns as suggested from (b).

number of false positive and false negative predictions were two each (equivalent to about 22% overall wrong assignments). With reference to Table 15.1 it is evident that compounds from the "most" DILI concern category (namely the uptake-limited ketoconazole (ECM class 1), erythromycin (ECM class 3), and atazanavir (ECM class 3) (Figure 15.2)) were significantly impacted by the use of the unbound maximum intracellular hepatocyte concentration ($I_{\mathrm{hep,inlet,u}}$) (exception pravastatin (ECM class 4)).

Referring to Eq. (15.11) it is obvious that significant inhibition of canalicular secretion or intracellular metabolism as a single pathway, in contrast to distinct sinusoidal uptake inhibition, will maximally result in a "less" DILI concern (i.e. consequential $1/R$-value$\geq$0.7). Only the complete inhibition of two or more pathways in parallel would result in $1/R$-values$>$0.5. These relationships are illustrated in more detail in Table 15.2.

Recent in-house data analyses provided evidence that ECM-based DILI prediction is largely driven by reversible OATP1B1 (for sinusoidal uptake) and BSEP (for canalicular efflux) as well as irreversible CYP3A4 (for metabolism) inhibition. Accordingly, Eq. (15.12) could be simplified as follows:

$$\frac{\mathrm{CL}_{\mathrm{int,all,i}}}{\mathrm{CL}_{\mathrm{int,all}}} = \frac{1}{R} \propto \frac{\left(\dfrac{1}{1+\dfrac{I_{\mathrm{u}}}{K_{\mathrm{i,u,OATP1B1}}}}+1\right)\cdot\left[\left(\dfrac{1}{1+\dfrac{I_{\mathrm{u}}}{K_{\mathrm{i,u,BSEP}}}}\right)+\left(\dfrac{1}{1+\dfrac{k_{\mathrm{inact,CYP3A4}}\cdot I_{\mathrm{u}}}{k_{\mathrm{deg,CYP3A4}}\cdot\left(K_{\mathrm{I,u,CYP3A4}}+I_{\mathrm{u}}\right)}}\right)\right]}{2+\left(\dfrac{1}{1+\dfrac{I_{\mathrm{u}}}{K_{\mathrm{i,u,BSEP}}}}\right)+\left(\dfrac{1}{1+\dfrac{k_{\mathrm{inact,CYP3A4}}\cdot I_{\mathrm{u}}}{k_{\mathrm{deg,CYP3A4}}\cdot\left(K_{\mathrm{I,u,CYP3A4}}+I_{\mathrm{u}}\right)}}\right)} \tag{15.13}$$

Applying Eq. (15.13) to our dataset (Table 15.1), with I_{u} equal to $I_{\mathrm{inlet,u}}$, and referring to the horizontal line at the $1/R$-value of 0.7, the number of false positive compounds in the "no-less" DILI concern category was two, while the number of false negative compounds in the "most" DILI concern category was one (Figure 15.4d). As such, about 83% of the entire dataset is predicted correctly with inhibition information for these three processes only. This is the same overall prediction accuracy as seen with the entire inhibition assessment according to Eq. (15.12) (Figure 15.4b)). The erroneously predicted compounds, independent of the applied approach, were identical (namely the "no-less" DILI concern compounds verapamil and rosuvastatin as well as the "most" DILI concern compound ketoconazole). However, the only compound in the dataset significantly impacted (i.e. predicted $1/R$-value of 0.64 and 0.84 using Eqs. (15.12) and (15.13), respectively) was ketoconazole for which the DILI concern potential is highly underestimated by only referring to OATP1B1, BSEP, and CYP3A4 inhibition data. With reference to Table 15.1 this is mainly owed to the fact that ketoconazole at therapeutic concentrations also acts as a potent UGT1A1 inhibitor (adding UGT1A1 inhibition into Eq. (15.13) provides a $1/R$-value of 0.78).

15.5.2 1/R vs Safety Margin Relationship

Based on Eq. (15.10), assuming just reversible enzyme and/or transporter inhibition, an apparent safety margin (SM_{app}) can be calculated as follows:

$$SM_{app} = \Sigma \frac{K_{i,u}}{I_u} = \frac{\frac{1}{R}}{1 - \frac{1}{R}} = \frac{1}{R-1} \tag{15.14}$$

Eq. (15.14) is the inherent attempt to convert ECM-based $1/R$-values into more tangible safety margins (equivalent to the ratio of process inhibition (e.g. Ki_u) constants and exposure (e.g. I_u)). Using the hyperbolic relationship represented by Eq. (15.14) a $1/R$-value of 0.7, which according to present analysis represents the limit between "no-less" and "most" DILI concern, translates into an apparent safety margin (SM_{app}) of about 2.3 (Figure 15.4 and Table 15.2). The higher the SM_{app}-value the less sensitive a compound's DILI risk is toward an exposure change. That said, compounds with a $SM_{app} \geq 4$ (i.e. $1/R > 0.8$) are rather insensitive toward an exposure change while for compounds with a $SM_{app} \leq 1.5$ (i.e. $1/R < 0.6$) a small change in I_u can have a drastic effect on the ECM-based DILI risk assessment (i.e. on the resulting $1/R$-value). For compounds in-between (i.e. $1.5 < SM_{app} < 4$), in Figure 15.4 representing the separation zone between drug classes tagged with "no-less" and "most" DILI concerns, a duplication and/or bisection of exposure correlates more or less directly with the $1/R$-value assessment (i.e. a duplication or bisection of exposure moves the $1/R$-value from 0.7 to 0.6 or 0.8, respectively). Not surprisingly, the chances for a DILI category misassignment are largest in this separation zone. Referring to the gray highlighted area in Figure 15.4, there were no false positives (i.e. number of compounds below a $1/R$-value of 0.6) for the "no-less" DILI concern categories for either prediction attempt (panel A, B, C, and D). Similarly, no false positives (i.e. number of compounds above a $1/R$-value of 0.8) were determined for the assessment using Eq. (15.12) and $I_{inlet,u}$, as a basis (Figure 15.4b). The number of false negatives for all other prediction approaches (either using Eq. (15.10) with $I_{sys,u}$ and $I_{hep,inlet,u}$ or Eq. (15.13) with $I_{inlet,u}$) was one (Figure 15.4a,d). In all three cases, the "most" DILI concern drug ketoconazole was again identified as the most difficult to predict compound.

Table 15.2 1/*R*-assessment analysis following single or parallel worst-case pathway(s) inhibition (i.e. $CL_{int,i}/CL_{int} \rightarrow 0$).

Affected pathways	None	Canalicular secretion	Intracellular metabolism	Sinusoidal uptake	Sinusoidal uptake and canalicular secretion	Sinusoidal uptake and intracellular metabolism	Intracellular metabolism and canalicular secretion	Sinusoidal uptake, intracellular metabolism, and canalicular secretion
1/*R*-value[a]	1	0.7	0.7	0.5	0.3	0.3	0	0
SM_{app}[b]	~100	2.3	2.3	1	0.4	0.4	0	0
Anticipated DILI severity[c]	No	Less	Less	Most	Most	Most	Most	Most

[a] Calculated according to Eq. (15.11).
[b] Calculated according to Eq. (15.14).
[c] Assignment according to limits identified in Figure 15.4 (i.e. 1/*R*-value of about 0.7 or apparent SM of about 2.3).

15.6 Discussion and Outlook

The ECM describes compound trafficking in the liver as a consequence of different pathways and processes, namely sinusoidal uptake, canalicular efflux, and enzymatic metabolism. Based on the ECM, we have derived several integral DILI prediction concepts to describe the overall inhibition potency of drug candidates on active processes involved in the transport and metabolism of endogenous and safety-relevant compounds. Particularly, the $1/R$-value approach represents a promising prediction concept as it allows an integral (all currently known enzyme- and transporter-derived mechanisms have been incorporated in the model) and quantitative (the lower the $1/R$-value determined for a drug candidate the higher is its likelihood to cause DILI) assessment of the cell stress derived DILI risk. By applying these concepts, our mechanistic understanding for the underlying processes and interdependencies is constantly growing. However, at the same rate new questions arise, which warrant further investigations.

In contrast to the DIC risk approach (Section 15.4), and independent of the assessment method (Eq. (15.12) vs Eq. (15.13)), the use of total intracellular concentrations instead of $I_{sys,u}$ did not improve predictions with the $1/R$-value method. The reason for this observation is unknown. Based on the rather small number of compounds with less and most DILI concern, however, it is difficult to draw strong conclusions regarding the best matrix, especially since different risk assignments for only few compounds have a big impact of the predictive power of a model. Since the unbound intrahepatic concentrations were calculated using the ECM-based approach for Kp_{uu}, a reliable determination of this parameter is critical. Validation of human Kp_{uu} is challenging, however, the successful IVIVC for Kp_{uu} in rat using the same approach provided confidence that Kp_{uu} represents a relevant estimation of the drug disposition in human liver. Moreover, for all $1/R$-value DILI risk assessments, the same inhibitor matrix was used for all inhibition processes, irrespective of the location of the enzyme. However, some drugs act simultaneously as inhibitors of intracellular targets (e.g. BSEP, CYP3A4) and as sinusoidal uptake carriers (such as the OATPs) where $I_{inlet,u}$ is likely more relevant than $I_{hep,inlet,u}$ [31]. Indeed, compared to using $I_{hep,inlet,u}$ only, applying such a mixed assessment approach (i.e. use of $I_{inlet,u}$ for all sinusoidal uptake pathways and of $I_{hep,inlet,u}$ for all other hepatic clearance processes) significantly improved the overall predictability. Referring to highlighted area in Figure 15.4e, there were neither false positives nor false negatives using this approach. The overall separation power was similar to the use of $I_{inlet,u}$ as a reference matrix (Figure 15.4b). In this context it is worth mentioning that many investigators have shown that total I_{sys} values are preferred over free $I_{sys,u}$ values for better DILI predictions [24, 47]. Shah et al. systematically compared the performance of total versus free maximal plasma concentrations for DILI assessments on a test set of about 100 compounds and obtained better results with an I_{sys}-based assessment [48]. Referring to our dataset (Table 15.1) and the DIC model, this interpretation could however not be confirmed. While with I_{sys}, compared to $I_{sys,u}$, the apparent Safety Margin (SM_{app}) thresholds between the DIC classes seemed to decrease significantly the separation of the three frequency classes remained incomplete and even got worse (Figure 15.3a,d). Similarly, applying I_{sys} to the $1/R$-value model, the two drug classes (i.e. "no-less" vs. "most" DILI concerns) could largely not be separated anymore (data not shown). Referring to the free-drug hypothesis, above observations are certainly intuitively correct and make a lot of sense ($I_{inlet,u}$ values are often much closer to I_{sys} than $I_{sys,u}$). However, the particular communalities behind these observations remain to be explored.

Kp_{uu} values are rarely accessible in early development projects. Similarly, inhibition data on NTCP, OATP1B3, MRP2, and/or UGT1A1 are almost never generated as part of compound screening efforts in the discovery phase. However, present analysis reveals that a DILI risk assessment according to Eq. (15.13) seems to be largely sufficient. Therefore, in light of cost-efficiency and development speed, inhibition data on OATP1B1, BSEP, and CYP3A4 paired with unbound systemic exposure estimates at the highest projected clinical daily dose might provide adequate overall DILI risk prognoses to reduce significantly the risk of post-approval market withdrawals. Additional investigations on the inhibition pathways of primary relevance, particularly using a complete reference dataset with entirely in-house generated numbers, are currently ongoing in our facilities.

Based on the current assessment and in line with observations from other groups, irreversible CYP3A4 inhibition is certainly a key driver for DILI [49]. So far, time-dependent enzyme inhibition was clearly associated with covalent adduct formation due to metabolic activation [47, 49]. CYP3A4 has been proposed to play a major role in the biotransformation and detoxification of bile acids [9]. As such, its exact association with cholestatic liver injury remains to be explored. Therefore, future in-house work will be tailored to verify the multifaceted relationship(s) between CYP3A4 inhibition (irreversible as well as reversible) and the pathogenesis of cholestasis.

The importance of bilirubin with regard to DILI is discussed controversially [19]. On one hand, increased bilirubin levels seem to result in enhanced DNA instability, interrupted protein synthesis and/or reduction of mitochondrial enzyme activities [50, 51]. On the other hand, bilirubin is reported to function as a potent radical scavenger due to its antioxidant function and is demonst, rated to exert anti-inflammatory effects on T helper type 17-mediated immune responses [52–54]. With reference to liver function, it seems that a systemic bilirubin increase is currently rather regarded as a sensitive biomarker for DILI than as a potential underlying mechanistic principle. This belief is supported by recent case examples that demonstrated systemic (conjugated and/or non-conjugated) bilirubin rises while transaminase levels and liver histology remained unchanged [55]. In the light of this controversy, additional investigations on the role of bilirubin in DILI are required.

DILI is a multifactorial pathophysiological process that cannot (yet) be recapitulated in a single *in vitro* model [56]. The current ECM-based 1/*R*-value strategy considers multiple pathways but remains limited to enzyme- and/or transporter-based cell stress perturbations (so-called intrinsic activation pathway). There is an emerging body of evidence that initial injury, besides through direct cell stress, is exerted by additional mechanisms (so-called extrinsic activation pathways) such as direct mitochondrial inhibition and/or specific immune reactions [4]. Consequently, with the aim of assisting in the design and development of safer and more effective medicines, pharmaceutical companies are moving towards the development of more complementary and tiered DILI screening platforms. These platforms, also often referred as DILI roadmaps, are combining different (often mainly *in vitro)* model systems using various statistical modeling or risk matrixes as data evaluation methods [24, 47, 56]. To develop such de-risking strategies further, additional research on the best model selection data integration approaches is certainly warranted.

References

1 Lasser KE, Allen PD, Woolhandler SJ, Himmelstein DU, Wolfe SM, Bor DH. Timing of new black box warnings and withdrawals for prescription medications. *JAMA* 2002;287(17):2215–2220.

2 Lee WM. Drug-induced hepatotoxicity. *The New England Journal of Medicine* 2003;349(5):474–485.

3 Food and Drug Administration (2009). Guidance for industry. Drug-induced liver injury: premarketing clinical evaluation. FDA Drug Safety.

4 Russmann S, Kullak-Ublick G, Grattagliano I. Current concepts of mechanisms in drug-induced hepatotoxicity. *Current Medicinal Chemistry* 2009;16(23):3041–3053.

5 Chen M, Vijay V, Shi Q, Liu Z, Fang H, Tong W. FDA-approved drug labeling for the study of drug-induced liver injury. *Drug Discovery Today* 2011;16(15–16):697–703.

6 Björnsson E, Olsson R. Outcome and prognostic markers in severe drug-induced liver disease. *Hepatology* 2005;42(2):481–489.

7 Hussaini SH, Farrington EA. Idiosyncratic drug-induced liver injury: an overview. *Expert Opinion on Drug Safety* 2007;6(6):673–684.

8 Stieger B. Role of the bile salt export pump, BSEP, in acquired forms of cholestasis. *Drug Metabolism Reviews* 2010. 42(3) p. 437–445.

9 Chen J, Zhao K-N, Chen C. The role of CYP3A4 in the biotransformation of bile acids and therapeutic implication for cholestasis. *Annals of Translational Medicine* 2014;2(1):7.

10 Bodin K, Lindbom U, Diczfalusy U. Novel pathways of bile acid metabolism involving CYP3A4. *BBA – Molecular and Cell Biology of Lipids* 2005;1687(1–3):84–93.

11 Gustafsson J, Andersson S, Sjövall J. Bile acid metabolism during development: metabolism of taurodeoxycholic acid in human fetal liver. *Biology of the Neonate* 1985;47(1):26.

12 Perreault M, Wunsch E, Bialek A, Trottier J, Verreault M, Caron P, et al. Urinary elimination of bile acid glucuronides under severe cholestatic situations: contribution of hepatic and renal glucuronidation reactions. *Canadian Journal of Gastroenterology and Hepatology*. 2018;2018: 8096314.

13 Trottier J, Milkiewicz P, Kaeding J, Verreault M, Barbier O. Coordinate regulation of hepatic bile acid oxidation and conjugation by nuclear receptors. *Molecular Pharmaceutics* 2006;3(3):212.

14 Akita H, Suzuki H, Ito K, Kinoshita S, Sato N, Takikawa H, et al. Characterization of bile acid transport mediated by multidrug resistance associated protein 2 and bile salt export pump. *Biochimica et Biophysica Acta* 2001;1511(1):7.

15 Keppler D. The roles of MRP2, MRP3, OATP1B1, and OATP1B3 in conjugated hyperbilirubinemia. *Drug Metabolism and Disposition: The Biological Fate of Chemicals* 2014;42(4):561.

16 Schinkel AH, Mol CA, Majoor D, Mooi WJ, Jongsma AP. Tissue distribution of the human MDR3 P-glycoprotein. *Laboratory Investigation* 1994;71(5):638–649.

17 van Helvoort A, Smith AJ, Sprong H, Fritzsche I, Schinkel AH, Borst P, et al. MDR1 P-glycoprotein is a lipid translocase of broad specificity, while MDR3 P-glycoprotein specifically translocates phosphatidylcholine. *Cell* 1996;87(3):507–517.

18 Ballatori N, Christian WV, Wheeler SG, Hammond CL. The heteromeric organic solute transporter, OSTα–OSTβ/SLC51: a transporter for steroid-derived molecules. *Molecular Aspects of Medicine* 2013;34(2–3):683–692.

19 Chang JH, Plise E, Cheong J, Ho Q, Lin M. Evaluating the in vitro inhibition of UGT1A1, OATP1B1, OATP1B3, MRP2, and BSEP in predicting drug-induced hyperbilirubinemia. *Molecular Pharmaceutics* 2013;10(8):3067.

20 Kiuchi Y, Suzuki H, Hirohashi T, Tyson CA, Sugiyama Y. cDNA cloning and inducible expression of human multidrug resistance associated protein 3 (MRP3). *FEBS Letters* 1998;433(1–2):149–152.

21 Zelcer N, van de Wetering K, de Waart R, Scheffer GL, Marschall H-U, Wielinga PR, et al. Mice lacking Mrp3 (Abcc3) have normal bile salt transport, but altered hepatic transport of endogenous glucuronides. *Journal of Hepatology* 2006;44(4):768–775.

22 Pauli-Magnus C, Meier PJ. Hepatobiliary transporters and drug-induced cholestasis. *Hepatology* 2006. 44 (4) p. 778–787.

23 Camenisch G. Drug-induced liver injury predictions: extended clearance model and its use for prospective transporter and enzyme-based hepatic cell stress grading. *International Journal of Pharmacy and Chemistry*; 5(2) 2019. p. 15–19.

24 Aleo MD, Shah F, Allen S, Barton HA, Costales C, Lazzaro S, et al. Moving beyond binary predictions of human drug-induced liver injury (DILI) toward contrasting relative risk potential. *Chemical Research in Toxicology* 2020;33(1):223.

25 Sirianni G, Pang K. Organ clearance concepts: new perspectives on old principles. *Journal of Pharmacokinetics and Biopharmaceutics* 1997;25(4):449–470.

26 Shitara Y, Sato H, Sugiyama Y. Evaluation of drug–drug interaction in the hepatobiliary and renal transport of drugs. *Annual Review of Pharmacology and Toxicology* 2005;45:689–723.

27 Camenisch G, Umehara KI. Predicting human hepatic clearance from in vitro drug metabolism and transport data: a scientific and pharmaceutical perspective for assessing drug–drug interactions. *Biopharmaceutics and Drug Disposition* 2012;33(4):179–194.

28 Camenisch G, Riede J, Kunze A, Huwyler J, Poller B, Umehara K. The extended clearance model and its use for the interpretation of hepatobiliary elimination data. *ADMET and DMPK* 2015;3(1):1–14.

29 Umehara K, Camenisch G. Novel in vitro-in vivo extrapolation (IVIVE) method to predict hepatic organ clearance in rat. *Pharmaceutical Research* 2012;29(2): 603–617.

30 Pfeifer ND, Bridges AS, Ferslew BC, Hardwick RN, Brouwer KLR. Hepatic basolateral efflux contributes significantly to rosuvastatin disposition II: characterization of hepatic elimination by basolateral, biliary, and metabolic clearance pathways in rat isolated perfused liver. *The Journal of Pharmacology and Experimental Therapeutics* 2013;347(3):737.

31 Riede J, Poller B, Huwyler J, Camenisch G. Assessing the risk of drug-induced cholestasis using unbound intrahepatic concentrations. *Drug Metabolism and Disposition: The Biological Fate of Chemicals* 2017;45(5):523.

32 Camenisch G. Drug disposition classification systems in discovery and development: a comparative review of the BDDCS, ECCS and ECCCS concepts. *Pharmaceutical Research*. 2016;33(11):2583–2593.

33 Kunze A, Poller B, Huwyler J, Camenisch G. Application of the extended clearance concept classification system (ECCCS) to predict the victim drug–drug interaction potential of statins. *Drug Metabolism and Personalized Therapy* 2015;30(3):175–188.

34 Yabe Y, Galetin A, Houston JB. Kinetic characterization of rat hepatic uptake of 16 actively transported drugs. *Drug Metabolism and Disposition: The Biological Fate of Chemicals* 2011;39(10):1808.

35 Shitara Y, Maeda K, Ikejiri K, Yoshida K, Horie T, Sugiyama Y. Clinical significance of organic anion transporting polypeptides (OATPs) in drug disposition: their roles in hepatic clearance and intestinal absorption. *Biopharmaceutics and Drug Disposition* 2013. 34(1) p. 45–78.

36 Guo Y, Chu X, Parrott NJ, Brouwer KLR, Hsu V, Nagar S, et al. Advancing predictions of tissue and intracellular drug concentrations using in vitro, imaging and physiologically based pharmacokinetic modeling approaches. *Clinical Pharmacology and Therapeutics* 2018;104(5):865–889.

37 Giacomini KM, Huang S-M, Tweedie DJ, Benet LZ, Brouwer KLR, Chu X, et al. Membrane transporters in drug development. *Nature Reviews Drug Discovery* 2010;9(3):215.

38 Fahmi OA, Maurer TS, Kish M, Cardenas E, Boldt S, Nettleton D. A combined model for predicting CYP3A4 clinical net drug-drug interaction based on CYP3A4 inhibition, inactivation, and induction determined in vitro. *Drug Metabolism and Disposition: The Biological Fate of Chemicals* 2008;36(8):1698.

39 Food and Drug Administration (2020). in vitro drug interaction studies—Cytochrome P450 Enzymee and transporter-mediated drug interactions guidance for industry. FDA Clinical Pharmacology.

40 European Medicines Agency (2012). Guideline on the investigation of drug interactions. EMA London, UK.

41 Stieger B, Fattinger K, Madon J, Kullak-Ublick GA, Meier PJ. Drug- and estrogen-induced cholestasis through inhibition of the hepatocellular bile salt export pump (Bsep) of rat liver. *Gastroenterology* 2000;118(2):422–430.

42 Fattinger K, Funk C, Pantze M, Weber C, Reichen J, Stieger B, et al. The endothelin antagonist bosentan inhibits the canalicular bile salt export pump: a potential mechanism for hepatic adverse reactions. *Clinical Pharmacology and Therapeutics* 2001;69(4):223.

43 Funk C, Pantze M, Jehle L, Ponelle C, Scheuermann G, Lazendic M, et al. Troglitazone-induced intrahepatic cholestasis by an interference with the hepatobiliary export of bile acids in male and female rats. Correlation with the gender difference in troglitazone sulfate formation and the inhibition of the canalicular bile salt export pump (Bsep) by troglitazone and troglitazone sulfate. *Toxicology* 2001;167(1):83–98.

44 Pauli-Magnus CJ, Meier PJ, Stieger BJ. Genetic determinants of drug-induced cholestasis and intrahepatic cholestasis of pregnancy. *Seminars In Liver Disease* 2010;30(2):147–159.

45 Dietrich CG, Geier A. Effect of drug transporter pharmacogenetics on cholestasis. *Expert Opinion on Drug Metabolism and Toxicology* 2014. 10(11) p. 1533–1551.

46 Morgan RE, Trauner M, van Staden CJ, Lee PH, Ramachandran B, Eschenberg M, et al. Interference with bile salt export pump function is a susceptibility factor for human liver injury in drug development. *Toxicological Sciences* 2010;118(2):485–500.

47 Schadt S, Simon S, Kustermann S, Boess F, McGinnis C, Brink A, et al. Minimizing DILI risk in drug discovery — a screening tool for drug candidates. *Toxicology in vitro* 2015;30(1):429–437.

48 Shah F, Leung L, Barton HA, Will Y, Rodrigues AD, Greene N, et al. Setting clinical exposure levels of concern for drug-induced liver injury (DILI) using mechanistic in vitro assays. *Toxicological Sciences* 2015;147(2):500–514.

49 Utkarsh D, Loretz C, Li AP. in vitro evaluation of hepatotoxic drugs in human hepatocytes from multiple donors: identification of P450 activity as a potential risk factor for drug-induced liver injuries. *Chemico-Biological Interactions* 2016;255:12–22.

50 Hansen TWR, Bratlid D. Bilirubin and brain toxicity. *Acta Pædiatrica* 1986;75(4): 513–522.

51 Dennery PA, Weng YH, Stevenson DK, Yang G. The biology of bilirubin production. *Journal of Perinatology: Official Journal of the California Perinatal Association*. 2001;21 Suppl 1:S17.

52 Min-Sun K, Donghee K, Goh Eun C, Seung Joo K, Min Jung P, Yoon Jun K, et al. Serum bilirubin levels are inversely associated with nonalcoholic fatty liver disease. *Clinical and Molecular Hepatology* 2012;18(4):383–390.

53 Hamoud A-R, Weaver L, Stec DE, Hinds TD. Bilirubin in the liver–gut signaling axis. *Trends in Endocrinology and Metabolism* 2018;29(3):140–150.

54 Longhi MS, Vuerich M, Kalbasi A, Kenison JE, Yeste A, Csizmadia E, et al. Bilirubin suppresses Th17 immunity in colitis by upregulating CD39. *JCI Insight*. 2017;2(9): e92791.

55 Nagar S, Argikar UA, Tweedie DJ. Enzyme kinetics in drug metabolism: fundamentals and applications. *Methods in Molecular Biology (Clifton, NJ)*. 2014;1113:1.

56 Weaver RJ, Blomme EA, Chadwick AE, Copple IM, Gerets HHJ, Goldring CE, et al. Managing the challenge of drug-induced liver injury: a roadmap for the development and deployment of preclinical predictive models. *Nature Reviews Drug Discovery* 2020;19(2):131.

Index

Transporters and Drug-Metabolizing Enzymes in Drug Toxicity, First Edition. Edited by Albert P. Li.

b

g

q

Printed and bound by CPI Group (UK) Ltd, Croydon, CR0 4YY